Advances in
PARASITOLOGY

VOLUME 41

Editorial Board

C. Bryant Division of Biochemistry and Molecular Biology, The Australian National University, Canberra, ACT 0200, Australia

M. Coluzzi Director, Istituto di Parassitologia, Università Degli Studi di Roma "La Sapienza", P. le A. Moro 5, 00185 Roma, Italy

C. Combes Laboratoire de Biologie Animale, Université de Perpignan, Centre de Biologie et d'Ecologie Tropicale et Méditerranéenne, Avenue de Villeneuve, 66860 Perpignan Cedex, France

S.L. James Chief, Parasitology and Tropical Diseases Branch, Division of Microbiology and Infectious Diseases, National Institute for Allergy and Infectious Diseases, Bethesda, MD 20892-7630, USA

W.H.R. Lumsden 16A Merchiston Crescent, Edinburgh EH10 5AX, UK

Lord Soulsby of Swaffham Prior Department of Clinical Veterinary Medicine, University of Cambridge, Madingley Road, Cambridge CB3 0ES, UK

K. Tanabe Laboratory of Biology, Osaka Institute of Technology, 5-16-1 Ohmiya, Asahi-Ku, Osaka 535, Japan

P. Wenk Falkenweg 69, D-72076 Tübingen, Germany

Advances in
PARASITOLOGY

Edited by

J.R. BAKER

*Royal Society of Tropical Medicine and Hygiene,
London, England*

R. MULLER

*International Institute of Parasitology,
St Albans, England*

and

D. ROLLINSON

*The Natural History Museum,
London, England*

VOLUME 41

ACADEMIC PRESS

San Diego   London   Boston
New York   Sydney   Tokyo   Toronto

ACADEMIC PRESS
525 B Street, Suite 1900, San Diego,
California 92101-4495, USA
http://www.apnet.com

ACADEMIC PRESS
24-28 Oval Road
LONDON NW1 7DX
http://www.hbuk.co.uk/ap/

Copyright © 1998, by
ACADEMIC PRESS

This book is printed on acid-free paper

All Rights Reserved
No part of this publication may be reproduced or transmitted in any form
or by any means, electronic or mechanical, including photocopy,
recording, or any information storage and retrieval system,
without permission in writing from the publisher

A catalogue record for this book is available from the British Library

ISBN 0-12-031741-9

Typeset by J&L Composition Ltd, Filey, North Yorkshire
Printed in Great Britain by MPG, Bodmin, Cornwall

98 99 00 01 02 03 MP 9 8 7 6 5 4 3 2 1

CONTRIBUTORS TO VOLUME 41

T.J.C. ANDERSON, *Wellcome Trust Centre for Epidemiology of Infectious Disease, Department of Zoology, South Parks Road, Oxford, OX1 3PS, UK*

R.M. BEECH, *Institute for Parasitology, 21111 Lakeshore Road, Ste-Anne de Bellevue, Quebec, H9X 3V9, Canada*

R.G. BELL, *James A. Baker Institute for Animal Health, College of Veterinary Medicine, Cornell University, Ithaca, NY 14853, USA*

M.S. BLOUIN, *Department of Zoology, Cordley Hall 3029, Oregon State University, Corvallis, OR 97331-2714, USA*

D.W.T. CROMPTON, *WHO Collaborating Centre for Soil-transmitted Helminthiases, University of Glasgow, Glasgow, G12 8QQ, Scotland, UK*

J. DE BONT, *Department of Parasitology, Faculty of Veterinary Medicine, University of Gent, Salisburylaan 133, 9820 Merelbeke, Belgium*

Y. NAKAMURA, *Department of Tumor Biology, Institute of Medical Science, University of Tokyo, 4-6-1 Shirokanedai, Minato-ku, Tokyo 108, Japan*

W. PETERS, *CABI Bioscience, Tropical Parasitic Diseases Unit, 395a Hatfield Road, St Albans, Hertfordshire, AL4 0XU, UK*

J. VERCRUYSSE, *Department of Parasitology, Faculty of Veterinary Medicine, University of Gent, Salisburylaan 133, 9820 Merelbeke, Belgium*

M. WADA, *Department of Tumor Biology, Institute of Medical Science, University of Tokyo, 4-6-1 Shirokanedai, Minato-ku, Tokyo 108, Japan*

PENG WEIDONG, *Department of Parasitology, Jiangxi Medical College, Nanchang, Jiangxi 330006, People's Republic of China*

ZHOU XIANMIN, *Department of Parasitology, Jiangxi Medical College, Nanchang, Jiangxi 330006, People's Republic of China*

PREFACE

This volume commences with a review of the current situation concerning drug-resistant malaria by Wallace Peters, currently working in the Tropical Parasitic Diseases Unit of CABI Bioscience in St Albans, UK (formerly the Institute of Parasitology). Professor Peters is universally acknowledged to be one of the, if not the, world expert(s) on this topic, and his monograph *Chemotherapy and Drug Resistance in Malaria* (published by Academic Press in 1970 and, as the revised second edition, again in 1987) is the malaria chemotherapists' 'bible'. In the first edition, Peters cited Paul Ehrlich's warning that trypanosomes would probably become resistant to any new drug and presciently predicted that 'The same maxim may equally well be applied to the malaria parasite.' This prediction has proved only too true, and it is entirely appropriate that Peters should bring malariologists up to date with this scholarly and wide-ranging review of the current situation which includes among its many references several dated 1997 and even, in the addendum, one published in 1998. The author concludes that 'The goal of eradicating malaria is probably no longer realistic . . . all we can do is to try to keep one step ahead of the parasites.' This review should contribute to the realization of that hope.

The second review in this volume, by Yoshikazu Nakamura and Miki Wada of the Tokyo University Institute of Medical Science, deals with that still rather enigmatic organism, *Pneumocystis carinii*. Propelled, sad to say, in the course of a decade or two into the front rank of pathogenic microorganisms by the advent of the AIDS pandemic, *P. carinii* remains taxonomically somewhere in limbo between protista and fungi, being probably (as Nakamura and Wada argue) 'much more closely related' to the latter. The review covers all aspects of the molecular biology of the organism, to our knowledge of which the authors have made very considerable contributions, with emphasis on the surface glycoproteins and the genes controlling their production. *P. carinii* was originally thought by Carlos Chagas to be a stage in the life cycle of the trypanosomes subsequently named after him, and it is perhaps ironic that the parasite shares with some other species of trypanosomes the ability to switch these antigens and thus, to some extent, escape the host's immune response (a feature not known in other fungi). The authors predict rapid advances in our understanding of this and other aspects of the pathobiology of the parasite in the not-too-distant future as a result of the hoped-for achievement of a reliable *in vitro* culture system and the sequencing of the organism's entire genome.

China has about one-quarter of the world's population but little is known about the present status of ascariasis in the country. This is due to two reasons: the first in that until recently there were considered to be more important priorities in health care; the second is that all the literature has been in Chinese. This review by Peng Weidong and Zhou Xianmin (Jianxi Medical College, China) and David Crompton (University of Glasgow, UK) attempts to answer the questions of what is the scale of disease in China, how is it distributed geographically and demographically, what is the pattern of infection and whether it follows that in other countries. Most of the literature cited is in Chinese. More that half the population is estimated to be infected, mostly in rural populations and particularly in the south-east. There is a close relationship with pigs in China and the possible degree of cross-infection is discussed. The need for a test to identify the source of eggs in the soil is stressed.

There have been numerous studies on immunity to *Trichinella spiralis* in laboratory rodents and this field is comprehensively reviewed by Robin Bell (Cornell University, USA). He separates immunological work into two broadly defined periods, that before 1970 and the wealth of more recent studies. However, as he points out, some of the earlier work was virtually ignored until rediscovered later. The most recent researches have concentrated particularly on helping to explain the basic immunological precepts governing the Th1 and Th2 models of CD4 cell function. Antibody responses, which have been neglected in recent years, are now being given the most important role in protection. The author believes that there is a need to define the novel parasite-specific mechanisms of protection by investigating the parasite's biology and host responsiveness, rather than relying completely on basic immunological knowledge and applying it to immunoparasitology.

Drawing on their extensive experience of animal parasitic nematodes, Tim Anderson (Wellcome Trust Centre for Epidemiology of Infectious Disease, Oxford, UK), Michael Blouin (Oregon State University, USA) and Robin Beech (Institute of Parasitology, Quebec, Canada) consider the use of genetic markers to unravel the population biology of parasitic nematodes. This is an area of parasitology where there has been considerable progress in recent years. The authors first pose the question: How variable are nematode parasites? The answer can be found by comparing patterns of variation in different genes and by making comparisons with other groups of organisms. The population structure of nematode populations has wide implications concerning the epidemiology of disease and interesting insights are presented into aspects of transmission and mating systems. Further sections deal with the use of genetic markers in the study of sibling species, host affiliation and hybridization and the ways in which genetic markers can contribute to the unravelling of parasite life cycles. The uses of genetic markers in the study of drug resistance are carefully reviewed and emphasis given to the monitoring of drug resistance in populations. Finally research areas, such as antigen evolution and intra-host dynamics, have been identified which are likely to benefit from increasing genetic analysis in future years.

The final chapter in this volume is a comprehensive review of cattle schistosomiasis by Jan De Bont and Jozef Vercruysse (Faculty of Veterinary Medicine, University of Gent, Belgium). Both authors are well qualified to review this topic and the chapter reflects their extensive field and laboratory experience in relation to cattle schistosomes. It is difficult to quantify losses caused by cattle schistosomiasis, but it is well recognized that the disease can be of veterinary significance in certain parts of the world and there are problems in terms of treatment and control. The authors consider many aspects of the biology and epidemiology of cattle schistosomiasis, the pathology and pathophysiology of infection, diagnosis, and treatment and control. It is increasingly being realized that cattle schistosomiasis may provide a natural animal model for the study of immunity against human schistosome infection and the dynamics of transmission, and this contribution should be of value to all workers on schistosomiasis.

J.R. Baker
R. Muller
D. Rollinson

CONTENTS

CONTRIBUTORS TO VOLUME 41 v
PREFACE vii

Drug Resistance in Malaria Parasites of Animals and Man

W. Peters

1. Introduction 1
2. Drug Resistance in Clinical Practice 4
3. Experimental Drug Resistance in Laboratory Models 15
4. The Biochemical and Genetic Bases of Drug Resistance . . . 21
5. Development of New Antimalarial Drugs and
   Drug Combinations 28
6. Conclusions 40
   Acknowledgements 40
   References 40

Molecular Pathobiology and Antigenic Variation of *Pneumocystis carinii*

Y. Nakamura and M. Wada

1. Introduction 64
2. Major Surface Glycoproteins 70
3. Major Surface Glycoprotein Genes 75
4. Conclusion 97
   Acknowledgements 98
   References 98

Ascariasis in China

Peng Weidong, Zhou Xianmin and D.W.T. Crompton

1. Introduction 110
2. Sources of Information 114
3. National Distribution of Ascariasis in China 114
4. Estimation of Prevalence 119
5. Distribution and Prevalence within Regions 120
6. Factors Influencing the Distribution of Ascariasis in China . 120
7. Intensity of Ascariasis in China 133
8. Observations on the Origin of the Human–*Ascaris* Association 136
9. Traditional Chinese Medicine for the Treatment of Ascariasis 138
10. Conclusions 140
    Acknowledgements 140
    References 141

The Generation and Expression of Immunity to *Trichinella spiralis* in Laboratory Rodents

R.G. Bell

1. Introduction 150
2. The Development of Mechanistic Theories of Rejection of *T. spiralis* in the Period up to 1970 150
3. Analysis of the Cellular and Humoral Immune Response to Infection Since 1970 155
4. Variation in the Host Response to Infection 173
5. The Response of Granulocytic Cell Populations to Infection . 180
6. Assessment of Proposed Mechanisms of Protection 190
7. Synthesis 197
8. Conclusions 201
   Acknowledgement 202
   References 202

Population Biology of Parasitic Nematodes: Applications of Genetic Markers

T.J.C. Anderson, M.S. Blouin and R. M. Beech

1. Introduction 220
2. How Variable are Nematode Parasites? 221

3. Population Structure — How is Genetic Variation Arranged in Populations? 237
4. Sibling Species, Host Affiliation, and Hybridization 245
5. Nematode Life-histories 260
6. Genetic Markers and Drug Resistance 262
7. Other Applications of Genetic Markers 270
   Acknowledgements 273
   References 273

Schistosomiasis in Cattle

J. De Bont and J. Vercruysse

1. Introduction 286
2. Schistosomes and Their Life Cycles 286
3. Pathology and Pathophysiology 303
4. Diagnosis 315
5. The Epidemiology of Cattle Schistosomiasis 320
6. Treatment and Control 337
7. Conclusion and Perspectives 345
   Acknowledgements 345
   References 345
   Index 365

   Colour Plate located between 18-19

Drug Resistance in Malaria Parasites of Animals and Man

W. Peters

CABI Bioscience, Tropical Parasitic Diseases Unit,* 395a Hatfield Road, St Albans, Hertfordshire, AL4 0XU, UK

1. Introduction 1
2. Drug Resistance in Clinical Practice 4
   2.1. *Plasmodium falciparum* 4
   2.2. *Plasmodium vivax* 13
   2.3. *Plasmodium ovale* and *Plasmodium malariae* 15
3. Experimental Drug Resistance in Laboratory Models 15
   3.1. *In vitro* models — *Plasmodium falciparum* 15
   3.2. Rodent malaria 16
4. The Biochemical and Genetic Bases of Drug Resistance 21
   4.1. Chloroquine and other 4-aminoquinolines 21
   4.2. Dihydrofolate reductase inhibitors 25
   4.3. Quinine, mefloquine and halofantrine 26
   4.4. Naphthoquinones 28
5. Development of New Antimalarial Drugs and Drug Combinations 28
   5.1. Single compounds 29
   5.2. Drug combinations 37
6. Conclusions 40
Acknowledgements 40
References 40

1. INTRODUCTION

'A great deal of progress has been made in the field of protozoal chemotherapy since the days of Paul Ehrlich. However, nature is as wily and

* Formerly the International Institute of Parasitology

unpredictable now as she was then and it would be unwise to forget . . . [his] . . . warning that trypanosomes would be likely to develop resistance to any new type of drug that could be designed to attack them. The same maxim may equally well be applied to the malaria parasite. The best we can hope to do probably is to keep one jump ahead of nature.' (Peters, 1970a).

In the quarter of a century since these lines were written, this forecast, sadly, has proved to be correct. Moreover, in this period very few new antimalarial drugs have been added to our limited repertoire and, of those, several are closely related analogues of a single compound, artemisinin, that was isolated in 1972 from a traditional Chinese medicinal herb, qing hao (*Artemisia annua*) (see review by Klayman, 1985). A high level of resistance has appeared to the most promising synthetic antimalarial of recent years, mefloquine, which is also now proving to be less safe than had been believed.

In this chapter, drug resistance is defined very simply as 'The ability of a parasite to survive in the presence of concentrations of a drug that normally destroy parasites of the same species or prevent their multiplication.' This definition must also be interpreted in clinical terms in relation to the maximum dose that is tolerated by the human host. In animal models the question of what dose is tolerated has to be gauged on careful observation of the host's response to the drug in question. In general, therefore, significantly higher levels of drug resistance can be attained experimentally in malaria-infected animals. The current widespread use of cultures of *Plasmodium falciparum* in relatively inert erythrocytes for testing levels of parasite response to antimalarial drugs *in vitro* has opened the road to the development of lines that can have a very high level of drug resistance — far higher than would ever be encountered in falciparum malaria in humans. In this type of model, relative host drug toxicity can be assessed by parallel exposure of mammalian cells in tissue culture (O'Neill *et al.*, 1985).

The ideal of eradicating malaria has long had to be abandoned and replaced by a more realistic target of malaria control. A global plan was drawn up by the World Health Organization to cover the period from 1993 to the end of the millenium and guidelines have been published for its implementation (WHO, 1993). The goal of this plan is simple: to reduce the morbidity and mortality due to malaria, which have changed little despite all the efforts of the past 50 years (Figure 1). There is no doubt that resistance of *P. falciparum* to the few currently available antimalarial drugs represents one of the major obstacles to the achievement of this simple objective. Much of the problem, as Verdrager (1995) has pointed out in an analysis of the genesis of multidrug-resistant falciparum malaria, especially in south-east Asia, is attributable to massive drug selection pressure in poorly immune or nonimmune people, working in highly endemic areas where efficient anopheline vectors are abundant — for example, gem

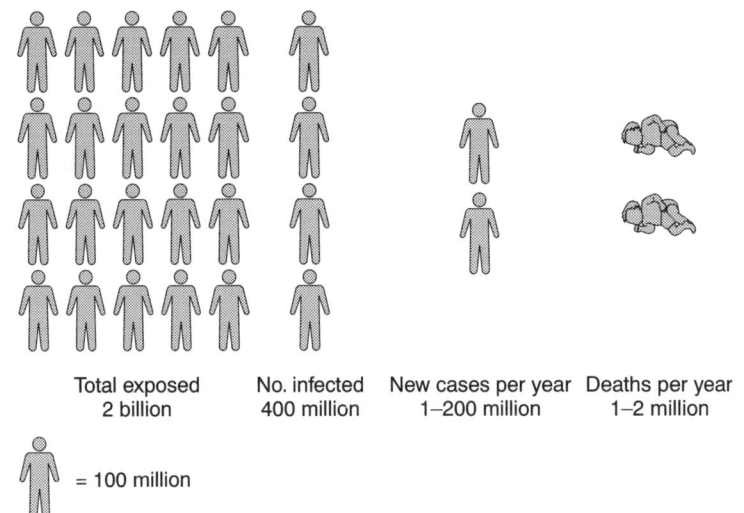

Figure 1 Graphic representation of the extent of the malaria problem — half of the world's population is exposed to malaria. Most of the deaths occur in infants and young children infected with *Plasmodium falciparum*, particularly on the African continent (1 billion = 10^9).

miners in the Pailin area of Kampuchea. To compound the problem, resistance to chloroquine of *Plasmodium vivax* is also being documented in an increasingly wide geographical area. Widespread resistance of malaria vectors to insecticides is a further major obstacle, but one that is outside the scope of the present review.

Much has been learned in recent years of the molecular basis of action of some of the older and widely used antimalarial compounds and, in a number of cases, of the molecular changes in the parasites that are associated with drug resistance. This wealth of academic knowledge, however, has not led to the development of any radically new drugs nor of better ways of deploying the older ones. One potentially valuable lead to possible ways of reversing resistance to chloroquine of *P. falciparum* was derived from the serendipitous application of experience gained in the field of cancer chemotherapy (Martin *et al.*, 1987) but, to date, no resistance modulator of practical clinical value has been reported.

In the following pages I shall review, firstly, the situation of antimalarial drug resistance as seen from the clinical point of view. I shall then examine the major developments in experimental drug resistance in laboratory models. To review adequately all the work of the past two decades on the biochemical aspects of the modes of action of different classes of drugs and the ways in which parasites overcome these actions, as well as the genetic bases of drug resistance, would demand at least twice as much space. These topics will therefore be presented in as succinct a form as possible, with references to major accounts that are available in the literature. The review closes with a summary of the relatively small number of novel drugs and drug combinations that have been introduced into the clinic in recent years or that are currently under development. Other major reviews of chemotherapy and drug resistance in malaria since that by Peters (1987) are those by WHO (1990, 1996a), Wernsdorfer and Payne (1991), Karbwang and Wernsdorfer (1993), Wernsdorfer (1994), and Basco et al. (1994). A valuable contribution was also made by Eaton et al. (1989). Plowe and Wellems (1995) have reviewed knowledge on the genetic basis of drug resistance. Cross and Singer (1991) made an interesting mathematical analysis of the epidemiological and genetic factors governing the selection of resistance under field conditions, taking well-documented reports of pyrimethamine resistance in *P. falciparum* as their examples. Using sophisticated mathematical models, Mackinnon (1997) has shown how increased drug selection pressure strongly favours the spread of multiple drug resistance, while Dye and Williams (1997) have explored the significance of multiple gene origins of resistance in a similar context.

2. DRUG RESISTANCE IN CLINICAL PRACTICE

2.1. *Plasmodium falciparum*

2.1.1. *Chloroquine and other 4-Aminoquinolines*

The first, simple procedure for assaying the clinical response to chloroquine, established over 30 years ago by the World Health Organization (WHO, 1965) and was most recently revised in 1973 (WHO, 1973), distinguished between an abbreviated '7-day test', which was considered to be the most practical for use in malaria endemic areas, and an 'extended' test that required freedom from reinfection over a period of 28 days. Both tests divided the clinical response to a standard dose of 25mg kg^{-1} chloroquine base administered over a period of 3 days into sensitive (S) or resistant at the RI, RII or RIII levels. Although these tests have, by and large, served a very useful purpose, they were based essentially on observations of parasitaemia and were most suitable for use in nonimmune subjects. While it has always been recognized that the responses were strongly influenced by the immune status of the test subjects, it is only recently that the conventional WHO test has been modified to place the emphasis on the clinical, rather than the parasitiological, response. The revised procedure has the advantage that it can be applied to evaluate the response to any antimalarial drug used for therapy of acute falciparum malaria. The response is now classified as 'early treatment failure' (ETF), 'late treatment failure' (LTF), or 'adequate clinical response' (ACR), the criteria for which are laid down in a document (WHO, 1996b) that contains provision, as before, for the adaptation of the procedure to an observation period of 7 or 28 days.

In parallel with drug sensitivity monitoring in humans, the *in vitro* procedure established by WHO (see Bruce-Chwatt *et al.*, 1986) has proved invaluable, especially since that organization has supported the production of standardized test kits, not only for chloroquine but also for a number of other widely used drugs including amodiaquine, mefloquine, sulfadoxine, pyrimethamine and artemisinin. The identification of drug response at the molecular level by the use of the polymerase chain reaction (PCR) technique in association with specific deoxyribonucleic acid (DNA) probes now offers a highly sensitive diagnostic tool for use in epidemiological studies, although one that is of limited use in individual patients (see the review by Walliker, 1994). Babiker *et al.* (1994) and Kain *et al.* (1996), for example, have used a PCR technique to distinguish between reinfection and treatment failure in drug trials on patients with *P. falciparum* in areas of high malaria endemicity.

Clinical and parasitological resistance of falciparum malaria to chloroquine now extend to all regions, with the possible exception of certain countries of Central America. Despite the high prevalence of resistant strains, even up to RIII level in many areas, chloroquine is still considered to play a valuable role in the treatment of acute, uncomplicated *P. falciparum* infection in a number of countries where transmission is intense, notably in tropical Africa. Although the drug often fails to clear parasitaemia, a clinical cure of ACR type is acceptable in many patients and the fact that chloroquine is cheap, safe and readily available is an important practical consideration. Amodiaquine, originally considered to have similar efficacy to chloroquine, was largely discarded for many years on the basis that it was more costly, many drug failures occurred in patients who had originally failed to respond to chloroquine, and the shadow of severe drug toxicity remained. The latter, it must be noted, proved to be a problem mainly in nonimmune travellers who had been taking amodiaquine as a prophylactic drug. Side-effects when amodiaquine was deployed for therapy in the field were rare. Interest has recently been revived in the potential value of amodiaquine as a substitute for chloroquine in view of its demonstrably superior activity (or, rather, that of its desbutyl metabolite), at least *in vitro* (Olliaro *et al.*, 1996), although a cautious line is taken by many experienced malariologists.

In addition to its strictly blood schizontocidal action, the broader pharmacological properties of chloroquine are being increasingly recognized as knowledge of the pathophysiology of severe falciparum malaria increases. It has been suggested, for example, that chloroquine may antagonize the deleterious action of certain cytokines such as tumour necrosis factor (TNF) (Picot *et al.*, 1993; Kwiatkowski and Bate, 1995) or nitric oxide (Balmer *et al.*, 1995), excessive production of which may be an important underlying factor in the pathophysiological changes associated with cerebral malaria (Clark and Rockett, 1996; Kremsner *et al.*, 1996). (See also comments on desferrioxamine, Section 5.1.7.) The possibility of resurrecting chloroquine as a therapeutic agent for severe malaria by combining it with a 'reversing agent' (see Section 5.2.1) is, therefore, a very attractive, if still somewhat theoretical, proposition. (See also the comments on the effect of withdrawing chloroquine in Section 2.1.3.)

At the time of writing no other 4-aminoquinolines has been shown to be of clinical value for the treatment of infections resistant to chloroquine although a number of such compounds are effective in experimental malaria (see Section 5.1.4).

2.1.2. *Dihydrofolate Reductase Inhibitors, Alone or in Combination*

Two of the earliest dihydrofolate reductase (DHFR) inhibitors ('antifols'), proguanil and pyrimethamine, which were introduced in 1948 and 1951, respectively, still play a useful role in various combinations in the prevention and treatment of malaria, in spite of widespread resistance to both compounds, especially in *P. falciparum*. In fact, pyrimethamine has for many years been used only in a synergistic combination with sulfadoxine or sulfalene and essentially for treatment, rather than prophylaxis, because of the hazard of severe toxic skin reactions to the sulphonamide component when given in repeated doses. Although these two combinations have been invaluable for the treatment of multidrug-resistant falciparum malaria and still play a major role as cheap, effective and relatively safe preparations in some countries, resistance even to these potent combinations is extending rapidly. In south-east Asia, for example in Thailand, the cure rate with pyrimethamine/sulfadoxine (Fansidar) has declined in recent years to the point where it now has little practical value. Conversely, in East and West Africa, although the response to this combination is now diminishing (e.g., only 62% curative in a recent study in Kenya (Anabwani *et al.*, 1996) and even less in Tanzania (Rønn *et al.*, 1996)), Fansidar still plays a valuable role in the treatment of uncomplicated falciparum malaria in semi-immune patients. Fansidar was, for example, very effective in protecting pregnant women (Schultz *et al.*, 1996) and children under five years old in Malawi (Nwanyanwu *et al.*, 1996), and also for treating uncomplicated falciparum malaria in Gambian children (Müller *et al.*, 1996). Moreover, a pyrimethamine/dapsone combination (Maloprim) has been shown to give useful prophylactic cover, for example, to schoolchildren in trials conducted in The Gambia (Greenwood, 1991; Greenwood *et al.*, 1995) although, in principle, chemoprophylaxis is no longer considered suitable for general deployment among semi-immune populations of areas with intense malaria transmission. Undoubtedly, resistance to such combinations will continue to spread rapidly in Africa as long as the population has unlimited access to them (Plowe *et al.*, 1996).

Mutation-specific primers to identify the sector of the mutant *DHFR–TS* (thymidilate synthase) gene responsible for pyrimethamine resistance were described by Zolg *et al.* (1990). A PCR technique for use with simple, filter paper samples for epidemiological studies on the response of *P. falciparum* to pyrimethamine has been published by Zindrou *et al.* (1996).

Proguanil, unlike pyrimethamine, still plays a useful role when administered as a prophylactic drug, for example to non-immune travellers, together with chloroquine in areas where the overall level of chloroquine resistance is moderate. Cross-resistance between pyrimethamine and proguanil in *P. falciparum* is a common, but by no means universal, phenomenon (Björkman, 1991) and is further discussed below. However, breakthroughs of falciparum parasitaemia in the face of prophylaxis with the recommended regimen of proguanil plus chloroquine are being observed on an increasing scale, especially in tropical African countries.

At present, no other antifol or antifol combination is in general use, although some are being investigated in experimental models or in clinical trials (see Section 5.1.5).

2.1.3. *Quinine, Mefloquine and Halofantrine*

When the clinical failure of chloroquine first became apparent in the 1960s it was rapidly recognized that quinine was the first drug of choice for the treatment of acute falciparum malaria (Peters, 1987). However, the toxicity, cost and relative scarcity of this compound necessitated a search for an equally effective alternative. The United States army, which urgently required a replacement for chloroquine, initiated an extensive drug research and development programme from which emerged two potent blood schizontocides, mefloquine, a quinolinemethanol, and halofantrine, a phenanthrenemethanol with structural analogies to quinine. Of the two, mefloquine proved to be the more satisfactory in early clinical trials and, following extended field studies under the auspices of the UNDP/World Bank/WHO Special Programme for Research and Training in Tropical Diseases (the TDR programme), this compound came to be widely used from the late 1970s for the treatment of chloroquine-resistant *P. falciparum* and for prophylaxis, largely in nonimmune travellers, military forces, etc. Clinical results initially were excellent although experimental investigations showed, as early as 1977, that mefloquine resistance was very likely to emerge rapidly in areas of chloroquine-resistant *P. falciparum* once it came to be widely deployed in monotherapy (Peters *et al.*, 1977). In the hope of minimizing this risk, a triple combination containing mefloquine, sulfadoxine and pyrimethamine (MSP) was developed under TDR auspices and introduced into the areas with the highest risk.

In the event, the introduction of MSP was too late, especially in a number of countries of south-east Asia, among which Kampuchea and Thailand are probably the most notorious. From over 97% in the early 1980s the cure rate fell to less than 30%, even with MSP, in Thai gem miners who had been infected with *P. falciparum* in Kampuchea by 1991 (Thimasarn *et al.*, 1995). Clinical failures were paralleled by increases in inhibitory levels *in vitro*. A similar situation was reported by Nosten *et al.* (1991) among Karen hill tribesmen on the Myanmar–Thai border where, by 1994, mefloquine failed to cure nearly half of the patients with falciparum malaria (Price *et al.*, 1995). It has even been suggested that such foci arose there from parasites imported from Kampuchea by gem miners. Bygbjerg *et al.* (1983) had demonstrated a decrease in the response *in vitro* of *P. falciparum* to mefloquine in a Danish man, infected in Tanzania, whose infection recrudesced. Increases in inhibitory concentrations with successive passages were recorded also in clones isolated from Thai patients (Webster *et al.*, 1985) and in isolates made following recrudescence *in vivo* (Childs *et al.*, 1991).

Despite the evident problem of rapidly emerging clinical resistance to mefloquine, the acute response of many patients severely infected with multidrug-resistant *P. falciparum* to quinine, when administered parenterally in adequate dosage, on the whole remained satisfactory in most countries. However, once again in the epicentres of resistance in south-east Asia resistance to quinine itelf has become an increasingly serious clinical problem. In Thailand, for example, the 50% inhibitory concentration (IC$_{50}$) of quinine for *P. falciparum in vitro* increased from about 130 ng mL^{-1} in 1984 to over 350 ng mL^{-1} in 1990 in one village bordering Kampuchea (Wongsrichanalai *et al.*, 1992). Of particular concern is the not unexpected cross-resistance that has become apparent between quinine and mefloquine. Although the latter compound is undoubtedly a more potent blood schizontocide than quinine, mefloquine cannot be guaranteed completely to clear falciparum parasitaemia from a significant proportion of patients with *P. falciparum* infections that recrudesce or even fail to clear after quinine alone, i.e., an LTF or even ETF response. Fortunately, quinine-resistant *P. falciparum* infections still respond rapidly to treatment with compounds of the artemisinin family such as artemether (Tran *et* al., 1996). In the northern part of Cameroon in West Africa, Brasseur *et al.*

(1992) detected a high level of resistance to mefloquine which they attributed to cross-resistance with quinine which had been widely deployed for therapy in that area (mefloquine itself not having been used there). Until recently, mefloquine was widely recommended for prophylaxis in non-immune travellers to areas with a high risk of infection with chloroquine or multidrug-resistant *P. falciparum*. Increasing reports of severe neurotoxicity with this compound, while undoubtedly exaggerated in their frequency, have nevertheless imposed the need for caution in the prescribing of this drug for prophylaxis but this, in turn, stresses the serious problem of suitable, effective alternatives which, at the moment, are almost entirely lacking.

The development of halofantrine, which has a similar spectrum of activity to mefloquine against multidrug-resistant *P. falciparum*, was delayed for several years because of problems of bioavailability. These were, however, largely overcome and this compound is now used for the treatment of falciparum malaria that has been acquired in areas of known multidrug resistance as an alternative to mefloquine. Halofantrine resistance, manifested both by failure to achieve a radical cure and by enhanced inhibitory levels *in vitro*, is now a widely recognized phenomenon and, again not surprisingly, cross-resistance with mefloquine has been reported. However, it is rather the high cost of halofantrine and questions regarding its cardiotoxicity that are currently the main impediments to its wider clinical deployment. The main metabolite of halofantrine in humans is N-monodesbutyl halofantrine. This may have an important advantage over the parent compound. While it is equipotent with halofantrine against *P. falciparum* (see Basco *et al.*, 1992), it lacks the cardiotoxicity of the parent compound (Wesche *et al.*, 1996). Because halofantrine is now one of the few compounds that yield a high cure rate in multidrug-resistant falciparum malaria, its manufacturers and WHO (1995) quite correctly insist that it should be conserved for therapy and not be deployed for prophylaxis at the present time (but see comments in Section 3.2.2 regarding its resistance potential).

Until recently it was uncertain whether chloroquine resistance in nature is immutable or whether, like resistance to antifols, the withdrawal of drug selection pressure may permit sensitive parasites to return. Jacquier *et al.* (1985) and Thaithong *et al.* (1988) suggested that the withdrawal of chloroquine for the treatment of multidrug-resistant *P. falciparum* might lead to a reversion of the parasites to chloroquine sensitivity. Data published from a long-term surveillance programme on the Chinese island of Hainan (Liu *et al.*, 1995) have now shown that the complete withdrawal of chloroquine from 1979 has been followed by a progressive return of the responsiveness of *P. falciparum* to that drug. Instead of chloroquine, patients were treated with piperaquine or, more recently, artemether. Monitoring *in vitro* has revealed a decrease from 97.9% resistance in 1981 to 60.9% in 1991. The 28-day *in vivo* test showed a fall in resistance from 84.2%

to 40% in the corresponding period, with almost half the numbers of cases at the RII and RIII levels in 1991. These data are in marked contrast to the responses recorded by Liu *et al.* (1996) from a wider area of Hainan and Yunnan Provinces, where the 1981 levels persisted. While the Chinese data imply that a complete withdrawal of 4-aminoquinolines and compounds with the same mode of action could be followed by a restoration of chloroquine to its former role as the drug of choice for therapy, in practical terms it is hard to envisage such action being taken on an adequate scale, other than in a tightly controlled island population such as that of Hainan. Another phenomenon that has been widely observed is a paradoxical restoration of the sensitivity of *P. falciparum* to chloroquine with increasing resistance to mefloquine or quinine (see review by Mockenhaupt, 1995) or halofantrine (Ritchie *et al.*, 1996). This is also reflected at the molecular level (Cowman *et al.*, 1994). Neither the biochemical nor molecular basis of this inverse relationship has been determined.

2.1.4. *Naphthoquinones*

A number of naphthoquinones has been demonstrated to inhibit the growth of intraerythrocytic malaria parasites that are resistant to 4-aminoquinolines, quinoline- and phenanthrenemethanols, antifols and antifol/sulphonamide combinations. As one of the first of these compounds to reach clinical trial, menoctone, proved to be too poorly bioavailable, a large series of analogues derived from it were investigated. Of these, atovaquone (Figure 2) is the first naphthoquinone to have passed all the hurdles from preliminary experimental studies to clinical trials in patients naturally infected with multidrug-resistant *P. falciparum*. Unfortunately, like its precursor menoctone in which resistance could be selected in a rodent malaria parasite in a single passage *in vivo* (Peters, 1987), resistance to atovaquone was observed in Thailand in the very first patients infected with *P. falciparum* who were treated with the new compound (Looareesuwan *et* al., 1996b). It was, however, rapidly demonstrated that similar patients who received atovaquone plus proguanil, tetracycline or doxycycline were completely cured. (Incidentally, this clinical observation once again confirmed the value of rodent malaria models for forecasting the role of antimalarials in humans — Peters (1970b) reported strong synergism between menoctone and cycloguanil (the active metabolite of proguanil) against *Plasmodium berghei*.) In Gabon, a combination of atovaquone with proguanil cured 87% of 71 semi-immune African patients with falciparum malaria, compared with only 72% of a comparison group of 71 who received atovaquone plus amodiaquine (Radloff *et al.*, 1996a). The prospects for the future of the combination of atovaquone with proguanil, currently being introduced under the name of Malarone (Figure 2), are uncertain

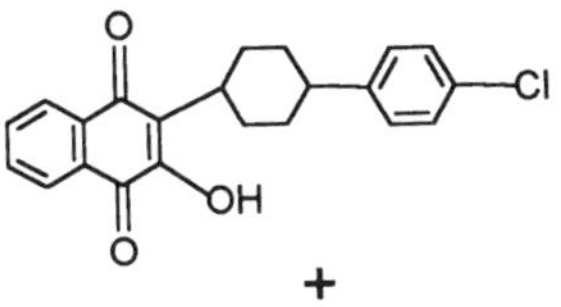

I

+

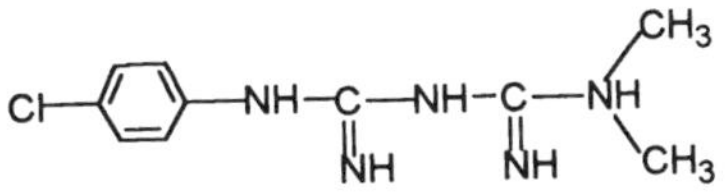

II

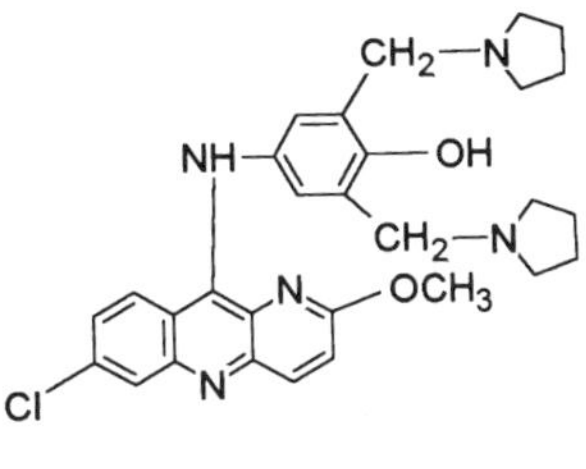

III

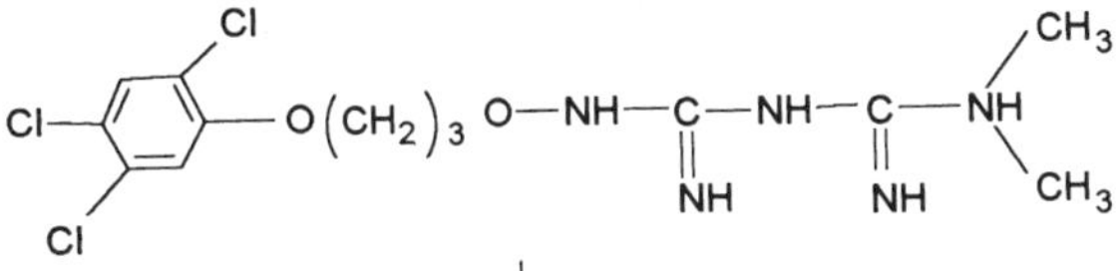

IV

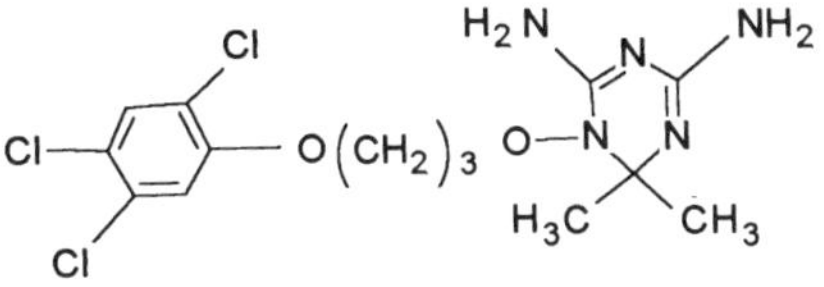

V

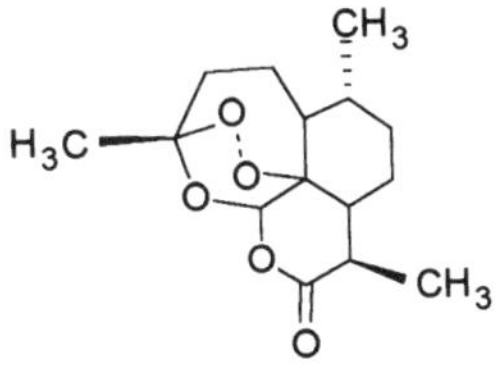

VI

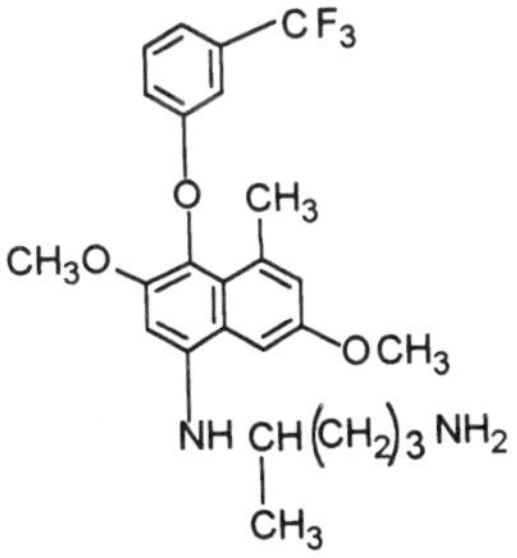

VII

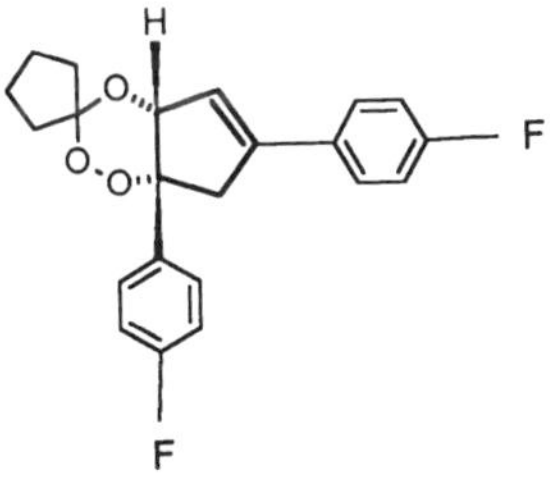

VIII

Figure 2 Structures of some novel antimalarial compounds. I, atovaquone; II, proguanil (the combination of I + II is known as Malarone); III, pyronaridine; IV, PS-15, which is metabolized in the mammal to the triazine; V, WR 99,210; VI, artemisinin (qinghaosu); VII, WR 238,605; VIII, Fenozan B07.

at this time, partly because of the high cost of the preparation and partly because of the lack of studies of the potential for the parasites to develop resistance to it, as has happened in the case of Fansidar. Another potentially valuable compound, which significantly impedes the selection of resistance to atovaquone and to itself in cultures of *P. falciparum* exposed to drug selection pressure, is 5-fluoroorotate. Gassis and Rathod (1996) have shown that the rate at which resistance is selected to the compounds used individually, i.e., a frequency of 10^{-6} for atovaquone (in a concentration of 10^{-8}) and 10^{-5} for 5-fluoroorotate (in a concentration of 10^{-7}), is reduced to below 5×10^{-10} when the parasites are exposed to the drugs simultaneously. If ten times these concentrations were to be applied, it was estimated that the frequency would be reduced to 10^{-17}. Although these combinations were not toxic to mammalian cells in culture, no data are available yet from any *in vivo* model.

2.1.5. *Sesquiterpene Lactones*

Artemisinin (Figure 2) is a sesquiterpene lactone derived from a widely distributed member of the Compositae, *Artemisia annua* Linn., which has been extensively used as a febrifuge in traditional Chinese medicine for two millennia. Isolated from the wild plant by Chinese scientists in the late 1970s, artemisinin has been shown to have a potent and rapid blood schizontocidal action against *Plasmodium*. Moreover, it is as active against strains that are resistant to virtually all other antimalarial drugs in clinical use as against drug-sensitive parasites both in experimental and human hosts, and *in vitro*. Artemisinin and a number of semisynthetic analogues, especially artemether and the water-soluble artesunate, are coming increasingly into clinical use for the treatment of acute, multidrug-resistant falciparum malaria (WHO, 1990; Anonymous, 1994), especially in south-east Asia but also prematurely in other parts of the world. Unfortunately, these include several African countries in which multidrug-resistant *P. falciparum* has not yet reached serious proportions but where, despite dire warnings from WHO and other authorities, the commercial exploitation of the artemisinins is rapidly being expanded.

To date, there has been no confirmed report to indicate that clinical resistance to any of these novel blood schizontocides has emerged, although Liu *et al.* (1996) noted that sensitivity tests *in vitro* with artesunate, artemether and dihydroartemisinin and isolates of *P. falciparum* from Hainan and Yunnan Provinces of China indicated resistance to these compounds in 5.8% (of 56 cases), 2.4% (of 204 cases) and 2.3% (of 88 cases), respectively. Experimentally, a number of drug-resistant lines has been developed, e.g., in the chloroquine-sensitive *P. berghei* N strain and in the chloroquine-resistant parasite *Plasmodium yoelii* ssp. NS (see Chawira *et al.*, 1986). It is essential to monitor closely the clinical and *in vitro* response to compounds of the artemisinin family which currently offer some of the few available drugs with which to cure otherwise highly refractory, severe falciparum malaria.

One property of the artemisinins that is of potentially great epidemiological significance is their gametocytocidal action (Dutta *et al.*, 1989; Chen *et al.*, 1994). This appears to be limiting transmission when they are widely used for therapy of multidrug-resistant *P. falciparum* in western Thailand (Price *et al.*, 1996).

2.2. *Plasmodium vivax*

Until recently it was generally understood that chloroquine could be guaranteed to produce clinical cure of an acute attack of vivax malaria, even though this compound would not prevent the emergence of relapses caused by the subsequent maturation of crops of hypnozoites in the hepatocytes. To eliminate these it was necessary to follow chloroquine therapy with appropriate dosage of the 8-aminoquinoline, primaquine. Regional variations in the sensitivity of *P. vivax* to primaquine were well established, with infections acquired in the south-west Pacific being notoriously hard to eliminate (see the review by Collins and Jeffrey, 1996). Although some degree of cross-resistance was known between chloroquine and primaquine in the asexual, intraerythrocytic stages of rodent malaria (Peters, 1970a), this phenomenon had never been tested in *P. vivax* since primaquine was never used as a blood schizontocide because of its inherent toxicity. For the same reason, despite its recognized causal prophylactic potential, primaquine was never deployed for prophylaxis. Recent studies by Fryauff *et al.* (1995) in Indonesia (see Section 5.2.2) and Weiss *et al.* (1995) in Kenya, suggest that primaquine may finally have a place in prophylaxis for individuals with normal levels of glucose-6-phosphate dehydrogenase.

The infallible ability of chloroquine to cure clinical attacks of *P. vivax* was first seriously put in doubt when Schuurkamp *et al.* (1989) and Rieckmann *et al.* (1989) recorded apparent failures in patients infected in Papua New Guinea. This report was succeeded by a minor avalanche of confirmatory reports, initially from West Irian (Irian Jaya) (Baird *et al.*, 1991; Schwartz *et al.*, 1991; Murphy *et al.*, 1993) and, more recently, from a number of countries of south-east Asia. In order to distinguish at community level between true chloroquine resistance of asexual parasites and relapses caused by hypnozoites, Baird *et al.* (1995) developed a simple *in vivo* procedure. Based on many earlier observations that a standard dose of 1.5 g of chloroquine base given over 3 days was an effective blood schizontocide against *P. vivax*, and that relapses from maturing hypnozoites would not appear earlier than 28 days after the start of treatment, they treated patients suffering from primary attacks of vivax malaria with the standard dose of chloroquine then monitored their parasitaemia up to 28 days from the start of treatment. The intake of chloroquine was also monitored by either a simple urine test or, when possible, analysis of blood samples. Using this procedure they were able to demonstrate, in nonimmune, immigrant labourers transferred from mainland Indonesia to the Arso region of eastern Irian Jaya, that nearly half of the newly acquired *P. vivax* infections recrudesced within 14 days and 78% did so within 28 days. Chloroquine resistance was also identified in 14% of *P. vivax* infections in an island population off the coast of north-west Sumatra (Baird *et al.*, 1996). Fortunately, these blood infections responded satisfactorily to other compounds such as mefloquine or halofantrine, although primaquine must still be administered to produce a radical cure since neither mefloquine nor halofantrine destroys hypnozoites. It must now be accepted that the geographical distribution of chloroquine-resistant *P. vivax* is extending, for example to India (Garg *et al.*, 1995; Dua *et* al., 1996), Myanmar (previously Burma) (Marlar-Than *et al.* (1995), Brazil (Garavelli and Corti, 1992) and Guyana (Phillips *et al.*, 1996). If the example of *P. falciparum* is followed, this spread can be expected to be rapid.

The conundrum is, why has it taken so long for chloroquine resistance to emerge in *P. vivax*? Epidemiological factors as well as genetic factors no doubt play a role. So far, both the genetic and biochemical bases of chloroquine resistance in this parasite are unknown, and there are still large gaps in our knowledge of these aspects of chloroquine resistance in *P. falciparum*. It will be interesting to observe, in coming years, whether there is any evidence of cross-resistance of the asexual blood stages between chloroquine and primaquine (or, perhaps, newer 8-aminoquinolines that are currently being developed, such as WR 238,605) or, indeed, whether the hypnozoites in chloroquine-resistant infections show any decrease in their sensitivity to primaquine. There are already indications that the standard radical curative (i.e., hypnozoitocidal) regimen is inadequate in up to 20% of cases originating over a wide geographical area, not just from the southwest Pacific (Doherty *et al.*, 1997).

Resistance of *P. vivax* to antifols is well known, particularly against pyrimethamine (see the review by Peters, 1987). The cure rate of *P. vivax* with Fansidar is also apparently less than total in many areas, but few critical analyses have been made to distinguish recrudescences due to resistance from relapses caused by surviving hypnozoites. Nevertheless, it is noteworthy that (i) pyrimethamine does not eliminate hypnozoites even of sensitive parasites and (ii) resistance to pyrimethamine (and probably other antifols) appears to occur at all stages of the malaria parasite's life cycle.

2.3. Plasmodium ovale and Plasmodium malariae

Resistance of *P. malariae* to pyrimethamine was first recorded in neurosyphilitic patients receiving malaria therapy. Since that time, few observations on the antimalarial resistance of this parasite have been published; neither has there been any report of note of resistance of the blood stages of *P. ovale* (see Peters, 1987).

3. EXPERIMENTAL DRUG RESISTANCE IN LABORATORY MODELS

3.1. In Vitro Models — Plasmodium falciparum

Numerous isolates and clones of *P. falciparum* with different levels of drug sensitivity have been employed for investigations of the modes of drug action, the mechanisms and the genetic bases of drug resistance. The seminal culture technique of Trager and Jensen (1976) and the drug sensitivity test of Desjardins *et al.* (1979) have remained the basic procedures used by most workers for the last two decades. However, a simpler model for many investigations is still required. To produce lines resistant to drugs, two approaches have been used. The first is simply to expose cultured parasites in successive passages to stepwise increases in drug concentration. In this way, for example, Nateghpour *et al.* (1993) developed ninefold resistance to halofantrine starting with the chloroquine-resistant K1 line of *P. falciparum*, but only three- to five-fold resistance in a chloroquine-sensitive clone. The response to chloroquine of the halofantrine-resistant K1 line was found to have reverted towards normal, but the sensitivity to mefloquine and quinine had decreased. Peel *et al.* (1994) produced a number of lines resistant to mefloquine or chloroquine, the former having an amplified *Pfmdr1* gene and a point mutation (tyrosine-86 to phenylalanine). The latter showed no amplification and reversion to tyrosine-86. The second approach is to expose the cultured parasites to various mutagens as well as the antimalarial drugs. For example, Thaithong *et al.* (1992), using chemical mutagens to select a series of pyrimethamine-resistant mutants, selected some in which the gene for the DHFR–TS enzyme complex was amplified and others with a single point mutation from isoleucine (Ile)-164 to methionine (Met)-164 in the DHFR domain. Rathod *et al.* (1994), after a single exposure of the parasites to ultraviolet light, rapidly selected, in a single step, a line highly resistant to 5-fluoroorotate. (See also Section 5.1.5 for comments on resistance to atovaquone and 5-fluoroorotate.)

The yeast *Saccharomyces cerevisiae* has been used as a surrogate host for genes of *P. falciparum* that are associated with resistance to chloroquine and to antifols. Ruetz *et al.* (1996) have inserted the *Pfmdr1* gene of *P. falciparum* from both chloroquine-sensitive and chloroquine-resistant isolates into this yeast. The transformed organisms containing the gene from the sensitive parasite proved to be resistant to the aminoalcohols quinine, mefloquine and halofantrine as well as, rather surprisingly, to the aminoacridine compound, mepacrine. This was associated with decreased cellular drug accumulation as well as an increased drug efflux from preloaded yeast cells (see also Section 4.1). The gene coding for the DHFR domain of the double enzyme, DHFR–TS, from *P. falciparum* has also been inserted into *S. cerevisiae* by Wooden *et al.* (1997). Their model is potentially valuable for the study of antifolate resistance.

3.2. Rodent Malaria

Although their value as models for the human malaria parasites has been adequately demonstrated, relatively little work appears to have been carried out during the past decade on antimalarial drug resistance in rodent malarias (other than molecular biological investigations of the genetic and molecular basis of resistance to antifols). When interpreting data on drug resistance induced in such parasites as *P. berghei* or *P. yoelii,* it is essential to note carefully the technique that has been employed. The exposure of the former to slowly increasing drug selection pressure by chloroquine, for example, resulted in the production of lines such as *P. berghei* RC that have a high level of resistance but totally abnormal morphology (Peters, 1987). The investigation of biochemical or other changes in such parasites may lead to conclusions that are irrelevant to the situation in chloroquine-resistant *P. falciparum*. Mahmalgi *et al.* (1989), for example, noted that 4-aminoquinolines were found in diffuse endocytotic vesicles scattered throughout the cytoplasm in parasites of the *P. berghei* RC type, in contrast to being concentrated in the area of the larger food vacuoles of sensitive organisms. *P. yoelii*, on the other hand, has a low level of inherent chloroquine resistance which can be increased by exposure to the drug without unduly modifying the morphology. Two different mechanisms are probably involved, but they have yet to be identified. A similar phenomenon has been recorded by van Dijk *et al.* (1994) in relation to pyrimethamine resistance in *P. berghei* NYU-2. Much of the research by my colleagues and myself has been carried out on mice infected with either the drug-sensitive *P. berghei* N strain or the chloroquine-resistant *P. yoelii* ssp.NS or, in a few cases, *P. yoelii nigeriensis*. The last two parasites produce abundant gametocytes so that cyclical transmission can be maintained and studies on the stability of resistance after mosquito transmission can be conducted.

These models have the disadvantage of yielding asynchronous parasitaemias. A few investigators have made use of synchronous infections in mice with *P. chabaudi*, *P. vinckei vinckei* or *P. vinckei petteri*. In this Section I shall consider new studies on the development of drug resistance in rodent models, while research aimed at explaining the biochemical and genetic mechanisms underlying resistance will be reviewed in the following Section. (See also Section 5.1.6.)

When a novel antimalarial compound appears, it is obviously important to explore the risk that its deployment may select parasites that are resistant to it. For this purpose, the use of a relapse technique using chloroquine-resistant *P. yoelii* ssp.NS in mice has proved particularly valuable since the principal target for new blood schizontocides is the discovery of compounds that will inhibit chloroquine- or multidrug-resistant *P. falciparum*. It was mentioned above that, starting with this parasite, resistance to mefloquine and halofantrine is readily selected by the procedure described as the '2% relapse technique' (Peters, 1987). Fonseca *et al.* (1995) were able to select a cyclically transmissible, mefloquine-resistant line from a clone of *P. berghei* ANKA by progressively increasing the drug selection pressure in gametocyte-carrying mice on which the anopheline vectors were fed. De *et al.* (1996) and Ridley *et al.* (1996) recently described a number of 4-aminoquinolines and bisquinolines (Ridley *et al.*, 1997) which are significantly more active as blood schizontocides against chloroquine-resistant rodent malaria parasites than chloroquine itself. Cogswell *et al.* (1997) pursued the matter further to show that several of the 4-aminoquinolines were also effective in *Saimiri* monkeys infected with a chloroquine-resistant strain of *P. falciparum*, as well as against *P. cynomolgi* in the rhesus monkey. However, there was a clear correlation between the IC_{50} values for chloroquine and some of the most active compounds reported by Ridley *et al.* (1996) against a large battery of *P. falciparum* isolates *in vitro*. Moreover, using the '2% relapse technique' it was shown (Figure 3) that resistance to one of the most active of the bisquinolines was almost as rapidly developed as resistance to mefloquine (Peters *et al.*, 1977) or to halofantrine (Peters *et al.*, 1987). Such observations lead to the tentative conclusion that the further development of such compounds has little chance of yielding a blood schizontocide that could replace chloroquine for the treatment of multidrug-resistant *P. falciparum*.

Another blood schizontocide currently under development is pyronaridine, a compound that resembles a hybrid between mepacrine and a Mannich base (Figure 2). This compound is highly active *in vivo* against asexual intraerythrocytic stages of rodent malarias, including *P. yoelii* ssp. NS and the N/1100 line of this parasite which is resistant to mefloquine. However, resistance was selected readily to this compound using the '2% relapse technique', although less readily with *P. berghei* N (see Peters and

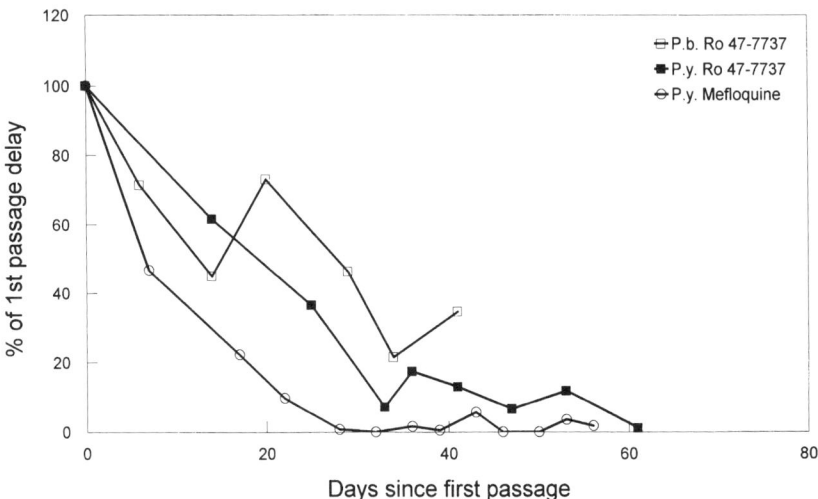

Figure 3 The rate of selection of resistance to the bisquinoline, Ro 47–7737, and to mefloquine in the '2% relapse test' in mice infected with chloroquine-sensitive *Plasmodium berghei* N strain (P.b.) or chloroquine-resistant *Plasmodium yoelii* ssp. NS (P.y.). Note that the time to complete resistance for the bisquinoline in the chloroquine-resistant parasite (indicated by a fall of the 2% delay time to zero) was twice that required for mefloquine. The chloroquine-sensitive parasite developed resistance to Ro 47–7737, but at a slower rate. (Based on data given by Ridley *et al.*, 1997.)

Robinson, 1992). When compared with the responses of its parent line, the SPN line developed from *P. yoelii* NS showed a high level of cross-resistance to amodiaquine and mepacrine but a lower level of resistance to chloroquine, mefloquine and halofantrine. A similar pattern was seen in the NPN line developed from *P. berghei* N when compared with its parent strain. In addition to our own experiments, studies in China have confirmed the ease with which resistance to pyronaridine can be developed by rodent malaria parasites (see the review by Fu and Xiao, 1991).These, and more recent experiments led to the conclusion that the eventual deployment of such a compound for monotherapy of multidrug-resistant *P. falciparum* is fraught with the danger that pyronaridine-resistant parasites may rapidly be selected, as has happened with mefloquine. Underlining this warning is a report by Looareesuwan *et al.* (1996a) from Thailand, who observed a relatively high recrudescence rate when patients infected with *P. falciparum* received treatment with pyronaridine alone. (See, however, the comment on pyronaridine combinations in Section 5.1.4.)

With the increasing popularity of blood schizontocides of the artemisinin series for the treatment of multidrug-resistant *P. falciparum*, it was

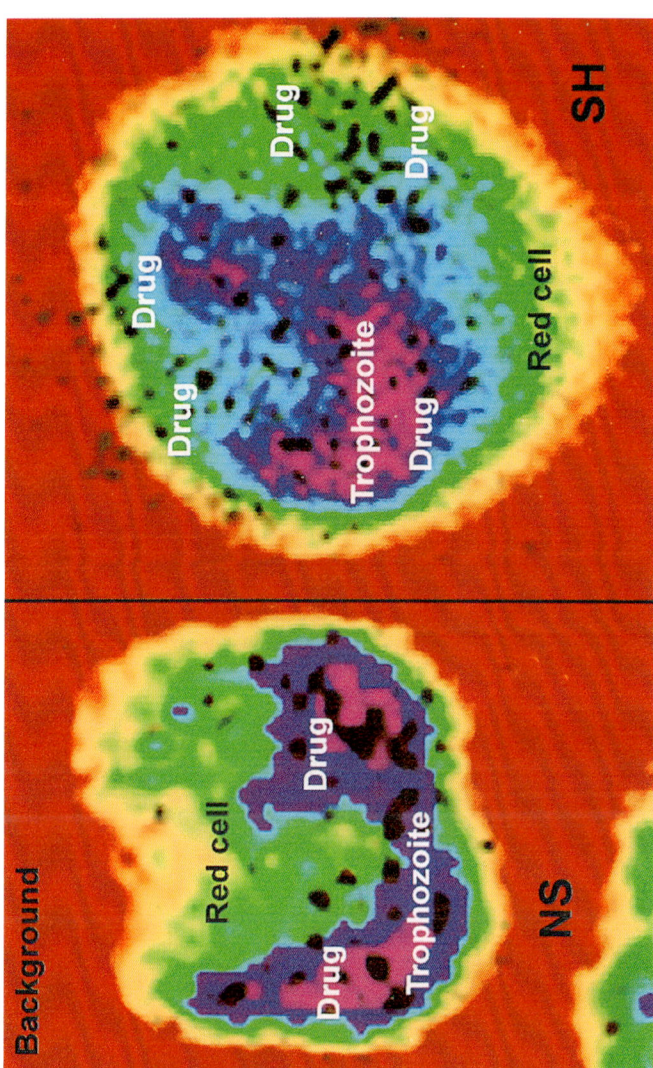

Plate 1. The localization of halofantrine in intraerythrocytic trophozoites of (left) *Plasmodium yoelii* spp. NS (sensitive to halofantrine) and (right) *Plasmodium yoelii* ssp. SH line (resistant to halofantrine). These electron micrographs, with computer-enhanced colouring were produced by a combination of scanning ion microscopy and mass spectrometry, the images from each of these being superimposed. The localization of halofantrine is based on the detection of fluorine in the compound and it appears as black granules, concentrated mainly over the trophozoite of the sensitive line but widely dispersed over the parasite and host erythrocyte of the SH line. Slight differences in the image produced by each technique make it appear that some drug is lying outside the erythrocyte containing the SH trophozoite. (Modified from micrographs kindly supplied by Dr Yves Boulard.)

obviously an urgent matter to explore the ability of parasites to become resistant to them. Between 1986 and 1992, Chinese workers reported the development of lines of *P. berghei* resistant to artemisinin, sodium artesunate and artemether, and a line of *P. falciparum* resistant to sodium artesunate. Fedorova (1991) also recorded a line of *P. berghei* with a high level of resistance to artemisinin. As already mentioned (Section 2.1.5), Chawira *et al.* (1986) developed two artemisinin-resistant lines using the '2% relapse technique', one initiated from *P. berghei* N and the other from *P. yoelii* NS (QS). In the latter line, resistance was first detected by the fifth passage and, by passage 50, 35-fold resistance had developed. However, resistance developed more slowly and to a lower level in *P. berghei* N. Interestingly, resistance in the QS line diminished when the drug selection pressure was withdrawn for 21 passages, but reappeared again after only two passages under renewed drug selection pressure. Recently, we repeated our earlier experiments, but using artesunate instead of artemisinin, and have observed a similar, relatively slow, emergence of resistance in the line derived from *P. yoelii* ssp. NS. If a parallel can be drawn with our previous experience with other compounds discussed above, it is to be hoped that resistance in *P. falciparum*, will also be slow to emerge in areas where the artemisinins are widely used.

Several studies have pointed to an apparent biological advantage that resistance to chloroquine confers on both rodent malaria (Ramkaran and Peters, 1969; Ichimori *et al.*, 1990) and *P. falciparum*, namely enhanced asexual schizogony (Wernsdorfer *et al.*, 1995) and transmissibility (see the review by Peters, 1987; also Handunnetti *et al.*, 1996). Robert *et al.* (1996) found that both gametocyte density and prevalence were higher in chloroquine-resistant than in chloroquine-sensitive infections of *P. falciparum* in a hypoendemic area of Senegal, implying that treatment with chloroquine would favour transmission of the resistant parasites. Conversely, another factor favouring the sesquiterpene lactones (as well, perhaps, as the synthetic endoperoxide, Fenozan B07; Peters *et al.*, 1993b) is that their gametocytocidal effect may contribute to a reduction of *P. falciparum* transmission in areas of high endemicity (Price *et al.*, 1996).

Rodent malaria models have also been widely used in the past, and continue to be used, to explore resistance to antifols, sulphonamides and naphthoquinones. Earlier work has been reviewed by Peters (1987). Puri and Dutta (1989) showed that the combination of mefloquine with dapsone impeded the emergence of resistance to these compounds in mice infected with *P. berghei*. More recently, my colleagues and I have investigated the rate at which resistance can be selected (in *P. berghei* N) to atovaquone and have succeeded in selecting a high level of resistance in a single passage, much as was found with menoctone (Peters, 1987). With equal rapidity we selected a new line of this parasite with a high level of resistance to cycloguanil. We are currently investigating whether resistance can be obtained in a further line that is exposed to a synergistic mixture of these two compounds.

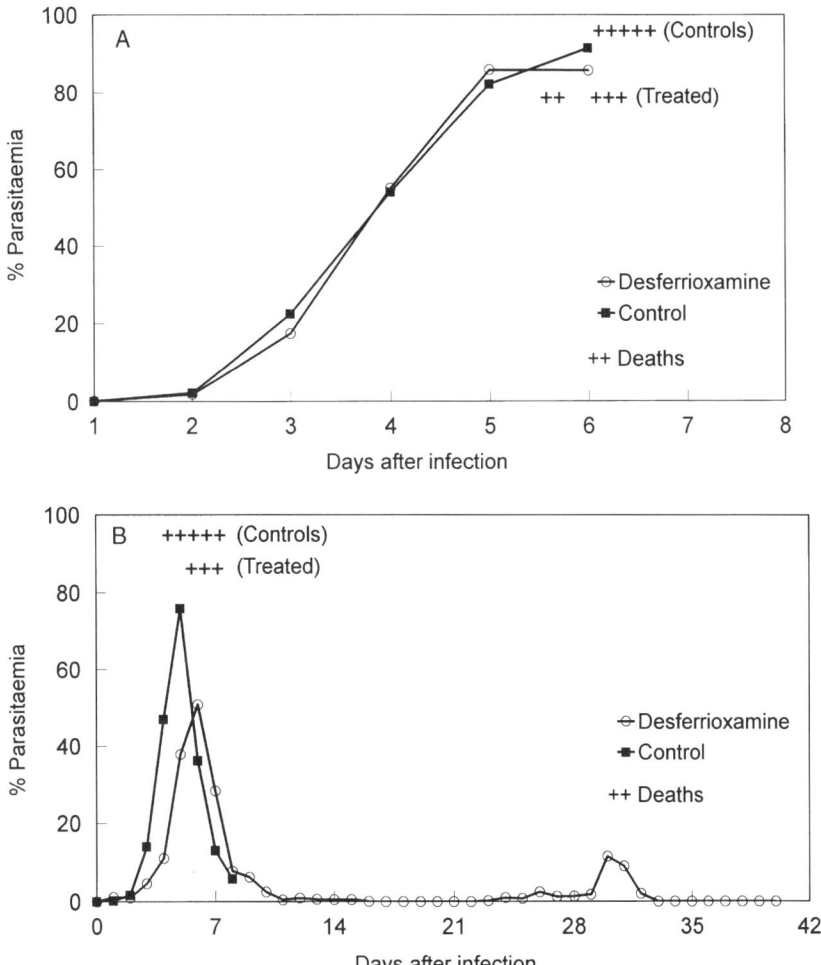

Figure 4 The response of *Plasmodium berghei* N strain (A) and *Plasmodium chabaudi* ASS strain (B) in mice to desferrioxamine in the '4-day test'. The animals (five per group) received a twice daily dose of 150 mg kg^{-1} body weight of drug subcutaneously for up to 7 days after infection. Two of five mice infected with *P. chabaudi* survived more than 35 days but none of those infected with *P. berghei* did so. Controls with either infection died in 6–8 days. This response pattern was consistent in repeat experiments with higher doses. (Author's unpublished data.)

Recently, we further investigated the value of rodent malaria models for evaluating potential chelating agents as antimalarial drugs, by examining the activity of desferrioxamine and newer iron chelators against rodent malaria parasites *in vivo*. Two interesting observations (so far unpublished) have emerged from our experiments. The first was that desferrioxamine must be given in high dose and at least twice daily for seven or more days to demonstrate an inhibitory action on parasitaemia. This necessitates modifying our standard '4-day test' into an 'extended test' for such a compound. The second observation was that, whereas *P. berghei* N strain (a non-synchronous infection in mice) was not inhibited by desferrioxamine (and other iron chelators), a synchronous infection with *P. chabaudi* ASS strain was almost completely suppressed (Figure 4).

These observations indicated two important principles: first, that one may need to use more than one species in screening for certain types of antimalarial activity, and second that the rodent system can be of value in determining the action of iron chelators (as well as giving a guide to their selective toxicity).

4. THE BIOCHEMICAL AND GENETIC BASES OF DRUG RESISTANCE

It is self evident that an exploration of the ways in which parasites become resistant to drugs implies a sound knowledge of the ways in which the drugs exert their antiparasitic action on the sensitive organisms. This knowledge, in relation to antimalarial agents, is reasonably solid for some groups, for example, the antifols and naphthoquinones, but still grossly inadequate for others, notably the 4-aminoquinolines. Three general directions have been pursued in the analysis of the mechanisms and genetics of resistance, biological (i.e., studies on the relationship between the parasite in all its developmental stages and its hosts, an approach that is considered rather 'old-fashioned' by some contemporary scientists), biochemical, and genetic, the last employing the tools of molecular biology. Reviews of these three aspects have been published by Landau and Chabaud (1994a), Basco *et al.* (1994) and Plowe and Wellems (1995), respectively. The rapid progress now being made in sequencing the genome of *P. falciparum* by Dame *et al.* (1996) should greatly facilitate these studies.

4.1. Chloroquine and other 4-Aminoquinolines

Free merozoites in the circulation are believed not to be vulnerable to the action of antimalarial drugs. In a broad comparison of the responses to chloroquine of a wide range of rodent malarias, Landau and Chabaud (1994b) have observed a clear correlation between the sensitivity of a particular species (or subspecies) and the time during which merozoites released during asexual, intraerythrocytic schizogony remain free, either in the circulation or possibly in the lymphatic vessels (i.e., 'merozoite latency') (Beauté-Lafitte et al., 1994). The presence of latent merozoites is also a factor related to the asynchrony of intraerythrocytic schizogony in chloroquine-resistant parasites. Coquelin et al. (1997) have found that, under chloroquine selection pressure, a normally synchronous infection of P. chabaudi in mice becomes progressively asynchronous as the resistance to chloroquine develops. In an elegant series of experiments, Cambie et al. (1991) were able to pinpoint the stages upon which different drugs exert their action on the intraerythrocytic parasites: chloroquine, for example, acted essentially upon 'middle-aged' trophozoites, while Caillard et al. (1992) found that arteether acted primarily on young ring forms of P. vinckei petteri. Landau et al. (1991) proposed the term 'chronotherapy' to describe the optimal timing of therapy in order to impose the maximal effect of a given compound, dependent upon the period of maximum drug sensitivity during the cycle. Landau et al. (1992) provided further evidence, demonstrating the practical application of this principle during chloroquine treatment of patients infected with P. falciparum. However, no explanation of the correlation between chloroquine response, merozoite latency and the duration of the intraerythrocytic cycle has been proposed in biochemical terms. Possibly of some relevance to this work is the observation by Dei-Cas et al. (1984) that the 50% effective dose (ED_{50}) for chloroquine within P. berghei inhabiting mature erythrocytes is lower than that found in parasites in reticulocytes since, in general, there is a direct correlation between the proportions of parasites found in reticulocytes and decreasing chloroquine sensitivity.

Deactivation of chloroquine by resistant parasites as a mechanism for resistance against it has been excluded by Berger et al. (1995) and other mechanisms are still being explored. Whereas chloroquine reaches a high concentration in parasitized erythrocytes, it has long been recognized that the concentration in erythrocytes that contain chloroquine-resistant parasites such as P. berghei is significantly lower (Macomber et al., 1966; see also review by Peters, 1987). In studies with cultures of P. falciparum in vitro, Krogstad et al. (1987) proposed that the mechanism of resistance to chloroquine is enhanced efflux of the drug from the infected host–parasite complex. This hypothesis initially gained wide acceptance but was challenged by Ginsburg and Stein (1991) who, in a mathematical analysis, showed that it failed to fit the kinetics of drug efflux. Subsequently, Bray et al. (1994) showed that changes in initial uptake of chloroquine more closely paralleled the level of drug response in P. falciparum in vitro than the rate of efflux. Further support for this argument was provided by Hawley et al. (1996), who obtained similar data with amodiaquine which has, nevertheless, a higher binding affinity than chloroquine for the parasite receptor in chloroquine-resistant *P. falciparum in vitro*. The uptake of both chloroquine and amodiaquine is increased by verapamil (Bray et al., 1996). The Krogstad hypothesis was, nevertheless, a spur to molecular biologists who, drawing a parallel with drug efflux from various types of cancer cells, sought to show whether a similar genetic mechanism was involved, namely the presence of a surface protein equivalent to the multiple drug-resistance (mdr) protein of mammalian cells that is amplified in resistant parasites.

In 1992 Slater and Cerami described an enzyme in intraerythocytic *P. falciparum* that they considered to be a haem polymerase which converts potentially toxic haem (formed during digestion of haemoglobin by the parasite) into insoluble haemozoin, malaria pigment. They also produced data to show that the activity of this enzyme was inhibited by a number of blood schizontocides including chloroquine, while Raynes et al. (1996) showed also that a series of novel antimalarial bisquinolines inhibited haem polymerization. Slater (1993) postulated that the mechanism of chloroquine resistance was the selection of a mutant gene coding for a modified polymerase. This would imply that, in chloroquine-resistant parasites, the mutant polymerase would still be able to detoxify haem, thus permitting the parasites to survive. Attractive as this hypothesis appeared, doubt was cast on it when Dorn et al. (1995) showed that this polymerase is not essential for the polymerization of haem in the parasite food vacuoles since haemozoin (β-haematin) forms spontaneously in the presence of haem in the acid conditions that prevail within the parasite's digestive vacuole. Histidine-rich proteins in the digestive vacuole also can bind haem and mediate haemozoin formation (Sullivan et al., 1996a). Haem polymerization is inhibited by quinine and 4-aminoquinolines (Egan et al., 1994), probably because the drugs form a complex with haem, which is first incorporated into the polymer, then blocks any further growth (Sullivan et al., 1996b). The free drug–haem complex that remains is toxic to the parasites, as first proposed by Fitch et al. (1982).

Surprisingly few studies have been published on the molecular basis of chloroquine resistance in any of the rodent malarias, in view of their ready availability and range of chloroquine responses. One of the exceptions is a report by Li et al. (1995), who detected overexpression of a 54 kDa protein in the serum of mice infected with lines of *P. berghei* ANKA that were resistant to chloroquine or to pyronaridine. Resistance in this parasite was induced by the technique of slowly increasing drug selection pressure. Immunoelectron microscopy indicated that the protein was widely distributed in the parasite cytoplasm but the function of the protein is unknown.

Carlton and her associates (J. Carlton, personal communication, 1997) found a single copy of the *Pfmdr1* homologue, *Pcmdr1*, on chromosome 12 of *P. chabaudi*. *Pcmdr1* was duplicated in a mefloquine-resistant line and translocated to chromosome 4. Crossing experiments of this line with a mefloquine-sensitive line produced evidence suggesting that mefloquine resistance may be multigenic in *P. chabaudi*. Most other studies have been carried out in cultured *P. falciparum*. A number of investigators have now provided clear evidence that a homologue of mammalian *p*-glycoprotein, now termed Pgh1, is present on the surface of the digestive vacuole of *P. falciparum* (Cowman *et al*., 1991) and two genes, *Pfmdr1* and *Pfmdr2*, that code for it have now been identified on chromosome 5 and sequenced. Two types of change in these genes have been investigated: their amplification and the existence of point mutations in association with drug resistance. Multiple copies of *Pfmdr1* have been identified in some chloroquineresistant isolates of this parasite (Foote *et* al., 1989; Wilson *et al*., 1989), but *Pfmdr2* does not appear to bear any relation to chloroquine resistance (Rubio and Cowman, 1994). However, the examination of a wider spectrum of newly isolated, chloroquine-sensitive and chloroquine-resistant isolates has negated the suggestion that amplification of *Pfmdr1* is responsible for chloroquine resistance since this has not been demonstrable in many of them (Wellems *et al*., 1990; Basco *et al*., 1995). Moreover, studies of *Pcmdr1* in *P. chabaudi* also failed to provide evidence of amplification in chloroquine-resistant lines which appeared to be multigenic in origin (J. Carlton, personal communication). Conversely, a direct link has now been established between the amplification of the *Pfmdr1* gene and resistance to mefloquine, halofantrine and quinine (Wilson *et al*., 1993; Cowman *et al*., 1994). Of particular interest was the observation that chloroquine resistance was not linked to the *Pfmdr* genes in a line obtained by crossing two isolates of *P. falciparum*, one chloroquine-sensitive and the other resistant, in an anopheline vector (Wellems *et al*., 1990). Although the evidence to date suggests that mutations in the *Pfmdr1* gene are not responsible for chloroquine resistance (see reviews by Cowman, 1991 and Wellems, 1991), several point mutations have been associated with this state, for example a change from codon Asn-86 to codon Tyr-86 in African isolates (Foote *et al*., 1990). However, marked geographical differences have since been identified in the point mutations that are present in isolates sensitive or resistant to chloroquine or to mefloquine (Basco *et al*., 1995, 1996) and the current belief is that resistance to chloroquine (and, perhaps, also to mefloquine) in *P. falciparum* is almost certainly a multigenic phenomenon, as was originally suggested by Padua (1981) on the basis of her studies on *P. chabaudi*. Further support for this hypothesis has been presented recently by Coquelin *et al*. (1997) for the same rodent parasite and by Triglia *et al*. (1991) for *P. falciparum*.

4.2. Dihydrofolate Reductase Inhibitors

Unlike those of chloroquine, both the molecular sites of action and the genetic changes underlying drug resistance are well documented in the case of the antifols (using this term to include sulphonamides and sulphones). Malaria parasites are unable to utilize preformed folate and have to synthesize tetrahydrofolate *de novo* from guanosine triphosphate (GTP) (see review by Gutteridge and Coombes, 1977). At an early stage in this metabolic path, sulphonamides and sulphones block the action of dihydropteroate synthase (DHPS) in binding pteridine pyrophosphate (derived from GTP) to *p*-aminobenzoic acid (PABA) to form dihydropteroate, by mimicking PABA. DHPS is now known to be one part of a double enzyme, the other domain of which is hydroxymethylpterine pyrophosphokinase. The absence of this enzyme from the human organism provides the basis for the highly selective antiparasitic (and antibacterial) action of sulphonamides and sulphones. Pyrimethamine and proguanil block a later step in folate metabolism (the latter essentially through its triazine metabolite, cycloguanil) from dihydrofolate to tetrahydrofolate which is mediated through the double enzyme DHFR–TS. The selective action of these compounds against *Plasmodium* depends upon a far higher binding affinity of the parasite enzyme to them compared with their binding to the DHFR of the host tissues.

It has long been recognized that cross-resistance between proguanil and pyrimethamine was a common but inconsistent finding in *P. falciparum*. With the identification, cloning and sequencing of the gene coding for the synthesis of DHFR, the road was opened to the discovery of the association of point mutations in the gene to sensitivity or resistance to antifols. Based on the examination of large numbers of isolates of *P. falciparum* from different geographical areas, various workers have been able to identify specific alleles on several codons of the DHFR domain of the DHFR–TS gene on chromosome 4 (see reviews by Hyde, 1989 and Plowe and Wellems, 1995). Moreover, in some rodent malaria parasites, gene amplification has also been identified, notably when resistance to pyrimethamine was induced in *P. berghei* NYU-2 by slowly increasing the drug selection pressure *in vivo* (van Dijk *et al.*, 1994). In this work, exposure of *P. berghei* ANKA to a high drug dose resulted in the selection of a resistant parasite with the single point mutation Ser-110 to Asn-110. Van Dijk *et al.* (1995) succeeded in transfecting *P. berghei* merozoites by electroporation with a portion of the gene coding for pyrimethamine-resistant DHFR–TS. The transcribed gene sequence was found to be located in the subtelomeric region of chromosome 7 (van Dijk *et al.*, 1996). The DHFR–TS gene was also amplified in a pyrimethamine-resistant line of *P. chabaudi* developed by Cowman and Lew (1989).

A single point mutation on codon 108 from Ser to Asn on chromosome 4 has been identified as the major change underlying resistance to pyrimethamine and to cycloguanil in *P. falciparum* (see review by Hyde, 1989). Cycloguanil resistance of a very high level is also associated with a change from Iso-164 to Leu-164. High levels of pyrimethamine resistance may also be associated with changes in codons 51 (Asn to Ile) and 59 (Cys to Arg). These mutations have been described in *P. falciparum* isolates derived from geographical areas as divergent as Latin America (see, for example, Peterson *et al.*, 1991), sub-saharan Africa, Kampuchea (Basco *et al.*, 1995) and Papua New Guinea (Reeder *et al.*, 1996). Sirawaraporn *et al.* (1990) have shown how the binding affinity of pyrimethamine for the DHFR of *P. falciparum* is reduced by the structural modification in the enzyme associated with the change from Ser-108 to Asn-108. These workers noted that, in isolates from humans, mutations at residues 51, 59, 108 and 164 were associated with cross-resistance between pyrimethamine and cycloguanil, whereas that at residue 16 conferred resistance only to cycloguanil (Sirawaraporn *et al.*, 1997). They suggested that multiple mutations have evolved from a primary mutant, namely at residue 108, and that new drugs targeted at both wild-type DHFR and at the Asn-108 mutant should minimize the further selection of antifol resistance. Curtis *et al.* (1996) found that the Asn-108 mutant was rapidly selected in a group of Tanzanian children who were treated with either Fansidar or a chlorproguanil/dapsone combination. Pyrimethamine resistance may also be associated with increased expression of the DHFR gene in *P. falciparum* (see Thaithong *et al.*, 1992).

Resistance of a number of Kampuchean clones of *P. falciparum* to sulfadoxine has now been associated with several point mutations in the DHPS domain of the gene on chromosome 8, notably Ala-581 to Gly-581 (L.K. Basco, personal communication). In a group of isolates from different continents, several alleles have been identified in codons 436, 437 and 613, but any association of these with sensitivity to sulphonamides remains to be defined. Wang *et al.* (1995) have developed a PCR procedure to detect these mutants.

4.3. Quinine, Mefloquine and Halofantrine

As with chloroquine, the mode of action at the molecular level of these aminoalcohol blood schizontocides is still not fully understood. While their site of action is almost certainly at the level of the digestive vacuole of the intraerythrocytic parasites, they do not produce the clumping of haemozoin that is characteristic of chloroquine and related 4-aminoquinolines but, nevertheless, do induce qualitative changes in the pigment granules that are visible at the electron microscope level (see Peters, 1987). Surprisingly, there are few reports on the ultrastructure of parasites that are resistant to these compounds. Boulard *et al.* (1996) used scanning ion microscopy to identify the location of halofantrine in the erythrocytes of mice that had been infected with either *P. yoelii* ssp. NS or a halofantrine-resistant line (SH) derived from it (Peters *et al.*, 1987). The drug was seen to be localized within the NS parasites, whereas there was diffuse distribution of drug in host cell cytoplasm as well as parasites in erythrocytes infected with the resistant SH line (see Plate 1). On the whole, research on resistance to the aminoalcohols appears to have started rather with the application of molecular biological techniques than by less sophisticated biological or biochemical routes. One striking difference has been observed between chloroquine and mefloquine-resistant parasites, namely their response to resistance-modulating agents. Whereas verapamil has a marked resistance-reversing effect on chloroquine-resistant *P. falciparum* or *P. yoelii* ssp. NS, this compound is ineffective against mefloquine-resistant parasites. However, the resistance of these is modulated by other compounds such as penfluridol (see Section 5.2.1).

Unfortunately, many of the chloroquine-resistant *P. falciparum* isolates referred to above, on which genetic studies have been carried out, were not resistant simply to chloroquine but also to mefloquine, especially those from south-east Asia. Responses *in vitro* to mefloquine have not always been reported, so that it is not always clear which point mutation is related to which compound. In 1993 Wilson *et al.* found that the *Pfmdr1* gene was amplified in a number of isolates of *P. falciparum* obtained from Thailand that were resistant to mefloquine and halofantrine. Basco *et al.* (1996), working with freshly isolated clones from Kampuchea, have, however, cast doubt on the hypothesis that resistance to mefloquine is directly associated with amplification of *Pfmdr1*; a possible explanation of these observations was that the level of resistance to mefloquine in their isolates may have been less than that of those from Thailand and that they were not cross-resistant to halofantrine, as were the latter. (See also comments on the inverse relationship between mefloquine or halofantrine resistance and the sensitivity to chloroquine in Sections 2.1.3. and 3.1.) Peel *et al.* (1994), who developed a line of *P. falciparum* resistant to mefloquine *in vitro*, observed a point mutation in amino acid 86 of the *Pfmdr1* molecule from tyrosine to phenylalanine. Similarly, Ritchie *et al.* (1996) developed a halofantrine resistant line from the chloroquine-resistant K1 clone of *P. falciparum* but found no change in Tyr-86; nor did their K1HF line show any amplification of *Pfmdr1*.

4.4. Naphthoquinones

The site of action of atovaquone and, presumably, that of its naphthoquinone precursor menoctone, has been identified as the metabolic step that is catalysed by the enzyme dihydroorotate dehydrogenase, which mediates the reduction of parasite dihydroorotate to orotate. The drug is believed to mimic the ubiquinones 7 and 8 which are specific to *Plasmodium* (the mammalian respiratory chain utilizing ubiquinone 10) and which play an essential role in electron transport at the level of the parasite mitochondrion. The genetic basis of atovaquone resistance has not yet been described. However, the speed at which resistance to menoctone or to atovaquone can be selected suggests that a point mutation is involved and that it is one that occurs at high frequency in wild populations of both *P. falciparum* and *P. berghei*. As with the antifols, it is likely that such a mutant yields a structural change in dihydroorotate dehydrogenase that diminishes the affinity of the enzyme for the drug.

5. DEVELOPMENT OF NEW ANTIMALARIAL DRUGS AND DRUG COMBINATIONS

To present a review of all the many compounds that have been examined in the search for novel antimalarial drugs during the past decade is beyond the scope of this chapter. However, a few general features of the approaches that are being taken do merit consideration. Several trends are especially notable. The first is a change from seeking only drugs that act upon the malaria parasite to identifying compounds that exert their effects through an influence on the host. Increasing understanding of the pathophysiological changes that underlie the more severe manifestations of clinical malaria, for example the overproduction of certain cytokines such as TNF with the cascade of secondary biochemical events that can succeed this (see reviews by Warrell *et al.*, 1990, Pasvol *et al.*, 1995 and Clark and Rockett, 1996), have led to the suggestion that drugs should be sought which can antagonize these host responses. To date, however, no specific drug has been proposed for this purpose, although chloroquine itself has been identified as a compound that antagonizes both TNF and nitric oxide (see Section 2.1.1). A second trend is the focus on seeking novel antimalarial compounds that may occur in plants. Encouragement for this approach derives from the outstanding success of the sesquiterpene artemisinin that occurs in *Artemisia annua*, as a potent blood schizontocide, and of yingzhaosu which has also been shown by Chinese workers to have potent antimalarial properties. From these two leads, medicinal chemists have been able to develop numerous analogues, some of which have proved invaluable in the treatment of multidrug-resistant *P. falciparum*. A further trend is the increasing change to the deployment of 'combinatorial chemistry' in association with mass, automated drug screening by certain sections of the pharmaceutical industry. For this, it is best if specific targets such as essential, parasite-specific enzymes can be identified for use as screening targets. A number of such enzymes, for example parasite proteases, have been identified (Vander Jagt *et al.*, 1989) but, to date, no commercial organization appears to have taken up the challenge. Progress in the development of new antimalarial drugs is hindered by a conspicuous lack of interest in this field by most of the pharmaceutical industry. However, two biochemical approaches that are currently receiving attention are the search for, firstly, inhibitors of parasite phospholipid synthesis (Vial, 1996) and, secondly, parasite-selective chelating agents. In the following pages I shall concentrate mainly on compounds that have achieved the stage of clinical trial or beyond into clinical deployment.

5.1. Single Compounds

5.1.1. *Sesquiterpene Lactone and Yingzhaosu Derivatives*

Of the numerous semisynthetic analogues of artemisinin that have been described, two have been introduced in a number of countries for the treatment of multidrug-resistant *P. falciparum*. Artemether, which has very low aqueous solubility, is administered either intramuscularly in an oily solution or orally in the form of tablets. Artesunate, a sodium salt, which can be administered in aqueous solution, has proved invaluable in the treatment of severe falciparum malaria, especially against parasites that have acquired some measure of resistance to quinine. It can also be administered orally in tablet form (WHO, 1995). Artesunate has also proved its value when administered in a suppository formulation (Looareesuwan *et al.*, 1995). Artemisinin functions through an active metabolite, dihydroartemisinin, which has also proved effective by oral administration in Thai patients infected with multidrug-resistant *P. falciparum* (see Looareesuwan *et al.*, 1996d). These compounds exert their action mainly on young ring forms of the intraerythrocytic parasites of *P. falciparum* (see Jiang *et al.*, 1982) and murine malarias (Caillard *et al.*, 1992) and their use in humans leads to rapid reduction of parasitaemia. Although the use of the artemisinins, especially artesunate, also results in rapid resolution of fever, direct comparisons of artemether with quinine for the treatment of cerebral malaria in Gambian children (van Hensbroek *et al.*, 1996) or Vietnamese adults (Hien *et al.*, 1996) have failed to show any advantage in terms of survival rates. However, since treatment with artemisinin or its analogues is followed by an undesirably high recrudescence rate (which, so far, has not been associated with the emergence of drug resistance), a second antimalarial drug such as mefloquine is frequently given in association with it to ensure a radical cure (see Section 5.2.2).

The synthetic compound known as Ro 42–1611 or Arteflene, derived from another naturally occurring endoperoxide, yingzhaosu, which occurs in *Artabotrys uncinatus*, was found to be curative in West African patients infected with *P. falciparum* (see Salako *et al.*, 1994; Radloff *et al.*, 1996b) but, unfortunately, problems of formulation and poor bioavailability led to its withdrawal from clinical trial.

5.1.2. Synthetic Endoperoxides and other 'Oxidant' Compounds

Several series of novel antimalarial drugs have been developed based on the identification of the endoperoxide bridge as the pharmacophore of artemisinin, the action of which may be related to the potentiation of the toxic activity of intraparasitic haem (Meshnick *et al.*, 1991; Jefford *et al.*, 1993; Meshnick *et al.*, 1996). The most advanced of the synthetic endoperoxides is a 1,2,4-trioxane known as Fenozan 50-F (also known as Fenozan B07, Figure 2) which is currently in the stage of preclinical development. This compound, like the artemisinins, is active against a broad spectrum of drug-resistant parasites but, in contrast, exerts its action against all stages of intraerythrocytic, asexual parasites except mature schizonts (Fleck *et al.*, 1997a). Like the artemisinins, it also inhibits the development of gametocytes, a property that may well prove to be important if and when the endoperoxides come to be widely used in endemic areas (Price *et al.*, 1996). Several other classes of endoperoxides have been synthesized. These include a series of tricyclic 1,2,4-trioxanes that, like Fenozan B07, cure multidrug-resistant *P. falciparum* infection in *Aotus* monkeys (Posner *et al.*, 1994), a fluorinated bicyclic endoperoxide which has blood schizontocidal activity against rodent malaria (Posner *et al.*, 1996), and a series of dispiro-1,2,4,5-tetroxanes which are active against multidrug-resistant *P. falciparum in vitro* as well as *P. berghei in vivo* (Vennerstrom *et al.*, 1992).

Marked synergism occurs between a benzophenone compound, exifone, and a structurally related compound, rufigallol, against sensitive and multidrug-resistant lines of *P. falciparum in vitro*. Winter *et al.* (1996) postulated that the compounds interact, with the generation of active hydroxyl radicals, to form 2,3,4,5,6-pentahydroxyxanthone which they suggested is a potent antimalarial. They also drew attention to the presence of a number of xanthone derivatives in higher plants that are reputed to possess antipyretic properties. It is striking to note a recent return to the exploration of such oxidant compounds as methylene blue (see, for example, Atamna *et al.*, 1996), which was a seminal drug in the evolution of synthetic antimalarial compounds and the origin of research leading to the 8-aminoquinolines, pamaquine and primaquine (see Peters, 1987).

5.1.3. *Phytochemical Leads*

Although many plant remedies are used by traditional healers to treat fevers, relatively few of them have been submitted to chemical and biological analysis in order to determine the nature of their active constituents, artemisinin itself being the most notable success story to date as regards antimalarial properties. In a valuable review Lampe (1994) drew up a systematic list of plants that have been investigated as sources of active blood schizontocides. Other reviews are those by Phillipson *et al.* (1995) and Kirby (1996). Most of the plant screening has been carried out on *P. falciparum in vitro*. Techniques are available by which a preliminary judgement can be made both of activity and of host toxicity (see, for example, O'Neill *et al.*, 1985). One reason for using *in vitro* screening is the difficulty of producing sufficient pure substance from the relatively crude extracts that are available from most plants for testing in animal models. The result has been that, of the relatively few natural products that have been found to be active and sufficiently nontoxic to merit testing *in vivo*, even fewer have been examined even against rodent malaria. Included in these are bisbenzylisoquinoline alkaloids such as tetrandine (Marshall *et al.*, 1994) and various quassinoids such as bruceantin (O'Neill *et al.*, 1986). A number of novel compounds derived from plant sources have been shown to be immunomodulators, a property that can, in turn, contribute both to antimalarial activity and to drug toxicity. As Kirby (1996) has pointed out, many problems impede the advance of phytochemistry in the antimalarial field. The next steps in the development of such compounds include an attempt to produce semisynthetic analogues with superior activity, to devise means of increasing the yield of an active compound from its natural source, or cultivation of that source on a large scale (as is being done with *A. annua*). Chemical development also includes the determination of the active pharmacophore of novel structures and the total synthesis *de novo* of new, active compounds based on it, as has been accomplished with Fenozan B07 and other endoperoxides (Jefford *et al.*, 1993). The interest of the pharmaceutical industry tends to lie somewhere other than in the problem of malaria.

5.1.4. *Novel 4-Aminoquinolines, Bisquinolines and Mannich Bases*

A number of monoquinolines analogous to chloroquine but with different sidechains have proved to be far more active than chloroquine itself against multidrug-resistant isolates of *P. falciparum in vitro* (Ridley *et al.*, 1996; De *et al.*, 1996), as have certain bisquinolines (Raynes *et al.*, 1996; Ridley *et al.*, 1997). Ridley *et al.* (1996) were, however, able to demonstrate, using their '2% relapse technique', that resistance to a number of these compounds develops very rapidly in the chloroquine-resistant parasite *P. yoelii* ssp. NS, even though, initially, this parasite is very susceptible to them. In the light of our experience with other compounds, e.g. mefloquine, in this system, we believe that this indicates a poor prognosis for the new compounds and places considerable doubt on the value of developing them as novel replacements for chloroquine. Basco and Le Bras (1993a) noted that amopyroquine, a Mannich base related to amodiaquine, shows significant cross-resistance *in vitro* with chloroquine in fresh isolates and clones of *P. falciparum* from Africa.

Related to both the aminoacridine drug, mepacrine and the Mannich base, amodiaquine (both of which show a high level of cross-resistance to chloroquine), is a relatively new Mannich base, pyronaridine (see review by Fu and Xiao, 1991). Reference has been made to this compound above. While it is far more active than chloroquine against multidrug-resistant *P. falciparum* and chloroquine-resistant *P. yoelii* ssp.NS, Liu *et al.* (1996) noted that 8.3% of 156 isolates of *P. falciparum* from Hainan and Yunnan Provinces of China were resistant to pyronaridine *in vitro*. Moreover, clinical experience has already indicated that multidrug-resistant *P. falciparum* infections in humans may prove to have a high recrudescence rate in patients receiving monotherapy with pyronaridine (Looareesuwan *et al.*, 1996a), although it was completely curative in semi-immune West African patients (Ringwald *et al.*, 1996). In my view, the future of this compound used alone is in considerable doubt. In the pyronaridine-resistant SPN line derived from *P. yoelii* ssp. NS, however, strong potentiation has been demonstrated between this compound and artemisinin (Peters and Robinson, 1997). This may indicate that a combination of these two compounds could impede the selection of resistance to one or both of the components should pyronaridine eventually find a place in the clinic. In China, pyronaridine has been used extensively in combination with sulfadoxine and pyrimethamine to treat patients infected with chloroquine-resistant *P. falciparum*. In part of Hainan Province where this combination was the only antimalarial drug employed from 1986, it retained complete activity for at least five years (Shao *et al.*, 1991).

5.1.5. *Compounds Interrupting Pyrimidine Metabolism, Including Antifols*

The fact that intraerythrocytic malaria parasites are dependent on pyrimidine biosynthesis *de novo* makes them particularly vulnerable to selective chemotherapy. While naphthoquinones such as atovaquone and antifols block different steps in thymidylate synthesis, few studies had been made on the blocking of thymidylate synthase itself. Rathod et al. (1992) found that this enzyme was very sensitive to 5-fluoroorotate. Moreover, 5-fluoroorotate was also curative in mice infected with *P. berghei* at a relatively low dose (Krungkrai et al., 1992). Subsequently Rathod et al. (1994) rapidly developed, with the prior use of ultraviolet irradiation, a line of *P. falciparum* that was highly resistant to this compound *in vitro*. They also developed a line that was resistant to atovaquone and found that resistance developed far less readily to a mixture of the two compounds (Gassis and Rathod, 1996). This dramatic example of the protection of two compounds that act on sequential steps in a crucial and specific parasite metabolic pathway could offer a new route to combination therapy for the future.

In 1973 it was reported that WR 99,210 (Figure 2), a triazine antifol analogous to cycloguanil, was an extremely potent inhibitor of *P. falciparum in vitro*, including strains that were resistant to chloroquine, to pyrimethamine and to cycloguanil itself (WHO, 1973). Subsequently this compound was examined in human volunteers but it proved to be very poorly tolerated and was dropped. Recently, a novel biguanide compound, code-named PS-15 (Figure 2), analogous to proguanil, has been shown to act as a prodrug for WR 99,210 and has proved very effective when administered orally to *Aotus* or *Saimiri* monkeys infected with multi-drug-resistant *P. falciparum* (see Canfield et al., 1993; Rieckmann et al., 1996). PS-15 is, moreover, synergistic with atovaquone, at least *in vitro* (Canfield et al., 1995), just as is proguanil *in vivo* (Looareesuwan et al., 1996b). Interestingly, the human phenotypic status as regards extensive or poor metabolic conversion of proguanil to its active metabolite, cycloguanil, does not affect the level of synergism (Edstein et al., 1996), implying that it is proguanil itself that is synergistic. The same may apply to PS-15; it would therefore be a very interesting candidate for clinical trials against resistant *P. falciparum* infections, first alone and, second, together with atovaquone. PS-15 is also synergistic with sulphamethoxazole against *P. falciparum* in *Saimiri* monkeys (Yeo and Rieckmann, 1994a).

Chlorproguanil (Lapudrine), the *p*-chloro analogue of proguanil, which has a somewhat longer half-life in humans than proguanil itself, was introduced briefly in the 1950s in the hope that it would prove of value as a once-weekly prophylactic, instead of proguanil which is given daily. This hope was not realized, but interest in chlorproguanil has been regenerated recently. The combination of dapsone and proguanil, which is synergistic, has long been known to give good protection as a chemoprophylactic against chloroquine/pyrimethamine-resistant *P. falciparum* in humans, just as do combinations of sulphonamides with pyrimethamine (see Peters, 1987). Since the war in Vietnam in the 1960s, however, the dapsone/proguanil combination has not been used, perhaps partly because of the (largely unwarranted) fear of toxicity from dapsone that has been generated by its use as a prophylactic in combination with pyrimethamine (Maloprim), although Yeo *et al.* (1992) showed that a lower dose of dapsone with proguanil was also effective. Chlorproguanil is once again undergoing clinical trials in East Africa, this time in combination with dapsone in areas where many *P. falciparum* infections are resistant to proguanil or pyrimethamine used alone (for prophylaxis — they are no longer deployed for therapy). The initial indications (Watkins *et al.*, 1988; Keuter *et al.*, 1990) are that this combination may prove to be a useful alternative to pyrimethamine/sulfadoxine (Fansidar), which has two potential disadvantages: first, that the sulfadoxine component has inherent skin toxicity and, second, that resistance to the combination is beginning to appear in parts of Africa. Unfortunately, neither proguanil/dapsone nor chlorproguanil/dapsone were curative in a clinical trial in Thailand where *P. falciparum* is known to be resistant to both cycloguanil and dapsone (Wilairatana *et al.*, 1997).

5.1.6. *Antibiotics*

A few reports have appeared indicating that some quinolone antibiotics, such as ciprofloxacin, have a modest but slow blood schizontocidal action against *P. falciparum*, both *in vitro* (Divo *et al.*, 1988; Tripathi *et al.*, 1993c; Yeo and Rieckmann, 1994b) and in humans (Watt *et al.*, 1991; Stromberg and Björkman, 1992). These compounds act against certain bacteria by blocking the metabolism of topoisomerases and it is possible that they have a similar mode of action against intraerythrocytic malaria parasites. Norfloxacin was also only partially effective against *P. vivax* (see Tripathi *et al.*, 1993a) and *P. falciparum* in Indian patients (Tripathi *et al.*, 1993b). Their future role, however, in the treatment of malaria seems doubtful.

Perhaps of greater potential interest are compounds such as thiostrepton and rifampicin that may interfere with protein synthesis in the plastid-like organelles that have recently been identified in *P. falciparum* (see Gardner *et al.*, 1991 and McConkey *et al.*, 1997).

The azalide antibiotic, azithromycin, has good blood schizontocidal activity against *P. berghei* in mice, *P. falciparum* in *Aotus* monkeys, and *P. falciparum* in humans (Andersen *et al.*, 1995a), but it is only partly effective as a causal prophylactic (Andersen *et al.*, 1995b); *in vitro*, it shows only an additive effect with chloroquine (Gingras and Jensen, 1992). Gingras and Jensen (1993) also reported that a combination of azithromycin with chloroquine is additive in a 4-day test in mice infected with *P. berghei*. In the course of a clinical study of its action against trachoma in Gambian schoolchildren, azithromycin was also found to halve the parasite rate and fever associated with *P. falciparum* over an observation period of 28 days, but it did not appear to prevent reinfection (Sadiq *et al.*, 1995).

Clindamycin, which also has long been known to be a slowly acting blood schizontocide, can be used together with quinine as an alternative to one of the tetracyclines to produce radical cure in patients infected with multidrug-resistant *P. falciparum*, but its gastrointestinal and other side-effects preclude its general use for this purpose. Yeo and Rieckmann (1994c) have suggested that chloramphenicol or azithromycin (Yeo and Rieckmann, 1995) could replace tetracycline to eliminate residual *P. falciparum* infections following initial treatment with a rapidly acting blood schizontocide, but no clinical report of their use in this way has appeared to date. Tetracycline which, in association with quinine, has a long-established role in the treatment of acute, multidrug-resistant *P. falciparum*, has now been largely superseded by doxycycline, which is equally active as a blood schizontocide against chloroquine-sensitive and multidrug-resistant *P. falciparum* (see Basco and Le Bras, 1993b). It is also valuable as a causal prophylactic in humans (see, for example, Pang *et al.*, 1987; Gras *et al.*, 1993; WHO, 1995). (See also combinations with tetracyclines in Section 5.2.2.)

5.1.7. *Chelating Agents*

Intraerythrocytic *P. falciparum* requires iron for several vital metabolic processes and obtains it in the form of transferrin, which is endocytosed from the host plasma (Pollack, 1989). Atkinson *et al.* (1991) showed that desferrioxamine, a potent, iron-chelating drug that is used to treat such conditions as haemochromatosis, prevents nuclear division in late trophozoites of *P. falciparum in vitro*, possibly by inhibiting iron-dependent ribonucleotide reductase. There have been numerous and conflicting reports suggesting that desferrioxamine could find a place as an antimalarial drug by blocking the access of intraerythrocytic *Plasmodium* to iron (Gordeuk *et al.*, 1993). However, this may not be the explanation for the apparently beneficial effects observed with desferrioxamine in patients with cerebral malaria (Mabeza *et al.*, 1996). Weiss *et al.* (1997) have shown that desferrioxamine, given together with quinine to Zambian children with cerebral malaria, stimulated a beneficial increase in the production of reactive nitrogen intermediates from nitric oxide in the brain. Data from both experimental studies and the few clinical data are, however, still less than convincing (Bunnag *et al.*, 1992). In *P. falciparum in vitro*, desferrioxamine and a number of hydroxypyridinone iron chelators showed a modest, inherent, growth inhibitory effect which was not potentiated by combining them with several classical antimalarial substances (Pattanapanyasat *et al.*,

1997). The further development of this class of iron chelators does now appear to show promise (Hider and Liu, 1996) (see also Figure 4).

5.1.8. *Compounds Influencing Lipid Metabolism*

The major component of the membrane phospholipids of malaria parasites is phosphatidylcholine. A choline analogue code-named G25, which is believed to block the biosynthesis of phosphatidylcholine (the major component of malaria-infected erythrocytes) by impeding choline transport, was found to be poorly active against *P. berghei* N and *P. yoelii* ssp. NS *in vivo*, but highly active against both *P. chabaudi* and *P. vinckei petteri in vivo*, as well as against *P. falciparum* both *in vitro* and in *Aotus* monkeys (Ancelin *et al.*, 1994). Although the toxicity of the choline analogues tested to date has been relatively high, the area of phospholipid synthesis inhibitors has been little explored and is clearly of considerable potential interest (Vial, 1996). Moreover, the presence of malaria parasites within erythrocytes induces numerous changes in the host cell membranes that may offer points for selective attack by novel compounds that affect both membrane structure and transport functions (Elford *et al.*, 1995). Parasite lysophospholipase, for example, which is distinct from the mammalian enzyme, may offer a potential target (Zidovetzki *et al.*, 1994), as may sphingomyelin synthase (Lauer *et al.*, 1995).

5.1.9. *Protease Inhibitors and other Parasite Enzymes as Targets for Automated Screening*

Intraerythrocytic malaria parasites synthesize a number of specific proteases for such essential activities as the digestion of host cell haemoglobin and the penetration of new host cell membranes by invading merozoites. One of these enzymes, cysteine protease, which has been identified in *P. falciparum* and *P. vinckei*, is readily inhibited by a chloroquine-haemin complex (Gluzman *et al.*, 1994). Another specific parasite enzyme in *P. falciparum* is aspartic haemoglobinase (Francis *et al.*, 1994), but cysteine proteinase inhibitors at present seem to offer more promise (Rosenthal, 1995). However, the cysteine protease inhibitor E64, and the aspartic protease inhibitor pepstatin A, act synergistically against *P. falciparum in vitro* (Bailly *et al.*, 1992). A logical approach to the development of novel antimalarial drugs is the search for protease inhibitors that would function selectively against the parasite enzymes but not affect those of the host. This target, however, is far from being met since many potential parasite protease inhibitors also have high host toxicity.

Nevertheless, the identification of specific parasite proteases (as well as other enzymes), the identification and sequencing of the genes that code for them, and the insertion of such genes into, for example, yeast would provide targets that could be employed for mass automated drug screening that is coming increasingly into use in association with combinatorial chemistry in the pharmaceutical industry. As noted above (Section 3.1), such a development has been reported by Wooden *et al.* (1997), who have succeeded in inserting the DHFR domain of the *DHFR-TS* gene of *P. falciparum* into a strain of *S. cerivisiae* that lacks endogenous DHFR activity.

Parasite proteinases also play a role in merozoite invasion of new host erythrocytes, a function that is strongly blocked by calpain inhibitors (Olaya and Wasserman, 1991) and may offer yet another target for chemotherapy.

5.2. Drug Combinations

It required several decades of hard experience for the principle of deploying combinations of antimalarial drugs to impede the selection of the drug resistance that inevitably followed the widescale use of monotherapy against malaria to become accepted. Antimalarial combinations can be of several types (Peters, 1987), from those in which the individual components are mutually synergistic to a high degree (e.g., sulphonamides with DHFR inhibitors) to combinations that provide one rapidly acting component (e.g., artemisinin) with a second, longer half-life compound that destroys any residual parasites (e.g., mefloquine) (see, for example, the review by White and Olliaro, 1996). The serendipitous discovery by Martin *et al.* (1987) that verapamil can 'reverse' chloroquine resistance in cultures of *P. falciparum*, just as it does with certain anticancer drugs in mammalian cell lines, opened a new route to the exploration of drug combinations (Milhous *et al.*, 1989) but one, unfortunately, that has failed to yield results of significant value. In my opinion, the difference between drugs that are 'synergistic', 'potentiating' or 'resistance-reversing' (or 'resistance-modulating') is a semantic rather than a real one, with the exception that, in the last case, one of the pair may possess no inherent antimalarial activity.

5.2.1. *Drug Resistance Modulators (Resistance-reversing Agents)*

The report by Martin *et al.* (1987) stimulated a search for the molecular mechanism underlying the phenomenon of resistance to chloroquine and other antimalarial drugs, particularly in *P. falciparum*, and resulted, as discussed above, in the identification of two multiple drug-resistance genes, *Pfmdr1* and *Pfmdr2*, the former of which may be concerned in resistance to aminoalcohols such as mefloquine, although probably not to chloroquine. It is not surprising, therefore, that a number of compounds that, at least *in vitro* and, in some cases, also *in vivo*, modulate resistance to chloroquine have no effect on mefloquine resistance and vice versa.

In addition to verapamil, a calcium-channel blocker employed in cardiology, several psychotropic drugs such as desipramine (Bitonti *et al.*, 1988), chlorpromazine and prochlorperazine have been reported to reverse resistance to chloroquine in *P. falciparum in vitro* and, in certain cases, also in *Aotus* monkeys (Kyle *et al.*, 1993). None, however, has been considered sufficiently safe to put into clinical trial because the joint toxicity of the combinations with chloroquine has been prohibitive (Watt *et al.*, 1990). Other neuronal monoamine re-uptake inhibitors, including fluoxetine, that reverse chloroquine resistance *in vitro* also have not been assessed in humans (Coutaux *et al.*, 1994). However, preliminary clinical studies were carried out with a combination of chloroquine and cyproheptadine, one of a series of antihistaminic agents that Peters *et al.* (1990) found to modulate chloroquine resistance in both *P. falciparum in vitro* and *P. yoelii* ssp. NS *in vivo*. In patients naturally infected with multidrug-resistant *P. falciparum*, however, no beneficial action was observed when chloroquine therapy was supplemented by the administration of cyproheptadine, possibly because inadequate blood concentrations were attainable at a tolerated dose level (Björkman *et al.*, 1990), although Sowunmi *et al.* (1997) considered that chlorpheniramine, another antihistaminic drug, enhanced the efficacy of chloroquine in Nigerian children.

Penfluridol, another psychotropic drug which was reported by Kyle *et al.* (1989) and Oduola *et al.* (1993) to reverse resistance to mefloquine in *P. falciparum in vitro*, also had this effect in a mefloquine-resistant strain derived from *P. yoelii* ssp. NS in mice (Peters and Robinson, 1991). No report has appeared, however, to indicate whether this combination has been studied in humans.

5.2.2. *Other Combinations*

Reference has already been made to the high recrudescence rate that follows the treatment of malaria with drugs of the artemisinin family. Since there is no evidence, so far, to suggest that recrudescing parasites are less sensitive to these drugs than those treated initially, it is possible that the cause of this phenomenon is rather a failure to achieve an optimum dosage regimen. A practical consequence has been the increasing use of combinations of the artemisinins with partner compounds to ensure the removal of residual or recrudescing parasites. Synergism has been shown between artemisinin and mefloquine in mice infected with mefloquine or artemisinin-resistant strains of *P. berghei* by Chawira *et al.* (1987) and against chloroquine-resistant strains of *P. falciparum in vitro* by Chawira and Warhurst (1987). The latter workers also demonstrated synergism between artemisinin and tetracycline against chloroquine-sensitive and chloroquine-resistant *P. falciparum in vitro*. In the clinic, various combinations with one or other of the artemisinins have been administered including artesunate with mefloquine (Luxemburger *et al.*, 1994; Nosten *et al.*, 1994; Cardoso *et al.*, 1996; Looareesuwan *et al.*, 1996c), doxycycline (Looareesuwan *et al.*, 1994), or tetracycline (Duarte *et al.*, 1996), artemether with mefloquine (Karbwang *et al.*, 1995; Bunnag *et al.*, 1996), and artemisinin with quinine or doxycycline (Bich *et al.*, 1996). Looareesuwan *et al.* (1996e), in a pilot study of safety and tolerability in Thai patients with severe falciparum malaria, all of whom were cured, obtained similar clinical responses in those who received artesunate alone and those who received a combination of artesunate with desferrioxamine. More recently, artemether has been examined in combination with a novel aminoalcohol, benflumetol, in clinical trials (Olliaro and Trigg, 1995). In areas where quinine is still effective and is used as a first line treatment for multidrug-resistant *P. falciparum*, tetracycline, doxycycline or Fansidar is often given in addition to ensure a radical cure.

The synthetic endoperoxide, Fenozan B07, has also a synergistic action with mefloquine and with halofantrine against chloroquine-resistant *P. yoelii* ssp. NS in rodents (Fleck *et al.*, 1997b). Unlike the artemisinins, Fenozan B07 is active against all the intraerythrocytic stages except mature schizonts (Fleck *et al.*, 1997a).

Chloroquine-resistant *P. vivax* presents another challenge in areas such as Irian Jaya, where resistant strains are prevalent among Indonesian immigrants. Peters *et al.* (1993a) demonstrated synergism between chloroquine and primaquine as well as a novel 8-aminoquinoline, WR 238,605 (Figure 2), against chloroquine-resistant *P. yoelii* ssp. NS. Subsequently, Baird *et al.* (1995) obtained evidence to indicate that synergism also exists between chloroquine and primaquine when they are administered in combination as blood schizontocides against chloroquine-resistant *P. vivax*. A combination of halofantrine with primaquine has proved to be a valuable replacement for chloroquine with primaquine to effect a radical cure of *P. vivax* in Irian Jaya (Fryauff *et al.*, 1997). (In that area it was necessary to use an increased dosage of primaquine because the hypnozoites of *P. vivax* are relatively insensitive to this compound.) The combined regimen was also curative in patients with falciparum malaria, which is commonly associated with vivax malaria in the south-west Pacific and is chloroquine-resistant. Daily doxycycline with primaquine also gave complete protection to nonimmune soldiers against both *P. falciparum* and *P. vivax* in a country of south-east Asia where the response to primaquine alone of *P. vivax* hypnozoites is poor (Rieckmann *et al.*, 1993).

Reference has already been made above to novel combinations of compounds that act on folate metabolism and to the potent combination of atovaquone with proguanil or 5-fluoroorotate. One practical problem that may arise with such combinations is that of cost since, if prohibitively high, this may deter either commercial producers from developing them or the potential recipients from purchasing them.

6. CONCLUSIONS

The goal of eradicating malaria is probably no longer realistic. The nature of the *Plasmodium* genome is extremely complex but rapid progress is being made on characterizing that of *P. falciparum* (see Dame *et al.*, 1996) and it is to be hoped that this will, in time, provide new avenues for the rationalization of novel antimalarial drug development. Meanwhile all we can do is try to keep one step ahead of the parasites by producing new compounds with which to treat, at a reasonable cost, people infected with multidrug-resistant parasites. It is crucial, moreover, that we should protect new drugs for as long as possible by deploying rationally designed combinations. Antimalarial vaccines are still a hope but, even if and when good vaccines do become available, there is no doubt that chemotherapy will still be required for the foreseeable future.

ACKNOWLEDGEMENTS

I am indebted to the librarians of the London School of Hygiene and Tropical Medicine for assistance in obtaining reference material. Original studies referred to were made with the invaluable assistance of Mr Brian Robinson. Dr Yves Boulard generously provided the electron micrographs for Figure 5.

REFERENCES

Anabwani, G.M., Esamai, F.O. and Menya, D.A. (1996). A randomised controlled trial to assess the relative efficacy of chloroquine, amodiaquine, halofantrine and Fansidar in the treatment of uncomplicated malaria in children. *East African Medical Journal* **73**, 155–158.

Ancelin, M., Vial, H.J., Calas, M., Giral, L., Piquet, G., Rubi, E., Thomas, A., Peters, W., Slomianny, C., Herrera, S., Louis, F., Mouanda, V., Herrera, S. and

Jepsen, S. (1994). Present development concerning antimalarial activity of phospholipid metabolism inhibitors with special reference to *in vivo* activity. *Memorias do Instituto Oswaldo Cruz* **89** (Supplement 2), 85–90.
Andersen, S.L., Ager, A., McGreevy, P., Schuster, B.G., Wesche, D., Kuschner, R., Ohrt, C., Ellis, W., Rossan, R. and Berman, J. (1995a). Activity of azithromycin as a blood schizontocide against rodent and human plasmodia *in vivo*. *American Journal of Tropical Medicine and Hygiene* **52**, 159–161.
Andersen, S.L., Berman, J., Kuschner, R., Wesche, D., Magil, A., Wellde, B., Schneider, I., Dunne, M. and Schuster, B.G. (1995b). Prophylaxis of *Plasmodium falciparum* malaria with azithromycin administered to volunteers. *Annals of Internal Medicine* **123**, 771–773.
Anonymous (1994). Artemisinin. *Transactions of the Royal Society of Tropical Medicine and Hygiene* **88** (Supplement, 1).
Atamna, H., Krugliaj, M., Shalmiev, G., Deharo, E., Pescarmona, G. and Ginsburg, H. (1996). Mode of antimalarial effect of methylene blue and some of its analogues on *Plasmodium falciparum* in culture and their inhibition of *P. vinckei petteri* and *P. yoelii nigeriensis in vivo*. *Biochemical Pharmacology* **51**, 693–700.
Atkinson, C.T., Bayne, M.T., Gordeuk, V.R., Brittenham, G.M. and Aikawa, M. (1991). Stage-specific ultrastructural effects of desferrioxamine on *Plasmodium falciparum in vitro*. *American Journal of Tropical Medicine and Hygiene* **45**, 593–601.
Babiker, H., Ranford-Cartwright, L., Sultan, A., Satti, G. and Walliker, D. (1994). Genetic evidence that RI chloroquine resistance of *Plasmodium falciparum* is caused by recrudescence of resistant parasites. *Transactions of the Royal Society of Tropical Medicine and Hygiene* **88**, 328–331.
Bailly, E., Jambou, R., Savel, J. and Jaureguiberry, G. (1992). *Plasmodium falciparum*: differential sensitivity *in vitro* to E-64 (cysteine protease inhibitor) and Pepstatin A (aspartyl protease inhibitor). *Journal of Protozoology* **39**, 593–599.
Baird, J.K., Basri, H., Purnomo, Bangs, M.J., Subianto, B., Patchen, L.C. and Hoffman, S.L. (1991). Resistance to chloroquine by *Plasmodium vivax* in Irian Jaya, Indonesia. *American Journal of Tropical Medicine and Hygiene* **44**, 547–552.
Baird, J.K., Basri, H., Subianto, B., Fryauff, D.J., McElroy, P.D., Leksana, B., Richie, T.L., Masbar, S., Wignall, F.S. and Hoffman, S.L. (1995). Treatment of chloroquine-resistant *Plasmodium vivax* with chloroquine and primaquine or halofantrine. *Journal of Infectious Diseases* **171**, 1678–1682.
Baird, J.K., Nalim, M.F.S., Basri, H., Masbar, S., Leksana, B., Tjitra, E., Dewi, R.M., Khairani, M. and Wigsall, F.S. (1996). Survey of resistance to chloroquine by *Plasmodium vivax* in Indonesia. *Transactions of the Royal Society of Tropical Medicine and Hygiene* **90**, 409–411.
Balmer, P., Mathers, K.E., Davidson, C. and Phillips, R.S. (1995). Chloroquine inhibition of nitric oxide production. *Transactions of the Royal Society of Tropical Medicine and Hygiene* **89**, 588–595.
Basco, L.K. and Le Bras, J. (1993a). *In vitro* activity of monodesethylamodiaquine and amopyroquine against African isolates and clones of *Plasmodium falciparum*. *American Journal of Tropical Medicine and Hygiene* **48**, 120–125.
Basco, L.K. and Le Bras, J. (1993b). Activity *in vitro* of doxycycline against multidrug-resistant *Plasmodium falciparum*. *Transactions of the Royal Society of Tropical Medicine and Hygiene* **87**, 469–470.
Basco, L.K., Gillotin, C., Gimenez, F., Farinotti, R. and Le Bras, J. (1992). Antimalarial activity *in vitro* of the N-desbutyl derivative of halofantrine. *Transactions of the Royal Society of Tropical Medicine and Hygiene* **86**, 12–13.

Basco, L.K., Ruggeri,C. and Le Bras, J. (1994). *Molécules Antipaludique. Mécanismes d'Action, Mécanismes de Résistance et Relations Structure–Activité des Schizontocides sanguins.* Paris: Masson.

Basco, L.K., Le Bras, J., Rhoades, Z. and Wilson, C.M. (1995). Analysis of *pfmdr1* and drug susceptibility in fresh isolates of *Plasmodium falciparum* from Subsaharan Africa. *Molecular and Biochemical Parasitology* **74**, 157–166.

Basco, L.K., Pécoulas, P.E., Le Bras, J. and Wilson, C.M. (1996). *Plasmodium falciparum*: molecular characterization of multidrug-resistant Cambodian isolates. *Experimental Parasitology* **82**, 97–103.

Beauté-Lafitte, A., Altmeyer-Caillard, V., Gonnet-Gonzalez, F., Ramiaramanana, L., Chabaud, A.G. and Landau, I. (1994). The chemosensitivity of the rodent malarias — relationships with the biology of merozoites. *International Journal for Parasitology* **24**, 981–986.

Berger, B.J., Martiney, J., Slater, A.F.G., Fairlamb, A.H. and Cerami, A. (1995). Chloroquine resistance is not associated with drug metabolism in *Plasmodium falciparum*. *Journal of Parasitology* **81**, 1004–1008.

Bich, N.N., de Vries, P., Thien, H.V., Phong, T.H., Hung, L.N., Eggelte, T.A., Anh, T.K. and Kager, P.A. (1996). Efficacy and tolerance of artemisinin in short combination regimens for the treatment of uncomplicated falciparum malaria. *American Journal of Tropical Medicine and Hygiene* **55**, 438–443.

Bitonti, A.J., Sjoerdsma, A., McCann, P.P., Kyle, D.E., Oduola, A.M.J., Rossan, R.N., Milhous, W.K. and Davidson, D.E., jr (1988). Reversal of chloroquine resistance in malaria parasite *Plasmodium falciparum* by desipramine. *Science* **242**, 1301–1303.

Björkman, A. (1991). Drug resistance — changing patterns. In: *Malaria. Waiting for the Vaccine* (G.A.T. Targett, ed.), pp. 105–120. Chichester: John Wiley.

Björkman, A., Willcox, M., Kihamia, C.M., Mahikawano, L.F., Phillips-Howard, P.A., Hakansson, A. and Warhurst, D.C. (1990). Field study of cyproheptadine/chloroquine synergism in falciparum malaria. *Lancet* **336**, 59–60.

Boulard, Y., Adovelande, J., Dennebouy, R., Peters, W., Galle, P., Slodzian, G. and Schrevel, J. (1996). Differential localization of halofantrine in drug sensitive and resistant rodent *Plasmodium* strains, a scanning ion microscopy and mass spectrometry study. *Biology of the Cell* **86**, 192.

Brasseur, P., Kouamouo, J., Moyou-Somo, R. and Druilhe, P. (1992). Multi-drug resistant falciparum malaria in Cameroon in 1987–1988. II. Mefloquine resistance confirmed *in vivo* and *in vitro* and its correlation with quinine resistance. *American Journal of Tropical Medicine and Hygiene* **46**, 8–14.

Bray, P.G., Boulter, M.K., Ritchie, G.Y. and Ward. S.A. (1994). Relationship of global chloroquine transport and reversal of chloroquine resistance in *P. falciparum*. *Molecular and Biochemical Parasitology* **63**, 87–94.

Bray, P.G., Hawley, S.R., Mungthin, M. and Ward, S.A. (1996). Physicochemical properties correlated with drug resistance and the reversal of drug resistance in *Plasmodium falciparum*. *Molecular Pharmacology* **50**, 1559–1566.

Bruce-Chwatt, L.J., Black, R.H., Canfield, C.J., Clyde, D.F., Peters, W. and Wernsdorfer, W.H. (1986). *Chemotherapy of Malaria*, 2nd edn. Geneva: World Health Organization.

Bunnag, D., Poltera, A.A., Viravan, C., Looareesuwan, S., Harinasuta, K.T. and Schindlery, C. (1992). Plasmodicidal effect of desferrioxamine B in human vivax or falciparum malaria from Thailand. *Acta Tropica* **52**, 59–67.

Bunnag, D., Kanda, T., Karbwang, J., Thimasarn, K., Pungpak, S. and Harinasuta T. (1996). Artemether or artesunate followed by mefloquine as a possible treat-

ment for multidrug resistant falciparum malaria. *Transactions of the Royal Society of Tropical Medicine and Hygiene* **90**, 415–417.
Bygbjerg, I.C., Shapira, A., Flachs, H., Gomme, G. and Jepsen, S. (1983). Mefloquine resistance of falciparum malaria from Tanzania enhanced by treatment. *Lancet* **i**, 774–775.
Caillard, V., Beauté-Lafitte, A., Chabaud, A. and Landau, I. (1992). *Plasmodium vinckei petteri*: identification of the stages sensitive to arteether. *Experimental Parasitology* **75**, 449–456.
Cambie, G., Caillard, V., Beauté-Lafitte, A., Ginsburg, H., Chabaud, A. and Landau, I. (1991). Chronotherapy of malaria: identification of drug-sensitive stage of parasite and timing of drug delivery for improved therapy. *Annales de Parasitologie Humaine et Comparée* **66**, 14–21.
Canfield, C.J., Milhous, W.K., Ager, A.L., Rossan, R.N., Sweeney, T.R., Lewis, N.J. and Jacobus, D.P. (1993). PS-15: a potent, orally active antimalarial from a new class of folic acid antagonist. *American Journal of Tropical Medicine and Hygiene* **49**, 121–126.
Canfield, C.J., Pudney, M. and Gutteridge, W.E. (1995). Interactions of atovaquone with other antimalarial drugs against *Plasmodium falciparum in vitro*. *Experimental Parasitology* **80**, 373–381.
Cardoso, B. da S., Dourado, H.V., Pinheiro, M. da C., Crescente, J.A., Amoras, W.W., Baena, J. and Saraty, S. (1996). Estudo da eficácia do artesunato oral isolado e em associação com mefloquina, no tratamento da malária falciparum não complicado em área endêmica do Para, Brasil. *Revista da Sociedade Brasileira de Medicina Tropical* **29**, 251–257.
Chawira, A.N. and Warhurst, D.C. (1987). The effect of artemisinin combined with standard antimalarials against chloroquine-sensitive and chloroquine-resistant strains of *Plasmodium falciparum in vitro*. *Journal of Tropical Medicine and Hygiene* **90**, 1–8.
Chawira, A.N., Warhurst, D.C. and Peters, W. (1986). Qinghaosu resistance in rodent malaria. *Transactions of the Royal Society of Tropical Medicine and Hygiene* **80**, 477–480.
Chawira, A.N., Warhurst, D.C., Robinson, B.L. and Peters, W. (1987). The effect of combinations of qinghaosu (artemisinin) with standard antimalarial drugs in the suppressive treatment of malaria in mice. *Transactions of the Royal Society of Tropical Medicine and Hygiene* **81**, 554–558.
Chen, P.Q., Li, G.Q., He, K.R., Fu, Y.X., Fu, L.C. and Song, Y.Z. (1994). The infectivity of gametocytes of *Plasmodium falciparum* from patients treated with artemisinin. *Chinese Medical Journal* **107**, 709–711.
Childs, G.E., Boudreau, E.F., Wimonwattratee, T., Pang, L. and Milhous, W.K. (1991). *In vitro* and clinical correlates of mefloquine resistance of *Plasmodium falciparum* in eastern Thailand. *American Journal of Tropical Medicine and Hygiene* **44**, 553–559.
Clark, I.A. and Rockett, K.A. (1996). Nitric oxide and parasitic disease. *Advances in Parasitology* **37**, 1–56.
Cogswell, F.B., Litterst, C., Riccio, E.S., Mirsalis, J.C., De, D., Krogstad, F.M. and Krogstad, D.J. (1997). Pre-clinical studies of aminoquinolines active against chloroquine- and mefloquine-resistant *Plasmodium falciparum*. Abstract No. 109, 46th annual meeting of the American Society of Tropical Medicine and Hygiene, Lake Buena Vista, Florida, December 7–11, 1997. *American Journal of Tropical Medicine and Hygiene* **57** (Supplement), 139.

Collins, W.E. and Jeffery, G.M. (1996). Primaquine resistance in *Plasmodium vivax*. *American Journal of Tropical Medicine and Hygiene* **55**, 243–249.

Coquelin, F., Biarnais, T., Deharo, E., Peters, W., Chabaud, A. and Landau, I. (1997). Modifications in the rhythm of evolution of the schizogony in *Plasmodium chabaudi chabaudi* associated with the selection of drug resistance. *Parasitology Research* **83**, 504–509.

Coutaux, A.F., Mooney, J.J. and Wirth, D.F. (1994). Neuronal monoamine reuptake inhibitors enhance *in vitro* susceptibility to chloroquine in resistant *Plasmodium falciparum*. *Antimicrobial Agents and Chemotherapy* **38**, 1419–1421.

Cowman, A.F. (1991). The P-glycoprotein homologues of *Plasmodium falciparum*: are they involved in chloroquine resistance? *Parasitology Today* **7**, 70–76.

Cowman, A.F. and Lew, A.M. (1989). Antifolate drug selection results in duplication and rearrangement of chromosome 7 in *Plasmodium chabaudi*. *Molecular and Cellular Biology* **9**, 5182–5188.

Cowman, A.F., Karcz, S., Galatis, D. and Culvenor, J.G. (1991). A P-glycoprotein homologue of *Plasmodium falciparum* is localized on the digestive vacuole. *Journal of Cell Biology* **113**, 1033–1042.

Cowman, A.F., Galatis, D. and Thompson, J.K. (1994). Selection for mefloquine resistance in *Plasmodium falciparum* is linked to amplification of the pfmdr1 gene and cross-resistance to halofantrine and quinine. *Proceedings of the National Academy of Sciences of the United States of America* **91**, 1143–1147.

Cross, A.P. and Singer, B. (1991). Modelling the development of resistance of *Plasmodium falciparum* to anti-malarial drugs. *Transactions of the Royal Society of Tropical Medicine and Hygiene* **85**, 349–355.

Curtis, J., Duraisingh, M.T., Trigg, J.K., Mbwana, H., Warhurst, D.C. and Curtis, C.F. (1996). Direct evidence that asparagine at position 108 of the *Plasmodium falciparum* dihydrofolate reductase is involved in resistance to antifolate drugs in Tanzania. *Transactions of the Royal Society of Tropical Medicine and Hygiene* **90**, 678–680.

Dame, J.B. *et al.* (1996). Current status of the *Plasmodium falciparum* genome project. *Molecular and Biochemical Parasitology* **79**, 1–12.

De, D., Krogstad, F.M., Cogswell, F.B. and Krogstad, D.J. (1996). Aminoquinolines that circumvent resistance in *Plasmodium falciparum in vitro*. *American Journal of Tropical Medicine and Hygiene* **55**, 579–583.

Dei-Cas, E., Slomianny, C., Prensier, G., Vernes, A., Colin, J.J., Verhaeghe, A., Savage, A. and Charet, P. (1984). Action preferentielle de la chloroquine sur les *Plasmodium* hebergés dans les hématies matures. *Pathologie Biologie* **32**, 1019–1023.

Desjardins, R.E., Canfield, C.J., Haynes, J.D. and Chulay, J.D. (1979). Quantitative assessment of antimalarial activity *in vitro* by a semiautomated microdilution technique. *Antimicrobial Agents and Chemotherapy* **16**, 710–718.

Divo, A.A., Sartorelli, A.C., Patton, C.L. and Bia, F.J. (1988). Activity of fluoroquinoline antibiotics against *Plasmodium falciparum in vitro*. *Antimicrobial Agents and Chemotherapy* **32**, 1182–1186.

Doherty, J.F., Day, J.H., Warhurst, D.C. and Chiodini, P.L. (1997). Treatment of *Plasmodium vivax* malaria — time for a change? *Transactions of the Royal Society of Tropical Medicine and Hygiene* **91**, 76.

Dorn, A.R., Stoffel, R., Matile, H., Bubendorf, A. and Ridley, R.G. (1995). Malarial haemozoin/β-haematin supports haem polymerisation in the absence of protein. *Nature* **374**, 269–271.

Dua, V.K., Kar, P.K. and Sharma, V.P. (1996). Chloroquine resistant *Plasmodium vivax* in India. *Tropical Medicine and International Health* **1**, 816–819.

Duarte, E.C., Fontes, C.J.F., Gyorkos, T.W. and Abrahamowicz, M. (1996). Randomized controlled trial of artesunate plus tetracycline versus standard treatment (quinine plus tetracycline) for uncomplicated *Plasmodium falciparum* malaria in Brazil. *American Journal of Tropical Medicine and Hygiene* **54**, 197–202.

Dutta, G.P., Bajpai, R. and Vishwakarma, R.A. (1989). Artemisinin (qinghaosu) — a new gametocytocidal drug for malaria. *Chemotherapy* **35**, 200–207.

Dye, C. and Williams, B.G. (1997). Multigenic drug resistance among inbred malaria parasites. *Proceedings of the Royal Society of London, B* **264**, 61–67.

Eaton, J.W., Meshnick, S.R. and Brewer, G.J. (eds). (1989). *Malaria and the Red Cell: 2*. New York: Alan R.Liss.

Edstein, M.D., Yeo, A.E.T., Kyle, D.E., Looareesuwan, S., Wilairatana, P. and Rieckmann, K.H. (1996). Proguanil polymorphism does not affect the antimalarial activity of proguanil combined with atovaquone *in vitro*. *Transactions of the Royal Society of Tropical Medicine and Hygiene* **90**, 418–421.

Egan T.J., Ross, D.C. and Adams, P.A. (1994). Quinoline anti-malarial drugs inhibit spontaneous formation of β-haematin (malaria pigment). *FEBS Letters* **352**, 54–57.

Elford, B.C., Cowan, G.M. and Ferguson, D.J.P. (1995). Parasite-regulated membrane transport processes and metabolic control in malaria-infected erythrocytes. *Biochemical Journal* **308**, 361–374.

Fedorova, O.V. (1991). [The sensitivity of a *Plasmodium berghei* strain resistant to artemisinin and other antimalarial preparations.] *Meditsinskaia Parazitologiia i Parazitarnye Bolezni* (1), 47–50. [In Russian.]

Fitch, C.S., Chevli, R., Banyal, H.S., Phillips, G., Pfaller, M.A. and Krogstad, D.J. (1982). Lysis of *Plasmodium falciparum* by ferriprotoporphyrin IX and a chloroquine–ferriprotoporphyrin complex. *Antimicrobial Agents and Chemotherapy* **21**, 819–822.

Fleck, S.L., Robinson, B.L., Peters, W., Thévin, F., Boulard, Y., Glénat, G., Caillard, V. and Landau, I. (1997a). The chemotherapy of rodent malaria. LIII. 'Fenozan B07' (Fenozan-50F), a difluorinated 3,3'-spirocyclopentane 1,2,4-trioxane: comparison with some compounds of the artemisinin series. *Annals of Tropical Medicine and Parasitology* **91**, 25–32.

Fleck, S.L., Robinson, B.L. and Peters, W. (1997b). The chemotherapy of rodent malaria. LIV. Combinations of 'Fenozan B07' (Fenozan-50F), a difluorinated 3,3'-spirocyclopentane 1,2,4-trioxane, with other drugs against drug-sensitive and drug-resistant parasites. *Annals of Tropical Medicine and Parasitology* **91**, 33–39.

Fonseca, L., Vigario, A.M., Seixas, E. and do Rosario, V.E. (1995). *Plasmodium berghei*: selection of mefloquine-resistant parasites through drug pressure in mosquitoes. *Experimental Parasitology* **81**, 55–62.

Foote, S.J., Thompson, J.K., Cowman, A.F. and Kemp, D.J. (1989). Amplification of the multidrug resistance gene in some chloroquine-resistant isolates of *P. falciparum*. *Cell* **57**, 921–930.

Foote, S.J., Kyle, D.E., Martin, R.K., Oduola, A.M.J., Forsyth, K., Kemp, D.J. and Cowman, A.F. (1990). Several alleles of the multidrug-resistance gene are closely linked to chloroquine resistance in *Plasmodium falciparum*. *Nature* **345**, 255–258.

Francis, S.E., Gluzman, I.Y., Oksman, A., Knickerbocker, A., Mueller, R., Bryant, M.L., Sherman, D.R., Russell, D.G. and Goldberg, D.E. (1994). Molecular characterization and inhibition of a *Plasmodium falciparum* aspartic hemoglobinase. *EMBO Journal* **13**, 306–317.

Fryauff, D.J., Baird, J.K., Basri, H., Sumawinata, I., Purnomo, Richie, T.L., Ohrt, C.K., Mouzin, E., Church, C.J., Richards, A.L., Subianto, B., Sandjaja, B., Wignall, F.S. and Hoffman, S.L. (1995). Randomised placebo-controlled trial of primaquine for prophylaxis of falciparum and vivax malaria. *Lancet* **346**, 1190–1193.

Fryauff, D.J., Baird, J.K., Basri, H., Wiady, I., Purnomo, Bangs, M.J., Subianto, B., Harjosuwarno, S., Tjitra, E., Richie, T.L.and Hoffman, S.L. (1997). Halofantrine and primaquine for radical cure of malaria in Irian Jaya, Indonesia. *Annals of Tropical Medicine and Parasitology* **91**, 7–16.

Fu, S. and Xiao, S.-H. (1991). Pyronaridine: a new antimalarial drug. *Parasitology Today* **11**, 310–313.

Garavelli, P.L. and Corti, E. (1992). Chloroquine resistance in *Plasmodium vivax*: the first case in Brazil. *Transactions of the Royal Society of Tropical Medicine and Hygiene* **86**, 128.

Gardner, M.J., Williamson, D.H. and Wilson, R.J.M. (1991). A circular DNA in malaria parasites encodes an RNA polymerase like that of prokaryotes and chloroplasts. *Molecular and Biochemical Parasitology* **44**, 115–124.

Garg, M., Gopinathan, N., Bodhe, P. and Kshirsagar, N.A. (1995). Vivax malaria resistant to chloroquine: case reports from Bombay. *Transactions of the Royal Society of Tropical Medicine and Hygiene* **89**, 656–657.

Gassis, S. and Rathod, P.K. (1996). Frequency of drug resistance in *Plasmodium falciparum*: a nonsynergistic combination of 5-fluorooroate and atovaquone suppresses in vitro resistance. *Antimicrobial Agents and Chemotherapy* **40**, 914–919.

Gingras, B.A. and Jensen, J.B. (1992). Activity of azithromycin (CP-62,993) and erythromycin against chloroquine-sensitive and chloroquine-resistant strains of *Plasmodium falciparum in vitro*. *American Journal of Tropical Medicine and Hygiene* **47**, 378–382.

Gingras, B.A. and Jensen, J.B. (1993). Antimalarial activity of azithromycin and erythromycin against *Plasmodium berghei*. *American Journal of Tropical Medicine and Hygiene* **49**, 101–105.

Ginsburg, H. and Stein, W.D. (1991). Kinetic modelling of chloroquine uptake by malaria-infected erythrocytes. *Biochemical Pharmacology* **41**, 1463–1470.

Gluzman, I.Y., Francis, S.E., Oksman, A., Smith, C., Duffin, K. and Goldberg, D.E. (1994). Order and specificity of the *Plasmodium* hemoglobin degradation pathways. *Journal of Clinical Investigation* **93**, 1602–1608.

Gordeuk, V.R., Thuma, P.E., Brittenham, G.M., Biemba, G., Zulu, S., Simwanza, G., Kalense, P., M'Hango, A., Parry, D., Poltera, A.A. and Aikawa, M. (1993). Iron chelation as a chemotherapeutic strategy for falciparum malaria. *American Journal of Tropical Medicine and Hygiene* **48**, 193–197.

Gras, C., Laroche, R., Guelian, J., Martet, G., Merlin, M., Pottier, G., Guisset, M. and Touze, J.E. (1993). Place actuelle de la doxycycline dans la chimioprophylaxie du paludisme à *Plasmodium falciparum*. *Bulletin de la Société de Pathologie Exotique* **86**, 52–55.

Greenwood, B.M. (1991). Malaria chemoprophylaxis in endemic regions. In: *Malaria. Waiting for the Vaccine* (G.A.T. Targett, ed.), pp. 83–102. Chichester: John Wiley.

Greenwood, B.M., David, P.H., Otoo-Forbes, L.N., Allen, S.J., Alonso, P.L. and Armstrong Schellenberg, J.R. (1995). Mortality and morbidity from malaria after

stopping malaria chemoprophylaxis. *Transactions of the Royal Society of Tropical Medicine and Hygiene* **89**, 629–633.
Gutteridge, W.E. and Coombs, G.H. (1977). *Biochemistry of Parasitic Protozoa*. London: Macmillan Press.
Handunnetti, S.M., Gunewardena, D.M., Pathirana, P.P.S.L., Ekanayake, K., Weerasinghe, S. and Mendis, K.N. (1996). Features of recrudescent chloroquine-resistant *Plasmodium falciparum* infections confer a survival advantage on parasites and have implications for disease control. *Transactions of the Royal Society of Tropical Medicine and Hygiene* **90**, 563–567.
Hawley, S.R., Bray, P.G., Park, K. and Ward, S.A. (1996). Amodiaquine accumulation in *Plasmodium falciparum* as a possible explanation for its superior antimalarial activity over chloroquine. *Molecular and Biochemical Parasitology* **80**, 15–25.
Hider, R.C. and Liu, Z. (1997). The treatment of malaria with iron chelators. *Journal of Pharmacy and Pharmacology* **49**, 59–64.
Hien, T.T., Day, N.P.J., Phu, N.H., Mai, N.T.H., Chau, T.T.H., Loc, P.P., Sinh, D.X., Chuong, L.V., Vinh, H., Waller, D., Peto, T.E.A. and White, N.J. (1996). A controlled trial of artemether or quinine in Vietnamese adults with severe falciparum malaria. *New England Journal of Medicine* **335**, 76–83.
Hyde, J.E. (1989). Point mutations and pyrimethamine resistance in *Plasmodium falciparum*. *Parasitology Today* **5**, 252–255.
Ichimori, K., Curtis, C.F. and Targett, G.A. (1990). The effects of chloroquine on the infectivity of chloroquine-sensitive and -resistant populations of *Plasmodium yoelii nigeriensis* to mosquitoes. *Parasitology* **100**, 377–381.
Jacquier, P., Druilhe, P., Felix, H., Diquet, B. and Djibo, L. (1985). Is *Plasmodium falciparum* resistance to chloroquine reversible in absence of drug pressure? *Lancet* **ii**, 270–271.
Jefford, C.W., Misra, D., Rossier, J.-C., Kamalaprija, P., Burger, U., Mareda, J., Bernardinelli, G., Peters, W., Robinson, B.L., Milhous, W.K., Zhang, F., Gosser, D.K. and Meshnick, S.R. (1993). Cyclopenteno-1,2,4-trioxanes as effective antimalarial surrogates of artemisinin. In: *Perspectives in Medicinal Chemistry* (B. Testa, E. Kyburz, W. Fuhrer and R. Giger, eds), pp. 460–472. Basel: Verlag Helvetica Chimica Acta.
Jiang, J.-B., Li, G.-Q., Guo, X.-B., Kong, Y.C. and Arnold, K. (1982). Antimalarial activity of mefloquine and quinhaosu. *Lancet* **ii**, 285–288.
Kain, K.C., Craig, A.A. and Ohrt, C. (1996). Single-strand conformational polymorphism analysis differentiates *Plasmodium falciparum* treatment failures from re-infections. *Molecular and Biochemical Parasitology* **79**, 167–175.
Karbwang, J. and Wernsdorfer, W.H. (eds) (1993). *Clinical Pharmacology of Antimalarials*. Bangkok: Faculty of Tropical Medicine, Mahidol University.
Karbwang, J., Na-Bangchang, K., Thanavibul, A., Ditta-in, M. and Harinasuta, T. (1995). A comparative clinical trial of two different regimens of artemether plus mefloquine in multidrug resistant falciparum malaria. *Transactions of the Royal Society of Tropical Medicine and Hygiene* **89**, 296–298.
Keuter, M., van Eijk, A., Hoogstrare, M., Raasveld, M., van der Ree, M., Ngwawe, W.A., Watkins, W.M., Were, J.B.O. and Brandling-Bennett, A.D. (1990). Comparison of chloroquine, pyrimethamine and sulfadoxine, and chlorproguanil and dapsone as treatment for falciparum malaria in pregnant and non-pregnant women, Kakamega district, Kenya. *British Medical Journal* **301**, 466–470.
Kirby, G.C. (1996). Medicinal plants and the control of protozoal disease, with

particular reference to malaria. *Transactions of the Royal Society of Tropical Medicine and Hygiene* **90**, 605–609.

Klayman, D.L. (1985). Qinghaosu (artemisinin): an antimalarial drug from China. *Science* **228**, 1049–1055.

Kremsner, P.G., Winkler, S., Wildling, E., Prada, J., Bienzle, U., Graninger, W. and Nussler, A.K. (1996). High plasma levels of nitrogen oxides are associated with severe disease and correlate with rapid parasitological and clinical cure in *Plasmodium falciparum* malaria. *Transactions of the Royal Society of Tropical Medicine and Hygiene* **90**, 44–47.

Krogstad, D.J., Gluzman, I.Y., Kyle, D.E., Oduola, A.M.J., Martin, S.K., Milhous, W. and Schlesinger, P.H. (1987). Efflux of chloroquine from *Plasmodium falciparum*: mechanism of chloroquine resistance. *Science* **238** 1283–1285.

Krungkai, J., Krungkrai, S.R. and Phakanont, K. (1992). Antimalarial activity of orotate analogs that inhibit dihydroorotase and dihydroorotate dehydrogenase. *Biochemical Pharmacology* **43**, 1295–1301.

Kwiatkowski, D. and Bate, C. (1995). Inhibition of tumour necrosis factor (TNF) production by antimalarial drugs used in cerebral malaria. *Transactions of the Royal Society of Tropical Medicine and Hygiene* **89**, 215–216.

Kyle, D.E., Webster, H.K. and Milhous, W.K. (1989). *In vitro* reversal of mefloquine and chloroquine resistance in multi-drug resistant *Plasmodium falciparum* in Thailand. Abstract no. 518, 38th Annual Meeting of the American Society of Tropical Medicine and Hygiene, Honolulu, Hawaii, 10–14 December 1989. *American Journal of Tropical Medicine and Hygiene* **41** (Supplement), 322.

Kyle, D.E., Milhous, W.K. and Rossan, R.N. (1993). Reversal of *Plasmodium falciparum* resistance to chloroquine in Panamanian *Aotus* monkeys. *American Journal of Tropical Medicine and Hygiene* **48**, 126–133.

Lampe, A. (1994) *Phytotherapeutika in der Anwendung gegen Malaria. Eine Literaturstudie.* Hannover: Tierarztliche Hochschule, inaugural dissertation.

Landau, I. and Chabaud, A. (1994a). *Plasmodium* species infecting *Thamnomys rutilans*: a zoological study. *Advances in Parasitology* **33**, 49–90.

Landau, I. and Chabaud, A. (1994b). Latency of *Plasmodium* merozoites and drug-resistance. A review. *Parasite* **1**, 105–114.

Landau, I., Chabaud, A., Cambie, G. and Ginsburg, H. (1991). Chronotherapy of malaria: an approach to malaria chemotherapy. *Parasitology Today* **7**, 350–352.

Landau, I., Lepers, J.-P., Ringwald, P., Rabarison, P., Ginsburg, H. and Chabaud, A. (1992). Chronotherapy of malaria: improved efficacy of timed chloroquine treatment of patients with *Plasmodium falciparum* infections. *Transactions of the Royal Society of Tropical Medicine and Hygiene* **86**, 374–375.

Lauer, S.A., Ghori,, N. and Haldar, K. (1995). Sphingolipid synthesis as a target for chemotherapy against malaria parasites. *Proceedings of the National Academy of Sciences of the United States of America* **92**, 9181–9185.

Li, G.D., Liu, S.Q., Ye, X.Y. and Qu, F.Y. (1995). Detection of 54–kDa protein overexpressed by chloroquine-resistant *Plasmodium berghei* ANKA strain in pyronaridine-resistant *P. berghei* ANKA. *Acta Pharmacologica Sinica* **16**, 17–20.

Liu, D., Liu, R., Ren, D., Gao, D., Zhang, C., Qiu, C., Cai, X., Ling, C., Song, A. and Tang, X. (1995). Changes in the resistance of *Plasmodium falciparum* to chloroquine in Hainan, China. *Bulletin of the World Health Organization* **73**, 483–486.

Liu, D., Liu, R., Zhang, C., Cai, X., Tang, X., Yang, H., Yang, P. and Dong, Y. (1996). Present status of the sensitivity of *Plasmodium falciparum* to antimalarials in China. *Chinese Journal of Parasitology and Parasitic Diseases* **14**, 37–41.

Looareesuwan, S., Viravan, C., Vanijanonta, S., Wilairatana, P., Charoenlarp, P., Canfield, C.J. and Kyle, D.E. (1994). Randomized trial of mefloquine–doxycycline, and artesunate–doxycycline for treatment of acute uncomplicated falciparum malaria. *American Journal of Tropical Medicine and Hygiene* **50**, 784–789.

Looareesuwan, S., Wilairatana, P., Vanijanonta, S., Viravan, C. and Andrial, M. (1995). Efficacy and tolerability of a sequential, artesunate suppository plus mefloquine, treatment of severe falciparum malaria. *Annals of Tropical Medicine and Parasitology* **89**, 469–475.

Looareesuwan, S., Kyle, D.E., Viravan, C., Vanijanonta, S., Wilairatana, P. and Wernsdorfer, W.H. (1996a). Clinical study of pyronaridine for the treatment of acute uncomplicated falciparum malaria in Thailand. *American Journal of Tropical Medicine and Hygiene* **54**, 205–209.

Looareesuwan, S., Viravan, C., Webster, H.K., Kyle, D.E., Hutchinson, D.B. and Canfield, C.J. (1996b). Clinical studies of atovaquone, alone or in combination with other antimalarial drugs, for treatment of acute uncomplicated malaria in Thailand. *American Journal of Tropical Medicine and Hygiene* **54**, 62–66.

Looareesuwan, S., Viravan, C., Vanijanonta, S., Wilairatana, P., Pitisuttithum, P. and Andrial, M. (1996c). Comparative clinical trial of artesunate followed by mefloquine in the treatment of acute uncomplicated falciparum malaria: two- and three-day regimens. *American Journal of Tropical Medicine and Hygiene* **54**, 210–213.

Looareesuwan, S., Wilairatana, P., Vanijanonta, S., Pitisuttithum, P., Viravan, C. and Kraisintu, K. (1996d). Treatment of acute, uncomplicated, falciparum malaria with oral dihydroartemisinin. *Annals of Tropical Medicine and Parasitology* **90**, 21–28.

Looareesuwan, S., Wilairatana, P., Vannaphan, SA., Gordeuk, V.R., Taylor, T.E., Meshnick, S.R. and Brittenham, G.M. (1996e). Co-administration of desferrioxamine B with artesunate in malaria: an assessment of safety and tolerance. *Annals of Tropical Medicine and Parasitology* **90**, 551–554.

Luxemburger, C., ter Kuile, F.O., Nosten, F., Dolan, G., Bradol, J.H., Phaipun, L., Chongsuphajaisiddhi, T. and White, N.J. (1994). Single day mefloquine-artesunate combination in the treatment of multi-drug resistant falciparum malaria. *Transactions of the Royal Society of Tropical Medicine and Hygiene* **88**, 213–217.

Mabeza, G.F., Biemba, G. and Gordeuk, V.R. (1996). Clinical studies of iron chelators in malaria. *Acta Haematologica* **95**, 78–86.

Mackinnon, M.J. (1997). Survival probability of drug resistant mutants in malaria parasites. *Proceedings of the Royal Society of London, (B* **264**, 53–59.

Macomber, P.B., O'Brien, R.L. and Hahn, F.E. (1966). Chloroquine: physiological basis of drug resistance in *Plasmodium berghei. Science* **152**, 1374–1375.

Mahmalgi, J., Veignie, E., Prensier, G. and Moreau, S. (1989). Relations between resistance to chloroquine and acidification of endocytotic vesicle of *Plasmodium berghei. Parasitology* **98**, 1–6.

Marlar-Than, Myat-Phone-Kyaw, Aye-Yu-Soe, Khaing-Khaing-Gyi, Ma-Sabai and Myint-Oo (1995). Development of resistance to chloroquine by *Plasmodium vivax* in Myanmar. *Transactions of the Royal Society of Tropical Medicine and Hygiene* **89**, 307–308.

Marshall, S.J., Russell, P.F., Wright, C.W., Anderson, M.M., Phillipson, J.D., Kirby, G.C., Warhurst, D.C. and Schiff, P.L., jr (1994). *In vitro* antiplasmodial, anti-amoebic, and cytotoxic activities of a series of bisbenzylisoquinoline alkaloids. *Antimicrobial Agents and Chemotherapy* **38**, 96–103.

Martin, S.K., Oduola, A.M.J. and Milhous, W.K. (1987). Reversal of chloroquine resistance in *Plasmodium falciparum* by verapamil. *Science* **235**, 899–901.

McConkey, G.A., Rogers, M.J. and McCutchan, T.F. (1997). Inhibition of *Plasmodium falciparum* protein synthesis — targeting the plastid-like organelle with thiostrepton. *Journal of Biological Chemistry* **272**, 2046–2049.

Meshnick, S.R., Thomas, A., Ranz, A., Xu, C. and Pan, H.Z. (1991). Artemisinin (qinghaosu): the role of intracellular hemin in its mechanism of antimalarial action. *Molecular and Biochemical Parasitology* **49**, 181–190.

Meshnick, S.R., Jefford, C.W., Posner, G.H., Avery, M.A. and Peters, W. (1996). Second-generation antimalarial endoperoxides. *Parasitology Today* **12**, 79–82.

Milhous, W.K., Gerena, L., Kyle, D.E. and Oduola, A.M.J. (1989). In vitro strategies for circumventing antimalarial drug resistance. In: *Malaria and the Red Cell: 2* (J.W. Eaton, S.R. Meshnick and G.J. Brewer, eds), pp. 61–72. New York: Alan R. Liss.

Mockenhaupt, F.P. (1995). Mefloquine resistance in *Plasmodium falciparum*. *Parasitology Today* **11**, 248–253.

Müller, O., van Hensbroek, M.B., Jaffar, S., Drakely, C., Okorie, C., Joof, D., Pinder, M. and Greenwood, B. (1996). A randomized trial of chloroquine, amodiaquine and pyrimethamine–sulphadoxine in Gambian children with uncomplicated malaria. *Tropical Medicine and International Health* **1**, 124–132.

Murphy, G.S., Basri, H., Purnomo, Andersen, E.M., Bangs, M.J., Mount, D.L., Gorden, J., Lal, A.A., Purwokusomo, A.R., Harjosuwarno, S., Sorensen, K. and Hoffman, S.L. (1993). Vivax malaria resistant to treatment and prophylaxis with chloroquine. *Lancet* **341**, 96–100.

Nateghpour, M., Ward, S.A. and Howells, R.E. (1993). Development of halofantrine resistance and determination of cross-resistance patterns in *Plasmodium falciparum*. *Antimicrobial Agents and Chemotherapy* **37**, 2337–2343.

Nosten, F., ter Kuile, F. O., Chongsuphajaisiddhi, T., Luxemburger, C., Webster, K., Edstein, H.K., Phaipun, L., Thew, K.L. and White, N.J. (1991). Mefloquine-resistant falciparum malaria on the Thai–Burmese border. *Lancet* **337**, 1140–1143.

Nosten, F., Luxemburger, C., ter Kuile, F.O., Woodrow, C., Eh, J.P., Chongsuphajaisiddhi, T. and White, N.J. (1994). Treatment of multidrug-resistant *Plasmodium falciparum* malaria with 3-day artesunate–mefloquine combination. *Journal of Infectious Diseases* **170**, 971–977.

Nwanyanwu, O.C., Ziba, C., Kazembe, P., Chitsulo, L., Wirima, J.J., Kumwenda, N. and Redd, S.C. (1996). Efficacy of sulphadoxine/pyrimethamine for *Plasmodium falciparum* malaria in Malawian children under five years of age. *Tropical Medicine and International Health* **1**, 231–235.

Oduola, A.M.J., Omitowoju, G.O., Gerena, L., Kyle, D.E., Milhous, W.K., Sowunmi, A. and Salako, L.A. (1993). Reversal of mefloquine resistance with penfluridol in isolates of *Plasmodium falciparum* from south-west Nigeria. *Transactions of the Royal Society of Tropical Medicine and Hygiene* **87**, 81–83.

Olaya, P. and Wasserman, M. (1991). Effect of calpain inhibitors on the invasion of human erythrocytes by the parasite *Plasmodium falciparum*. *Biochimica et Biophysica Acta* **1096**, 217–221.

Olliaro, P. and Trigg, P.I. (1995). Status of antimalarial drugs under development. *Bulletin of the World Health Organization* **73**, 565–571.

Olliaro, P., Nevill, C., Le Bras, J., Mussano, P., Garner, P. and Brasseur, P. (1996). Systematic review of amodiaquine treatment in uncomplicated malaria. *Lancet* **348**, 1196–1201.

O'Neill, M.J., Bray, D.H., Boardman, P., Phillipson, J.D. and Warhurst, D.C. (1985). Plants as sources of antimalarial drugs. Part 1. *In vitro* test method for the evaluation of crude extracts from plants. *Planta Medica* **47**, 394–398.

O'Neill, M.J., Bray, D.H., Boardman, P., Phillipson, J.D., Warhurst, D.C., Peters, W. and Suffness, M. (1986). Plants as sources of antimalarial drugs: *in vitro* antimalarial activities of some quassinoids. *Antimicrobial Agents and Chemotherapy* **30**, 101–104.

Padua, R.A. (1981). *Plasmodium chabaudi*: genetics of resistance to chloroquine. *Experimental Parasitology* **52**, 419–426.

Pang, L.W., Boudreau, E.F., Limsomwong, N. and Singharaj, P. (1987). Doxycycline prophylaxis for falciparum malaria. *Lancet* **i**, 1161–1164.

Pasvol, G., Clough, B., Carlsson, J. and Snounou, G. (1995). The pathogenesis of severe falciparum malaria. In: *Baillière's Clinical Infectious Diseases. Malaria.* (G. Pasvol, ed.). Vol. 2, no. 2, pp. 249–270. London: Baillière Tindall.

Pattanapanyasat, K., Thaithong, S., Kyle, D.E., Udomsangpetch, R., Yongvanitchit, K., Hider, R.C. and Webster, H.K. (1997). Flow cytometric assessment of hydroxypyridinone iron chelators on *in vitro* growth of drug-resistant malaria. *Cytometry* **27**, 84–91.

Peel, S.A., Bright, P., Yount, B., Handy, J. and Baric, R.S. (1994). A strong association between mefloquine and halofantrine resistance and amplification, overexpression, and mutation in the P-glycoprotein gene homolog (pfmdr) of *Plasmodium falciparum in vitro*. *American Journal of Tropical Medicine and Hygiene* **51**, 648–658.

Peters, W. (1970a). *Chemotherapy and Drug Resistance in Malaria.* London: Academic Press.

Peters, W. (1970b). A new type of antimalarial drug potentiation. *Transactions of the Royal Society of Tropical Medicine and Hygiene* **64**, 462–464.

Peters, W. (1987). *Chemotherapy and Drug Resistance in Malaria*, 2nd edn. London: Academic Press.

Peters, W. and Robinson, B.L. (1991). The chemotherapy of rodent malaria. XLVI. Reversal of mefloquine resistance in rodent *Plasmodium*. *Annals of Tropical Medicine and Parasitology* **85**, 5–10.

Peters, W. and Robinson, B.L. (1992). The chemotherapy of rodent malaria. XLVII. Studies on pyronaridine and other Mannich base antimalarials. *Annals of Tropical Medicine and Parasitology* **86**, 455–465.

Peters, W. and Robinson, B.L. (1997). The chemotherapy of rodent malaria. LV. Interactions between pyronaridine and artemisinin. *Annals of Tropical Medicine and Parasitology* **91**, 141–145.

Peters, W., Portus, J. and Robinson, B.L. (1977). The chemotherapy of rodent malaria. XXVIII. The development of resistance to mefloquine (WR 142,490). *Annals of Tropical Medicine and Parasitology* **71**, 419–427.

Peters, W., Robinson, B.L. and Ellis, D.S. (1987). The chemotherapy of rodent malaria. XLII. Halofantrine and halofantrine resistance. *Annals of Tropical Medicine and Parasitology* **81**, 639–646.

Peters, W., Ekong, R., Robinson, B.L., Warhurst, D.C. and Pan, X.-Q. (1990). The chemotherapy of rodent malaria. XLV. Reversal of chloroquine resistance in rodent and human *Plasmodium* by antihistaminic agents. *Annals of Tropical Medicine and Parasitology* **84**, 541–551.

Peters, W., Robinson, B.L. and Milhous, W.K. (1993a). The chemotherapy of rodent malaria. LI. Studies on a new 8-aminoquinoline, WR 238,605. *Annals of Tropical Medicine and Parasitology* **87**, 547–552.

Peters, W., Robinson, B.L., Tovey, G., Rossier, J.C. and Jefford, C.W. (1993b). The chemotherapy of rodent malaria. L. The activities of some synthetic 1,2,4-trioxanes against chloroquine-sensitive and chloroquine-resistant parasites. Part 3: Observations on 'Fenozan-50F', a difluorinated 3,3'-spirocyclopentane 1,2,4-trioxane. *Annals of Tropical Medicine and Parasitology* **87**, 111–123.

Peterson, D.S., Di Santi, S.M., Povoa, M., Calvosa, V.S., Do Rosario, V.E., and Wellems, T.E. (1991). Prevalence of the dihydrofolate reductase Asn-108 mutation as the basis for pyrimethamine-resistant falciparum malaria in the Brazilian Amazon. *American Journal of Tropical Medicine and Hygiene* **45**, 492–497.

Phillips, E.J., Keystone, J.S. and Kain, K.C. (1996). Failure of combined chloroquine and high-dose primaquine therapy for *Plasmodium vivax* malaria acquired in Guyana, South America. *Clinical Infectious Diseases* **23**, 1171–1173.

Phillipson, J.D., Wright, C.W., Kirby, G.C. and Warhurst, D.C. (1995). Phytochemistry of some plants used in traditional medicine for the treatment of protozoal diseases. In: *Phytochemistry of Plants used in Traditional Medicine* (K. Hostettman, A. Marston, M. Maillard and M. Hamburger, eds), pp. 95–135. Oxford: Clarendon Press.

Picot, S., Peyron, F., Donadille, A., Vuillez, J.P., Barbe, G. and Ambroise-Thomas, P. (1993). Chloroquine-induced inhibition of the production of TNF, but not of IL-6, is affected by disruption of iron metabolism. *Immunology* **80**, 127–133.

Plowe, C.V. and Wellems, T.E. (1995). Molecular approaches to the spreading problem of drug resistant malaria. *Advances in Experimental Medicine and Biology* **390**, 197–209.

Plowe, C.V., Djimde, A., Wellems, T.E., Diop, S., Kouriba, B. and Doumbo, O.K. (1996). Community pyrimethamine–sulfadoxine use and prevalence of resistant *Plasmodium falciparum* genotypes in Mali: a model for deterring resistance. *American Journal of Tropical Medicine and Hygiene* **55**, 467–471.

Pollack, S. (1989). *P. falciparum* iron metabolism. In: *Malaria and the Red Cell: 2* (J.W. Eaton, S.R. Meshnick and G.J. Brewer, eds), pp. 151–161. New York: Alan R. Liss.

Posner, G.H., Oh, C.H., Webster, H.K., Ager, A.L., jr and Rossan, R.N. (1994). New, antimalarial, tricyclic 1,2,4-trioxanes: evaluations in mice and monkeys. *American Journal of Tropical Medicine and Hygiene* **50**, 522–526.

Posner, G.H., Tao, X., Cumming, J.N., Klinedinst, D. and Shapiro, T.A. (1996). Antimalarially potent, easily prepared, fluorinated endoperoxides. *Tetrahedron Letters* **37**, 7225–7228.

Price, R.N., Nosten, F., Luxemburger, C., Kham, A., Brockman, A., Chongsuphajaisiddhi, T. and White, N.J. (1995). Artesunate versus artemether in combination with mefloquine for the treatment of multidrug-resistant falciparum malaria. *Transactions of the Royal Society of Tropical Medicine and Hygiene* **89**, 523–527.

Price, R.N., Nosten, F., Luxemburger, C., ter Kuile, F.O., Paiphun, L., Chongsuphajaisiddhi, T. and White, N.J. (1996). Effects of artemisinin derivatives on malaria transmissibility. *Lancet* **347**, 1654–1658.

Puri, S.K. and Dutta, G.P. (1989). Delay in emergence of mefloquine resistance in *Plasmodium berghei* by use of drug combinations. *Acta Tropica* **46**, 209–212.

Radloff, P.D., Philipps, J., Nkeyi, M., Hutchinson, D. and Kremsner, P.G. (1996a). Atovaquone and proguanil for *Plasmodium falciparum* malaria. *Lancet* **347**, 1511–1514.

Radloff, P.D., Philipps, J., Nkeyi, M., Sturchler, D., Mittelholzer, M.-L. and Kremsner, P.G. (1996b). Arteflene compared with mefloquine for treating *Plasmodium*

falciparum malaria in children. *American Journal of Tropical Medicine and Hygiene* **55**, 259–262.
Ramkaran, A.E. and Peters, W. (1969). Infectivity of chloroquine-resistant *Plasmodium berghei* to *Anopheles stephensi* enhanced by chloroquine. *Nature* **223**, 635–636.
Rathod, P.K., Leffers, N.P. and Young, R.D. (1992). Molecular targets of 5-fluoroorotate in the human malaria parasite, *Plasmodium falciparum*. *Antimicrobial Agents and Chemotherapy* **36**, 704–711.
Rathod, P.K., Khosla, M., Gassis, S., Young, R.D. and Lutz, C. (1994). Selection and characterization of 5-fluoroorotate-resistant *Plasmodium falciparum*. *Antimicrobial Agents and Chemotherapy* **38**, 2871–2876.
Raynes, K., Foley, M., Tilley, L. and Deady, L.W. (1996). Novel bisquinoline antimalarials: synthesis, antimalarial activity, and inhibition of haem polymerisation. *Biochemical Pharmacology* **52**, 551–559.
Reeder, J.C., Rieckmann, K.H., Genton, B., Lorry, K., Wines, B. and Cowman, A.F. (1996). Point mutations in the dihydrofolate reductase and dihydropteroate synthetase genes and *in vitro* susceptibility to pyrimethamine and cycloguanil of *Plasmodium falciparum* isolates from Papua New Guinea. *American Journal of Tropical Medicine and Hygiene* **55**, 209–213.
Ridley, R.G., Hofheinz, W., Matile, H., Jaquet, C., Dorn, A., Masciadri, R., Jolidon, S., Richter, W.F., Guenzi, A., Girometta, M.-A., Urwyler, H., Huber, W., Thaithong, S. and Peters, W. (1996). 4-Aminoquinoline analogs of chloroquine with shortened side chains retain activity against chloroquine-resistant *Plasmodium falciparum*. *Antimicrobial Agents and Chemotherapy* **40**, 1846–1854.
Ridley, R.G., Matile, H., Jaquet, C., Dorn, A., Hofheinz, W., Leupin, W., Masciadri, R., Theil, F.-P., Richter, W.F., Girometta, M.-A., Guenzi, A., Urwyler, H., Gocke, E., Potthast, J.-M., Csato, M., Thomas, A. and Peters, W. (1997). Antimalarial activity of the bisquinoline trans-N1,N2-bis(7-chloroquinolin-4-yl) cyclohexane-1,2-diamine. A comparison of two stereoisomers and a detailed evaluation of the (S,S)-enantiomer, Ro 47–7737. *Antimicrobial Agents and Chemotherapy* **41**, 677–686.
Rieckmann, K.H., Davis, D.R. and Hutton, D.C. (1989). *Plasmodium vivax* resistance to chloroquine? *Lancet* **334**, 1183–1184.
Rieckmann, K.H., Yeo, A.E.T., Davis, D.R., Hutton, D.C., Wheatley, P.F. and Simpson, R. (1993). Recent military experience with malaria chemoprophylaxis. *Medical Journal of Australia* **158**, 446–449.
Rieckmann, K.H., Yeo, A.E.T. and Edstein, M.D. (1996). Activity of PS-15 and its metabolite, WR99210, against *Plasmodium falciparum* in an *in vivo–in vitro* model. *Transactions of the Royal Society of Tropical Medicine and Hygiene* **90**, 568–571.
Ringwald, P., Bickii, J. and Basco, L. (1996). Randomised trial of pyronaridine versus chloroquine for acute uncomplicated falciparum malaria in Africa. *Lancet* **347**, 24–28.
Ritchie, G.Y., Mungthin, M., Green, J.E., Bray, P.G., Hawley, S.R. and Ward, S.A. (1996). *In vitro* selection of halofantrine resistance in *Plasmodium falciparum* is not associated with increased expression of Pgh1. *Molecular and Biochemical Parasitology* **83**, 35–46.
Robert, V., Molez, J.-F. and Trape, J.-F. (1996). Short report: gametocytes, chloroquine pressure, and the relative parasite survival advantage of resistant strains of falciparum malaria in West Africa. *American Journal of Tropical Medicine and Hygiene* **55**, 350–351.

Rønn, A.M., Msangeni, H.A., Mhina, J., Wernsdorfer, W.H. and Bygbjerg, I.C. (1996). High level of resistance of *Plasmodium falciparum* to sulfadoxine–pyrimethamine in children in Tanzania. *Transactions of the Royal Society of Tropical Medicine and Hygiene* **90**, 179–181.

Rosenthal, P.J. (1995). *Plasmodium falciparum*: effects of proteinase inhibitors on globin hydrolysis by cultured malaria parasites. *Experimental Parasitology* **80**, 272–281.

Rubio, J.P. and Cowman, A.F. (1994). *Plasmodium falciparum*: the pfmdr2 protein is not overexpressed in chloroquine-resistant isolates of the malaria parasite. *Experimental Parasitology* **79**, 137–147.

Ruetz, S., Delling, U., Brault, M., Scurr, E. and Gros, P. (1996). The *pfmdr1* gene of *Plasmodium falciparum* confers cellular resistance to antimalarial drugs in yeast cells. *Proceedings of the National Academy of Sciences of the United States of America* **93**, 9942–9947.

Sadiq, S.T., Glasgow, K.W., Drakeley, C.J., Muller, O., Greenwood, B.M., Mabey, D.C.W. and Bailey, R.L. (1995). Effects of azithromycin on malariometric indices in The Gambia. *Lancet* **346**, 881–882.

Salako, L.A., Guiguemde, R., Mittelholzer, M.L., Haller, L., Sorenson, F., Sturchler, D. and Bradley, D.J. (1994). Ro 42–1611 in the treatment of patients with mild malaria: a clinical trial in Nigeria and Burkina Faso. *Tropical Medicine and Parasitology* **45**, 284–287.

Schultz, L.J., Steketee, R.W., Chitsulo, L., Macheso, A., Kazembe, P. and Wirima, J.J. (1996). Evaluation of maternal practices, efficacy, and cost-effectiveness of alternative antimalarial regimens for use in pregnancy: chloroquine and sulfadoxine–pyrimethamine. *American Journal of Tropical Medicine and Hygiene* **55** (supplement 1), 87–94.

Schuurkamp, G.J., Spicer, P.E., Kereu, R.K. and Bulongol, P.L. (1989). A mixed infection of vivax and falciparum malaria apparently resistant to 4-aminoquinoline: a case report. *Transactions of the Royal Society of Tropical Medicine and Hygiene* **83**, 607–608.

Schwartz, I.K., Lacehitz, E.M. and Patchen, L.C. (1991). Chloroquine-resistant *Plasmodium vivax* from Indonesia. *New England Journal of Medicine* **324**, 927.

Shao, B.R., Huang, Z.S., Shi, X.H. and Meng, F. (1991). A 5-year surveillance of sensitivity *in vivo* of *Plasmodium falciparum* to pyronaridine/sulfadoxine/pyrimethamine in Diaoluo area, Hainan Province. *Southeast Asian Journal of Tropical Medicine and Public Health* **22**, 65–67.

Sirawaraporn, W., Sirawaraporn, R., Cowman, A.F., Yuthavong, Y. and Santi, D.V. (1990). Heterologous expression of active thymidylate synthase–dihydrofolate reductase from *Plasmodium falciparum*. *Biochemistry* **29**, 10779–10785.

Sirawaraporn, W., Sathitkul, T., Sirawaraporn, R., Yuthavong, Y. and Santi, D.V. (1997). Antifolate-resistant mutants of *Plasmodium falciparum* dihydrofolate reductase. *Proceedings of the National Academy of Sciences of the United States of America* **94**, 1124–1129.

Slater, A.F. (1993). Chloroquine: mechanism of drug action and resistance in *Plasmodium falciparum*. *Pharmacology and Therapeutics* **57**, 203–235.

Slater, A.F. and Cerami, A. (1992). Inhibition by chloroquine of a novel haem polymerase enzyme active in malaria trophozoites. *Nature* **355**, 167–169.

Sowunmi, A., Oduola, A.M.J., Ogundahunsi, O.A.T., Falade, C.O., Gbotosho, G.O. and Salako, L.A. (1997). Enhanced efficacy of chloroquine–chlorpheniramine combination in acute uncomplicated falciparum malaria in children. *Transactions of the Royal Society of Tropical Medicine and Hygiene* **91**, 63–67.

Stromberg, A. and Björkman, A. (1992). Ciprofloxacin does not achieve radical cure of *Plasmodium falciparum* infection in Sierra Leone. *Transactions of the Royal Society of Tropical Medicine and Hygiene* **86**, 373.
Sullivan, D.J., Gluzman, I.Y. and Goldberg, D.E. (1996a). *Plasmodium* hemozoin formation mediated by histidine-rich proteins. *Science* **271**, 219–222.
Sullivan, D.J., Gluzman, I.Y., Russell, D.G. and Goldberg, D.E. (1996b). On the molecular mechanism of chloroquine's antimalarial action. *Proceedings of the National Academy of Sciences of the United States of America* **93**, 11865–11870.
Thaithong, S., Suebsaeng, L., Rooney, W. and Beale, G.H. (1988). Evidence of increased chloroquine sensitivity in Thai isolates of *Plasmodium falciparum*. *Transactions of the Royal Society of Tropical Medicine and Hygiene* **82**, 37–38.
Thaithong, S., Chan, S.W., Songsomboon, S., Wilairat, P., Seesod, N., Sueblinwong, T., Goman, M., Ridley, R. and Beale, G. (1992). Pyrimethamine-resistant mutations in *Plasmodium falciparum*. *Molecular and Biochemical Parasitology* **52**, 149–157.
Thimasarn, K., Sirichaisinthop, J., Vijaykadga, S., Tansophalaks, S., Yamokgul, P., Laomiphol, A., Palananth, C., Thamewat, U., Thaithong, S. and Rooney, W. (1995). *In vivo* studies of the response of *Plasmodium falciparum* to standard mefloquine/sulfadoxine/pyrimethamine (MSP) treatment among gem miners returning from Cambodia. *Southeast Asian Journal of Tropical Medicine and Public Health* **26**, 204–212.
Trager, W. and Jensen, J.B. (1976). Human malaria parasites in continuous culture. *Science* **193**, 673–675.
Tran, T.H., Day, N.P., Nguyen, H.P., Nguyen, T.H., Tran, T.H., Pham, P.L., Dinh, X.S., Ly, V.C., Waller, D., Peto, T.E. and White, N.J. (1996). A controlled trial of artemether or quinine in Vietnamese adults with severe falciparum malaria. *New England Journal of Medicine* **335**, 76–83.
Triglia, T., Foote, S.J., Kemp, D.J. and Cowman, A.F. (1991). Amplification of the multidrug resistance gene pfmdr1 in *Plasmodium falciparum* has arisen as multiple independent events. *Molecular and Cellular Biology* **11**, 5244–5250.
Tripathi, K.D., Sharma, A.K. and Valecha, N. (1993a). Norfloxacin in the treatment of vivax malaria. *Medical Science Research* **21**, 159–160.
Tripathi, K.D., Sharma, A.K., Valecha, N. and Biswas, S. (1993b). *In vitro* activity of fluoroquinolones against chloroquine-sensitive and chloroquine-resistant *Plasmodium falciparum*. *Indian Journal of Malariology* **30**, 67–73.
Tripathi, K.D., Sharma, A.K., Valecha, N. and Kulpati, D.D.S. (1993c). Curative efficacy of norfloxacin in falciparum malaria. *Indian Journal of Medical Research, Section A: Infectious Diseases* **97**, 176–178.
Vander Jagt, D.L., Caughey, W.S., Campos, N.M., Hunsaker, L.A. and Zanner, M.A. (1989). Parasite proteases and antimalarial activities of protease inhibitors. In: *Malaria and the Red Cell: 2* (J.W. Eaton, S.R. Meshnick and G.J. Brewer, eds), pp. 105–118. New York: Alan R. Liss.
van Dijk, M.R., McConkey, G.A., Vinkenoog, R., Water, A.J. and Janse, C.J. (1994). Mechanisms of pyrimethamine resistance in two different strains of *Plasmodium berghei*. *Molecular and Biochemical Parasitology* **68**, 167–171.
van Dijk, M.R., Waters, A.P. and Janse, C.J. (1995). Stable transfection of malaria parasite blood stages. *Science* **268**, 1358–1362.
van Dijk, M.R., Janse, C.J. and Waters, A.P. (1996). Expression of a *Plasmodium* gene introduced into subtelomeric regions of *P. berghei* chromosomes. *Science* **271**, 662–665.
van Hensbroek, M.B., Onyiorah, E., Jaffar, S., Schneider, G., Palmer, A., Frenkel,

J., Enwere, G., Forck, S., Nusmeijer, A., Bennett, S., Greenwood, B. and Kwiatkowski, D. (1996). A trial of artemether or quinine in children with cerebral malaria. *New England Journal of Medicine* **335**, 69–75.

Vennerstrom, J.L., Fu, H.-N., Ellis, W.Y., Ager, A.L., jr, Wood, J.K., Andersen, S.L., Gerena, L. and Milhous, W.K. (1992). Dispiro-1,2,4,5-tetroxanes: a new class of antimalarial peroxides. *Journal of Medicinal Chemistry* **35**, 3023–3027.

Verdrager, J. (1995). Localized permanent epidemics: the genesis of chloroquine resistance in *Plasmodium falciparum*. *Southeast Asian Journal of Tropical Medicine and Public Health* **26**, 23–28

Vial, H. (1996). Recent developments and rationale towards new strategies for malarial chemotherapy. *Parasite* **3**, 3–23.

Walliker, D. (1994). The role of molecular genetics in field studies on malaria parasites. *International Journal for Parasitology* **24**, 799–808.

Wang, P., Brooks, D.R., Sims, P.F. and Hyde, J.E. (1995). A mutation-specific PCR system to detect sequence variation in the dihydropteroate synthetase gene of *Plasmodium falciparum*. *Molecular and Biochemical Parasitology* **71**, 115–125.

Warrell, D.A., Molyneux, M.E. and Beales, P.F. (1990). Severe and complicated malaria, 2nd edition. *Transactions of the Royal Society of Tropical Medicine and Hygiene* **84** (Supplement 2).

Watkins, W.M., Brandling-Bennett, A.D., Nevill, C.G., Carter, J.Y., Boriga, D.A., Howells, R.E. and Koech, D.K. (1988). Chlorproguanil/dapsone for the treatment of non-severe *Plasmodium falciparum* malaria in Kenya: a pilot study. *Transactions of the Royal Society of Tropical Medicine and Hygiene* **82**, 398–403.

Watt, G., Long, G.W., Grogl, M. and Martin, S.K. (1990). *Reversal of drug-resistant falciparum malaria by calcium antagonists: potential for host cell toxicity*. Geneva: World Health Organization, mimeographed document WHO/MAL 90.1056.

Watt, G., Shanks, G.D., Edstein, M.D., Pavanand, K., Webster, H.K. and Wechgritaya, S. (1991). Ciprofloxacin treatment of drug-resistant falciparum malaria. *Journal of Infectious Diseases* **164**, 602–604.

Webster, H.K., Thaithong, S., Pavanand, K., Yongvanitchit, K., Pinswasdi, C. and Boudreau, E.F. (1985). Cloning and characterization of mefloquine-resistant *Plasmodium falciparum* from Thailand. *American Journal of Tropical Medicine and Hygiene* **34**, 1022–1027.

Weiss, G., Thuma, P.E., Mabeza, G., Werner, E.R., Herold, M. and Gordeuk, V.R. (1997). Modulatory potential of iron chelation therapy on nitric oxide formation in cerebral malaria. *Journal of Infectious Diseases* **175**, 226–230.

Weiss, W.R., Oloo, A.J., Johnson, A., Koech, D. and Hoffman, S.L. (1995). Daily primaquine is effective for prophylaxis against falciparum malaria in Kenya: comparison with mefloquine, doxycycline, and chloroquine plus proguanil. *Journal of Infectious Diseases* **171**, 1569–1575.

Wellems, T.E. (1991). Molecular genetics of drug resistance in *Plasmodium falciparum* malaria. *Parasitology Today* **7**, 110–112.

Wellems, T.E., Panton, L.J., Gluzman, I.Y., Rosario, V.E., Gwadz, R.W., Walker-Jonah, A. and Krogstad, D.J. (1990). Chloroquine resistance not linked to mdr-like genes in *Plasmodium falciparum* cross. *Nature* **345**, 253–255.

Wernsdorfer, W.H. (1994). Epidemiology of drug resistance in malaria. *Acta Tropica* **56**, 143–156.

Wernsdorfer, W.H. and Payne, D. (1991). The dynamics of drug resistance in *Plasmodium falciparum*. *Pharmacology and Therapeutics* **50**, 95–121.

Wernsdorfer, W.H., Landgraf, B., Wiedermann, G. and Kollaritsch, H. (1995).

Chloroquine resistance of *Plasmodium falciparum*: a biological advantage. *Transactions of the Royal Society of Tropical Medicine and Hygiene* **89**, 90–91.
Wesche, D.L., Woosley, R.L., Chen, Y., Wang, W., Miller, R.E., Kyle, D.E., Ngampochjana, M., Nuzum, E.O., Rosaan, R.N. and Schuster, B.G. (1996). Desbutylhalofantrine as a potentially safer alternative to halofantrine. *American Journal of Tropical Medicine and Hygiene* **55** (2, Supplement), abstract no. 110, p. 135.
White, N.J. and Olliaro, P.L. (1996). Strategies for the prevention of antimalarial drug resistance: rationale for combination chemotherapy for malaria. *Parasitology Today* **12**, 399–401.
WHO (1965). *Resistance of Malaria Parasites to Drugs*. Geneva: World Health Organization, Technical Report Series, no. 296.
WHO (1973). *Chemotherapy of Malaria and Resistance to Antimalarials*. Geneva: World Health Organization, Technical Report Series, no. 529.
WHO (1990). *Practical Chemotherapy of Malaria*. Geneva: World Health Organization, Technical Report Series, no. 805.
WHO (1993). *Implementation of the Global Malaria Control Strategy. Report of a WHO Study Group on the Implementation of the Global Plan of Action for Malaria Control 1993–2000*. Geneva: World Health Organization, Technical Report Series, no. 839.
WHO (1995). *WHO Model Prescribing Information. Drugs Used in Parasitic Diseases*, 2nd edition. Geneva: World Health Organization.
WHO (1996a). *Management of Uncomplicated Malaria and the use of Antimalarial Drugs for the Protection of Travellers*. Geneva: World Health Organization, mimeographed document WHO/MAL/96.1075.
WHO (1996b). *Assessment of Therapeutic Efficacy of Antimalarial Drugs for Uncomplicated Falciparum Malaria in areas with Intense Transmission*. Geneva: World Health Organization, mimeographed document WHO/MAL/96.1077.
Wilairatana. P., Kyle, D.E., Looareesuwan, S., Chinwongprom, K., Amradee, S., White, N.J. and Watkins, W.M. (1997). Poor efficacy of antimalarial biguanide–dapsone combinations in the treatment of acute, uncomplicated, falciparum malaria in Thailand. *Annals of Tropical Medicine and Parasitology* **91**, 125–132.
Wilson, C.M., Serrano, A.E., Wasley, A., Bogenschutz, M.P., Shankar, A.H. and Wirth, D.F. (1989). Amplification of a gene related to mammalian mdr genes in drug-resistant *Plasmodium falciparum*. *Science* **244**, 1184–1186.
Wilson, C.M., Volkman, S.K., Thaithong, S., Martin, R.K., Kyle, D.E., Milhous, W.K. and Wirth, D.F. (1993). Amplification of pfmdr1 associated with mefloquine and halofantrine resistance in *Plasmodium falciparum* from Thailand. *Molecular and Biochemical Parasitology* **57**, 151–160.
Winter, R.W., Cornell, K.A., Johnson, L.L., Ignatushchenko, M., Hinrichs, D.J. and Riscoe, M.K. (1996). Potentiation of the antimalarial agent rufigallol. *Antimicrobial Agents and Chemotherapy* **40**, 1408–1411.
Wongsrichanalai, C., Webster, H.K., Wimonwattrawatee, T., Sookto, P., Chuanak, N., Timasarn, K. and Wernsdorfer, W.H. (1992). In vitro sensitivity of *Plasmodium falciparum* isolates in Thailand to quinine and chloroquine, 1984–1990. *Southeast Asian Journal of Tropical Medicine and Public Health* **23**, 533–536.
Wooden, J.M., Hartwell, L.H., Vasquez, B. and Sibley, C.H. (1997). Analysis in yeast of antimalaria drugs that target dihydrofolate reductase of *Plasmodium falciparum*. *Molecular and Biochemical Parasitology* **85**, 25–40.
Yeo, A.E.T. and Rieckmann, K.H. (1994a). The activity of PS-15 in combination with sulfamethoxazole. *Tropical Medicine and Parasitology* **45**, 136–137.
Yeo, A.E.T. and Rieckmann, K.H. (1994b). Prolonged exposure of *Plasmodium*

falciparum to ciprofloxacin increases anti-malarial activity. *Journal of Parasitology* **80**, 158–160.

Yeo, A.E.T. and Rieckmann, K.H. (1994c). The *in vitro* antimalarial activity of chloramphenicol against *Plasmodium falciparum*. *Acta Tropica* **56**, 51–54.

Yeo, A.E.T. and Rieckmann, K.H. (1995). Increased antimalarial activity of azithromycin during prolonged exposure of *Plasmodium falciparum in vitro*. *International Journal for Parasitology* **25**, 531–532.

Yeo, A.E.T., Edstein, M.D. and Shanks, G.D. (1992). Proguanil combined with dapsone chemoprophylaxis for malaria. *Medical Journal of Australia* **156**, 883.

Zidovetzki, R., Sherman, I.W., Prudhomme, J. and Crawford, J. (1994). Inhibition of *Plasmodium falciparum* lysophospholipase by anti-malarial drugs and sulphydryl reagents. *Parasitology* **108**, 249–255.

Zindrou, S., Dao, L.D., Xuyen, P.T., Dung, N.P., Sy, N.D., Skold, O. and Swedberg, G. (1996). Rapid detection of pyrimethamine susceptibility of *Plasmodium falciparum* by restriction endonuclease digestion of dihydrofolate gene. *American Journal of Tropical Medicine and Hygiene* **54**, 185–188.

Zolg, J.W., Chen, G.-X. and Plitt, J.R. (1990). Detection of pyrimethamine resistance in *Plasmodium falciparum* by mutation-specific polymerase chain reaction. *Molecular and Biochemical Parasitology* **39**, 257–266.

ADDENDUM

After submission of the manuscript for this chapter, a number of key publications appeared.

Section 3.1. Reynolds and Roos (1998) have developed an elegant surrogate model for folate-resistant *P. falciparum* by transforming the DHFR sequences of *Toxoplasma gondii* to produce mutants that are identical to those found in the malaria *DHFR–TS* gene. The mutant parasites are readily studied either *in vitro* or *in vivo*.

Section 3.2. Buckling *et al.* (1997) have shown in *P. chabaudi* that sub-curative treatment of mice with chloroquine leads to an enhanced rate of gametocyte production and an increase in infectivity to mosquitoes. These findings reinforce the hypothesis, first proposed by Ramkaran and Peters in 1969, that the continued deployment of chloroquine in areas where *P. falciparum* is highly resistant to this compound may increase transmission.

Section 4.1. New light has been thrown on the elusive genetic basis of chloroquine resistance in *P. falciparum* which Su *et al.* (1997) have now shown is linked to a gene named *cg*2 which occurs on chromosome 7. This work was made possible by the examination of clones obtained from the genetic cross of the chloroquine-sensitive HB3 and the chloroquine-resistant Dd2 lines of *P. falciparum* described by Wellems *et al.* (1990). (These clones have also been invaluable in studies on the genetics of sulfadoxine resistance – see below.) The gene encodes for a protein of approximately 330 kDa (named CG2) which has complex polymorphisms and is located at the periphery of the intraerythrocytic parasites, as well as in association with the haemozoin of the digestive vacuole where chloroquine is believed to exert its antiparasitic action. Their data suggest that the chloroquine-resistant parasites of south-east Asia and Africa originated over 40 years ago in the Indochina region, whereas those seen now in South America are of independent origin. The CG2 protein is believed to be associated with chloroquine transport, either into or out of the parasite. The initial slow emergence of chloroquine resistance, in contrast with the rapid selection of point mutations in the DHFR gene, is probably due to the fact that a complex of multiple mutations involving 12 or more codons was necessary for the generation of the resistant *cg2* alleles. Furthermore, the data suggest that another, so far unidentified gene with a 'complimentary or permissive role' may be involved in the mechanism of resistance to chloroquine.

Section 4.2. Resistance to Fansidar is rapidly increasing in several countries where it has been used as a replacement for chloroquine. Knowledge of the distribution and prevalence, in nature, of mutants of the genes coding for DHPS and DHFR has been greatly enriched through the work of several groups who have correlated the various mutants or mutant combinations with the clinical responses to Fansidar in patients infected with *P. falciparum*. A new and very sensitive *in vitro* procedure, involving culture in totally folate- and PABA-free medium, has enabled Wang *et al.* (1997a) to show that four of 16 clones obtained from a genetic cross of the sulfadoxine-sensitive HB3 line of this parasite with the sulfadoxine-resistant Dd2 line were highly resistant to this compound and 12 were sensitive. Resistant mutants were associated with a marked reduction of the affinity of parasite DHPS for sulfadoxine. Moreover, their data suggested that a hitherto unrecognized genetic factor linked to the DHFR gene may control the ability of intraerythrocytic *P. falciparum* to utilize host cell folate. In this and a subsequent paper (Wang *et al.*, 1997b) based on the examination of 141 field samples (from Mali, several Middle East countries, Vietnam, Kenya and Tanzania) they concluded that their material (in which 60% of isolates contained mutant *DHPS*) included 10 genotypes of the *DHPS* gene with 13 alleles, while there were 11 variants of the *DHFR* gene. One Pakistan isolate revealed the novel *DHFR* mutation 16 (Ala–Ser):59 (Cys–Arg). In total, they identified 25 different combinations of the two enzymes. Their data indicate that the addition of folic acid to patients receiving Fansidar therapy is especially liable to reduce the activity of the combination if multiple *DHFR* mutants are already present. The data, coupled with those from a large survey of drug responses matched to the genetic determination of the *DHFR* and DHPS mutant frequencies in Mali, Kenya, Malawi and Bolivia by Plowe *et al.* (1997), show that the level of pyrimethamine or sulfadoxine resistance is associated with a stepwise accumulation of mutants in the *DHFR* or *DHPS* gene respectively. Two new *DHFR* mutations (50 Cys–Arg, and a 15 bp repeat between 30 and 31), as well as a novel *DHPS* (540 Lys–Glu), were present in Bolivian isolates. (The last has also been detected in Malawi and Thai isolates.) The latter investigators found a clear association between the prevalence of resistance to Fansidar and the extent of the use either of the individual compounds or, especially, this drug combination in a given area. The prior widespread deployment of other antifolate drugs and sulphonamides (e.g. co-trimoxazole, sulphaguanidine) is probably also responsible for the selection of mutant *DHFR* and *DHPS* genes. Moreover, serial data showed an enhanced number of mutants in matched samples taken before and after treatment of individual patients. Djimde *et al.* (1997) in Mali observed an increase of resistant mutants from 10% to 90% of infections in 109 residents of an endemic area who received pyrimethamine alone prophylactically for five weeks. The Asn-108 mutant of *DHFR*, which is the commonest found in association with pyrimethamine in nature, is apparently the ideal mutant. When Sirawaraporn *et al.* (1997) replaced the wild Ser-108 codon for *P. falciparum* DHFR with those for 19 other amino acids, most failed to lead to pyrimethamine resistance or resulted in an enzyme with diminished activity.

Section 5.1.5. Fidock and Wellems (1997) found that the inhibitory action on parasite growth of WR 99,210 and cycloguanil, but not that of proguanil, is completely blocked in *P. falciparum* that has been transfected with the gene for folate (methotrexate)-resistant DHFR. This confirms earlier suggestions that the target for proguanil (and probably PS-15) is not DHFR. Further evidence for the specific action of WR 99,210 on parasite DHFR was provided by kinetic interaction studies of this compound, which Hekmat-Nejad and Rathod (1997) found to bind almost equally to both pyrimethamine-sensitive and pyrimethamine-resistant *P. falciparum* DHFR. The synergistic action of atovaquone with proguanil against *P. falciparum in vitro* is further enhanced when dapsone is added to the medium. Yeo *et al.* (1997) suggested that the triple combination, in principle, may be of advantage in inhibiting the selection of resistance to the atovaquone–proguanil combination.

Section 5.1.9. Considerable progress has been made in identifying the structure and genetic determinants of the plasmalepsins that initiate the cleavage of haemoglobin by intracellular *Plasmodium* and a number of inhibitors have been synthesized to bind the enzyme of different species (Westling *et al.*, 1997).

Section 5.2.2. Obaldia *et al.* (1997) have shown that a combination of chloroquine with the novel 8-aminoquinoline, WR 238,605 (now named etaquine), is more effective against chloroquine-resistant *P. vivax* in *Aotus* monkeys than either compound alone. Moreover, preliminary clinical trials of this long-acting compound indicate that it may prove to be more effective than primaquine with activity against intraerythrocytic as well as intrahepatic stages of *P. falciparum* (see, e.g., Shanks *et al.*, 1997).

References

Buckling, A.G.J., Taylor, L.H., Carlton, J.M.R. and Read, A.M. (1997). Adaptive changes in *Plasmodium* transmission strategies following chloroquine chemotherapy. *Proceedings of the Royal Society of London, B* **264**, 553–559.

Djimde, A., Cortese, J.F., Kayentao, K., Diourte, Y., Doumbo, O. and Plowe, C.V. (1997). Rapid selection of DHFR mutations *in vivo* by pyrimethamine prophylaxis. *American Journal of Tropical Medicine and Hygiene* **57**, (supplement), abstract no. 379, p. 229.

Fidock, D.A. and Wellems, T.E. (1997). Transformation with human dihydrofolate reductase renders malaria parasites insensitive to WR 99210 but does not affect the intrinsic activity of proguanil. *Proceedings of the National Academy of Sciences of the United States of America* **94**, 10931–10936.

Hekmat-Nejad, M. and Rathod, P.K. (1997). *Plasmodium falciparum*: kinetic interactions of WR 99,210 with pyrimethamine-sensitive and pyrimethamine-resistant dihydrofolate reductase. *Experimental Parasitology* **87**, 222–228.

Obaldia, N., Rossan, R.N., Cooper, R.D., Kyle, D.E., Nuzum, E.O., Rieckmann, K.H. and Shanks, G.D. (1997). WR 238,605, chloroquine, and their combinations as blood schizontocides against a chloroquine-resistant strain of *Plasmodium vivax* in *Aotus* monkeys. *American Journal of Tropical Medicine and Hygiene* **56**, 508–510.

Plowe, C.V., Cortese, J.F., Djimde, A., Nwanyanwu, O.C., Watkins, W.M., Winstanley, P.A., Estrada-Franco, J.G., Mollinedo, R.E., Avila, J.C., Cespedes, J.L., Carter, D. and Doumbo, O.K. (1997). Mutations in *Plasmodium falciparum* dihydrofolate reductase and dihydropteroate synthase and epidemiologic patterns of pyrimethamine–sulfadoxine use and resistance. *Journal of Infectious Diseases*, **176**, 1590–1596.

Reynolds, M.G. and Roos, D.S. (1998). A biochemical and genetic model for parasite resistance to antifolates. *Toxoplasma gondii* provides insights into pyrimethamine and cycloguanil resistance in *Plasmodium falciparum*. *Journal of Biological Chemistry* **273**, 3461–3469.

Shanks, G.D., Oloo, A., Klotz, F.W., Aleman, G.M., Wesche, D., Brueckner, R. and Horton, J. (1997). Evaluation of weekly etaquine (WR 238605) compared to placebo for chemosuppression of *Plasmodium falciparum* in adult volunteers in

Western Kenya. *American Journal of Tropical Medicine and Hygiene* **57** (supplement) Abstract no. 518, p. 277.

Sirawaraporn, W., Yongkiettrakul, S., Sirawaraporn, R., Yuthavong, Y. and Santi, D.V. (1997). *Plasmodium falciparum*: asparagine mutant at residue 108 of dihydrofolate reductase is an optimal antifolate-resistant single mutant. *Experimental Parasitology* **87**, 245–252.

Su, X.-Z., Kirkman, L.A., Fujioka, H. and Wellems, T.E. (1997). Complex polymorphisms in an ~ 330 kDa protein are linked to chloroquine-resistant *P. falciparum* in Southeast Asia and Africa. *Cell* **91**, 593–603.

Wang, P., Read, M., Sims, P.F.G. and Hyde, J.E. (1997a). Sulfadoxine resistance in the human malaria parasite *Plasmodium falciparum* is determined by mutations in dihydropteroate synthetase and an additional factor associated with folate utilization. *Molecular Microbiology* **23**, 979–986.

Wang, P., Lees, C.-S., Bayoumi, R., Djimde, A., Doumbo, O., Swedberg, G., Dao, L.D., Mshinda, H., Tanner, M., Watkins, W.M., Sims, P.F.G. and Hyde, J.E. (1997b). Resistance to antifolates in *Plasmodium falciparum* monitored by sequence analysis of dihydropteroate synthetase and dihydrofolate reductase alleles in a large number of field samples of diverse origins. *Molecular and Biochemical Parasitology* **89**, 161–177.

Westling, J., Yowell, C.A., Majer, P., Erickson, J.W., Dame, J.B. and Dunn, B.M. (1997). *Plasmodium falciparum*, *P. vivax*, and *P. malariae*: a comparison of the active site properties of plasmalepsins cloned and expressed from three different species of the malaria parasite. *Experimental Parasitology* **87**, 185–193.

Yeo, A.E.T., Edstein, M.D. and Rieckmann, K.H. (1997). Antimalarial activity of the triple combination of proguanil, atovaquone and dapsone. *Acta Tropica* **67**, 207–214.

Molecular Pathobiology and Antigenic Variation of *Pneumocystis carinii*

Yoshikazu Nakamura and Miki Wada

Department of Tumor Biology, Institute of Medical Science, University of Tokyo, Minato-ku, Tokyo 108, Japan

1. Introduction 64
   1.1. Organism 64
   1.2. Life cycle 64
   1.3. Molecular taxonomy 67
   1.4. Molecular biology 67
   1.5. Molecular diagnosis 68
2. Major Surface Glycoproteins 70
   2.1. Biochemistry 70
   2.2. Pathobiology 73
3. Major Surface Glycoprotein Genes 75
   3.1. Gene hunting 75
   3.2. Major surface glycoprotein cDNA 76
   3.3. Silent genomic repertoire 80
   3.4. Genomic expression site 82
   3.5. Chromosomal organization 85
   3.6. Genetic control of antigenic variation 89
   3.7. Predicted protein features 93
   3.8. Subtilisin-like protease 96
4. Conclusion 97
Acknowledgements 98
References 98

1. INTRODUCTION

1.1. Organism

Pneumocystis carinii is an opportunistic pathogen which often causes fatal pneumonia in patients who are immunosuppressed or immunologically compromised due to the acquired immune deficiency syndrome (AIDS), cancer chemotherapy or immunosuppressive therapy for organ transplantation (Selik *et al.*, 1987). *Pneumocystis carinii*, first described by Chagas in 1909, was thought to be a stage in the life cycle of *Trypanosoma* but was recognized as a distinct and separate organism in rats in 1914 by Delanöe and Delanöe (1912, 1914). Although there were outbreaks of *P. carinii* infection in Europe during the Second World War (1939–1945) and antimicrobial agents such as pentamidine were found to be therapeutically effective in *P. carinii* pneumonia, it was not until the 1980s that people became seriously aware of *P. carinii* when it emerged as a leading cause of opportunistic pulmonary infection and mortality in AIDS patients.

More than 60% of AIDS patients suffer from *P. carinii* pneumonia (PCP) at some time in the course of the disease (Selik *et al.*, 1987). It was estimated that *c.* 52000 AIDS patients in the United States of America alone developed PCP between 1981 and 1988 (Telzak *et al.*, 1990), and that there were about two million cases of PCP among AIDS patients around the world. The clinical importance of the organism was accentuated by the fact that anti-*P. carinii* drugs had major problems of efficacy and toxicity. Solutions to these problems, however, were not easily found because basic research on the organism had been ignored. Nevertheless, investigators succeeded in isolating nucleic acids from this organism in 1988 (Edman *et al.*, 1988; Tanabe *et al.*, 1988) and there have been marked advances in the molecular biology of *P. carinii*, which led to a resurgence of pathobiological and clinical studies (for reviews, see Walzer, 1993; Su and Martin, 1994).

1.2. Life Cycle

Pneumocystis carinii is a eukaryotic microorganism that can infect many mammalian hosts, including mice, rats, guinea pigs, rabbits, ferrets, dogs, and horses. The difficulty of continuous cultivation *in vitro* of *P. carinii* impeded the fundamental study of this organism. Despite many years of investigation, researchers have been able to achieve less than a 10-fold increase of *P. carinii* isolated from rats (Cushion, 1989; Sloand *et al.*, 1993). Therefore, most of our knowledge has emerged from ultrastructural studies of the parasites in infected rodents or humans (Ruffolo, 1993). *Pneumocystis carinii* develops naturally in lung alveoli, and has two predominant forms in the life cycle (Figure 1): one, a vegetative form, i.e. a small unicellular trophozoite; and the other, a resting form or cyst containing eight intracystic bodies (Figure 2). Although there is no evidence which form is infective, trophozoites are probably responsible for infection; cysts may also represent the infective stage, from which daughter trophozoites emerge (see the review by Yoshida, 1989) to infect alveolar cells.

According to the 'textbook' view, inhalation of airborne organisms in early childhood exposes humans to *P. carinii*. When the organism enters the lower respiratory tract, it causes a mild subclinical infection in hosts with normal humoral and cellular immunity. It is widely believed that the organism evades clearance and remains quiescent in the host for long periods of time despite the host's fully functional immunity. Under conditions of immune deficiency or suppression, the latent infection is reactivated and

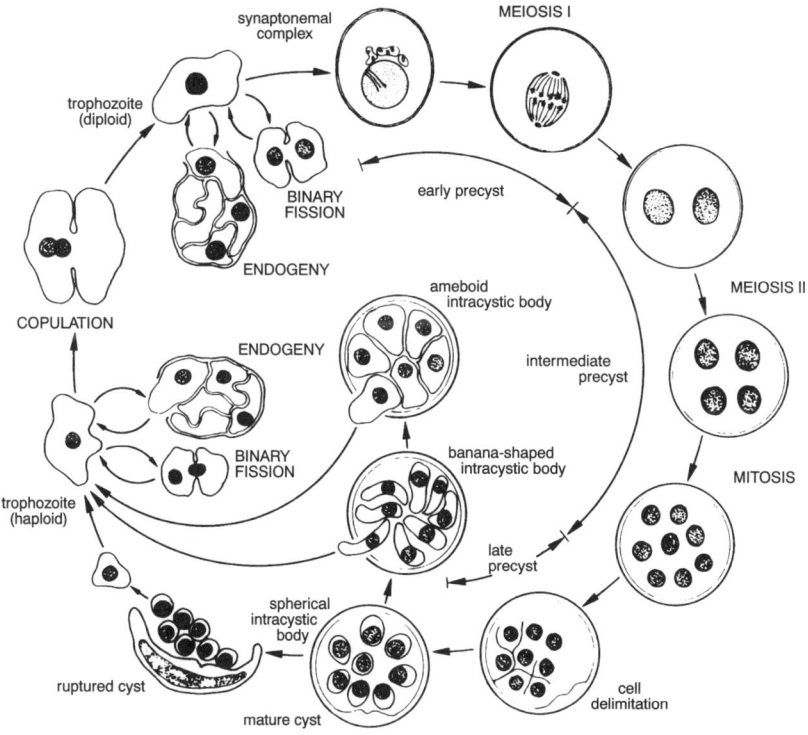

Figure 1 A proposed scheme of the life cycle of *Pneumocystis carinii* (Yoshida, 1989). Reprinted with permission of The Society of Protozoologists.

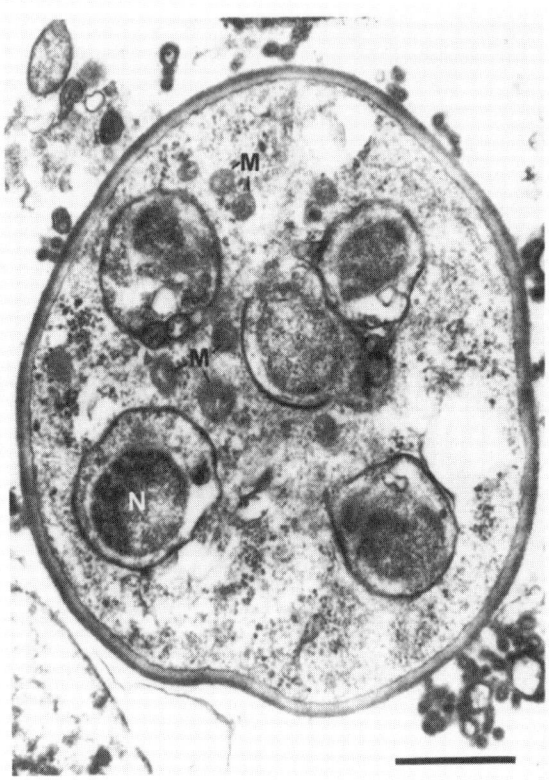

Figure 2 Ultrastructure of *Pneumocystis carinii* cyst (Yoshida, 1981). N, nucleus; M, mitochondria. Bar indicates 1 μm. Reprinted with permission of Nanzando Co., Ltd.

P. carinii begins to proliferate and causes pneumonia. On the other hand, the occurrence of miniepidemics of PCP has recently raised the possibility that PCP results from reinfection rather than relapse, since the infection may be transferable from one immunosuppressed patient to another (Jacobs *et al.*, 1991). Animal studies of PCP have demonstrated animal-to-animal transfer and appear to support this possibility (Walzer *et al.*, 1989). Current application of the polymerase chain reaction (PCR) to diagnosis indicated that *P. carinii* disappears during the course of therapeutic treatment with anti-*P. carinii* drugs and is not detected in normal lungs or sputum (Kitada *et al.*, 1991a,b). It remains to be definitively demonstrated whether PCP results from relapse or reinfection, although the two explanations may not be mutually exclusive.

1.3. Molecular Taxonomy

Although *P. carinii* was discovered more than 80 years ago, the taxonomic classification of the organism has not yet been finalized. Chagas (1909) misidentified it as a stage in the life cycle of trypanosomes, and so *P. carinii* was traditionally thought to be a protozoan. This misconception was perpetuated by its response to antiprotozoan agents such as pentamidine and trimethoprim/sulfamethoxazole. However, several *P. carinii* genes have been cloned and sequenced, facilitating molecular taxonomy. Ribosomal ribonucleic acid (rRNA) genes have been widely used to study evolutionary relationships among microorganisms (Edman and Sogin, 1993). Phylogenic trees have been constructed based on the degree of similarity in nucleotide sequences. Analysis of the 5S RNA or 18S RNA of *P. carinii* isolated from rats revealed that the organism is much more closely related taxonomically to fungi such as *Saccharomyces cerevisiae* or *Schizosaccharomyces pombe* than to protozoa (Edman *et al.*, 1988; Stringer *et al.*, 1989; Watanabe *et al.*, 1989). Analysis of other ribosomal RNA genes (5.8S, 26S) supported a close relationship to fungi (Pixley *et al.*, 1991; Liu *et al.*, 1992). This was further confirmed by the presence of the translation elongation factor 3 gene, which is specific to fungi, in *P. carinii* (see Ypma-Wong *et al.*, 1992). Ultrastructural and biochemical studies of the cyst wall of *P. carinii* (Bedrossian, 1989; Matsumoto *et al.*, 1989, 1991; Williams *et al.*, 1991), together with the cross-reactivity with fungi of monoclonal antibodies raised against *P. carinii* (see Lundgren *et al.*, 1992), provided further evidence of this close relationship.

Although there is increasing evidence of its close relationship to fungi, it is unknown to what degree this molecular taxonomy reflects the true nature of *P. carinii*. As described below, *P. carinii* has a genetic system for switching cell surface antigens, like that found in the protozoan *Trypanosoma*, but which is not otherwise known among fungi. Several criteria may be required to determine the true taxonomic status of *P. carinii*.

Because most of the gene cloning has been done using *P. carinii* from rats, some caution is required when extrapolating this information to *P. carinii* of humans. Nevertheless, analysis of the rat *P. carinii* genes will provide useful information on the basic biology of the parasite, and information for clinical application (Walzer *et al.*, 1992).

1.4. Molecular Biology

There have been striking advances in recent years in the area of molecular biology and the related pathobiology of *P. carinii* (for reviews, see Walzer,

1993; Su and Martin, 1994). As described above, several *P. carinii* genes encoding proteins, enzymes and rRNAs were cloned and sequenced. For taxonomic analysis, and for research into the development of anti-*P. carinii* drugs, several metabolic enzyme genes were cloned, including those for dihydrofolate reductase (DHFR) and thymidylate synthase (TS), which are encoded on separate chromosomes (Edman *et al.*, 1989a,b). Studies of the base composition of the genome of rat *P. carinii* have shown a high content of adenine and thymine (Edman *et al.*, 1989a,b; Smulian *et al.*, 1992, 1993; Zhang *et al.* 1993). These studies not only significantly improved our basic understanding of *P. carinii*, but provided a powerful technology for clinical investigation, as will be described shortly.

The genome of *P. carinii* has been studied by electrophoretic karyotyping, i.e., the separation of chromosomes by pulse field gel electrophoresis. These studies showed that the haploid genome consists of about 13–16 chromosomes containing 700–1000 kb of DNA, which is approximately two-thirds the size of the *Saccharomyces* genome (Yoganathan *et al.*, 1989; Hong *et al.*, 1990; Lundgren *et al.*, 1990). Differences in the karyotype pattern have appeared in *P. carinii* isolates from different strains of rat and from animals supplied by different commercial vendors (Hong *et al.*, 1990; Cushion *et al.*, 1993a,b). Diversity of the karyotype pattern has also been observed among human *P. carinii* isolates, which appear to produce similar numbers of karyotype bands between 200 and 700 kb in size, as well as among *P. carinii* isolates from other animals (Stringer, 1993), showing the heterogeneity of *P. carinii* isolates. Another study has indicated that some laboratory rats can contain two different genetic variants of *P. carinii* that differ in electrophoretic karyotype (Cushion *et al.*, 1993b).

1.5. Molecular Diagnosis

A definitive diagnosis of PCP can be made only by detecting the organism in clinical specimens such as sputum (Zaman *et al.*, 1988; Leigh *et al.*, 1989), bronchoalveolar lavage fluid (Hopkin *et al.*, 1983), or transbronchial biopsy samples (Peters and Prakash, 1987). Conventionally, Grocott's stain, toluidine blue O, or Giemsa's stain is usually used to identify *P. carinii* in these materials. These histochemical methods have, however, an inherent lack of sensitivity and/or specificity and the organism cannot always be detected, especially in noninvasive specimens. Therefore, because of the uncertainty of the diagnosis, patients are sometimes routinely treated with anti-*P. carinii* therapy coincident with ganciclovir and/or other antibiotic therapy. In contrast to the uncertainty of standard diagnostic methods, the PCR method has better sensitivity and specificity in detecting

P. carinii (see Wakefield *et al.*, 1990; Kitada *et al.*, 1991a,b). There has been an increasing number of applications of the PCR to detect *P. carinii* genes, including ribosomal DNA in the respiratory tract of patients and animal models with pneumocystosis (Leigh *et al.*, 1992; Lipschik *et al.*, 1992; Reddy *et al.*, 1992; Peters *et al.*, 1992; Schlugen *et al.*, 1992; Olsson *et al.*, 1993). *Pneumocystis carinii* has also been detected by hybridization *in situ* (Hayashi *et al.*, 1990; Haidaris *et al.*, 1993). Because PCR is highly sensitive, attention must be given to the assay conditions and control of contamination by host tissues or other microbes to ensure specificity. The PCR data must be interpreted in the light of other clinical and laboratory data because a positive PCR indicates the presence of *P. carinii* DNA but does not necessarily indicate the presence of active disease. Obviously, PCR diagnosis can reduce the need for unnecessary invasive procedures (Wakefield *et al.*, 1991; Lipschik *et al.*, 1992; Kitada *et al.*, 1993; Oka *et al.*, 1993). Improved diagnoses by PCR have made bronchoscopy unnecessary in many cases.

PCP in patients with AIDS is usually treated for 3 weeks (Wharton *et al.*, 1986; Kovacs and Masur, 1988; Davey and Masur, 1990), and in those with other types of immunodeficiency for 2 weeks (Sattler and Remington, 1981; Peters and Prakash, 1987), both periods having been determined empirically. In other words, the duration of therapy is not based on a valid index of morbidity, because there is no reliable therapeutic marker. Nevertheless, the PCR allows early diagnosis of PCP and close therapeutic monitoring using noninvasive specimens such as sputum (Figure 3; see Nakamura, 1993).

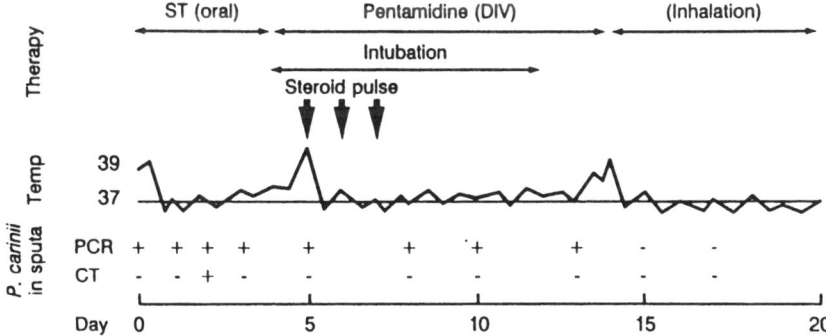

Figure 3 Clinical course of an AIDS patient who suffered from *Pneumocystis carinii* pneumonia. Administration of anti-*P. carinii* drugs began on day 0, and *P. carinii* shedding in sputum was monitored by 5S rDNA PCR and the cytological test (CT) (Nakamura, 1993). Reprinted with the permission of the American Society for Microbiology.

2. MAJOR SURFACE GLYCOPROTEINS

2.1. Biochemistry

2.1.1. Molecular Mass

Pneumocystis carinii organisms are coated by abundant surface proteins that are of great interest to investigators (Gigliotti *et al.*, 1986, 1988, 1991, 1992; Graves *et al.*, 1986a,b; Lee *et al.*, 1986; Walzer and Linke, 1987; Kovacs *et al.*, 1988, 1989, 1993; Pesanti and Shanley, 1988; Linke and Walzer, 1989, 1991; Linke *et al.*, 1989; Nakamura *et al.*, 1989, 1991; Radding *et al.*, 1989; Tanabe *et al.*, 1989; Lundgren *et al.*, 1991; Haidaris *et al.*, 1992; Smulian *et al.*, 1992, 1993; Wada *et al.*, 1993). The major protein components of both cysts and trophozoites are the major surface glycoproteins (MSGs) (Kovacs *et al.*, 1993; Wada *et al.*, 1993; Walzer, 1993), referred to as P115 (Tanabe *et al.*, 1989), gp120 (Radding *et al.*, 1989) or gpA (Gigliotti, 1992; Haidaris *et al.*, 1992), with an apparent molecular mass ranging from 95 to 140 kDa, depending on host species, degree of glycosylation, and isolation procedures. The detailed nomenclature of this antigen complex is somewhat confusing, and we shall refer to it here inclusively as MSG.

MSG accounted for three-quarters of the total cellular protein when analysed by sodium dodecyl sulphate-polyacrylamide gel electrophoresis (SDS–PAGE) (Figure 4) using *P. carinii* cells isolated from rats (Tanabe *et al.*, 1989; Wada *et al.*, 1993). Although these surface antigens are not readily available in a purified form because of the difficulty of mass cultivation, the use of monoclonal antibodies has facilitated the study of the surface structure of the cells. Several investigators reported the preparation of monoclonal antibodies against *P. carinii* in the late 1980s. Antigenic molecules with masses of 90–95 kDa (Gigliotti *et al.*, 1986), 110–116, 90, 55–60 and 35 kDa (Graves *et al.*, 1986a,b), and 110, 65 and 35 kDa (Lee *et al.*, 1986), were identified by immunoblotting. The largest molecules reported in these investigations corresponded to MSG. There is clearly a large protein complex on the surface of *P. carinii* isolates from humans, rats and other animals. Under nonreducing conditions, this antigen has an estimated molecular mass ranging from about 300 to more than 2000 kDa, while it migrates as a prominent band of 95–140 kDa when run under reducing conditions.

2.1.2. Surface Glycoprotein

The ease of generation of anti-MSG monoclonal antibodies indicates the strong antigenicity of MSG molecules, presumably, in part, because of their

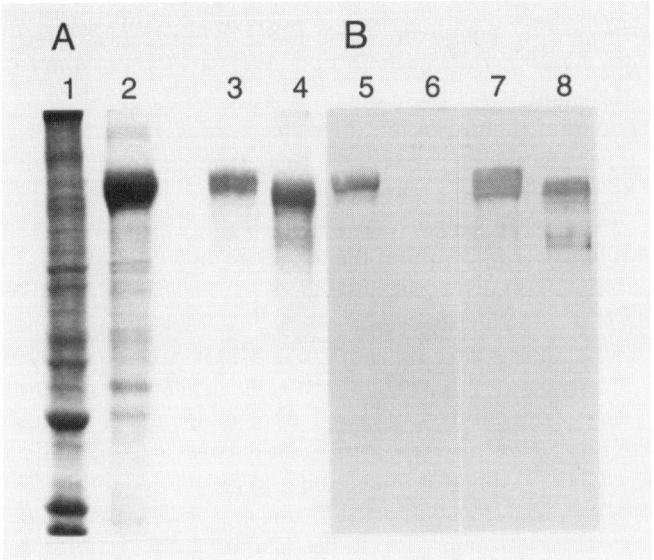

Figure 4. SDS–polyacrylamide gel electrophoresis and immunoblotting analyses of *Pneumocystis carinii* MSG (Wada *et al.*, 1993). A. Gels stained with Coomassie blue: lane 1, uninfected rat lung homogenate; lane 2, homogenate of *P. carinii* organisms purified from infected rat lungs; lane 3, purified MSG; lane 4, MSG treated with endoglycosidase-F. B. Immunoblots of MSG antigens before (lanes 5 and 7) and after (lanes 6 and 8) endoglycosidase-F treatment using anti-MSG monoclonal antibody (lanes 5 and 6) or polyclonal anti-deglycosylated-MSG antibody (lanes 7 and 8). Reprinted with the permission of The University of Chicago Press.

unusual abundance on the surface of *P. carinii*. Immunohistochemical analysis using anti-MSG monoclonal antibodies demonstrated that ferritin granules conjugated with these antibodies were bound to the surface of the cyst and to that of the trophozoite of *P. carinii* (see Tanabe *et al.*, 1989). Immunoblotting analysis of *P. carinii* proteins separated by SDS–PAGE clearly showed that a major part of the total protein of both cysts and trophozoites formed a broad band at 105–120 kDa, which corresponds to a common epitope recognized by the anti-MSG monoclonal antibody (Tanabe *et al.*, 1989; Radding *et al.*, 1989).

The *P. carinii* cyst has a thick cell wall composed of carbohydrates, consisting primarily of glucose, mannose and galactose, with lesser amounts of *N*-acetyl-D-glucosamine, ribose, and sialic acid (De Stefano *et al.*, 1990). MSG contains mannosyl, glycosyl, and *N*-acetylglucosamine residues, displays species-specific antigenic variation, and possesses collagenase sensitivity. Tanabe *et al.* (1989) first demonstrated biochemically that MSGs are glycoproteins containing mannose-rich oligosaccharides. They analysed the monosaccharide composition of MSG by high-performance liquid chromatography after hydrolysis to monosaccharides, revealing that mannose was the main sugar component (Figure 5). The cell walls of many organisms contain mannose-rich oligosaccharides. In several cases, high-mannose-type sugar chains are known to contribute to cell–cell interaction or recognition. Mammalian alveolar macrophages express a receptor that binds mannosylated glycoproteins (Stahl et al., 1978, 1980; Sung et al., 1983) and it has been shown that the mannose moiety of MSGs is attacked by the macrophage during the uptake of P. carinii, as described below (Ezekowitz et al., 1991).

Biochemical studies have revealed that MSGs are acidic glycoproteins with several isoelectric points (Gigliotti et al., 1988; Tanabe et al., 1989). Deglycosylation with endoglycosidase F or α-mannosidase increased migration of MSG molecules in SDS–PAGE, and decreased isoelectric variation (Tanabe et al., 1989). Most monoclonal and polyclonal antibodies reacted with all of these isoelectric variants. However, these deglycosylated MSGs no longer reacted with most antibodies against authentic MSGs, showing that the sugar moiety participated in the strong immunodeterminant structure of P. carinii. Limiting proteolysis of the isoelectric variants generated common major fragments, showing that they shared a common polypep-

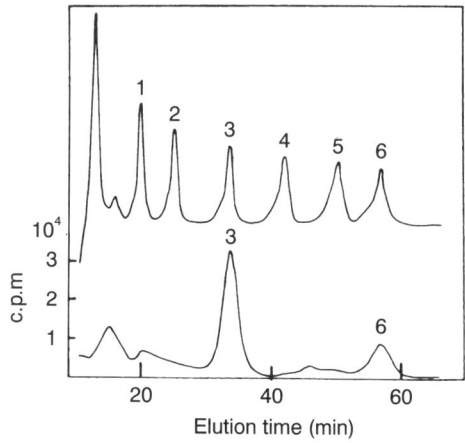

Figure 5. HPLC analysis of sugar composition of MSG (Tanabe et al., 1989). Monosaccharides of acid hydrolysates of MSG were converted into ^{3}H-labeled sugar alcohols, and analyzed by HPLC with sugar alcohol internal standards (upper curve). The comparative amounts of ^{3}H-labeled monosaccharide components of MSG are shown in the lower curve. Peak 1, N-acetylgalactosaminitol; peak 2, N-acetylglucosaminitol; peak 3, mannitol; peak 4, fucitol; peak 5, galactitol; peak 6, glucitol or sorbitol; c.p.m., counts per minute. Reproduced with permission of the American Society for Microbiology.

tide portion. Isoelectric variation, therefore, seems to be caused mainly by heterogeneity in the carbohydrate chains.

MSG has been purified by several techniques. Zymolyase treatment has been favored by some investigators as the initial step in this process because it liberates MSG from the *P. carinii* cell wall (Linke and Walzer, 1991; Lundgren *et al.*, 1991). Approximately 10% of the molecular mass of MSG is composed of *N*-linked carbohydrates, which are rich in glucose, mannose and *N*-acetylglucosamine residues.

The other major antigen complex consists of a glycoprotein with a molecular mass of 45–55 kDa in *P. carinii* isolated from rats and other animal species, including humans (Walzer and Linke, 1987; Kovacs *et al.*, 1988; Linke *et al.*, 1989; Walzer *et al.*, 1989). The gene coding for this complex has been cloned and sequenced (Smulian *et al.*, 1992, 1993). The predicted amino acid sequence of this antigen shows a repeated motif rich in glutamic acid residues, which seems to be highly immunogenic. This low molecular mass antigen does not show any sequence diversity among *P. carinii* populations in contrast to MSG (see below). The function of the 45–55 kDa antigen complex of *P. carinii*, including perhaps evasion of the host immune response, has yet to be investigated.

2.2. Pathobiology

2.2.1. *Attachment*

When *P. carinii* infects a new host, the organism passes through the initial host defences and is deposited within lung alveoli. The initial event in the pathogenesis of PCP is the preferential attachment of the organism to type I alveolar epithelial cells (for review, see Su and Martin, 1994). Although the mechanisms of attachment are not fully understood, MSG is thought to play an important role in the interaction of *P. carinii* with host cells, involving other components such as cell adhesive proteins and lectins (Yoshikawa *et al.*, 1987; Cushion *et al.*, 1988; Pesanti and Shanley, 1988; Pottratz and Martin, 1990a; Fishman *et al.*, 1991; Pottratz *et al.*, 1991; Wisniowski and Martin, 1992; Aliouat *et al.*, 1993; Limper *et al.*, 1993; Narasimhan *et al.*, 1994). *Pneumocystis carinii* trophozoites can bind to alveolar epithelial cells by using fibronectin as a ligand, and MSG has been implicated as the binding site for fibronectin. Other extracellular matrix adhesive proteins that may be involved include laminin and vitronectin. Vitronectin binds to *P. carinii* and mediates attachment of the organism to cultured lung epithelial cells (Limper *et al.*, 1993). *Pneumocystis carinii* may also bind to epithelial cell surface glycoproteins by means of a mannose-dependent mechanism (Limper *et al.*, 1991). Thus, attachment to type I alveolar epithelial cells may be mediated by multiple mechanisms. Consistent with the crucial role of MSG in attachment to host cells, hybridization studies *in situ* showed that transcription of MSG occurs in the developmental stages of *P. carinii* that are actively replicating, and in close proximity to alveolar epithelial cells (Haidaris *et al.*, 1993).

As *P. carinii* infection evolves, the alveolar–capillary membrane barrier is disrupted, resulting in epithelial cell death. This process may involve the release of degradative enzymes, such as a specific serine protease, from *P. carinii*, and mediates injury to the epithelial cells (Breite *et al.*, 1993). The precise mechanism of this injury awaits elucidation.

2.2.2. *Interaction With Surfactant*

The type II alveolar cell may also be involved in the pathogenesis of *P. carinii* (for review, see Su and Martin, 1994). As the source of pulmonary surfactant, the type II cell maintains the structural integrity of the alveolus necessary to conduct gas exchange, and impairment of the pulmonary surfactant can lead to hypoxemia. *Pneumocystis carinii* can interact with pulmonary surfactant by binding to surfactant proteins and can interfere with its function. This binding is in part due to specific binding of MSG to surfactant proteins A, the major surfactant protein (Zimmerman *et al.*, 1992), and D (O'Riordan *et al.*, 1995). It is possible that, by becoming coated with surfactant proteins, *P. carinii* may avoid detection by alveolar macrophages, thus allowing PCP to develop (Phelps and Rose, 1991; Sternberg *et al.*, 1993).

2.2.3. *Host Response*

There is evidence that *P. carinii* infection stimulates both humoral and cellular immune responses in the host. For example, passive immunization with an anti-MSG monoclonal antibody partially protected against the progression of PCP in animal models (Gigliotti and Hughes, 1988), and a specific T cell response to MSG molecules occurred after both immunization and natural infection (Fisher *et al.*, 1991; Theus *et al.*, 1993). Administration of hyperimmune serum to severe combined immune deficiency (SCID) mice with PCP resulted in clearing of the infection (Roths and Sidman, 1992). Cases of pneumocystosis have also been described in patients and mice with B cell defects (Esolen *et al.*, 1992; Sidman and Roths, 1993; Marcotte *et al.*, 1996).

Alveolar macrophages have generally been considered the principal effector cells in the host cell-mediated immune response to *P. carinii*. The alveolar macrophages play a central role in the clearance of microorganisms from the lung through migration, attachment and phagocytosis. Binding of *P. carinii* to macrophages is mediated by adhesive proteins such as fibronectin (Pottratz and Martin, 1990b) and by mannose-dependent mechanisms in which the macrophage mannose-receptor clearly targets the mannose moiety of MSG to initiate uptake (Ezekowitz *et al.*, 1991). Nevertheless, *P. carinii* organisms tend to avoid phagocytosis by macrophages. Apparently, the balance between the ability of *P. carinii* to avoid phagocytosis and the ability of the macrophages to engulf the organism is critical in host defence against PCP.

3. MAJOR SURFACE GLYCOPROTEIN GENES

3.1. Gene Hunting

Analysis of MSG genes has been hampered for several reasons. First, MSG molecules contain highly immunogenic sugar moieties with which most of the existing anti-MSG antibodies react, and which do not permit the immunoscreening of complementary DNAs (cDNAs). Second, analysis of the protein portion of MSG is impeded by a blocked amino terminus (Nakamura *et al.*, 1991). Third, as revealed by cDNA analysis, MSG genes are highly polymorphic, preventing us from viewing them simplistically. A breakthrough in immunoscreening was made with antibodies prepared against deglycosylated MSG molecules of *P. carinii* isolated from rats (Wada *et al.*, 1993; Kitada *et al.*, 1994). These antibodies reacted well with the endoglycosylated MSG molecules that migrate slightly faster than intact MSG in SDS–PAGE (Wada *et al.*, 1993). Partial peptide sequences of proteolytic MSG fragments have been determined (Paulsrud *et al.*, 1991), but, to our knowledge, no cloning work based on the peptide sequence has yet been reported.

After some years of investigation, three groups have succeeded in cloning MSG cDNAs. Haidaris *et al.* (1992) reported the partial cloning and sequencing of a cDNA encoding a glycoprotein A (gpA) antigen, analogous to MSG, in *P. carinii* from ferrets. Wada *et al.* (1993) and Kovacs *et al.* (1993) reported the cloning and sequencing of a cDNA encoding MSG in *P. carinii* from rats. The ferret *P. carinii* gpA sequence was about 60% and 30% identical to the rat *P. carinii* MSG sequence for nucleotides and amino acids, respectively. In these initial reports, sequence diversity was demonstrated for rat *P. carinii* MSGs, although not for the ferret *P. carinii* gpA. Nevertheless, sequence diversity of the ferret gpA was subsequently reported (Wright *et al.*, 1995a).

MSG sequences were also found among repetitive sequence elements. Stringer *et al.* (1991) isolated repeated, polymorphic, DNA elements from rat *P. carinii* which hybridized to most chromosomes. It was shown later that these DNAs encode MSG (Sunkin *et al.*, 1994).

Extensive studies have followed these early investigations. Consequently, the chromosomal structure and expression of MSG genes have been uncovered, giving rise to the surprising discovery of antigenic variation in *P. carinii*. This will now be reviewed in some detail.

3.2. Major Surface Glycoprotein cDNA

3.2.1. *Heterogeneity*

The cDNA expression library of *P. carinii* was prepared in a λ phage expression vector from poly(A)$^+$ RNA isolated from rat-derived organisms, and screened using polyclonal antibody against deglycosylated MSG. The frequency of appearance of positive clones was unusually high (10^{-2}–10^{-3}), reflecting the abundance of MSG in the vegetative cycle of *P. carinii*. These cDNA clones encoded a single open reading frame, shared identical or homologous sequences, and were classified into different subtypes (referred to as MSG1, MSG2, MSG3, etc.) Their amino acid sequences are 65–80% identical (see Figure 6). Codon usage is strongly biased towards codons with A or T in the third position. A similar bias is seen in the previously sequenced protein-encoding genes of *P. carinii* (see Edman *et al.*, 1989a,b; Dyer *et al.*, 1992; Edlind *et al.*, 1992; Haidaris *et al.*, 1992; Smulian *et al.*, 1992; Volpe *et al.*, 1992; Fletcher *et al.*, 1993).

Despite extensive cloning, a complete MSG-cDNA sequence was not easily obtained. Instead, new members of the MSG family were continuously discovered, suggesting a high degree of sequence diversity among MSG-cDNAs. Finally, Kitada *et al.* (1994) reported a cDNA fragment encoding a large MSG polypeptide isolated from an oligo(dT)-primed cDNA library. The deduced MSG, referred to as the MSG5 subtype, was a 120 765 Da protein (composed of 1076 amino acids), seemingly equivalent in size to MSG and thus thought to be a full-length product. Kovacs *et al.* (1993) have also reported the analogous sequence of an MSG-cDNA from rat *P. carinii*, determined by expression library screening combined with PCR amplification of junction sequences between isolated cDNA clones. However, it was shown later that these sequences lack the *N*-terminal part of MSG, including a signal peptide sequence necessary for protein secretion (see below).

As a result of these cDNA analyses, it became apparent that the MSG subtypes of rat *P. carinii* are homologous over the entire length of the proteins, except for some variable regions. The variable regions map to the

```
MSG1  EIDEKHLLAFIVKDKYKEEQKCKEELEKYCKELKEADKNLENVDDKVKGLCDDKKRDEKCKDVKKKVEDEL    71
MSG2  D··········A·E······N···Q······E···KI·GGSD··NKN·····E·G·QQD···L·GE··KV·
MSG3  D··············E··SN··Q·T··K···E······-G·K··N···EI··T···G··EL·D··KK··
gpA   KVE·ADV··LL·GKE·NNKDQ·EK··Q···DG··VE··IP·GI·PL··NI·QKDNGQK··T·L·NNIQQKC  180

MSG1  KDFEEELQKVLNN------IKDENCEKYEEKCILLEETDYDVIKDNCVKLREGCYELKRKKVAEELLLRALG 137
MSG2  ·A··G···EA·KD-----------·····················IE······K···E···········
MSG3  ET·K···E·A·KD····················NH·DV·K··········K····R··D·········
gpA   NA·KTK·DEKFPKNGSVTLESKD·PG··VQ·F···KA--SSL·E··N·V·NT··GI··VL·IDTFA··L·-  249

MSG1  KEAKEEVKCQAEMKKVCPVLSRESDELMFLCLD---SDGTCQALKKKSEEVCQLLKEKLK------DGELKE 205
MSG2  GD····A··KGK·NT··········SF······AK··GD···LGT··EP··KE···-----·N··A·
MSG3  ·DV··NGE·EKK··D··S············SF······AK··GE··T·LDT··EA··T··----AKDFEK
gpA   ·GHLDGAQ·NNKL·EI··SI·G······KT··EAQTGG·S·TD·PQS·VTK·NA··DEVEKALGSNTT··· 320
```

Figure 6 Amino acid alignment of homologous regions of MSG1, MSG2 and MSG3 of rat *Pneumocystis carinii* and gpA of ferret *P. carinii* (single-letter code). Only amino acids that differ from MSG1 are indicated; dashes denote absence of an amino acid (Wada *et al.*, 1993).

middle of the proteins and to the *C*-terminal region (Kitada *et al.*, 1994), except for the *C*-terminal end which is highly conserved (Linke *et al.*, 1994). The MSG family of rat *P. carinii* contains 65–80% identical amino acids among the subtypes and shares 32–36% identical amino acids with gpA of ferret-derived *P. carinii*.

Pneumocystis carinii has not yet been successfully cultured *in vitro* and therefore a genetically homogeneous strain has not been established. This raises the possibility that some of the sequence diversity revealed in these studies was due to the heterogeneity of *P. carinii* organisms derived from rats. However, it is very unlikely that this was the major cause of diversity, as was shown by the results of the Southern blot, PCR or karyotype analyses described below.

3.2.2. *Upstream Conserved Sequence*

The 5' upstream regions of MSG-cDNAs were investigated by 5' rapid amplification of cDNA ends (RACE) (Wada *et al.*, 1995). 5' RACE products were synthesized from bulk *P. carinii* RNA using an antisense primer complementary to a coding sequence of the MSG variable region. Two discrete bands were produced, which sequencing showed to contain two types of conserved sequences, namely the upstream conserved sequences (UCS). Northern blot hybridization showed that the UCS in the longer RACE product (type I) was present in the majority of the MSG transcripts, while RNAs containing the UCS in the shorter RACE product (type II) were present in only a small percentage of the total number of MSG transcripts. To date, only the type I transcripts have been characterized in detail.

Sequence analysis of type I MSG transcripts showed them all to contain the same UCS. The sequences of the three longest type I MSG clones are shown in Figure 7 (Wada *et al.*, 1995). MSG mRNAs initiate at one of two

Figure 7 (opposite) The upstream conserved sequence (UCS) region of MSG cDNAs containing full-length 5' ends plus extra guanine residues derived from the cap structure (Wada *et al.*, 1995). The numbers of the nucleotide positions are indicated from the 5' end of the clone A24; those of the amino acid positions refer to the initial methionine of the deduced amino acid sequence (single-letter code). The boxed region of the nucleotide sequence represents the type I UCS and the amino acid sequence shown as white letters on black represents the *N*-terminal signal peptide. The position of the intron is marked by an inverted triangle. The amino acid sequence downstream of UCS is noted only when the three clones conserve the same amino acids; those nonconserved are indicated by dashes. Dots denote the absence of a nucleotide. The conserved recombination junction element (CRJE) sequence is underlined. Reprinted with permission of The University of Chicago Press.

UCS

```
                                                                                                        100
MSG A11  1          .         .         .         .         .         .         .         .         .         .
MSG A15             GAGTTTGTTGTGCAATAATGAGGATTGCATTTTTGCGCAACTAGTTGTATTTTAGTTTATTCAATAGCAG
MSG A24             GAGTTTGTTGTGCAATAATGAGGATTGCATTTTTGCGCAACTAGTTGTATTTTAGTTTATTCAATAGCAG
         GGATATCCCTCGTCGTTCTTCAGTTGTTGTGTCAATAATGAGGATTGCATTTTTGCGCAACTTAGTTGTATTTTAGTTTATTCAATAGCAG
                                                                                                   E  22
                    .         .         .         .         .         .         .         .         .         .
                                                                                                        200
MSG A11  AAAGGGATTTCATGTCATTAGATGAAATATGAAGGAGGCGATATAAGTTTGATCATGAAAACTCGAATTTAACGAATATAATCAAGTTTACAAT
MSG A15  AAAGGGATTTCATGTCATTAGATGAAATATGAAGGAGGCGATATAAGTTTGATCATGAAAACTCGAATTTAACGAATATAATCAAGTTTACAAT
MSG A24  AAAGGGATTTCATGTCATTAGATGAAATATGAAGGAGGCGATATAAGTTTGATCATGAAAACTCGAATATAATCAAGTTTACAAT
         R   D   F   M   S   L   D   E   I   Y   E   G   G   D   I   S   F   D   H   E   K   L   E   F   N   E   Y   N   Q   V   L   Q   M
         23                                                                                             55
                    .         .         .         .         .         .         .         .         .         .
                                                                                                        300
MSG A11  GCCTGAAAAGGCAAAAAAATTGGAACCGGCTTTGTTGATAGAACCAAAGATTTTTCTAATAGACGATATGAAGGAGAATTGAGTTAAATCATTTGGG
MSG A15  GCCTGAAAAGGCAAAAAAATTGGAACCGGCTTTGTTGATAGAACCAAAGATTTTTCTAATAGACGATATGAAGGAGAATTGAGTTAAATCATTTGGG
MSG A24  GCCTGAAAAGGCAAAAAAATTGGAACCGGCTTTGTTGATAGAACCAAAGATTTTCTAATAGACGATATGAAGGAGAATTGAGTTAAATCATTTGGG
         L   E   K   A   K   K   L   G   T   G   F   V   D   R   T   K   D   F   S   N   R   R   Y   E   G   R   I   E   L   N   H   L   G
         56                                                                                              88
                    .         .         .         .         .         .         .         .         .         .
                                                                                                        400
MSG A11  AGACGCCCAGGAGTCGACTATTTTAGGAAAGGTGGGATGTTTTTACTGATGGTTATCCTCGTGGAGGTCATTTGATCGAGGATGAGTTGTCCGAAGAGG
MSG A15  AGACGCCCAGGAGTCGACTATTTTAGGAAAGGTGGGATGTTTTTACTGATGGTTATCCTCGTGGAGGTCATTGATGAGGATGAGTTGTCCGAAGAGG
MSG A24  AGACGCCCAGGAGTCGACTATTTTAGGAAAGGTGGGATGTTTTTACTGATGGTTATCCTCGTGGAGGTCATTGATGAGGATGAGTTGTCCGAAGAGG
         R   R   P   G   V   D   Y   F   R   K   G   G   D   V   F   T   D   G   Y   P   R   G   G   H   L   I   E   D   E   L   S   E   E   V
         89                                                                                             122
                    .         .         .         .         .         .         .         .         .         .
                                                                                                        500
MSG A11  TGGCAATGCACGGCCGGTTAAGAGGCAAAATCAA...GCAGCACCAGCAGCGATGGAATTAAGGACGGAACACCTTTTGCCTTTCATTGCGAAGGAGAA
MSG A15  TGGCAATGCACGGCCGGTTAAGAGGCAACA...GCAGCACCGGCAGTAGCAGAGCAGAACACCTTTTGCCTTTCATTGTGAAGGACAA
MSG A24  TGGCAATGCACGGCCGGTTAAGAGGCACAAGGAGCAGGGGACCAGCAGCGATGACATTGATGAGAACACCTTTTGCCTTTCATTGTGAAGGACAA
         A   M   A   R   P   V   K   R   Q   -   Q   -   A   -   -   -   A   -   -   I   -   E   -   H   L   L   A   F   I   -   K   -   K
         123                                                                                            155
                    .         .         .         .         .         .         .         .         .         .
                                                                                                        600
MSG A11  ATACAAAGAAGTACAACAATGCAAAGAAGAACTCGAAGAAGAACTCGAAGAATATTGTGAAGAGTTGAAGAAATGATGGTTCCCATGTGAATAAAAATGTAAAGAA
MSG A15  ACATGGTGATGAATGAATGCAAAAAAGCTCGAAGAATATTGTAAGGAGTTGAAGACTTGAAGTTCAGTGTAATGAGAAGTTAAAGAA
MSG A24  A......TATGAAGAAGATTGCAAAGGAAGAACTCGAAGAATATTGTCAAGACTTGAAGGAAATAGATGGTCAATTCAAAGTGAATGATAAAGTTAAAGAA
         -   -   -   -   C   K   -   -   L   E   -   Y   C   -   -   L   K   -   -   -   D   -   -   -   V   N   -   -   V   K   -
         .         .         .         .         .         .         .         .         .         .
                                                                                                        700
MSG A11  CTTTGTGAAGATGA........AAACAACAGATAAATCAA...CTGAAAGGCGAAGTTGAAAAGTATTGAAGGGGAACTTCAAG
MSG A15  CTTTGTGGTGGTTGGTGTGATAAAACAAACGATAAAATGCACGACCTG.....CAAGTTGAAGATGAATTGCACACTTTGATGCGAAATTCAAA
MSG A24  CTTTGTGGTGGTTGGTGTGATAAAACAAGAGAAAAATGCCAAGAACTGAAAGACGAAGTTAAAAAAAATTGGAAGCTTTAAAAAGAACTGAAG
         L   C   -   -   -   G   -   -   -   K   -   -   -   K   C   -   L   -   -   -   V   -   -   L   -   -   F   -   E   -   -
```

sites, P1 or P2. All the clones were identical in sequence from P2 (Figure 7). Sequence analysis of these RACE clones revealed a translation start site (see Figure 7), and initiation from this AUG codon would produce a putative signal peptide which may be used in deposition of MSGs on the cell surface.

3.3. Silent Genomic Repertoire

3.3.1. *Tandem Repeat*

Genomic hybridization analysis clearly indicated that the cDNA diversity reflects a multiple MSG-gene family in the *P. carinii* chromosome. As shown in Figure 8A, an MSG cDNA probe hybridized to multiple restriction fragments, while the same probes did not hybridize to DNAs from uninfected rat lungs. Therefore, a complex group of MSG genes must indeed exist in *P. carinii*. Moreover, PCR analysis demonstrated that at least some MSG genes cluster in tandem fashion in the chromosome. When primers were designed with *C*-terminal sense and *N*-terminal antisense sequences, so as to amplify a spacer sequence between tandem MSG genes, multiple products were amplified by PCR (Figure 9A), indicating that the gene family was closely spaced. Therefore, sequence diversity of MSG-cDNAs is probably caused by the diversity and the multiplicity of genomic MSG genes.

Several genomic fragments carrying two or three MSG genes have been cloned (Wada and Nakamura, 1994a,b; Sunkin *et al.*, 1994; Wright *et al.*, 1995a). Among those, an 11 kb fragment was sequenced and shown to carry three open reading frames (ORFs) in tandem repeat (Figure 9B). The first ORF protein (MSG99) was composed of 774 amino acids and shared 37% protein sequence identity (and 61% protein sequence similarity) with MSG5, but it was shorter than MSG5 in the *C*-terminal region. The second ORF protein (MSG100) contained 1083 amino acids, with a deduced mass of 120 921 Da, equivalent to MSG5. The protein shared 76% protein sequence identity with MSG5. Consistent with the PCR amplification of spacer sequences, MSG99 and MSG100 are spaced 311 bp apart. Sunkin *et al.* (1994) also cloned a tandem repeat of rat-derived *P. carinii* genes encoding MSG.

3.3.2. *Multiplicity*

Each *P. carinii* cell has the potential to express a number of different MSG isoforms, because the genome of *P. carinii* encodes at least 100 different MSG genes, distributed among all of the 14–15 discernible chromosomes

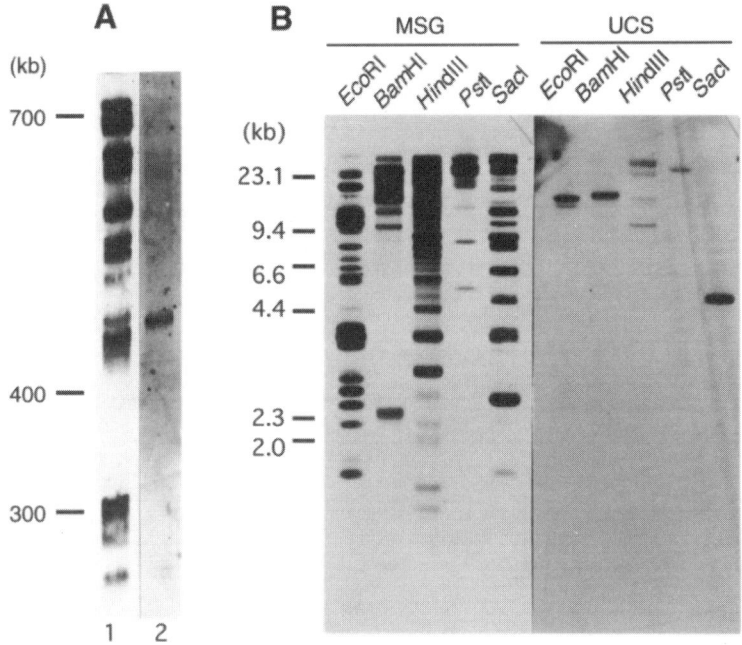

Figure 8 Southern blot and karyotype hybridization analyses of MSG genes (Wada *et al.*, 1995). A. Electrophoretic karyotype hybridization. Rat *Pneumocystis carinii* chromosomes were separated by pulse-field gel electrophoresis, and the blot was analyzed by hybridization using probes specific to MSG (lane 1) or type I UCS (lane 2). B. Southern blot hybridization. Rat *P. carinii* genomic DNA was digested with the endonucleases indicated above the lanes and fragments were separated by 0.8% agarose-gel electrophoresis and analyzed by Southern blot hybridization using probes specific to MSG (left) or type-I UCS (right). Reprinted with permission of The University of Chicago Press.

(see Figure 8B) (Kovacs *et al.*, 1993; Wada *et al.*, 1993; Linke *et al.*, 1994; Sunkin *et al.*, 1994). The multiplicity and diversity of MSG genes were first discovered in *P. carinii* from rats. Cloning and sequencing analyses, using organisms from other animals such as ferrets, mice, and humans, have subsequently shown a similar sequence diversity and multiplicity of MSGs (Stringer *et al.*, 1993; Garbe and Stringer, 1994; Wright *et al.*, 1995a). The precise degree of multiplicity has not been determined.

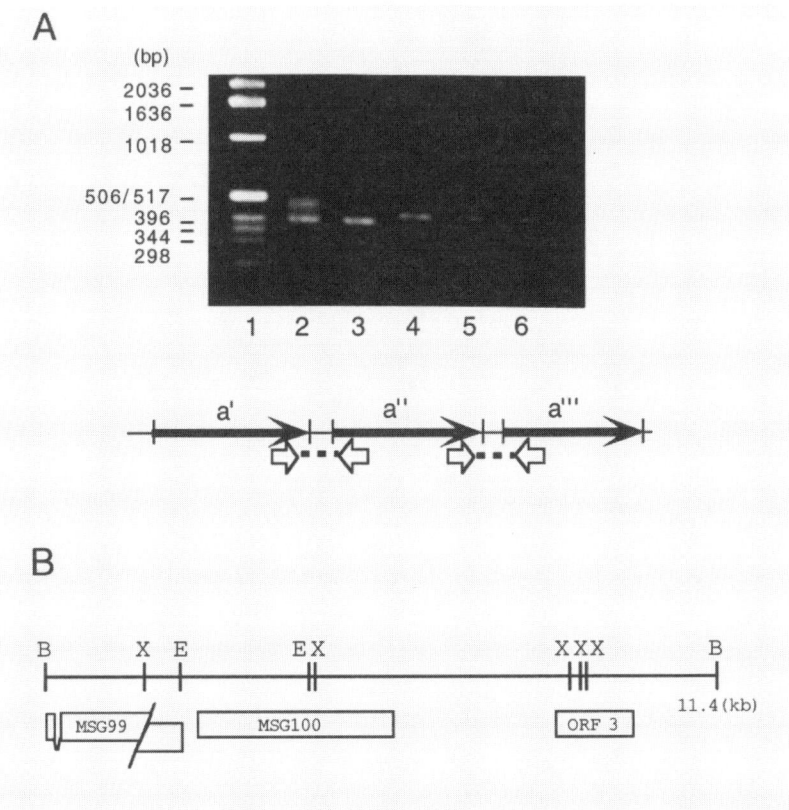

Figure 9 Tandem clusters of MSG genes. A. PCR amplification of spacer sequences between tandem MSG genes from *Pneumocystis carinii* bulk DNA (lane 2) and four genomic MSG clones (lanes 3 to 6) using a set of primers (open arrows in the lower panel) (Wada *et al.*, 1993). Lane 1, size markers. (Reprinted with permission of Chicago University Press.) B. Schematic representation of ORFs encoded by the 11370 bp *P. carinii* DNA cloned in λMW111 (Wada and Nakamura, 1994b). An intron is shown in the *N*-terminal region of MSG99. Restriction endonuclease sites are indicated thus: B, *Bam*HI; E, *Eco*RI; X, *Xba*I. Reprinted with the permission of the Society of Protozoologists.

3.4. Genomic Expression Site

3.4.1. *Genomic Upstream Conserved Sequence*

Comparison of RACE clones with genomic clones (Sunkin *et al.*, 1994; Wada and Nakamura, 1994a,b) showed that they diverged just upstream of position 400 (Figure 7). Southern blot hybridization clarified the reason for this. The UCS probe hybridized to a single band (approximately 500 kb) in the electrophoretic karyotype, while an MSG probe hybridized to all the bands (Figure 8B). Similarly, the UCS hybridized to relatively few restriction fragments of *P. carinii* DNA compared with an MSG probe (Figure 8A). These experiments suggested that the UCS is located on one chromosome, in which case most MSG genes cannot be linked to the UCS.

Those MSG genes that were attached to a UCS were amplified by the PCR using an upstream sense primer of UCS, paired with a downstream antisense primer of the variable portion of MSG. The amplified products contained a mixture of DNA fragments that shared a UCS at one end, and contained a different MSG gene at the other (Wada *et al.*, 1995; Sunkin and Stringer, 1996). Some of these clones are illustrated in Figure 10. These clones, which were chosen at random, were all different in the MSG region, suggesting that the population of *P. carinii* genomes contained many different MSG genes attached to the UCS. Although the sequence of UCS is highly conserved, extensive amplification of UCS sequences from rat *P. carinii* populations showed some slight polymorphism within the UCS sequence (Sunkin and Stringer, 1996).

These results show that the most abundant MSG mRNAs are attached to UCS, a sequence that is found on only one of the *P. carinii* chromosomes

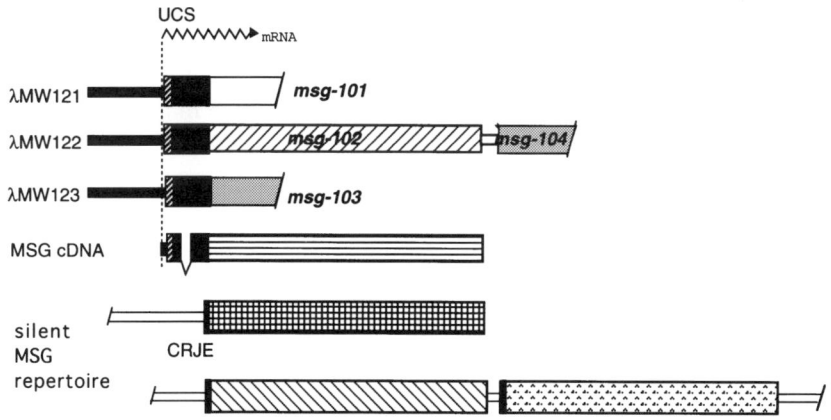

Figure 10 Genomic structure of the UCS element attached to MSG genes (Wada *et al.*, 1995). Structures of three genomic UCS clones whose nucleotide sequences were determined are represented schematically, together with those of cDNA and silent repertoires. The UCS are shown as black boxes and the signal peptide regions are hatched. Divergent coding sequences are shown by different patterns. The intron is indicated by a deletion within the UCS in the cDNA construct. CRJE is a conserved recombination junction element. Reprinted with permission of The University of Chicago Press.

resolved by electrophoresis. The UCS is attached to MSG genes, suggesting that it occupies a genome position that is permissive for MSG transcription. The most likely explanation is that the population is genetically heterogeneous at the UCS locus, and that this heterogeneity is brought about by DNA recombination, which installs different MSG coding sequences at a limited number of expression sites (perhaps only one), as will be discussed shortly. Until now, the UCS element has been studied only in rat *P. carinii*, and it remains to be investigated whether it exists universally in *P. carinii* organisms isolated from other animal hosts, including humans.

3.4.2. Upper Conserved Sequence Copy Number

Initial hybridization data have shown that the UCS was present on only one band in an electrophoretic karyotype, suggesting that the UCS locus is unique (Wada *et al.*, 1995). This prediction was further confirmed by quantitative hybridization of UCS compared with a single-copy gene probe (Sunkin and Stringer, 1996).

Wada *et al.* (1995) have roughly estimated the number of UCS copies attached to variant MSG genes per genome by screening a genomic library with the UCS probe. The number of plaques that hybridized to the UCS probe was only three times the number that hybridized to a single-copy gene probe and only 1/30 of the number which hybridized to an MSG gene. This result was consistent with the Southern blot data, showing that most of the restriction fragments that hybridized to an MSG probe did not hybridize to the UCS probe. Sunkin and Stringer (1996) estimated that at least 19 different MSG genes were linked to the UCS in a population of *P. carinii*.

3.4.3. Transcript Expression and Splicing

Genomic MSG clones were isolated, and each was found to contain a UCS, as revealed in cDNA, as well as a 150 bp intron located 98 bases downstream of the translation start site (Figure 10). These genomic clones had identical intron sequences and the ORF was maintained within the intron. The intron had the expected splicing junction motifs (Wada *et al.*, 1995). There are, apparently, some UCS clones that carry two or three MSGs, raising the possibility of transcription of tandem MSG genes from a single promoter. If both genes are transcribed, an MSG mRNA containing the UCS attached to the second MSG gene could be formed by alternative splicing.

Figure 11 shows that the genomic clones contained the two transcription start sites, P1 and P2. The sequence upstream of P1 contains a canonical TATA element and sequence motifs recognized by other transcription factors, such as AP-2, γ-IRE, CTF and Myb (Wada *et al.*, 1995). The

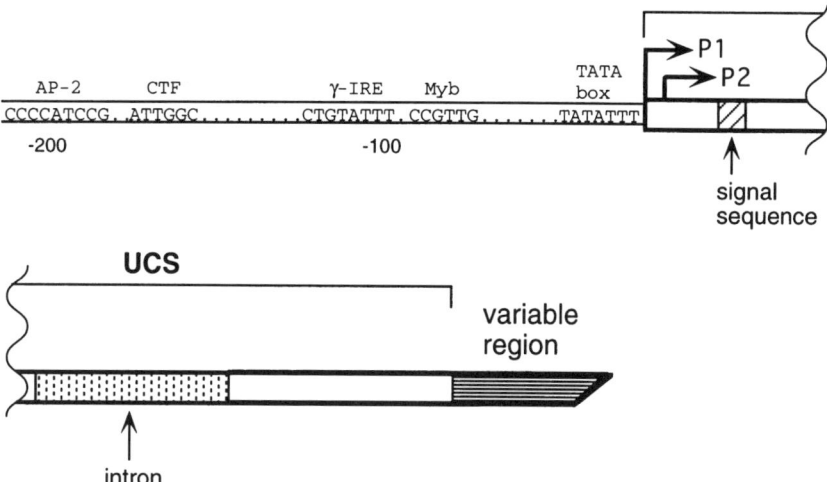

Figure 11 UCS expression site (Wada *et al.*, 1995). The numbers of the nucleotide positions are counted from the translation start site. Two transcription start sites are indicated by arrows, P1 and P2, and several transcription signal motifs are marked.

promoter activity of UCS was examined using a UCS-*lacZ* (β-galactosidase gene) fusion in the budding yeast *Sac. cerevisiae* and the fission yeast *S. pombe*, because *P. carinii* is phylogenically close to yeast. UCS allowed β-galactosidase synthesis in *S. pombe* but not in *Sac. cerevisiae* (M. Wada and Y. Nakamura, unpublished observations). The transcript start sites determined by 5′ RACE analysis are located slightly upstream of the two authentic start sites, P1 and P2, in *P. carinii*. The promoter activity of UCS itself is about one-fifth of that of a known strong promoter of *S. pombe*, *nmt1* (Maundrell, 1990).

3.5. Chromosomal Organization

3.5.1. *Telomeric Upper Conserved Sequence*

Several lines of evidence showed that a UCS site is telomeric. First, Southern blotting predominantly detected an 11–13 kb segment hybridizable to the UCS probe when the *P. carinii* DNA was cleaved with *Bam*HI. However, attempts to isolate the fragments were unsuccessful, suggesting that the 11–13 kb UCS fragment is not susceptible to conventional cloning using restriction endonucleases. Second, digestion of *P. carinii* DNA with

*Bal*31 before *Bam*HI cleavage increased the migration of the *Bam*HI UCS fragment, in a manner which was dependent on the duration of *Bal*31 digestion (Wada and Nakamura, 1996a). A similar shortening of the UCS fragment by *Bal*31 endonuclease was described by Sunkin and Stringer (1996). Finally, a telomeric UCS DNA was cloned from a genomic library of *P. carinii* constructed by prior digestion with mung bean nuclease to blunt the telomere ends, and telomeric 11 kb UCS fragments were isolated (Wada and Nakamura, 1996a). Sequence analysis of one of the clones, λMW124, showed that it contained the UCS and one MSG-coding sequence, but no other significant ORFs. This MSG sequence, MSG105, was highly homologous, but not identical, to the known MSG genes. At the 3′ terminus, there were seven tandem hexanucleotide repeats with the sequence of TTAGGG oriented 5′ to 3′ to the chromosomal end (Figure 12). TTAGGG is the sequence of telomeric DNA in all vertebrates, the protozoan *Trypanosoma*, and several slime molds and fungi, revealing that the UCS is telomeric.

In most organisms, the subtelomeric regions immediately internal to the simple repeats consist of moderately repetitive sequences, called telomere-associated DNA (for review, see Zakian, 1995). Between MSG105 and the TTAGGG repeats, there are distinct regions, I–IV, that have imperfect, but moderately repetitive, sequences, as revealed by the Macintosh software Dotty Plotter™ analysis (Figure 12). Most of these regions were composed of characteristic core sequences that appeared frequently within the respective regions: region I has 23 repeats of $T_{3-4}A_{5-8}$, region II has 17 repeats of $T_3A_5(A/T)$ as well as 12 tandem repeats of a highly homologous 160 bp sequence, region III has 18 repeats of $GA_{1-2}GAGA$, and region IV has eight repeats of TTCACAAAGTG. The nucleotide content in regions III and IV is highly biased toward GC, regardless of the high AT content of the *P. carinii* chromosome. These findings are consistent with the common features of subtelomeres.

3.5.2. *Telomeric Repertoire*

The TTAGGG telomere sequence and subtelomeric sequences are not unique to the 0.5 megabase chromosome containing UCS, but appear to be common to all *P. carinii* chromosomes. The 14–15 chromosome blot resolved by pulse field gel electrophoresis hybridized to this hexamer sequence and to the subtelomeric sequences, regions II and IV (Wada and Nakamura, 1996a). Telomeric fragments other than the UCS were cloned by screening with the (TTAGGG)$_5$-oligo DNA probe and the subtelomeric region IV probe. To date, two non-UCS telomeric clones have been sequenced and shown to contain TTAGGG repeats at the termini as well as the subtelomeric repetitive regions I–IV, common to the UCS-MSG

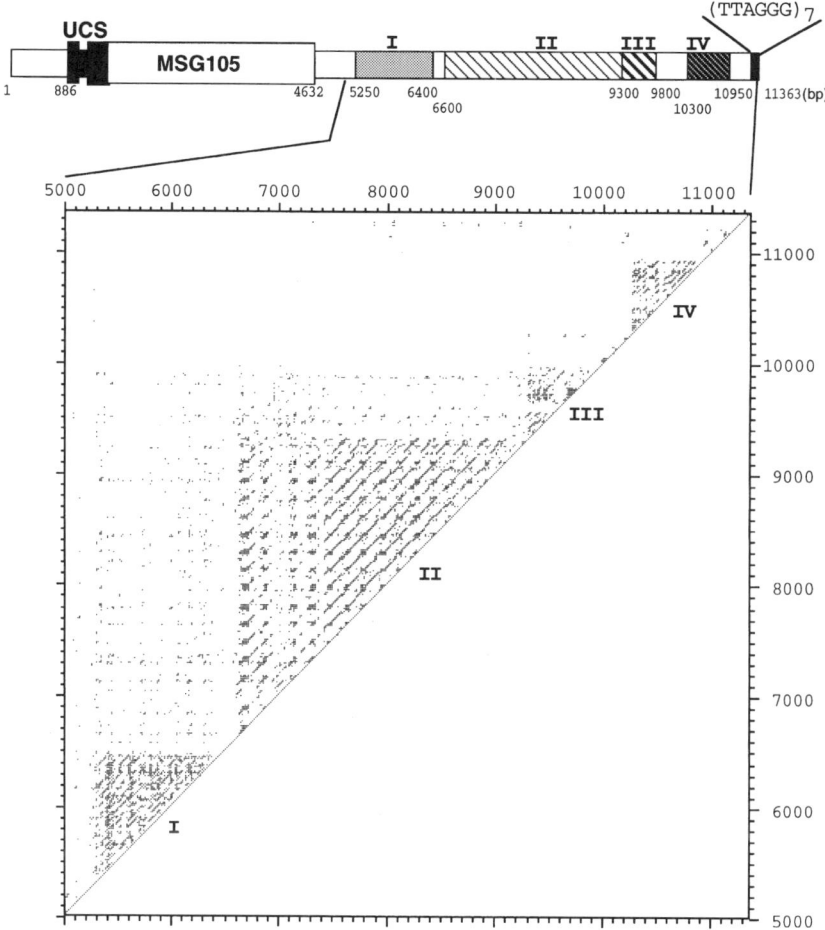

Figure 12 Schematic diagram of the UCS-MSG and telomere-associated sequences (Wada and Nakamura, 1996a). The moderately repetitive sequences of regions I–IV were defined using Macintosh software Dotty Plotter™ version 1.0 under the following conditions: match window 25 and stringency 18. Reprinted with permission of Universal Academy Press, Inc.

expression site (Figure 13). Moreover, these two clones, screened with non-MSG probes, also contained one or more MSG gene(s). The MSG sequences, MSG107 and MSG108, were not associated with the UCS, but attached to a spacer sequence found in the MSG gene cluster (Wada and Nakamura, 1994a). The frequent occurrences of these telomeric MSG clones (3×10^{-4}) indicate that they represent silent MSG genes localized in

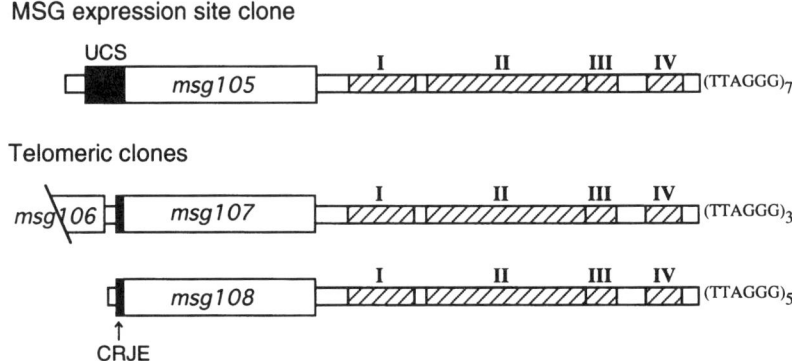

Figure 13 Schematic diagram of UCS-MSG and telomeric MSG gene clones (Wada and Nakamura, 1996b). Black boxes represent UCS and open boxes represent MSG genes. The hatched regions indicate subtelomeric repetitive regions I–IV. Reprinted with permission of the Society of Protozoologists.

the telomere regions. Underwood *et al.* (1996) have cloned a telomere and adjacent sequences from rat-derived *P. carinii* using the ability of foreign telomeres to complement a yeast artificial chromosome (YAC) deficient by one telomere in *Sac. cerevisiae*. They also found that one telomeric fragment of *P. carinii* contained an MSG gene.

3.5.3. *Telomeric DNA* Trans-*action*

In many organisms, including yeast and protozoa, telomere-associated DNA is repeated, can extend for tens or hundreds of kb, can include genes, and can be a source of chromosome polymorphism with different yeast strains or *Plasmodium* lines containing different combinations of repeated sequence elements at their telomeres (Charron *et al.*, 1988; Corcoran *et al.*, 1988; Zakian and Blanton, 1988; Brown *et al.*, 1990). The presence of these repeats and their polymorphic distribution is thought to reflect the exchange of DNA sequences between nonhomologous telomeres. Given that the majority of MSG genes are located at the chromosomal ends, the discovery of a UCS expression site within the telomere region suggests that MSG genes, or batteries of them, are translocated to the UCS expression site by means of reciprocal recombination or gene conversion. This seems to be true of the DNA rearrangement of the variant surface glycoprotein (VSG) genes of the protozoan parasite *Trypanosoma brucei* (see Pays and Steinert, 1988; Borst, 1991; Pays *et al.*, 1994). Trypanosome VSG undergoes antigenic variation, which enables the organism to evade the host's immune response. There are many VSG genes on the chromosomes and minichromosomes, but the VSG expression sites are restricted to telomeric regions and silent VSG genes can be translocated to the expression site by DNA recombination (Pays and Steinert, 1988; Borst, 1991; Pays *et al.*, 1994). A similar telomere-associated recombination model for *P. carinii* MSGs has been proposed by Sunkin and Stringer (1996) and Wada and Nakamura (1996b).

3.6. Genetic Control of Antigenic Variation

3.6.1. *Switch Model*

It has been demonstrated that MSGs play a crucial role in various aspects of the pathobiology of *P. carinii*, including infection, host–parasite interaction and host defence by T cells, alveolar macrophages or humoral immunity. In immunocompromised states in humans or experimental animals, an environment seems to be created that allows the proliferation of *P. carinii* and the establishment of an extracellular infectious focus. Therefore, it is likely that the surface structure of *P. carinii* may change by switching expression of the MSG genes to escape from cellular or humoral immunity (Figure 14) or alternatively, but less probably, the mannose-rich surface proteins are hyperproduced and shed during *P. carinii* infection to affect the expression and/or activity of the macrophage mannose receptor. Because the *P. carinii* cell surface is composed of highly immunogenic sugar moieties of MSG, it can be assumed that, if MSG polypeptide switching takes place to override the host immune defence, this polypeptide change may also cause altered glycosylation of MSG, leading to antigenic variation.

An alternative model, which is not necessarily exclusive of the recombination model, posits that the common UCS on MSG mRNAs is acquired by RNA splicing in the *cis* position (Figure 14). However, the existence of multiple MSG genes that are directly linked to the UCS shows that *trans* splicing is not necessary to explain the data. A role for *cis* splicing is suggested by the tandem linkage of MSG genes, and is attractive because it would allow an organism to express a battery of linked MSG genes from one expression site promoter (Wada *et al.*, 1995).

3.6.2. *Recombination*

(a) *Site-specific recombination.* If recombination is involved in expression of specific MSG genes, it could take place through either of two mechanisms. One possibility is recombination mediated by sequence homology (see below), as is thought to be the case in the

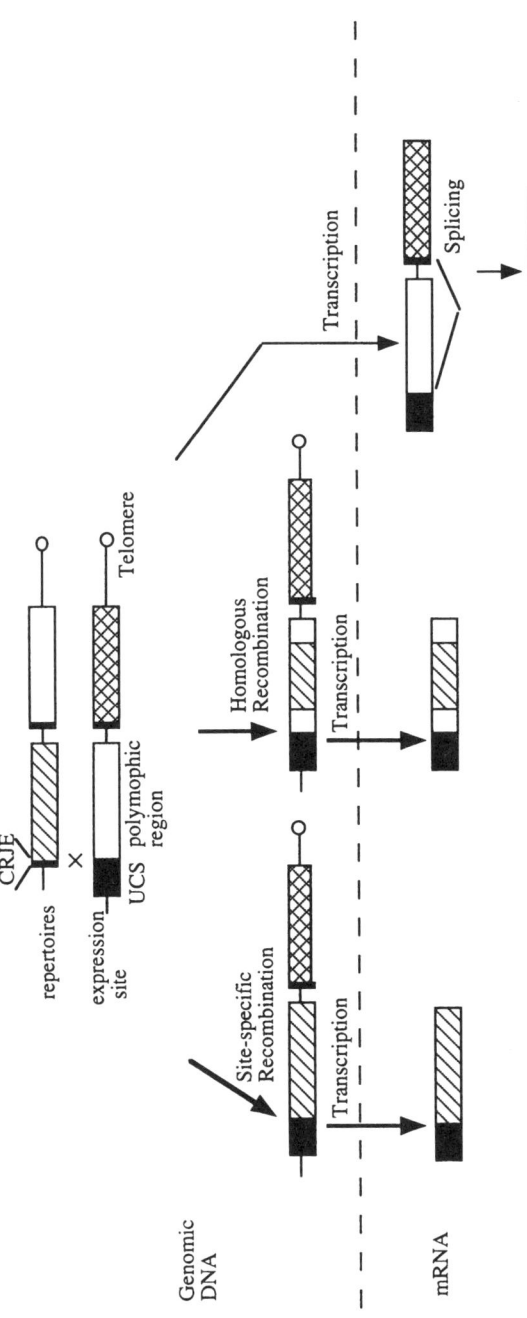

Figure 14 MSG switch model.

rearrangement of trypanosome VSG genes (Pays and Steinert, 1988; Borst, 1991; Pays *et al.*, 1994). Alternatively, a site-specific endonuclease (or recombinase) might be used. It is noteworthy that the 3′ terminus of the UCS shares a 28 bp sequence with the 5′ termini of silent MSG coding sequences (conserved recombination junction element, CRJE, in Figure 7). Recombination at this site is suggested by most UCS–MSG junctions, and each showed sequence differences within 50 bp downstream of the UCS (Wada *et al.*, 1995).

(b) *Homologous recombination.* Recombination could also occur between homologous regions within coding sequences other than the CRJE. In fact, when the full-length sequences of the known genomic or cDNA MSGs were compared, the *C*-terminal one-third of the sequence of MSG105 perfectly matched that of the MSG1cDNA (Figure 15), providing the first genetic evidence for homologous recombination within the MSG coding sequences (Wada and Nakamura, 1996a,b). The region where the recombination point of MSG105 and MSG1 maps is highly conserved in MSGs (see Figure 15) and may serve as a hot spot of homologous recombination. There is another 'hot spot' of recombination at the 3′-flanking sequence of the UCS-MSG gene, since it is highly homologous with the spacer region of tandem-linked MSG genes (Wada and Nakamura, 1994a). These homologous recombinations should increase the MSG repertoire and facilitate propagation of *P. carinii* organisms, given that MSG polymorphism is a manifestation of antigenic variation, which enables the parasite to evade the host immune system.

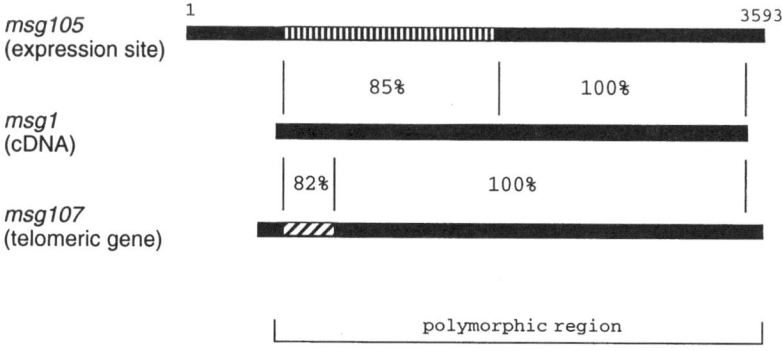

Figure 15 Sequence comparison of three MSG genes, MSG1, MSG105 and MSG107 (Wada and Nakamura, 1996b). The percentage identity between two relevant regions is indicated. The numbers of the nucleotide positions are indicated from the translation start site of MSG105. Reprinted with permission of the Society of Protozoologists.

3.6.3. *Antigenic diversity*

(a) *Interspecies diversity.* Early immunological studies demonstrated that *P. carinii* isolates from rats, mice and humans have species-specific, as well as shared, antigenic determinants. Kitada *et al.* (1994) examined whether a cloned MSG antigen was specific to *P. carinii* from rats, or contained a common determinant recognized by other species in immunoblotting analysis. The rabbit antibody against a recombinant MSG polypeptide reacted with an authentic MSG of rat *P. carinii*, and rabbit antibody or rat serum against an authentic MSG of rat *P. carinii* also recognized the recombinant MSG polypeptide. Similarly, mouse sera raised against mouse *P. carinii* recognized both authentic and recombinant MSG molecules from rat *P. carinii*. These results clearly demonstrated that the recombinant MSG polypeptide was not an immunodeterminant specific to *P. carinii* of rat origin, but also shared common epitopes with MSG from mouse *P. carinii*. Conversely, antiserum derived from an AIDS patient recovering from PCP reacted with the authentic MSG but not with the recombinant MSG (Kitada *et al.*, 1994). This indicated that an MSG subtype of rat *P. carinii* exists that differs antigenically from the cloned MSG but shares common epitopes with human *P. carinii*, or that the authentic glycosylation of the recombinant MSG provides the common immunodeterminant with human *P. carinii* MSG. As expected, the recombinant MSG proteins reacted consistently with the antiserum from rats infected with *P. carinii* and with antiserum generated in mice infected with *P. carinii*, indicating the existence of common determinants in MSG polypeptides (Kitada *et al.*, 1994). Antigenic differences were also demonstrated among genetically distinct (defined by karyotyping) *P. carinii* from rats using anti-MSG monoclonal antibody (Vasquez *et al.*, 1996).

(b) *Intraspecies diversity.* It is likely that the *P. carinii* population is genetically heterogeneous at the UCS locus (Wada *et al.*, 1995). Several lines of evidence suggest that MSG expression is regulated. Populations of *P. carinii* growing in immunosuppressed rats express numerous different MSG genes (Kovacs *et al.*, 1993; Wada *et al.*, 1993; Kitada *et al.*, 1994; Linke *et al.*, 1994), but individual *P. carinii* in a population express some MSGs but not others (Stringer, 1993). It was pointed out that a recombinant MSG protein does not always react with polyclonal antibodies raised against *P. carinii* from the same host (Garbe and Stringer, 1994; Linke *et al.*, 1994). Recently, immunohistochemical evidence for antigenic variation has been provided by using antisera raised against peptides whose sequences were determined from the deduced amino acid sequences of variants of rat-derived MSG (Angus *et al.*, 1996). This investigation revealed that several variants of MSG were differentially expressed in an individual lobe when serial sections of the lungs of infected rats were examined. There was a substantial difference in the fraction of organisms reacting with a specific antipeptide antiserum when comparing organisms isolated from rats raised in a single colony over a period of 2 years. An analogous observation also resulted from indirect immunofluorescent staining with anti-MSG polyclonal and monoclonal antibodies, which showed differential expression of an MSG epitope (Sunkin and Stringer, 1996). We also found, although rarely, a drastic decrease of the molecular mass of MSG among *P. carinii* populations during chronic infection in different rats (H. Yoshida and Y. Nakamura, unpublished observations). These observations support the hypothesis that *P. carinii* utilizes antigenic variation to evade host defence mechanisms.

3.6.4. *Antigenic Variation and AIDS*

Antigenic variation is a strategy used by a number of pathogenic bacteria and protozoa (Pays and Steinert, 1988; Meyer *et al.*, 1990), but it has not been previously described in the fungi. The complexity of MSG gene expression implies that antigenic polymorphism and variability play an important role in *P. carinii* pathobiology. Antigenic variation could have evolved in *P. carinii* because this feature allowed the microbe to colonize immunocompetent individuals often enough and long enough to allow replication and transmission to new hosts. Immunocompetent rats that have been exposed to *P. carinii* are known to harbor organisms, although it is not known for how long. Antigenic variation may also contribute to the prevalence of PCP in AIDS patients who have CD4 cell counts between 200 and 400 per µL (Corey and Coombs, 1993). These residual CD4 cells may furnish protection from other microbes, but not from the mercurial *P. carinii*.

3.7. Predicted Protein Features

3.7.1. *General Features*

MSG proteins contain at least 5–6 potential *N*-linked glycosylation sites (Asn-X-Ser/Thr) as well as 59 cysteine residues (Kitada *et al.*, 1994). These protein features enable complex structures of MSG molecules to be formed by glycosylation and by intra- or inter-polypeptide disulfide bond formation (Figure 16). The protein mass of the complex and dense cell walls of *P. carinii* is composed of these highly abundant MSG molecules. A hydrophobic domain exists at the *C*-terminus of the MSG protein that could be used for membrane anchorage (Kovacs *et al.*, 1993; Kitada *et al.*, 1994) (Figure 16). However, there is no transmembrane motif or hydrophilic

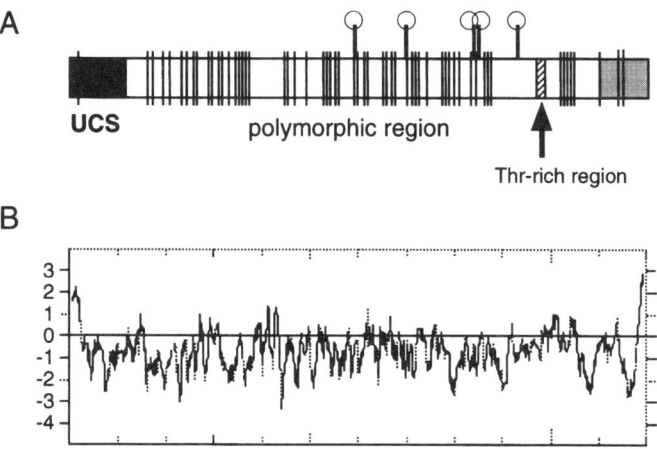

Figure 16 Predicted protein features of MSG. A. Schematic diagram of conserved protein features. Vertical bars represent conserved cysteine residues and open circles represent potential N-linked glycosylation sites. The N-terminal UCS (black box) is strictly conserved and the C-terminal region (shaded box) is also well conserved. The hatched region is rich in threonine (Thr) residues. B. Hydropathicity profile of a complete MSG protein composed of UCS and MSG5 sequences. The GCG program based on that of Kyte and Doolittle (1982) was used. The numbers below the Figure refer to amino acid residues. Areas above the median line are hydrophobic, and areas below are hydrophilic. There are two hydrophobic areas, one in the N-terminus and the other in the C-terminus.

region compatible with an intracytoplasmic domain. Analysis of hydropathicity and secondary structure revealed several areas that may be surface-exposed epitopes. It is noteworthy that some MSG species have a stretch of 25 threonine residues at the C-terminal region, although the significance of this is not known (Kitada *et al.*, 1994).

3.7.2. *Conserved Cysteines*

The complete MSG5 (Kitada *et al.*, 1994) and gp116 (Kovacs *et al.*, 1993) proteins each contain 59 cysteine residues, while the partial gpA polypeptide (Haidaris *et al.*, 1992) has 47. A striking observation is that these cysteines are almost perfectly conserved in all MSG subtypes, regardless of species specificity (Kitada *et al.*, 1994; Wright *et al.*, 1994). MSG5 from rat *P. carinii* shares only 32% identical amino acids with gpA from ferret *P. carinii*. However, they share 94% (44/47 in gpA) and 86% (44/51 in MSG5) of cysteine residues within the overlapped protein sequences. An extremely high conservation of cysteine residues was also found among the *P. carinii* organisms from other animals such as mice, rhesus monkeys and humans (Wright et al., 1994, 1995b).

The highly conserved nature of the cysteine residues in MSG may indicate that they play a key role in the functional structure of MSG on the cell surface. It appears that the cysteine residues do not occur at random, but appear at regular intervals in MSG. In most cases, adjacent cysteines are spaced six amino acids apart. Assuming that these regions are capable of forming α-helical structures, cysteines may be arranged on the same surface of the helix. Accordingly, one can speculate that the occurrence of intra- or inter-polypeptide disulfide bond formation may create or link the repetitive secondary structure on the *P. carinii* cell surface, which may be functionally important in host–organism interaction, immune response or other pathobiological activities of *P. carinii*.

3.7.3. *Potential glycosyl phosphatidylinositol anchorage*

There is a potential glycosyl phosphatidylinositol (GPI) anchorage feature, motifs 1 through 3 (Antony and Miller, 1994), in the C-terminal region of MSGs, as shown here (by single-letter code) using the two most distantly related MSG sequences derived from rat and ferret organisms (M. Wada and Y. Nakamura, unpublished data):

```
     [1]            [2]             [3]
     NDG            MKIRVPD         MIKIMLLGVIVMGM
     .:|            .: ::|          : .|:::||:|  :
     SGG            RR•WLP          SL•SIVIVGVVVALV
```

where [1] = 3 small amino acids region, [2] = putative hydrophilic amino acid region, and [3] = hydrophobic amino acid cluster. Identical and conserved amino acids are marked by vertical bars and colons, respectively.

A putative GPI anchor signal was also described in ferret gpA (Wright *et al.*, 1995a). GPI anchors are known to play an important role in attachment of surface proteins to cell glycans in many mammalian cells and in some parasitic protozoa (Englund, 1993; Kinoshita and Takeda, 1994). For example, in *Trypanosoma*, VSGs are attached to the cell surface by GPI anchorage and are probably removed by cleavage of the GPI anchor during VSG antigenic variation. The GPI anchor of MSG and its pathobiological significance have not been investigated in *P. carinii*.

3.7.4. *Other Functional Motifs*

It has been pointed out that the MSG sequences have significant sequence homology to tropomyosins and myosins, with 20–23% identities (Wada *et al.*, 1993). Because tropomyosins associate with the cortical actin cytoskeleton to regulate cell shape, cell adherence and cell movement, it is tempting to speculate that MSG proteins may also interact with this system and may transmit an environmental signal across the outer membrane to the membrane cytoskeleton during proliferation and establishment of an extracellular infectious focus of *P. carinii*. However, this is in apparent conflict with the prediction that MSG may not contain any transmembrane motif.

It has been reported that MSG participates in the attachment of the organism to host lung cells via fibronectin (Pottratz *et al.*, 1991) and that MSG may have an integrin receptor-like domain susceptible to fibronectin, since antiserum against a synthetic peptide for the cytoplasmic domain of the β_1 subunit of integrin cross-reacted with rat *P. carinii* MSG (Pottratz *et al.*, 1991). However, MSG protein sequences do not carry any putative integrin receptor-like sequence. Therefore, these molecules may have more than one method of interacting with cell surface proteins. The MSG–fibronectin interaction remains to be clarified at the molecular level by using recombinant MSG molecules.

3.8. Subtilisin-like Protease

The 11 kb chromosomal DNA encoding two tandem MSG repertoires contained a third ORF, ORF-3 (Wada and Nakamura, 1994a). ORF-3 encodes a protein that does not appear to be related to MSG molecules but which encodes a subtilisin protease motif (Barr, 1991; Rawlings and Barrett, 1994), as shown in Figure 17. The predicted ORF-3 protein is rich in proline residues as well as serine and threonine residues in its *C*-terminal region (ST-rich domain), and it is highly hydrophilic except for the *C*-terminus, which appears to be an anchoring sequence in the cell membrane. Northern blot hybridization revealed a 3 kb ORF-3 transcript in *P. carinii* cells, and cloning of a full-length cDNA segment confirmed that it encodes a 90 kDa protein (M. Wada and Y. Nakamura, unpublished observations). Given that ORF-3 encodes a subtilisin-like protease, questions immediately arise, such as what does this protease do and why is it linked to MSG?

An ORF-3 DNA probe hybridized to multiple restriction fragments in Southern blot hybridization (Wada and Nakamura, 1994a), as did the MSG probe, showing that ORF-3 is associated with a diverse and multiple gene family and, presumably, localized within the MSG clusters. In fact, all

```
PCSUB   VAEVDNASDYT.....HGTDCAGEVAA.....GNGG.LLD.....GSSASTAEAAG
HFUIRN  VSESDDGEKNH.....HGTRCAGEVAA.....GNGGREHD.....GTSASAPLAAG
KRP1    VAFVDDGIDFK.....HGTRCAGEVAA.....GNGGHYHD.....GTSAAAPLASA
```

Figure 17 Amino acid alignment of catalytic regions of subtilisin proteases and the subtilisin-like motif in the ORF-3 sequence (M. Wada and Y. Nakamura, unpublished data). Letters in bold type indicate active site residues. PCSUB, *Pneumocystis carinii* ORF-3; HFUIRN, human furin; KRP1, *Saccharomyces pombe* *KEX2*-related protease.

15 independent genomic MSG DNAs tested to date carried ORF-3. One might guess that this protease is involved in cleavage of MSGs or other related proteins. A subtilisin protease cleaves protein at pairs of basic amino acids, and such a cleavage site (a lysine–arginine sequence) is strictly conserved within the CRJE polypeptide sequence. Therefore, although it is entirely speculative, one might guess that a conserved *N*-terminal portion of MSG is processed by this subtilisin-like protease, and a variant MSG portion is exposed on the cell surface to aid in host immune evasion.

4. CONCLUSION

To our knowledge, *P. carinii* is the only fungus that uses a genetic system to switch cell surface determinants. The complexity of MSG gene expression implies that antigenic polymorphism and variability play an important role in *P. carinii* pathobiology. Antigenic variation may contribute to the prevalence of PCP in AIDS patients with a low number of $CD4^+$ T cells, compared with other microbes. Alternatively, we speculate that antigenic variability is used to maintain an opportunistic infection in a single animal host, rather than a series of relapses, because of the moderate alterations (about 70% sequence conservation) in the MSG structures. As is true in other cases of recombination-mediated events, site-specific endonuclease(s) may be an important target for further investigation into genetic control of MSG switching.

Establishment of a reliable culture system *in vitro* in the near future will greatly facilitate the study of *P. carinii*. Recent rapid progress in genome sequencing of pathogenic microorganisms should also enable us to provide a complete picture of the genome organization of *P. carinii*, including MSG genes. If a small amount of very pure *P. carinii* cells were available, the major research institutes should be able to complete the entire sequence of the *P. carinii* genome in less than a year. Further clues will also be given by the heterologous expression of *P. carinii* genes (including MSG) in its phylogenically close relatives, such as the fission yeast *S. pombe*. It may be interesting to build a heterologous expression system of MSG genes, then reconstitute variant MSG switching in *S. pombe* cells using artificial chromosomes derived from *P. carinii*. Much work remains to be done to uncover the mechanism of UCS-MSG switching, as well as its pathobiological relevance.

ACKNOWLEDGEMENTS

We acknowledge our current and former colleagues at the Institute of Medical Science of the University of Tokyo, especially Kazuhiro Kitada, Koichi Ito, Shinichi Oka, Satoshi Kimura, Kiyokatsu Tanabe, Kaoru Shimada and Kohji Egawa, for collaboration and discussion. We also extend thanks to John Hershey for reading and criticizing the whole of this manuscript, and Carolyn Nakamura for editing it. We are also grateful to the authors and publishers who allowed us to reproduce materials. The investigations on which this review is based received financial support from the Ministry of Education, Science, Sports and Culture, Japan, the Human Science Foundation, the Uehara Memorial Foundation, the Japan Health Sciences Foundation, the Tokyo Biochemical Research Foundation, and the Naito Foundation, Japan.

REFERENCES

Aliouat, E.M., Dei-Cas, E., Ouaissi, A., Palluault, F., Soulez, B. and Camus, D. (1993). *In vitro* attachment of *Pneumocystis carinii* from mouse and rat origin. *Biology of the Cell* **77**, 209–217.

Angus, C.W., Tu, A., Vogel, P., Qin, M. and Kovacs, J.A. (1996). Expression of variants of the major surface glycoprotein of *Pneumocystis carinii*. *Journal of Experimental Medicine* **183**, 1229–1234.

Antony, A.C. and Miller, M.E. (1994). Statistical prediction of the locus of endoproteolytic cleavage of the nascent polypeptide in glycosylphosphatidylinositol-anchored proteins. *Biochemical Journal* **298**, 9–16.

Barr, P.J. (1991). Mammalian subtilisins: the long-sought dibasic processing endoproteases. *Cell* **66**, 1–3.

Bedrossian, C.W. (1989). Ultrastructure of *Pneumocystis carinii*: a review of internal and surface characteristics. *Seminars in Diagnostic Pathology* **6**, 212–237.

Borst, P. (1991). Molecular genetics of antigenic variation. *Immunology Today* **12**, A29–A33.

Breite, W.M., Bailey, A.M. and Martin, W.J., II (1993). *Pneumocystis carinii* chymase is capable of altering epithelial cell permeability. *American Review of Respiratory Disease* **147**, A33.

Brown, W.R.A., MacKinnon, P.J., Villasanté, A., Spurr, N., Buckle, V.J. and

Dobson, M.J. (1990). Structure and polymorphism of human telomere-associated DNA. *Cell* **63**, 119–132.
Chagas, C. (1909). Nova tripanozomiaza. *Memorias do Instituto Oswaldo Cruz* **1**, 159–218.
Charron, M.J., Read, E., Haut, S.R. and Michels, C.A. (1988). Molecular evolution of the telomere associated *MAL* loci of *Saccharomyces. Genetics* **122**, 307–316.
Corcoran, L.M., Thompson, J.K., Walliker, D. and Kemp, D.J. (1988). Homologous recombination within subtelomeric repeat sequences generates chromosome size polymorphisms in *Plasmodium falciparum. Cell,* **53**, 807–813.
Corey, L. and Coombs, R.W. (1993). The natural history of HIV infection: implications for the assessment of antiretroviral therapy. *Clinical Infectious Diseases* **16** (Supplement 1), S2–S6.
Cushion, M.T. (1989). In vitro studies of *Pneumocystis carinii. Journal of Protozoology* **36**, 432–433.
Cushion, M.T., Destefano, J.A. and Walzer, P.D. (1988). *Pneumocystis carinii*: surface reactive carbohydrates detected by lectin probes. *Experimental Parasitology* **67**, 137–147.
Cushion, M.T., Kaselis, M., Stringer, S.L. and Stringer, J.R. (1993a). Genetic stability and diversity of *Pneumocystis carinii* infecting rat colonies. *Infection and Immunity* **61**, 4801–4813.
Cushion, M.T., Zhang, J., Kaselis, M., Giuntoli, D., Stringer, S.L. and Stringer, J.S. (1993b). Evidence for two genetic variants of *Pneumocystis carinii* coinfecting laboratory rats. *Journal of Clinical Microbiology* **31**, 1217–1223.
Davey, R.T. and Masur, H. (1990). Recent advances in the diagnosis, treatment, and prevention of *Pneumocystis carinii* pneumonia. *Antimicrobial Agents and Chemotherapy* **34**, 499–504.
Delanöe, P. and Delanöe, M. (1912). Sur les rapports des kystes de carinii du poumon des rats avec le *Trypanosoma lewisi. Comptes Rendus de l'Académie des Sciences* **155**, 658–660.
Delanöe, P. and Delanöe, M. (1914). De la rareté de *Pneumocystis carinii* chez les cobayes de la région de Paris. Absence de kystes chez d'autres animaux (lapin, grenouille, Zanguilles). *Bulletin de la Société de Pathologie Exotique* **7**, 271–274.
De Stefano, J.A., Cushion, M.T., Puvanesarajah, V. and Walter, P.D. (1990). Analysis of *Pneumocystis carinii* cyst wall. II. Sugar composition. *Journal of Protozoology* **37**, 436–441.
Dyer, M., Volpe, F., Delves, C.J., Somia, N., Burns, S. and Scaife, J.G. (1992). Cloning and sequence of a beta-tubulin cDNA from *Pneumocystis carinii*: possible implications for drug therapy. *Molecular Microbiology* **6**, 991–1001.
Edlind, T.D., Bartlett, M.S., Weinberg, G.A., Prah, G.N. and Smith, J.W. (1992). The β-tubulin gene from rat and human isolates of *Pneumocystis carinii. Molecular Microbiology* **6**, 3365–3373.
Edman, J.C. and Sogin, M.L. (1993). Molecular phylogeny of *Pneumocystis carinii*. In: Pneumocystis carinii *Pneumonia* (P.D. Walzer, ed.), pp. 91–106. New York: Marcel Dekker.
Edman, J.C., Kovacs, J.A., Masur, H., Santi, D.V., Elwood, H.J. and Sogin, M.L. (1988). Ribosomal RNA sequence shows *Pneumocystis carinii* to be a member of the fungi. *Nature* **334**, 519–522.
Edman, U., Edman, J.C., Lundgren, B. and Santi, D.V. (1989a). Isolation and expression of the *Pneumocystis carinii* thymidylate synthase gene. *Proceedings of the National Academy of Sciences of the United States of America* **86**, 6503–6507.
Edman, J.C., Edman, U., Cao, M., Lundgren, B, Kovacs, J.A. and Santi, D.V.

(1989b). Isolation and expression of the *Pneumocystis carinii* dihydrofolate reductase gene. *Proceedings of the National Academy of Sciences of the United States of America* **86**, 8625–8629.

Englund, P.T. (1993). The structure and biosynthesis of glycosyl phosphatidylinositol protein anchors. *Annual Review of Biochemistry* **62**, 121–138.

Esolen, L.M., Fasano, M.B., Flynn, J., Burton, A. and Lederman, H.M. (1992). Brief report: *Pneumocystis carinii* osteomyelitis in a patient with common variable immunodeficiency. *New England Journal of Medicine* **326**, 909–1001.

Ezekowitz, R.A.B., Williams, D.J., Koziel, H. Armstrong, M.Y., Warner, A., Richards, F.F. and Rose, R.M. (1991). Uptake of *Pneumocystis carinii* mediated by the macrophage mannose receptor. *Nature* **351**, 155–158.

Fisher, D.J., Gigliotti, F., Zauderer, M. and Harmsen, A.G. (1991). Specific T-cell response to a *Pneumocystis carinii* surface glycoprotein (gp120) after immunization and natural infection. *Infection and Immunity* **59**, 3372–3376.

Fishman, J.A., Samia, J.A., Fuglestad, J. and Rose, R.M. (1991). The effects of extracellular matrix (ECM) proteins on the attachment of *Pneumocystis carinii* to lung cell lines. *Journal of Protozoology* **38**, 34S-37S.

Fletcher, L.D., Berger, L.C., Peel, S.A., Baric, R.S., Tidwell, R.R. and Dykstra, C.C. (1993). Isolation and identification of six *Pneumocystis carinii* genes utilizing codon bias. *Gene* **129**, 167–174.

Garbe, T.R. and Stringer, J.R. (1994). Molecular characterization of clustered variants of genes encoding major surface antigens of human *Pneumocystis carinii Infection and Immunity* **62**, 3092–3101.

Gigliotti, F. (1991). Antigenic variation of a major surface glycoprotein of *Pneumocystis carinii*. *Journal of Protozoology* **38**, 4S-5S.

Gigliotti, F. (1992). Host species-specific antigenic variation of a mannosylated surface glycoprotein of *Pneumocystis carinii*. *Journal of Infectious Diseases* **165**, 329–336.

Gigliotti, F. and Hughes, W.T. (1988). Passive immunoprophylaxis with specific monoclonal antibody confers partial protection against *Pneumocystis carinii* pneumonitis in animal models. *Journal of Clinical Investigation* **81**, 1666–1668.

Gigliotti, F., Stokes, D.C., Cheatham, A.B., Davis, D.S. and Hughes, W.T. (1986). Development of murine monoclonal antibodies to *Pneumocystis carinii*. *Journal of Infectious Diseases* **154**, 315–322.

Gigliotti, F., Ballou, L.R., Hughes, W.T. and Mosley, B.D. (1988). Purification and initial characterization of a ferret *Pneumocystis carinii* surface antigen. *Journal of Infectious Diseases* **158**, 848–854.

Graves, D.C., McNabb, S.J.N., Ivey, M.H. and Worley, M.A. (1986a). Development and characterization of monoclonal antibodies to *Pneumocystis carinii*. *Infection and Immunity* **51**, 125–133.

Graves, D.C., McNabb, S.J.N., Worley, M.A., Downs, T.S. and Ivey, M.H. (1986b). Analyses of rat *Pneumocystis carinii* antigens recognized by human and rat antibodies by using Western immunoblotting. *Infection and Immunity* **54**, 96–103.

Haidaris, P.J., Wright, T.W., Gigliotti, F. and Haidaris, C.G. (1992). Expression and characterization of a cDNA clone encoding an immunodeterminant surface glycoprotein of *Pneumocystis carinii*. *Journal of Infectious Diseases* **166**, 1113–1123.

Haidaris, P.J., Wright, T.W., Gigliotti, F., Fallon, M.A., Whitbeck, A.A. and Haidaris, C.G. (1993). *In situ* hybridization analysis of developmental stages of *Pneumocystis carinii* strains that are transcriptionally active for a major surface glycoprotein gene. *Molecular Microbiology* **7**, 647–656.

Hayashi, Y., Watanabe, J., Nakata, K., Hukayama, M. and Ikeda, H. (1990). A

novel diagnostic method of *Pneumocystis carinii*. *In situ* hybridization of ribosomal ribonucleic acid with biotinylated oligonucleotide probes. *Laboratory Investigation* **63**, 576–580.
Hong, S. T., Steele, P.E., Cushion, M.T., Walzer, P.D., Stringer, S.L. and Stringer, J.R. (1990). *Pneumocystis carinii* karyotypes. *Journal of Clinical Microbiology* **28**, 1785–1795.
Hopkin, J.M., Turney, J.H., Young, J.A., Adu, D. and Michael, J. (1983). Rapid diagnosis of obscure pneumonia in immunosuppressed renal patients by cytology of alveolar lavage fluid. *Lancet* **ii**, 299–301.
Jacobs, J.L., Libby, D.M., Winters, R.A., Gelmont, D.M., Fried, E.D., Hartman, B.J. and Laurence, J. (1991). A cluster of *Pneumocystis carinii* pneumonia in adults without predisposing illnesses. *New England Journal of Medicine* **324**, 246–250.
Kinoshita, T. and Takeda, J. (1994). GPI-anchor synthesis. *Parasitology Today* **10**, 139–143.
Kitada, K., Oka, S., Kimura, S., Shimada, K., Serikawa, T., Yamada, J., Tsunoo, H., Egawa, K. and Nakamura, Y. (1991a). Detection of *Pneumocystis carinii* sequences by polymerase chain reaction: animal models and clinical application to noninvasive specimens. *Journal of Clinical Microbiology* **29**, 1985–1990.
Kitada, K., Oka, S., Kimura, S., Shimada, K. and Nakamura, Y. (1991b). Diagnosis of *Pneumocystis carinii* pneumonia by 5S ribosomal DNA amplification. *Journal of Protozoology* **38**, 90S–91S.
Kitada, K., Oka, S., Kojin, T., Kimura, S., Nakamura, Y. and Shimada, K. (1993). *Pneumocystis carinii* pneumonia monitored by *P. carinii* shedding in sputum by the polymerase chain reaction. *Internal Medicine* **32**, 370–373.
Kitada, K., Wada, M. and Nakamura, Y. (1994). Multi-gene family of major surface glycoproteins of *Pneumocystis carinii*: full-size cDNA cloning and expression. *DNA Research* **1**, 57–66.
Kovacs, J.A. and Masur, H. (1988). *Pneumocystis carinii* pneumonia: therapy and prophylaxis. *Journal of Infectious Diseases* **158**, 254–259.
Kovacs, J.A., Halpern, J.L., Swan, J.C., Moss, J., Parrillo, J.E. and Masur, H. (1988). Identification of antigens and antibodies specific for *Pneumocystis carinii*. *Journal of Immunology* **140**, 2023–2031.
Kovacs, J.A., Halpern, J.L., Lundgren, B., Swan, J.C., Parrillo, J.E., and Masur, H. (1989). Monoclonal antibodies to *Pneumocystis carinii*: identification of specific antigens and characterization of antigenic differences between rat and human isolates. *Journal of Infectious Diseases* **159**, 60–70.
Kovacs, J.A., Powell, F., Edman, J.C., Lundgren, B. and Martinez, A. (1993). Multiple genes encode the major surface glycoprotein of *Pneumocystis carinii*. *Journal of Biological Chemistry* **268**, 6034–6040.
Kyte, J. and Doolittle, R.F. (1982). A simple method for displaying the hydropathic character of a protein. *Journal of Molecular Biology* **157**, 105–132.
Lee, C.H., Bolinger, C.D., Bartlett, M.S., Kohler, R.B., Wide, C.E. and Smith, J. W. (1986). Production of monoclonal antibody against *Pneumocystis carinii* by using a hybrid of rat spleen and mouse myeloma cells. *Journal of Clinical Microbiology* **23**, 505–508.
Leigh, T.R., Parsons, P., Hume, C., Husain, O.A.N., Gazzard, B. and Collins, J.V. (1989). Sputum induction for diagnosis of *Pneumocystis carinii* pneumonia. *Lancet* **ii**, 205–206.
Leigh, T.R., Wakefield, A.E., Peters, S.E., Hopkin, J.M. and Collins, J.V. (1992). Comparison of DNA amplification and immunofluorescence for detecting

Pneumocystis carinii in patients receiving immunosuppressive therapy. *Transplantation* **54**, 468–470.

Limper, A.H., Pottratz, S.T. and Martin, W.J., II (1991). Modulation of *Pneumocystis carinii* adherence to cultured lung cells by a mannose-dependent mechanism. *Journal of Laboratory and Clinical Medicine* **118**, 492–499.

Limper, A.H., Standing, J.E., Hoffman, O.A., Castro, M. and Neese, L.W. (1993). Vitronectin binds to *Pneumocystis carinii* and mediates organism attachment to cultured lung epithelial cells. *Infection and Immunity* **61**, 4302–4309.

Linke, M.J. and Walzer, P.D. (1989). Analysis of a surface antigen of *Pneumocystis carinii*. *Journal of Protozoology* **36**, 60S-61S.

Linke, M.J. and Walzer, P.D. (1991). Identification and purification of a soluble species of gp120 released by zymolyase treatment of *Pneumocystis carinii*. *Journal of Protozoology* **38**, 176S-178S.

Linke, M.J., Cushion, M.T. and Walzer, P.D. (1989). Properties of the major antigens of rat and human *Pneumocystis carinii*. *Infection and Immunity* **57**, 1547–1555.

Linke, M.J., Smulian, A.G., Stringer, J.R. and Walzer, P.D. (1994). Characterization of multiple unique cDNAs encoding the major surface glycoprotein of rat-derived *Pneumocystis carinii*. *Parasitology Research* **80**, 478–486.

Lipschik, G.Y., Gill, V.J., Lundgren, J.D., Andrawis, V.A., Nelson, N.A., Nielsen, J.O., Ognibene, F.P. and Kovacs, J.A. (1992). Improved diagnosis of *Pneumocystis carinii* infection by polymerase chain reaction on induced sputum and blood. *Lancet* **340**, 203–206.

Liu, Y., Rocourt, M., Pan, S., Liu, C. and Leibowitz, M.J. (1992). Sequence and variability of the 5.8S and 26S rRNA genes of *Pneumocystis carinii*. *Nucleic Acids Research* **20**, 3763–3772.

Lundgren, B., Cotton, R., Lundgren, J.D., Edman, J.C. and Kovacs, J.A. (1990). Identification of *Pneumocystis carinii* chromosomes and mapping of five genes. *Infection and Immunity* **58**, 1705–1710.

Lundgren, B., Lipschik, G.Y. and Kovacs, J.A. (1991). Purification and characterization of a major human *Pneumocystis carinii* surface antigen. *Journal of Clinical Investigation* **87**, 163–170.

Lundgren, B., Kovacs, J.A., Nelson, N.N., Stock, F., Martinez, A. and Gill, V.J. (1992). *Pneumocystis carinii* and specific fungi have a common epitope, identified by a monoclonal antibody. *Journal of Clinical Microbiology* **30**, 391–395.

Marcotte, H., Levesque, D., Delanay, K., Bourgeault, A., de la Durantaye, R., Brochu, S. and Lavoie, M.C. (1996). *Pneumocystis carinii* infection in transgenic B cell-deficient mice. *Journal of Infectious Diseases* **173**, 1034–1037.

Matsumoto, Y., Matsuda, S. and Tegoshi, T. (1989). Yeast glucan in the cyst wall of *Pneumocystis carinii*. *Journal of Protozoology* **36**, 21S-22S.

Matsumoto, Y., Yamada, M. and Amagi, T. (1991). Yeast glucan of *Pneumocystis carinii* cyst wall: an excellent target for chemotherapy. *Journal of Protozoology* **38**, 6S-7S.

Maundrell, K. (1990). nmt1 of fission yeast. *Journal of Biological Chemistry* **265**, 10857–10864.

Meyer, T.F., Gibbs, C.P. and Haas, R. (1990). Variation and control of protein expression in *Neisseria*. *Annual Review of Microbiology* **44**, 451–477.

Nakamura, Y. (1993). PCR detection of *Pneumocystis carinii*: diagnosis and therapeutic monitoring. In: *Diagnostic Molecular Microbiology* (D.H. Persing, F. Tenover, T.J. White and T.F. Smith, eds), pp. 437–440. Washington, DC: American Society for Microbiology.

Nakamura, Y., Tanabe, K. and Egawa, K. (1989). Structure of major surface determinants and DNA diagnosis of *Pneumocystis carinii*. *Journal of Protozoology* **36**, 58S-60S.

Nakamura, Y., Kitada, K., Wada, M. and Saito, M. (1991). Epitope study and cDNA screening of major surface glycoprotein of *Pneumocystis carinii*. *Journal of Protozoology* **38**, 3S-4S.

Narasimhan, S., Armstrong, M.Y., Rhee, K., Edman, J.C., Richards, F.F. and Spicer E. (1994). Gene for an extracellular matrix receptor protein from *Pneumocytis carinii*. *Proceedings of the National Academy of Sciences of the United States of America* **91**, 7440-7444.

Oka, S., Kohjin, T., Kitada, K., Nakamura, Y., Kimura, S. and Shimada, K. (1993). Direct monitoring as well as sensitive diagnosis of *Pneumocystis carinii* pneumonia by the polymerase chain reaction on sputum samples. *Molecular and Cellular Probes* **9**, 419-424.

Olsson, M., Elvin, K., Lofdahl, S. and Linder, E. (1993). Detection of *Pneumocystis carinii* DNA in sputum and bronchoalveolar lavage samples by polymerase chain reaction. *Journal of Clinical Microbiology* **31**, 221-226.

O'Riordan, D.M., Standing, J.E., Kwon, K.Y., Chang, D., Crouch, E.C. and Limper, A.H. (1995). Surfactant protein D interacts with *Pneumocystis carinii* and mediates organism adherence to alveolar macrophages. *Journal of Clinical Investigation* **95**, 2699-2710.

Paulsrud, J.R., Queener, S.F., Bartlett, M.S. and Smith, J.W. (1991). Isolation and characterization of rat lung *Pneumocystis carinii* gp120. *Journal of Protozoology* **38**, 10S-11S.

Pays, E. and Steinert, M. (1988). Control of antigen gene expression in African trypanosomes. *Annual Review of Genetics* **22**, 107-126.

Pays, E., Vanhamme, L. and Berberof, M. (1994). Genetic controls for the expression of surface antigens in African trypanosomes. *Annual Review of Microbiology* **48**, 25-52.

Pesanti, E.L. and Shanley, J.D. (1988). Glycoproteins of *Pneumocystis carinii*: characterization by electrophoresis and microscopy. *Journal of Infectious Diseases* **158**, 1353-1359.

Peters, S.G. and Prakash, U.B.S. (1987). *Pneumocystis carinii* pneumonia. Review of 53 cases. *American Journal of Medicine* **82**, 73-78.

Peters, S.E., Wakefield, A.E., Banerji, S. and Hopkin, J.M. (1992). Quantification of the detection of *Pneumocystis carinii* by DNA amplification. *Molecular and Cellular Probes* **6**, 115-117.

Phelps, D.S. and Rose, R.M. (1991). Increased recovery of surfactant protein A in AIDS-related pneumonia. *American Review of Respiratory Disease* **143**, 1072-1075.

Pixley, F.J., Wakefield, A.E., Banerji, S. and Hopkin, J.M. (1991). Mitochondrial gene sequences show fungal homology for *Pneumocystis carinii*. *Molecular Microbiology* **5**, 1347-1351.

Pottratz, S.T. and Martin, W.J., II (1990a). Role of fibronectin in *Pneumocystis carinii* attachment to cultured lung cells. *Journal of Clinical Investigation* **86**, 351-356.

Pottratz, S.T. and Martin, W.J., II (1990b). Mechanism of *Pneumocystis carinii* attachment to cultured rat alveolar macrophages, *Journal of Clinical Investigation* **86**, 1678-1683.

Pottratz, S.T., Paulsrud, J., Smith, J.S. and Martin, W.J., II (1991). *Pneumocystis carinii* attachment to cultured lung cells by *Pneumocystis* gp120, a fibronectin binding protein, *Journal of Clinical Investigation* **88**, 403–407.

Radding, J.A., Armstrong, M.Y.K., Ullu, E. and Richards, F.F. (1989). Identification and isolation of a major cell surface glycoprotein of *Pneumocystis carinii*. *Infection and Immunity* **57**, 2149–2157.

Rawlings, N.D. and Barrett A.J. (1994). Families of serine peptidases. *Methods in Enzymology* **244**, 19–61.

Reddy, L.V., Zammit, C., Schuman, P. and Crane, L.R. (1992). Detection of *Pneumocystis carinii* in a rat model of infection by polymerase chain reaction. *Molecular and Cellular Probes* **6**, 137–143.

Roths, J.B. and Sidman, C.L. (1992). Both immunity and hyper-responsiveness to *Pneumocystis carinii* result from transfer of CD4+ but not CD8+ T cells into severe combined immunodeficiency mice. *Journal of Clinical Investigation* **90**, 673–678.

Ruffolo, J.J. (1993). *Pneumocystis carinii* cell structure. In: Pneumocystis carinii Pneumonia (P.D. Walzer, ed.) pp. 25–44. New York: Marcel Dekker.

Sattler, F.R. and Remington, J.S. (1981). Intravenous trimethoprim-sulfamethoxazole therapy for *Pneumocystis carinii* pneumonia. *American Journal of Medicine* **70**, 1215–1221.

Schlugen, N., Godwin, T., Sepkowitz, K., Armstrong, D., Bernard, E., Rifkin, M., Cerami, A. and Bucala, R. (1992). Application of DNA amplification to pneumocystosis: presence of serum *Pneumocystis carinii* DNA during human and experimentally induced *Pneumocystis carinii* pneumonia. *Journal of Experimental Medicine* **176**, 1327–1333.

Selik, R.M., Starcher, E.T. and Curran, J.W. (1987). Opportunistic diseases reported in AIDS patients: frequencies, associations, and trends. *AIDS* **1**, 175–182.

Sidman, C.L. and Roths, J.B. (1993). New animal models for *Pneumocystis carinii* research: immunodeficiency mice. In: Pneumocystis carinii Pneumonia (P.D. Walzer, ed.) pp. 223–236. New York: Marcel Dekker.

Sloand, E., Laughon, B., Armstrong, M., Bartlett, M.S., Blumenfeld, W., Cushion, M., Kalica, A., Kovacs, J.A., Martin, W., Pitt, E., Pesanti, E.L., Richards, F., Rose, R. and Walzer, P. (1993). The challenge of *Pneumocystis carinii* culture. *Journal of Eukaryotic Microbiology* **40**, 188–195.

Smulian, A.G., Stringer, J.R., Linke, M.J. and Walzer, P.D. (1992). Isolation and characterization of a recombinant antigen of *Pneumocystis carinii*. *Infection and Immunity* **60**, 907–915.

Smulian, A.G., Theus, S.A., Denko, N., Walzer, P.D. and Stringer, J.R. (1993). A 55 kDa antigen of *Pneumocystis carinii*: analysis of the cellular immune response and characterization of the gene. *Molecular Microbiology* **7**, 745–753.

Stahl, P., Rodman, J.S., Miller, J. and Schlesinger, P.H. (1978). Evidence for receptor-mediated binding of glycoproteins, glycoconjugates, and lysosomal glycosidases by alveolar macrophages. *Proceedings of the National Academy of Sciences of the United States of America* **75**, 1399–1403.

Stahl, P., Schlesinger, P.H., Sigardson, E., Rodman, J.S. and Lee, Y.C. (1980). Receptor-mediated pinocytosis of mannose glycoconjugates by macrophages: characterization and evidence for receptor recycling. *Cell* **19**, 207–215.

Sternberg, R.I., Whisett, J.A., Hull, W. and Baughman, R.P. (1993). *Pneumocystis carinii* alters surfactant protein-A concentrations found in bronchoalveolar lavage fluid. *American Review of Respiratory Disease* **147**, A33.

Stringer, J.R. (1993). Molecular genetics of *Pneumocystis*. In: Pneumocystis carinii Pneumonia (P.D. Walzer, ed.), pp. 73–90. New York: Marcel Dekker.

Stringer, S.L., Stringer, J.R., Blase, M.A., Walzer, P.D. and Cushion, M.T.(1989). *Pneumocystis carinii* sequences from ribosomal RNA implies a close relationship with fungi. *Experimental Parasitology* **68**, 450–461.

Stringer, S.L., Hong, S.T., Giuntoli, D. and Stringer, J.R. (1991). Repeated DNA in *Pneumocystis carinii*. *Journal of Clinical Investigation* **29**, 1194–1201.

Stringer, S.L., Garbe, T., Sunkin, S.M., Stringer, J.R. and Cushion M.T. (1993). Genes encoding antigenic surface glycoproteins in *Pneumocystis* from humans. *Journal of Eukaryotic Microbiology* **40**, 821–826.

Su, T.H. and Martin, W.J., II (1994). Pathogenesis and host response in *Pneumocystis carinii* pneumonia. *Annual Review of Medicine* **45**, 261–272.

Sung, S.S., Nelson, R.S. and Silverstein, S.C. (1983). Yeast mannans inhibit binding and phagocytosis of zymosan by mouse peritoneal macrophages. *Journal of Cell Biology* **96**, 160–166.

Sunkin, S.M. and Stringer, J.R. (1996). Translocation of surface antigen genes to a unique telomeric expression site in *Pneumocystis carinii*. *Molecular Microbiology* **19**, 283–295.

Sunkin, S.M., Stringer, S.L. and Stringer, J.R. (1994). A tandem repeat of rat-derived *Pneumocystis carinii* genes encoding the major surface glycoprotein. *Journal of Eukaryotic Microbiology* **41**, 292–300.

Tanabe, K., Fuchimoto, M., Egawa, K. and Nakamura, Y. (1988). Use of *Pneumocystis carinii* genomic DNA clones for DNA hybridization analysis of infected human lungs. *Journal of Infectious Diseases* **157**, 593–596.

Tanabe, K., Takasaki, S., Watanabe, J., Kobata, A., Egawa, K. and Nakamura, Y. (1989). Glycoproteins composed of major surface immunodeterminants of *Pneumocystis carinii*. *Infection and Immunity* **57**, 1363–1368.

Telzak, E.E., Cote, R.J., Gold, J.W.M., Campbell, S.W. and Armstrong, D. (1990). Extrapulmonary *Pneumocystis carinii* infections. *Reviews of Infectious Diseases* **12**, 380–386.

Theus, S.A., Linke, M.J., Andrews, R.P. and Walzer, P.D. (1993). Proliferative and cytokine responses to a major surface glycoprotein of *Pneumocystis carinii*. *Infection and Immunity* **61**, 4703–4709.

Underwood, A.P., Louis, E.J., Borts, R.H., Stringer, J.R. and Wakefield, A.E. (1996). *Pneumocystis carinii* telomere repeats are composed of TTAGGG and the subtelomeric sequence contains a gene encoding the major surface glycoprotein. *Molecular Microbiology* **19**, 273–281.

Vasquez, J., Smulian, G., Linke, M.J. and Cushion, M.T. (1996). Antigenic differences associated with genetically distinct *Pneumocystis carinii* from rats. *Infection and Immunity* **64**, 290–297.

Volpe, F., Dyer, M., Scaife, J.G., Darby, G., Stammers, D.K. and Delves, C.J. (1992). The multifunctional folic acid *fax* gene of *Pneumocystis carinii* appears to encode dihydtopteroate synthase and hydroxythyldihydropterin pyrophosphokinase. *Gene* **112**, 213–218.

Wada, M. and Nakamura, Y. (1994a). MSG gene cluster encoding major cell surface glycoproteins of rat *Pneumocystis carinii*. *DNA Research* **1**, 163–168.

Wada, M. and Nakamura, Y. (1994b). Chromosomal organization of *MSG* antigen genes of rat *Pneumocystis carinii*: tandem repeat and unique 5' UTR sequence encoding intron. *Journal of Eukaryotic Microbiology* **41**, 115S .

Wada, M. and Nakamura, Y. (1996a). Unique telomeric expression site of major-surface-glycoprotein genes of *Pneumocystis carinii*. *DNA Research* **3**, 55–64.

Wada, M. and Nakamura, Y. (1996b). Antigenic variation by telomeric recombination of major-surface-glycoprotein genes of *Pneumocystis carinii*. *Journal of Eukaryotic Microbiology* **43**, 8S.

Wada, M., Kitada, K., Saito, M., Egawa, K. and Nakamura, Y. (1993). cDNA sequence diversity and genomic clusters of major surface glycoprotein genes of *Pneumocystis carinii*. *Journal of Infectious Diseases* **168**, 979–985.

Wada, M., Sunkin, S. M., Stringer, J. R. and Nakamura, Y. (1995). Antigenic variation by positional control of major surface glycoprotein gene expression in *Pneumocystis carinii*. *Journal of Infectious Diseases* **171**, 1563–1568.

Wakefield, A.E., Pixley, F.J., Banerji, S., Sinclair, K., Miller, R.F., Moxon, E.R. and Hopkin, J. M. (1990). Detection of *Pneumocystis carinii* with DNA amplification. *Lancet* **336**, 451–453.

Wakefield, A.E., Guiver, L., Miller, R.F. and Hopkin, J.M. (1991). DNA amplification on induced sputum samples for diagnosis of *Pneumocystis carinii* pneumonia. *Lancet* **337**, 1378–1379.

Walzer, P.D. (1993). *Pneumocystis carinii*: recent advances in basic biology and their clinical application. *AIDS* **7**, 1293–1305.

Walzer, P.D. and Linke, M.J. (1987). A comparison of the antigenic characteristics of rat and human *Pneumocystis carinii* by immunoblotting. *Journal of Immunology* **138**, 2257–2265.

Walzer, P.D., Kim, C.K., Linke, M.J., Pogue, C.L., Huerkamp, M.J., Chrisp, C.E., Lerro, A.V. Wixson, S.K., Hall, E. and Shultz, L.D. (1989). Outbreaks of *Pneumocystis carinii*. *Infection and Immunity* **57**, 62–70.

Walzer, P.D., Foy, J., Steele, P. and White, M. (1992). Treatment of experimental pneumocystosis: review of 7 years of experience and development of a new system for classifying antimicrobial drugs. *Antimicrobial Agents and Chemotherapy* **36**, 1943–1950.

Watanabe, J., Hori, H., Tanabe, K. and Nakamura, Y. (1989). Phylogenetic association of *Pneumocystis carinii* with the 'Rhizopoda/Myxomycota/Zygomycota group' indicated by comparison of 5S ribosomal RNA sequences. *Molecular and Biochemical Parasitology* **32**, 163–168.

Wharton, J.M., Coleman, D.L., Wofsy, C.B., Luce, J.M., Blumenfeld, W., Hadley, W.K., Ingram-Drake, L., Volberding, P.A. and Hopewell, P.C. (1986). Trimethoprim-sulfamethoxazole or pentamidine for *Pneumocystis carinii* pneumonia in the acquired immune deficiency syndrome. A prospective randomized trial. *Annals of Internal Medicine* **105**, 37–44.

Williams, D.J., Radding, J.A., Dell, A.L., Khoo, K.H., Rogers, M.E., Richards, F.F. and Armstrong, M.Y. (1991). Glucan synthesis in *Pneumocystis carinii*. *Journal of Protozoology* **38**, 427–437.

Wisniowski, P. and Martin, W.J., II (1992). Interaction of vitronectin with *Pneumocystis carinii*: evidence for binding via the heparin binding domain. *Clinical Research* **40**, A697.

Wright, T.W., Simpson-Haidaris, P.J., Gigliotti, F., Harmsen, A.G. and Haidaris, C.G. (1994). Conserved sequence homology of cysteine-rich regions in genes encoding glycoprotein A in *Pneumocystis carinii* derived from different host species. *Infection and Immunity* **62**, 1513–1519.

Wright, T.W., Bissoondial, T.Y., Haidaris, C.G., Gigliotti, F. and Haidaris, P.J. (1995a). Isoform diversity and tandem duplication of the glycoprotein A gene in ferret *Pneumocystis carinii*. *DNA Research* **2**, 1977–1988.

Wright, T.W., Gigliotti, F., Haidaris, C.G. and Simpson-Haidaris, P.J. (1995b). Cloning and characterization of a conserved region of human and rhesus macaque *Pneumocystis carinii* gpA. *Gene* **167**, 185–189.

Yoganathan, T., Lin, H. and Buck, G.A. (1989). An electrophoretic karyotype and assignment of ribosomal genes to resolved chromosomes of *Pneumocystis carinii*. *Molecular Microbiology* **3**, 1473–1480.

Yoshida, Y. (1981). Pneumocystis carinii *Pneumonia*. Tokyo: Nanzando.

Yoshida, Y. (1989). Ultrastructural studies of *Pneumocystis carinii*. *Journal of Protozoology* **36**, 53–60.

Yoshikawa, H., Tegoshi, T. and Yoshida, Y. (1987). Detection of surface carbohydrates on *Pneumocystis carinii* by fluorescein-conjugated lectins. *Parasitology Research* **74**, 43–49.

Ypma-Wong, M.F., Fonzi, W.A. and Sypherd, P.S. (1992). Fungus-specific translation elongation factor 3 gene present in *Pneumocystis carinii*. *Infection and Immunity* **60**, 4140–4145.

Zakian, V.A. (1995). Telomeres: beginning to understand the end. *Science* **27**, 1601–1607.

Zakian, V.A. and Blanton, H. M. (1988). Distribution of telomere associated sequences on natural chromosomes in yeast *Saccharomyces cerevisiae*. *Molecular and Cellular Biology* **8**, 2257–2260.

Zaman, M.K., Wooten, O.J., Suprahmanya, B., Ankobiah, W., Finch, P.J.P. and Kamholz, S.L. (1988). Rapid noninvasive diagnosis of *Pneumocystis carinii* from induced liquefied sputum. *Annals of Internal Medicine* **109**, 7–10.

Zhang, J., Cushion, M.T. and Stringer, J.R. (1993). Molecular characterization of a novel repetitive element from *Pneumocystis carinii* from rats. *Journal of Clinical Microbiology* **31**, 244–248.

Zimmerman, P.E., Voelker, D.R., McCormack, F.X., Paulsrud, J.R. and Martin, W.J., II (1992). 120–kD surface glycoprotein of *Pneumocystis carinii* is a ligand for surfactant proteins A. *Journal of Clinical Investigation* **89**, 143–149.

Ascariasis in China

Peng Weidong[1], Zhou Xianmin[1] and D. W. T. Crompton[2]

[1] *Department of Parasitology, Jiangxi Medical College, Nanchang, Jiangxi 330006, P.R. China, and*
[2] *WHO Collaborating Centre for Soil-transmitted Helminthiases, University of Glasgow, Glasgow G12 8QQ, Scotland, UK*

1. Introduction 110
2. Sources of Information 114
3. National Distribution of Ascariasis in China 114
4. Estimation of Prevalence 119
5. Distribution and Prevalence within Regions 120
6. Factors Influencing the Distribution of Ascariasis in China 120
   6.1 Demographic factors 120
   6.2 Environmental factors 125
   6.3 Socio-economic factors 127
   6.4 Polyparasitism 130
   6.5 Pigs and ascariasis in China 131
7. Intensity of Ascariasis in China 133
8. Observations on the Origin of the Human–*Ascaris* Association 136
9. Traditional Chinese Medicine for the Treatment of Ascariasis 138
10. Conclusions 140
Acknowledgements 140
References 141

1. INTRODUCTION

Ascariasis, defined here for convenience as human infection with *Ascaris lumbricoides* regardless of whether disease is subsequently detected or not, is a highly prevalent human parasitic infection occurring in at least 150 countries (Crompton, 1989) and having an estimated global prevalence of about 1500 million cases (Chan *et al.*, 1994). There is evidence from many countries that people, especially children, with this infection may suffer from some degree of nutritional deficit, cognitive impairment, complications, and occasionally death (Thein Hlaing, 1993; Bundy and Guyatt, 1996). Experimental infections of *Ascaris suum* in pigs have shown that the parasite is positively linked to nutritional problems in that host. Briefly, some degree of migration of larval *A. suum* through the liver and lungs of pigs appears to be associated with decreased food intake, increased nitrogen loss and acute allergic asthma (Nesheim, 1989). Infection by adult *A. suum* of the intestine of pigs can cause chronic decreased food intake and increased nutrient excretion in addition to malabsorption, accompanying villus atrophy, impaired absorption of vitamin A and temporary lactose intolerance. Similar observations have been made in humans infected with *A. lumbricoides* (Thein Hlaing, 1993) and these impairments may stunt the growth of children; after deworming, children often showed improvement in weight gain and other variables associated with growth. Furthermore, the aggregation and migration of adult worms can give rise to acute diseases which are often life-threatening and require admission to hospital (Crompton, 1989; WHO, 1987; Nesheim, 1989; Thein Hlaing *et al.*, 1991). In Myanmar, for example, 1185 out of 2057 patients admitted with acute abdominal problems during 1981–83 to the surgical wards of the Yangoon Children's Hospital were found to be suffering from ascariasis (Thein Hlaing, 1985). In China, around 1977, of 16944 patients diagnosed as cases of acute intestinal obstruction and admitted to 79 hospitals, 2997 (17.6%) were judged to be due to infection with *A. lumbricoides*. Another longitudinal case analysis showed that of 8639 patients with cholangitis and cholelithiasis, admitted to the hospital affiliated to the Sichuan Medical College over a period of 44 years, 1607 (18.6%) of the complications were due to the presence of *A. lumbricoides* (Xu Rong-Qi *et al.*, 1991). The above example illustrates the fact that ascariasis remains a significant public health problem in many developing countries.

China, in modern economic terms, is a developing country with the largest population in the world. By 1990, the overall population in the country was about 1160 million. There are 56 nationalities in China of which the largest, the Han, accounts for 94% of the overall population. The population is unevenly distributed; the average density is about 121 people per km², with the western part of the country being sparsely populated while the eastern part is highly populated. More than 80% of the population live in rural areas (Bunge and Shinn, 1981; Beazley, 1982). Most of China has a monsoon climate, but there is considerable climatic variation owing to the complex topography and size of the country. Administratively, China comprises 23 provinces, five autonomous regions (ARs) and three municipalities (MCs) directly under the control of central government. More detailed information about population, climate and topography of China is summarized in Table 1.

Until recently, little information on the distribution and abundance of ascariasis in China has been available to scientists outside the country, with a few papers being quoted repeatedly when the topic has been discussed (Yu *et al.*, 1989). The lack of information can be attributed to two facts. First, almost all studies or surveys on ascariasis in China have been published in Chinese and are relatively inaccessible to the many people whose main language is not Chinese. Second, after the founding of the Peoples Republic of China (P.R.C.), great efforts have been made by the government to combat parasitic diseases judged to be more hazardous than ascariasis such as schistosomiasis japonicum, malaria, filariasis and kala-azar, and much progress has been achieved. For instance, kala-azar has been under overall control since 1958. The national annual number of cases of malaria has decreased from 33 million in the early 1950s to 3.3 million by 1980 and down to 0.14 million in recent years. Schistosomiasis japonicum has been declared to be eradicated in four of 12 provinces where it was endemic and the number of counties affected by the disease has been reduced from 380 to 129, with the number of cases falling from 12 million to 1.6 million. Filariasis has been eradicated in 14 out of 15 provinces (Wang Zhao and Xu Shu-Hui, 1991; Zhou Daren *et al.*, 1994). Action against other parasitic infections, including soil-transmitted helminthiases, had to be delayed owing to the constraints of funding and human resources (Yu *et al.*, 1989; Wang Zhao and Xu Shu-Hui, 1991). Accordingly, only scattered information about ascariasis had been gathered in China during the first 37 years of the People's Republic (Mao Shou-Bai, 1992). In recent years (1988–92), a nation-wide survey on the distribution of human parasites was undertaken under the direction of the Ministry of Public Health in cooperation with local governments. This survey covered 726 counties sampled at random from 22 provinces, five ARs and three MCs directly under the control of the central government. In total, 1 477 742 people from 2848 locations were examined (Wang Zhao and Xu Shu-Hui, 1991; Xu Shu-Hui, 1993; Yu Senhai *et al.*, 1994).

The main objective of this review is to present for western parasitologists and public health workers a general and concise description of the current state of ascariasis in China based on the most recent information. The

Table 1 Demography, climate and topography of the administrative regions of China (abstracted from *Atlas of Provinces of China*, 1992.)

No	Region	Population	Density	T	R	FF	Topography	MAU
1	Anhui	56.18	432	14–17	700–1700	200–250	Mountain 67%, plain 33%	18 c, 63 co
2	Beijing	10.81	643	10–12	>600	180–200	Mountain 62%, plain 38%	10 d, 8 co
3	Fujian	30.09	251	15–22	800–1900	240–330	Upland 90%, plain 10%	16 c, 54 co
4	Guangdong	62.82	350	>19	1500	yearly	Upland 67%, plain 33%	21 c, 78 co
5	Guangxi	42.24	184	17–23	1000–1800	>300	Upland 85%, plain 15%	12 c, 76 co
6	Gansu	22.37	57	−1–15	30–860	160–280	Mainly highland and mountain	13 c, 67 co
7	Guizhou	32.39	191	10–20	900–1500	210–300	Highland	9 c, 70 co
8	Hainan	6.55	193	22–27	>1600	yearly	Plain 67%, upland 33%	3 c, 16 co
9	Hebei	61.08	321	4–13	400–800	110–220	Upland 60%, plain 40%	25 c, 124 co
10	Heilongjiang	35.21	77	−6–4	250–700	90–120	Mountain and plain	25 c, 54 co
11	Henan	85.50	534	12–16	500–900	180–240	Upland 44%, plain and lowland 56%	27 c, 103 co
12	Hubei	53.96	300	13–18	750–1500	220–300	Upland 70%, plain 30%	30 c, 49 co
13	Hunan	60.65	289	16–18.5	1250–1750	260–300	Upland 80%, plain 20%	26 c, 78 co
14	Jiangsu	67.05	671	13–16	800–1200	200–240	Plain 95%, hill 5%	28 c, 47 co
15	Jiangxi	37.71	236	16–20	1200–1900	240–300	Upland 70%, plain 30%	16 c, 74 co
16	Jilin	24.65	140	−3–7	350–1000	120–150	Upland 60%, plain 40%	22 c, 25 co
17	Liaoning	39.45	263	4–10	400–1200	150–180	Upland 67%, plain 33%	22 c, 36 co
18	Nei Monggol	21.45	20	−1–10	50–450	60–160	Mainly highland	17 c, 17 co
19	Ningxia	4.65	71	5–10	190–700	100–162	Upland 75%, plain 25%	4 c, 16 co
20	Qinghai	4.45	6.2	−5–8	15–700	30	Highland	3 c, 37 co

No	Region	Population	Density	T	R	FF	MAU	Counties
21	Shaanxi	32.88	173	7–12	400–1000	180–220	Upland 81%, plain 19%	12 c, 85 co
22	Shandong	84.39	563	11–14.5	560–1170		Upland 35% Plain and lowland 65%	36 c, 74 co
23	Shanghai	13.34	2300	15–16	1000	250	Plain	12 d, 9 co
24	Shanxi	28.75	192	3–14	350–700	120–210	Upland 72%, basin 28%	13 c, 93 co
25	Sichuan	107.2	191	−1–19	500–1200	90–330	Highland 50%, basin 50%	24 c, 168 co
26	Taiwan	20.15	560	20–25	2000	yearly	Upland 67%, low and flatland 33%	–
27	Tianjing	8.78	798	12	550–560	210	Plain	13 d, 5 co
28	Xinjiang	15.15	9.5	−4–14	150	120–240	Highland 55%, basin 45%	16 c, 71 co
29	Xizang	2.16	1.83	−3–12	60–1000	<150	Mainly highland	2 c, 76 co
30	Yunnan	36.97	97	4–24	600–2300	>173	Upland 93%, basin 6%	11 c, 114 co
31	Zhejiang	41.44	414	15–19	850–1700	230–270	Upland 70%, plain and basin 30%	26 c, 50 co

No: see map of China (Figure 1) page 118.

Region: provinces, autonomous regions(ARs) and municipalities (MCs) directly under the central government.

Population: population in millions, from nationwide census in 1990. Although more recent data about population in China is available now, we chose this data because the nationwide survey of the distribution of human parasites was undertaken during 1988–1992.

Density: population density per km^2. T: mean annual temperature, °C.

R: mean annual rainfall, mm.

FF: frost-free days yearly.

MAU: main administrative units in a region; c, city; co, counties, usually rural areas; d, district in an MC.

following questions are addressed. (1) what is the scale of ascariasis in China? (2) how is ascariasis distributed geographically and demographically? (3) what is the pattern of ascariasis in China? (4) does it follow the pattern demonstrated in other countries? The discussion of these questions is important because it helps to form the basis for placing ascariasis in the order of health priorities for action in P.R.C. The discussion also raises issues likely to be important for planning, implementing and sustaining control activities intended to reduce morbidity resulting from ascariasis in China.

2. SOURCES OF INFORMATION

Most papers and reports consulted for this review are publications in Chinese. The main sources of information are *Excerpts of Papers of Human Parasitic Diseases in China from 1949–1986* (Mao Shou-Bai, 1992); *Tropical Medicine* (Zhong Huilan, 1986) – most information about ascariasis before 1986 in China was drawn from these two sources – the special issues of *Chinese Journal of Parasitology and Parasitic Diseases* (1991 and 1994); the supplement of the *Chinese Journal of Zoonoses* (1993). Both the special issues and supplement were published to present the results from the nation-wide survey on the distribution of human parasites in China.

3. NATIONAL DISTRIBUTION OF ASCARIASIS IN CHINA

Infection with *A. lumbricoides* was found to be widely distributed in all the administrative regions of China (Table 2). On the basis of the information in this table, the total number of infections in China was estimated to be about 532 million, which is in general agreement with the estimate of 531 million (range 523–539) made by Yu Senhai *et al.* (1994) and the estimate of 568 million published by Chan *et al.* (1994). Stool examination probably underestimates the actual prevalence, with the degrees of underestimation being dependent on the level of actual infection. It has been demonstrated that the error is likely to be largest when the actual prevalence fell in the range of 40% to 60% (Guyatt and Bundy, 1993).

Infection with *A. lumbricoides* in China is clearly unevenly distributed, regional prevalence estimates ranged from 0.2% to 71.2% (Table 2). Generally, ascariasis is more prevalent in the south and the east of the country and less in the west and the north. Detailed information can be obtained with the help of the coordinates set on the map of China (Figure 1).

Table 2 Nationwide distribution of *Ascaris lumbricoides* infection in China

No	Region	Before 1986					After 1986				
		Sites	Size	Subj[a]	P(%)	Ref[b]	Sites	Size	P(%)	References	Estimate[c] ($\times 10^6$)
1	Anhui	1 co, 2 c	1631	a,ch	84.76	1	24 co and c	54376	49.1	Hu Wanchueng et al., 1991; Xu Funiu et al., 1991	27.58
2	Beijing	co	863	a,ch	96.41	2	3 co	13730	30.7	Zhao Gui-Su et al., 1991	3.32
3	Fujian	1 c	617	a,ch	67.91	1	26 co and c	53416	57.11	Lin Jinxiang et al., 1993	17.18
4	Guangdong	1 c, 1 co	5465	ch	77.96	2	31 co and c	61517	46.44	Chen Xiqi et al., 1993	29.17
5	Guangxi						20 co and c	51883	65.98	Huang Jian et al., 1995	27.87
6	Gansu	—	—	—	>60.0	4	19 co and c	28700	37.5	Wei Shaokuan, 1989; Zhang Shouyi et al., 1994	8.39
7	Guizhou						25 co and c	52938	71.1	Chen Zhaoyi et al., 1994	23.03
8	Hainan						5 co and c	7958	61.8	Xu Fengsui et al., 1994	4.05
9	Hebei	3 c	2089	a,ch	45.56	2	31 co and c	65803	31.84	Li Yan-Bi et al., 1993	19.45
10	Heilongjiang	1 c	1431	ch	42.5	2	26 co and c	52131	10.3	Kang Qingde et al., 1994	3.63
11	Henan						40 co and c	85741	44.6	Human parasitic distribution survey team of Henan, 1991	38.13
12	Hubei	1 c	3026	ch	51.21	2	31 co and c	53284	39.96	Chen Shili et al., 1993	21.56
13	Hunan						30 co and c	63794	67.7	Zhang Xianjun et al., 1994	41.06
14	Jiangsu	2 c	53157	ch	52.81	2	32 co and c	63699	32.5	Hang Panyu et al., 1994	21.79

Table 2 (continued)

No	Region	Before 1986					After 1986				
		Sites	Size	Subj[a]	P(%)	Ref[b]	Sites	Size	P(%)	References	Estimate[c] (×10^6)
15	Jiangxi	1 c	537	ch	72.62	2	23 co and c	52046	71.2	Jiangxi Office for Schistosomiasis and Endemic Disease Control, 1991; Wang Weizhou et al., 1994	32.72
16	Jilin						5 co, 6c	35094	26.24	Tang Min et al., 1991; Zhang Zhongshan et al., 1991; Zhang Xuelin et al., 1991; Ma Wan-Hai et al., 1991; Yang Guohua et al., 1991; Xu Maolin and Xu Hongjun 1991; Zhang Xizhong et al., 1991	6.47
17	Liaoning						10 c	37978	57.5	Xu Jing-Tian et al., 1991	22.68
18	Nei Moggol	2 co, 2c	3015	ch	43.45	2	21 co	30713	16.6	Zhang Bin et al., 1994	3.56
19	Ningxia	—	9661	a,ch	50.42	2	17 co	7980	30.8	Ma Cheng-Ji et al., 1991	1.43
20	Qinghai						32 co and c	16079	33.9	Han Xiumin et al., 1994	1.51
21	Shaanxi	—	15291	a,ch	61.64	2	26 co, 1c	53324	41.6	Zhang Guangyuan et al., 1994	13.68
22	Shandong	1 co, 1 c	2226	a,ch	67.97	2	35 co	85417	38.77	Zhu Yu-Guang et al., 1991	32.72
23	Shanghai	c	1649	ch	53.29	2	10 co	47612	35.1	Xu Fengshui et al., 1994	4.68

No.	Region	Subjects[a]	n		Reference[b]	Total subjects	Mean		Estimate[c]
24	Shanxi	26 co and c				54506	25.9	Shanxi Sanitary and Anti-epidemic Station, 1991	7.45
25	Sichuan	3 co, 2 c	33435	a,ch	86.69	2	44 co and c	97159	68.5	Han Jiajun et al., 1994	73.43
26	Taiwan	—	—	ch	13.7	3	—	—	0.2	Chen and Hsieh, 1989	—
27	Tianjing	c	3584	a,ch	64.6	2	5 co	22144	28.3	Wang Qing et al., 1994	2.49
28	Xinjiang						23 co and c	26302	9.09	Li Baoshan et al., 1993	1.38
29	Xizang	3 co	867	ch	28.08	2	13 co	10315	5.99	Guo Wen-Min et al., 1994	0.13
30	Yunnan	1c	92911	ch	63.57	1	18 co	60865	59.6	Yang Jia-Lun et al., 1992	22.03
31	Zhejiang	1co	1344	ch	98.36	2	28 co and c	55291	60.03	Lei Changqui et al., 1993	24.99
	Total										531.69

[a] Subjects; a: adults, ch: children, subjects after 1986 in all regions were both adults and children except that of Taiwan where only children were sampled.
[b] References: 1, Mao Shou-Bai, 1992; 2, Zhong Huilan, 1986; 3, Chen and Hsieh, 1988; 4, Zhang Shou-Yi and Chen Ming, 1991.
[c] Estimate in millions, it was not made for Taiwan Province because the subjects were children only. Also see footnotes of Table 1.

Supposing the junction of the axes to be at the southern end of Ningxia AR (about 107°E, 35°N), then the distribution pattern is seen by reference to the prevalence values estimated for each main location on the map (Figure 1 and Table 2). It is clear that the most prevalent region is the south-east portion of the country, with the prevalence ranging from about 39% (Shandong province) to over 70% (Jiangxi and Guizhou Provinces). Nearly half of the regions in the south-east portion showed a high prevalence of around 60%. Assuming that the population and the infected fraction of it are evenly distributed (Figure 1), then it appears that roughly 60% of the all cases of ascariasis in China occur in the south-east of the country, where about 55% of total population lives.

In contrast, the Northwest portion (Figure 1) was found to have the lowest prevalence, with the third-lowest value of 9.09% in Xinjiang AR. The prevalence estimates in other parts of this portion were between about 20% and 40% (Table 2). The south-west was found to have a higher prevalence than the north-west and north-east portions, with high prevalences of ascariasis being reported from the provinces of Sichuan and

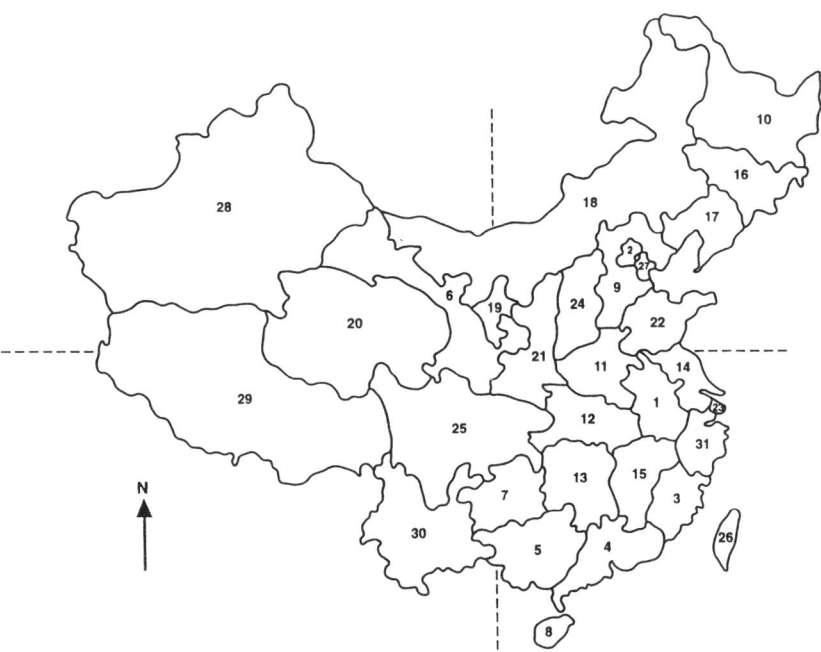

Figure 1 A diagrammatic representation of the map of China showing the principal regions which can identified by reference to Tables 1 and 2. The dashed lines may be used to divide the country into NW, NE, SW and SE quarters.

Yunnan (68.5% and about 60% respectively) and parts of Guizhou Province (71%) and Guangxi AR (66%). The other regions in this portion included much of Xizang AR with the second-lowest prevalence of 6%, and parts of Shaanxi, Qinghai and about one-third of Gansu Provinces with prevalences ranging from 34% to 42%. Much of the north-east portion is occupied by Nei Monggol AR and Heilongjiang Province with prevalences below 17%. The other areas in this portion showed higher prevalences, especially the value of 57.5% in Liaoning Province which is much higher than those reported in other provinces and MCs in the north-east (Figure 1). Further analysis of the national spatial distribution of soil-transmitted helminths in China has been undertaken recently by Lai and Hsi (1996).

4. ESTIMATION OF THE PREVALENCE

The prevalence estimates in Table 2 have been arranged for comparison into data obtained both before and after 1986. The prevalence before 1986 was calculated either by pooling sets of data, probably collected at different times, from different locations and from different populations and by means of different diagnostic methods or quoted from data in a publication when only one report was available. The method used to obtain the average prevalence estimate was the same as that used by Crompton (1989). The total number of subjects sampled and the number of those with *A. lumbricoides* were obtained from the various publications, giving an average prevalence expressed as a percentage. For the prevalence values after 1986, almost all (except that in Taiwan Province) were quoted from the original papers which had presented the average prevalence of ascariasis in related provinces, ARs and MCs. The total estimate of cases with ascariasis in a given region was calculated by multiplying the population in the region by the estimated prevalence value applying after 1986.

Some difficulties were encountered in attempts to estimate the average prevalence in a given region. For example, there were no estimates given for prevalence before 1986 in some provinces or ARs (Table 2). We were not sure whether any surveys had been conducted before 1986 in these regions or whether some investigations had taken place with the results having been published in local reports and distributed only within a county or a prefecture. Second, many papers or reports gave a prevalence value for ascariasis in the survey areas but did not supply information about sample size, population structure, diagnostic technique and other details which should be expected from an epidemiological study. The third difficulty was encountered when analysing some papers on the nation-wide survey.

Information presented in these papers was often pooled parasitic infection data rather than data about a specific parasitic infection. In these cases, information about ascariasis was not given. However, because there were a few papers with detailed information on ascariasis in the survey regions it is possible to apply this to the pooled data to suggest likely trends and patterns of distribution of the prevalence of ascariasis.

5. DISTRIBUTION AND PREVALENCE WITHIN REGIONS

It is well known that *A. lumbricoides* is not evenly distributed even within districts of a country, since climate, season, housing, socio-economic status, family and other factors influence the distribution and prevalence of ascariasis (WHO, 1987; Crompton, 1989). The uneven distribution pattern of ascariasis in China can be observed at different levels, i.e. within the country, within a province, AR or MC, within a prefecture or even within a county. In Xizang AR, for example, the reported prevalence values were varied, ranging widely from 0 (n = 436), 0.3% (n = 396), 1.7% (n = 358), 5.2% (n = 771), 50.8% (n = 1975) to 82.1% (n = 201) at different survey sites of different counties (Yu *et al.*, 1989; Guo Wen-Min *et al.*, 1991, 1994). A survey carried out in the suburbs of Zhangzhou city and Lunghai county, Fujian Province, also demonstrated a wide range of prevalence values for ascariasis in the areas, ranging from 10.0% to 60.1% (n = 4614, nine survey sites) (Chen Youzhu *et al.*, 1991). The patchy distribution pattern in some provinces is shown in Table 3. However, because prevalence values in most provinces and ARs were calculated by pooling data from several survey sites within one county and presented as a county prevalence, and in some provinces by further pooling data from several counties, this would narrow the range of the actual patchy distribution of ascariasis in these regions.

6. FACTORS INFLUENCING THE DISTRIBUTION OF ASCARIASIS IN CHINA

6.1 Demographic Factors

Some data demonstrated that the prevalence of ascariasis was somewhat low in young children under the age of 5 years, or even lower if the available data were related to age groups under 1, 2 and 3 years (Xu Feng-Hui *et al.*, 1991). Thereafter the prevalence was found to rise rapidly and to reach a peak value which was usually found in children aged between 5 and 9 years,

Table 3 Patchy distribution of infection of *Ascaris lumbricoides* in some regions of China

Regions	Sample size	Prevalence range (%)	Sites[a]	References
Fujian	53416	10.0–81.4	26 co	Chen Youzhu et al., 1993
Gansu	28700	0.8–80.9	19 co and c	Zhang Shouyi et al., 1994
Guangxi	51883	25.85–86.28	20 co and c	Huang Jian et al., 1995
Guizhou	52938	50.9–98.2	25 co and c	Chen Zhaoyi et al., 1994
Hebei	5898	8.6–52.0	7 v and t	Li Shu-Ben et al., 1991
Henan	85741	16.0–79.3	40 co and c	Human parasitic distribution survey team of Henan, 1991
Hubei	53284	10.0–77.8	31 co and c	Wu Chixu et al., 1994
Hunan	63794	41.7–85.4	30 co and c	Zhang Xiangjun et al., 1994
Jiangsu	21102	10.3–71.1	11 co	Zheng Chang-Qian et al., 1991
Jiangxi	9748	44.7–81.4	4 co	Fang Jixin et al., 1991
Jilin	35095	5.4–59.8	11 co	Tang Min et al., 1991
				Zhang Zhongshan et al., 1991
				Zhang Xuelin et al., 1991
				Ma Wan-Hai et al., 1991
				Yang Guohua et al., 1991
				Xu Maolin & Xu Hongjun 1991
				Zhang Xizhong et al., 1991
Liaoning	37978	33.4–79.7	10 c	Xu Jing-Tian et al., 1991
Shaanxi	53324	13.8–70.5	26 co	Zhang Guagyuan et al., 1994
Shanxi	1046	7.5–38.7	5 co, 1 c	Feng Bu-Wei et al., 1991
				Shanxi Sanitary and Anti-epidemic Station, 1991
Sichuan	94159	6.3–92.9	44 co and c	Han Jiajun et al., 1994
Tianjing	22144	12.2–42.8	5 co	Wang Qing et al., 1994
Xizanfg	4137	0–82.1	8 co	Yu et al., 1989
				Guo Wen-Min et al., 1991
Xinjiang	5630	0.23–61.8	6 co	Li Baoshan et al., 1993
Zhejiang	55291	33.48–83.7	28 co and c	Lei Changqui et al., 1993

[a] v, village; t, town; for others see footnotes of Table 1.

although was sometimes the peak observed in children aged between 10 and 14 years. The prevalence then tends to decline gradually until the oldest age group is reached. The general pattern of age-structured prevalence is shown in Figure 2 by pooling data from several survey reports (Guan Qin-Li et al., 1991; Hu Shao-Yuan et al., 1991; Huang Jian et al., 1995; Ma Cheng-Ji et al., 1991; Ma Xing-Bao et al., 1991; Tong Shu-Fen et al., 1991; Xu Feng-Hui et al., 1991; Xu Jing-Tian et al., 1991; Yang Jia-Lun et al., 1991; Zhao

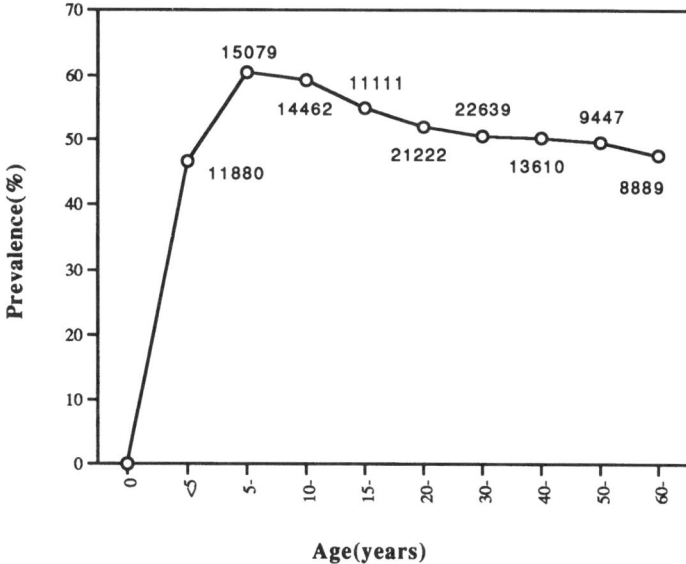

Figure 2 Age-stratified prevalence of *Ascaris lumbricoides* in China (size in each group shown)

Gui-Su *et al.*, 1991; Zheng Chang-Qian *et al.,* 1991). However, the trend shown in Figure 2 did not always apply if the results from individual reports were plotted. Occasionally, the greater prevalence was observed in the group aged between 15 and 19 years (Hu Shao-Yuan *et al.,* 1991), or even in the group between 30 and 39 years (Wang Guilin *et al.,* 1991). The distribution patterns of prevalence in different age groups between male and female hosts were usually found to be similar in most of the communities that had been surveyed. Some results from surveys suggested different patterns in the distribution of prevalence of ascariasis between males and females. In Dalian Prefecture, Heilongjang Province, 6590 (3234 male and 3356 female) people were sampled from 13 survey sites and the highest prevalence in males was seen in the group aged between 10 and 14 years, but in females the highest value was observed in the group aged between 40 and 44 years. The unusual distribution pattern of prevalence in females was attributed to a gender-related difference in occupation. In this region, females were more often engaged in vegetable planting and kitchen work, these being activities which might expose females more frequently than males to the infective eggs of the worm (Zhang Bei-Xiu and Chen Feng-Yi, 1991).

Most of the data indicated that females carried a significantly higher prevalence of ascariasis than males (Table 4). This result agrees with the evidence obtained from other countries (Crompton, 1989), but there were

Table 4 Prevalence(%) of *Ascaris lumbricoides* of males and females in some regions of China

Regions	Sample size	Males	Females	P	References
Fujian	53416	54.6	59.6	<0.01	Chen Youzhu et al., 1993
Guangxi	25900	63.90	68.04	<0.01	Huang Jian et al., 1995
	19341[a]	68.5	70.1	<0.05	Huang Jian et al., 1994
Changchueng, Jilin	11000	27.5	25.6	<0.05	Tang Min et al., 1991
Hebei	65803	30.95	32.74	<0.01	Li Yan-Bi et al., 1993
Heilongjiang	2526	5.4	7.5	<0.05	Tong Shu-Fen et al., 1991
Henan	22424	42.7	48.0	—	Feng Shixian et al., 1994
Jiangsu	63699	37.7	41.3	<0.005	Hang Panyu et al., 1994
Jiangxi	52042	69.0	73.4	—	Wang Weizhou et al., 1994
Liaoning	37978	53.2	58.9	<0.01	Xu Jing-Tian et al., 1991
Ningxia	7980	29.3	32.3	<0.05	Ma Cheng-Ji et al., 1991
Changchueng, Jilin	11000	27.5	25.6	<0.05	Tang Min et al., 1991
Qinghai	16079	31.3	36.5	<0.005	Han Xiumin et al., 1994
Shannxi	53324	41.0	42.2	<0.05	Zhang Guangyuan et al., 1994
Shandong	85417	37.64	41.29	—	Zhu Yu-Guang et al., 1991
Sichuan	18468	64.3	68.1	—	Han Jia-Jun et al., 1991
Xizang	2746	15.3	19.6	<0.05	Guo Wen-Min et al., 1991
Xinjiang	304	44.7	55.3	<0.05	Dong Qiang et al., 1991
Yunnan	60865	58.0	61.2	<0.001	Yang Jia-Lun et al., 1992
Zhejiang	55291	57.31	62.71	<0.01	Lei Changqiu et al., 1993
Qinghai	4838	12.1	13.5	>0.1	Wu Xianhong et al., 1994
Liaoyuan, Jilin	2013	56.2	58.0	>0.05	Yang Guohua et al., 1991
Jutai, Jilin	501	34.0	34.7	>0.05	Ma Wan-Hai et al., 1991
Ningxia	9460[b]	30.0	30.6	>0.05	Ma Chengji et al., 1994
Hainan	387[c]	66.0	68.3	>0.05	Chen Jizhang et al., 1994
Hubei	10149[a]	45.54	27.19	<0.01	Chen Sili et al., 1993

[a] ≤14 years old.
[b] <18 years old.
[c] 1–6 years old.

some contrasting reports presenting similar prevalence values between males and females or even significantly higher prevalences in males (Fong Zhishan et al., 1989; Din Jian-Guo, 1987; see Table 4).

The nation-wide survey of human parasites in China also covered populations from the different nationalities. Results from some reports are given in Table 5, but more information is needed to explain the differences in the prevalence of ascariasis among people of different nationalities in China.

Table 5 Prevalence(%) of *Ascaris lumbricoides* in people with different nationalities in China

Nationality	Yunnan	Qinghai	Hunan	Hainan	Guangxi	Ningxia	Harbin	Guangdong	Zhejiang	Hebei
Han	58.7	39.4	49.4	60.0	69.04	30.3	5.7	46.0	59.92	31.69
Man										50.09
Hui		44.8		73.5		32.0				9.44
Korea							7.7			
Yao					75.48			73.0		
Miao			61.1	73.5	61.54					
Li				56.5						
Tujia			60.6							
Dong			33.2							
She									63.49	
Zhuang					60.13					
Mulam					59.26					
Moggol		8.3								
Shala		44.1								
Tu		43.7								
Zang		8.4								
Dai	45.3									
Naxi	39.4									
Wa	80.3									
Hani	> 60.0									
Jingpo	> 60.0									
Lahu	> 60.0									
P value	—	—	< 0.001	< 0.001	—	> 0.05	< 0.05	< 0.01	< 0.01	< 0.01
Ref.	Yang Jia-Lun et al. (1993)	Han Xiumin et al. (1994)	Zhang Xiang-Jun et al. (1991)	Xu Feng-Shui et al. (1993)	Huang Jian et al. (1995)	Ma Chen-Ji et al. (1991)	Tong Shu-Fen et al. (1991)	Chen Xiqi et al. (1993)	Lei Chang-qui et al. (1993)	Li Yan-Bi et al. (1993)

Not only are many nationalities widely distributed in a variety of regions but there are different cultural traditions and customs which might influence the transmission of ascariasis.

6.2 Environmental Factors

The results from some of the surveys indicated that the prevalence of ascariasis was generally greater in rural areas than in urban communities. In Harbin, Liaoning Province, it was 7.5% versus 2.6% (n = 2526, $P < 0.01$) (Tong Shu-Fen et al., 1991), in Changchun, Jilin Province, it was 27.6% versus 21.8% (n = 11000, $P < 0.01$) (Tang Min et al., 1991), in Yingchuan, Ningxia AR, it was 63.9% versus 47.4% (n = 900, $P < 0.005$) (Din Jian-Guo 1987) and in Mojiang county, Yunnan Province, it was 63.9% versus 34.9% ($P < 0.001$) (Yang Baogui et al., 1991). However, information about the difference in prevalence of ascariasis between rural and urban communities was not as plentiful as might have been expected. Some reports included a comparison of the pooled prevalences of all parasitic infections between rural and urban communities; a high overall parasitic prevalence was usually observed in the rural communities. Because infection with *A. lumbricoides* was the most prevalent parasitic infection and consisted of a large part of all parasitic infections in these regions, it is perhaps reasonable to assume that the prevalence of ascariasis will be higher in rural communities than in urban communities in the same regions.

The impact of geographic and topographic characters on the distribution of ascariasis was investigated by reference to some reports. A survey conducted in Beijing suburbs suggested ascariasis to be more prevalent in mountain areas than in the plains, because the county located in the mountains showed a higher prevalence (41.3%) than in two counties located in the plains (31.6%, 17.4%) (Zhao Gui-Su et al., 1991). Similar results were also reported in Fuyuan county, Yunnan Province (n = 2518, 81.7% versus 64.8%) (Xu Yunhua and Tong Nuei, 1991). In Fujian Province, 27 counties were sampled by the nation-wide survey programme, of which in the 21 located in mountain and hill regions of the province, there was a higher prevalence (about 60-70%, and up to 81.4% in Zhouning county) compared with results from the other six counties located in the plains, where prevalences were less, ranging from 16.6% to 37.7% (Chen Youzhu et al., 1993). Results of an analysis related to topographic characters in Zhejiang Province also suggested a high prevalence of ascariasis in the plains (Lei Changqiu et al., 1993). However, because many other environmental factors such as climate, rainfall, latitude, longitude and elevation as well as socio-economic factors may relate to the prevalence of ascariasis, conclusions drawn from analyses based only on geographic or topographic factors may be misleading. In Sichuan Province, for example, the highest prevalence (74%) was observed in plains regions of Eastern Sichuan and the lowest (22.2%) was seen in the highland areas of Western Sichuan (Han Jia-Jun *et al.*, 1991). The difference in the prevalence of ascariasis in this province might also be attributed to differences in climate, to the distribution of population and to varied agricultural activity between the Eastern and Western Sichuan. As seen from data in Table 1, of all the administrative regions in China, Sichuan Province shows the greatest variety of mean annual temperature and number of frost-free days in a year. In the western part, for example, the frost-free period is less than 90 days but in the eastern part it is 280–330 days. In addition, this province has the largest population among all administrative regions in China, but the population is distributed mainly in the agricultural eastern part while the western part is sparsely populated with pastoral areas. In Henan Province, of the five topographic regions (including 40 counties and cities), the lowest prevalence was found in the mountain region (nine counties, 54.4%) rather than in other topographic regions (plains, hill land and basin, 63.3–75.3%, pooled parasitic prevalence). Similarly, the influence of the variety of climate between these five topographic regions was also considered to be important (Human Parasitic Distribution Survey Team of Henan, 1991).

Climatic factors were found to be related to the prevalence and distribution of ascariasis. Rainfall and annual mean humidity were demonstrated to be significantly positively correlated to prevalence of ascariasis in Anhui province, but the mean annual temperature was not correlated significantly (Hu Wanchueng *et al.*, 1991). In the summary paper of the results of the survey in Hebei Province, 24 environmental factors were analysed by using three statistical correlation and regression methods to investigate the relationship of these factors to the prevalence of ascariasis. The results revealed that 17 factors were identified by at least one method, while longitude, elevation of the survey sites, relative humidity and kinds of crop planted in counties were selected by two methods and soil condition, elevation of the sampled county and topographic characters were recognized by all three methods (Li Yan-Bi *et al.*, 1993). Another detailed analysis was made between environmental factors and the prevalence of all parasitic infections in Fujian Province. The results showed that elevation, latitude, longitude, rainfall and relative humidity were positively correlated to pooled prevalence while temperature and duration of sunshine were found to be negatively correlated to pooled parasite prevalence. Since *A. lumbricoides* infection consisted of about 70% of all parasitic infections detected in the province, a similar correlation of these environmental factors with ascariasis seems likely (Lin Jinxiang *et al.*, 1993).

6.3 Socio-economic Factors

With the steady improvement of socio-economic conditions since the establishment of the P.R.C., the prevalence of ascariasis in many regions has, in varying degrees, been seen to decrease on the basis of a comparison of prevalence values before and after 1986. This longitudinal comparison remains open to criticism, however, because it tends to ignore differences in sampling methods and sites, sample sizes and diagnostic techniques between the two periods. The findings, however, from many cross-sectional surveys showed that socio-economic status is a determinant of the prevalence of ascariasis in China. When all sites sampled in a given region were classified by socio-economic conditions (mainly using annual average income per capita), a relatively high prevalence was observed in those communities with relatively poor socio-economic status (Table 6). Similar results were also reported from Guiyang, Guizhou Province (Hu Shao-Yuan et al., 1991), Dalian, Liaoning Province (Zhang Bei-Xiu and Chen Feng-Yi, 1991), Jiamusi, Heilongjiang Province (Guo Shou-De et al., 1991) and Fujian Province (Lin Jinxiang et al., 1993). This conclusion is again

Table 6 Prevalence(%) of *Ascaris lumbricoides* in the population with different average annual incomes in some regions of China

Region	Prevalence (%) in population with			P	References
	Low income	Middle income	High income		
Gansu	47.1 (10609)	34.3 (13503)	17.0 (4475)	<0.005	Zhang Shou-Yi and Chen Min, 1991
Guizhou	73.7 (2330)	73.6 (13524)	70.0 (25367)	<0.01	Chen Zhaoyi et al., 1994
Hunan	67.0 (11276)	55.6 (10398)	50.4 (6208)	—	Zhang Xiang-Jun et al., 1991
Shanghai	45.2 (16260)	31.8 (22062)	25.1 (9290)	<0.001	Ma Xingbao et al., 1994
Shannxi	44.3 (28196)[a]		38.6 (11158)	<0.01	Zhang Guangyuan et al., 1994
Zhejiang	62.97 (18760)	61.91 (17299)	55.46 (19232)	<0.05	Lei Changqiu et al., 1993
	70.13 (5828)	62.55 (4032)	56.54 (4988)	<0.05	Huang Xuemin et al., 1993
Yunnan	67.0 (12283)	57.0 (40778)[b]		<0.005	Yang Jialun et al., 1994

[a] Middle and low incomes.
[b] Middle and high incomes.

based on the assumption that *A. lumbricoides* was the most prevalent infection in pooled data for all parasite infections in these regions.

Few analyses were made to examine any relationship between prevalence and the quality and use of water supplies. The summary report of Guizhou Province showed that the prevalence was found to be the lowest in communities where tap water was available, compared with other places where only river or other water sources (wells or ponds) were available (Chen Zhaoyi *et al.*, 1994).

The occupations of those people sampled was also found be related to the prevalence of ascariasis. Higher prevalence was usually observed in children of preschool age, students attending primary and secondary schools and farmers. Some prevalence values observed in populations with different occupations are listed in Table 7.

The types of crop planted, the method of treating and applying fertilizer, the extent of education and the personal hygiene of subjects were also revealed to be important subsidiary factors correlated with the prevalence of ascariasis. In Guangdong Province, a significant difference in the prevalence of ascariasis was found between people with and without the habitual practice of eating uncooked vegetables (49.11% versus 43.96%, $P < 0.01$) (Chen Xiqi *et al.*, 1993). In rural areas of Jiamusi, Heilongjiang province, where ascariasis was found to be the most prevalent of all intestinal parasitic infections, the overall prevalence of intestinal parasite infections was observed to be higher in residents with low family income, inferior educational level and the habitual practice of drinking unboiled water and/or eating uncooked vegetables ($P < 0.01$). In addition, a significantly higher prevalence of intestinal parasite infections was found in people living where toilets were located very close to their residences, where night soil was not properly treated before use as a fertilizer, and where domestic animals were not reared in pens ($P < 0.01$) (Guo Shou-De *et al.*, 1991). In Anshan Prefecture, Liaoning Province, different prevalences were observed between three areas where different crops were planted. The highest (60.5%) was found where vegetables were the main product and the lowest (45%) where grain was the main crop (Yu Hongbao *et al.*, 1991). However, a survey in Shanghai showed different results in that the prevalence in a vegetable-planted area was much lower than in grain- and cotton-planted areas (20.5% versus 35.4%, $P < 0.01$, n = 47612) (Ma Xingbao *et al.*, 1994). In Guangdong Province, the prevalence of ascariasis, and that of other soil-transmitted helminthiases, was found to be significantly lower in areas where dry land crops were main products compared with the prevalence in other areas ($P < 0.01$) (Chen Xiqi *et al.*, 1993). In the summary reports of surveys in Guizhou and Yunnan Provinces, prevalences were stratified by different areas where different types of fertilizer (organic, inorganic and mixed) were used. People using inorganic fertilizer

Table 7 Prevalence (%) of *Ascaris lumbricoides* in people with different occupations in some regions of China

Regions	School student	Farmer	Preschool children	Fisherman	Worker	Teacher	Administrator	Other[a]	References
Guangxi (50976)[b]	83.12 (11316)	65.26 (28821)	63.03 (8888)	[d]	46.50 (1028)[d]	46.06 (165)[d]	53.58 (292)[d]	55.40 (213)[d]	Huang Jian et al., 1995
Guangdong (61517)	50.4	47.3							Chen Xiqi et al., 1993
Hainan (7958)	68.9 (2251)	56.2 (3447)	66.8 (1493)	60.3 (209)	43.2 (111)	30.2 (43)[c]		71.3 (237)	Xu Fengshui et al., 1994
Hebei (64880)	39.85 (13926)	29.07 (38178)	35.15 (9891)	46.77 (248)	18.79 (1941)	25.69 (144)	6.81 (896)	18.87 (106)	Li Yan-Bi et al., 1993
Qinghai (16079)	47.5 (3485)	36.4 (6101)	40.0 (2127)	—	22.6 (677)	17.2 (134)	15.6 (435)	27.4 (1311)	Han Xiumin et al., 1994
Shannxi (53324)	49.8 (12090)	38 (11517)	44.5 (9454)	—	21.1 (502)	36.5 (63)	21.7 (350)	41.6 (517)	Zhang Guangyuan et al., 1994
Yunnan (60865)	65.8	59.3	54.6	[e]	[e]	[e]	[e]	[e]	Yang Jia-Lun et al., 1992
Zhejiang (55291)	66.82	60.32	58.72	4.48	41.43	36.71	35.04	25.77	Lei Changqiu et al., 1993

[a] People without permanent occupation.
[b] Sample size.
[c] Teachers and administrators
[d] Below 47.3%
[e] Below 54.6%

were found to have the lowest prevalence while those using organic fertilizer (mainly human and domestic animal faeces) showed the highest prevalences (53.2% versus 75.9%, $P < 0.001$; 44.1% versus 71.8%, $P < 0.01$, respectively) (Chen Zhaoyi et al., 1994; Yang Jia-Lun et al., 1994). In Ningxia AR, the low prevalence of ascariasis was attributed to climate and the manner of applying fertilizer. In the area investigated, night soil was applied after being properly treated by drying it under sunlight or composting it to ensure fermentation (Ma Cheng-Ji et al., 1991). Although night soil has been used as a fertilizer for a long time in China, it is not generally common for farmers to use fresh night soil. The problem is, however, that the treatment for night soil before its use as a fertilizer is not always as thorough as it should be to kill helminth eggs and fresh stools may also be mixed in from time to time.

6.4 Polyparasitism

One difficulty encountered in preparing this evaluation of the current state of ascariasis in China was the problem of separating information about ascariasis from data relevant to all the parasitic infections found in a given region. Polyparasitism is probably the normal condition for many people in China (Table 8). It was reported that a total of 56 species of parasite were detected during the period of the first nation-wide survey on human parasite distribution in China and that *Entamoeba histolytica, Giardia lamblia, A. lumbricoides, Trichuris trichiura* and *Enterobius vermicularis* were distributed nation-wide (Yu Senhai et al., 1994). In practice, people infected with at least two species of parasite usually comprised one-third to two-thirds of the total infections in these regions. Some individuals harboured eight or nine species of parasite (Yang Jia-Lun et al., 1992, Chen Sili et al., 1993; Chen Xiqi et al., 1993; Lin Jinxiang et al., 1993). The host–parasite interaction in this situation is inevitably more complicated than that occurring during monoparasitism. Hosts suffering from one parasitic infection might become more sensitive to another due to some impairment of immune system compared with those hosts without parasitic infections. Conversely, the presence in the host of one parasite species might influence the invasion of another species because of competition for limited living space and nutrients, especially for those parasite species living in the same or adjacent locations within the body of the host (see Crompton, 1973). It is possible that the prevalence of ascariasis might be influenced by the presence of other infections especially those occupying the alimentary tract.

Table 8 Polyparasitism in the population of some regions of China

Regions	No. of species	Proportion (%) of infection with				References
		1 species	2 species	3 species	>3 species	
Anhui	30[a]	50.83	34.03	12.41	2.73	Xu Funiu et al., 1991
Fujian	41	32.7	35.3	21.02	10.98	Lin Jinxiang et al., 1993
Guangdong	64	48.31	34.96	13.42	3.31	Chen Xiqi et al., 1993
Guizhou	21[a]	48.9	34.8	13.8	2.53	Chen Zhaoyi et al., 1994
Hainan	30	20.9	34	30.6	14.4	Xu Fengshui et al., 1994
Hebei	22[a]	81.71	16.41	1.62	0.26	Li Yan-Bi et al., 1993
Heilongjiang	17	93.3	6.5	0.3	0	Kang Qingde et al., 1994
Hubei	31	60.1	29.28	7.64	2.98	Chen Sili et al., 1993
Hunan	26	55.0	33.8	9.5	1.7	Zhang Xiangjun et al., 1994
Jiangsu	22	56.7	32.0	9.6	1.6	Hang Panyi et al., 1994
Nei Moggol	11[a]	88.7	11.0	0.3	1/7298	Zhang Bin et al., 1994
Shandong	19[a]	62.9	29.7	6.83	0.57	Zhu Yu-Guang et al., 1991
Shannxi	17[a]	82.3	16.1	1.5	0.1	Zhang Guangyuan et al., 1994
Xinjiang	29	64.32	23.84	8.82	2.94	Li Baoshan et al., 1993
Yunnan	25[a]	47.0	33.7	14.4	4.5	Yang Jia-Lun et al., 1992
Zhejiang	26	38.05	37.03	18.31	6.61	Lei Changqiu et al., 1993

[a] Only parasite species in alimentary tract of humans.

6.5 Pigs and Ascariasis in China

In rural areas of China, pigs are invariably reared as a household sideline production and people and pigs live in close proximity. Because both human and pig hosts can be infected with *Ascaris*, pigs might play an important role in the epidemiology of human *Ascaris* infection in these areas. Although this question has drawn attention since the early 1980s

(Kofies and Dipeolu, 1983), it is only in recent years that the question of cross infection has been investigated clearly with reference to communities in Guatemala where it was demonstrated that *Ascaris* from humans and those from pigs were involved in essentially separate transmission cycles (Anderson et al., 1993). In their study in rural villages in Guatemala, Anderson and colleagues investigated genetic variation in sympatric populations of *Ascaris* infecting children and pigs. Overall, 34% of the children and 32% of the pigs were infected. By means of enzyme electrophoresis and mitochondrial (mt) DNA analysis, it was concluded that gene flow between the population of worms in humans and the population in pigs was extremely limited. The results strongly suggested that two distinct, host-specific cycles of transmission occurred in this location in Latin America. Anderson et al. (1993) considered that people must ingest eggs originating from worms in pigs and *vice versa* and that some form of mating barrier within the host was probably responsible for the lack of gene flow. That proposition seems to merit further study, probably by means of experimental infection in *Ascaris*-free pigs bred for the purpose. Anderson et al. (1993) stressed that the situation they studied in Guatemala might not apply in other parts of the world. However, in areas such as North America and Western Europe where ascariasis is now thought to be nonendemic, cross-infections from pigs were not only inferred by clinical case histories, but also proved by molecular evidence (Anderson, 1995). Worms retrieved from eight infected people, five of whom had regular contact with pigs, were used to obtain preparations of genomic DNA. Molecular analysis of amplifications from this material showed that the *Ascaris* infections had originated from pigs; genomic DNA derived from *Ascaris* passed by humans in Guatemala (see Anderson et al., 1993) was significantly different. Anderson stated that cross-infection (that is *Ascaris* from pigs being acquired by humans) may occur naturally. In rural China, there would be plenty of opportunity for this to occur. The question of whether *A. lumbricoides* and *A. suum* are the same or separate species has still to be resolved (see Crompton, 1989).

However, the question of cross infectivity of *Ascaris* spp. between people and pigs in communities in China has not yet been investigated. According to our epidemiological study of ascariasis in a rural area of Jiangxi Province, China, the prevalence of ascariasis of pigs and humans in the study communities was about 60% and 70%, respectively (Peng Weidong et al., 1996). It appears that the two prevalence values are not much different considering there might be an underestimate of prevalence in pigs compared with that in humans because of the problems in preparing and examining stool smears from pigs. In both communities, mass chemotherapy was given to all villagers once, but to pigs, deworming was given only in one village at about two-month intervals. The reinfection rates of humans in both villages were measured at about two-month intervals following chemotherapy. The results did not reveal much difference between reinfection rates of people in the two villages. At about 2, 4 and 6 months of post treatment, the reinfection rates of humans were about 31% versus 33%, 56% versus 57% and 63% versus 59% in the communities, respectively.

Associated with the high prevalence of ascariasis in this area was a major contamination of soil with *Ascaris* eggs in the two villages. About 80% of samples taken from the sitting rooms of the houses and about 90% of samples taken from front yards and vegetable gardens contained *Ascaris* eggs. From June to September in the region, more than 50% of eggs developed into infective stage within two weeks. In these circumstances, the infection of *Ascaris* in the communities might be in a state of saturated infection. Unfortunately for health workers studying transmission or monitoring the progress of a control programme for ascariasis, there is no simple, reliable technology for identifying either the species or source of *Ascaris* whose eggs are found in soil. Peng Weidong *et al.* (1996) had no means of knowing from where the *Ascaris* eggs, found in the soil sampled from their study communities, had originated. Thus, their intervention with anthelminthic chemotherapy might not have been powerful enough to cause a significant decrease in transmission and reinfection. Further investigations, especially using molecular and genetic methods, are urgently needed to elucidate the role of pigs in transmission of ascariasis in human communities of rural areas of China. A useful tool for studying the epidemiology of ascariasis would be the development of a simple procedure for identifying *Ascaris* eggs retrieved from soil.

7. INTENSITY OF ASCARIASIS IN CHINA

Cross-sectional surveys from locations in China with a variety of climate and topography as well as diversity of socio-economic conditions showed that most cases with ascariasis fell into the categories of light and moderate infections according to the numbers of eggs per gram (epg) detected in stools (WHO, 1987). This trend, displayed in Table 9, agrees with the well-known aggregated distribution pattern of the numbers of helminth per host, where the majority of individuals harbouring a few or no worms while the minority of the host harbouring a great part of the worm population (Anderson, 1986).

A longitudinal investigation in Jiangxi Province also showed the same trend (Peng Weidong *et al.*, 1996). As measured by egg counts (epg), about 80% of individuals with ascariasis may be classified as having light and

Table 9 Constitution of intensity in populations with infections of *Ascaris lumbricoides* in some regions of China

Regions	No of infections	Prevalence (%)	Proportion (%) of epg counts			References
			epg^a <5000	epg <50000	epg >50000	
Anhui	—	74.0	56.7	38.3	0	Yu et al., 1989
Beijing	4215	30.7	100.0	0	0	Mao Shou-Bai, 1992
Fujian	5505	57.11	66.27	31.77	1.96	Lin Jinxiang et al., 1993
Guangdong	28570	46.44	82.96	16.26	0.78	Chen Xiqi et al., 1993
	10127^b	50.9	77.7	20.8	1.5	Liu Meizhen et al., 1993
Hainan	183	34.9	65.5	31.2	3.2	Xu Feng-Hui et al., 1991
	858	61.8	41.1	47.0	11.9	Xu Fengshui et al., 1994
	1029	76.4	44.8	47.6	7.6	Yu et al., 1989
Heilongjiang	5388	10.3	94.4	5.1	0.4	Kang Qingde et al., 1994
Jiangsu	2831	44.4	61.5	33.2	5.3	Zheng Chang-Qian et al., 1991
Jiangxi	5458	71.2	68.9	19.1	12.0	Wang Weizhou et al., 1994
Ningxia	2460	30.8	78.4	21.6	0	Ma Cheng-Ji et al., 1991
Qinghai	5452	33.9	78.0	20.5	1.6	Han Xiumin et al., 1994
Shaanxi	22182	41.6	83.4	16.3	0.3	Zhang Guangyuan et al., 1994
Shanghai	37	7.3	100.0	0	0	Ma Xing-Bao et al., 1991
	100	18.6	50.0	40.0	10.0	Ma Xing-Bao et al., 1991
	105	20.9	77.8	22.2	0	Ma Xing-Bao et al., 1991
	206	44.0	55.6	33.3	11.1	Ma Xing-Bao et al., 1991
	234	60.8	50.0	40.9	9.1	Ma Xing-Bao et al., 1991
Xizang	528	5.99	80.11	19.13	0.76	Guo Men-Min et al., 1994
	—	43.7	73.3	26.7	0	Yu, Jiang & Xu, 1989
	201	82.1	65.4	32.0	2.6	Yu, Jiang & Xu, 1989
Yunnan	561	49.2	85.1	14.9	0	Yu, Jiang & Xu, 1989
	58	71.2	56.9	43.1	0	Yan Jia-Lun et al., 1992
	77	92.1	41.6	54.5	3.9	Yan Jia-Lun et al., 1992

| Zhejiang | 5163 | 60.03 | 65.72 | 32.27 | 2.01 | Lei Changqiu et al., 1993 |
| | 1391[b] | 63.5 | 51.26 | 43.64 | 5.1 | Huang Xuemin et al., 1993 |

[a] epg = egg counts per gram
[b] population below 15 years old

moderate infections, with about 20% of the cases having a heavier infection. Furthermore, this trend continued to be observed during the study year in results from six cross-sectional surveys in the same community. However, the trend was observed with some fluctuation of the constituent ratios of different intensity levels (Table 10). The fluctuation observed might be caused by a complex combination of factors related to exposure to infection, such as environment and host behaviour, to factors relevant to host resistance to ascariasis and also to factors involving population processes of the parasite such as sex ratio of adults, density-dependent constraints on egg production and the development and survival of eggs in the surroundings.

There is still little detailed information available from China about age-related intensity, measured either by egg or worm counting. Based on our research data (Peng Weidong et al., 1996), the highest intensity of ascariasis was almost always found in the age group between 5 and 9 years old throughout the six periods of cross-sectional survey. The second-highest intensity usually occurred in children below 5 years old, then followed by older age groups, but with some variation. The difference in intensity among age groups was significant ($P < 0.001$). This distribution pattern of intensity (epg) in relation to host-age structure matched the pattern of intensity measured by worm counting after expulsion chemotherapy with pyrantal pamoate in the villages. The highest mean intensity was found to be about 14 worms in the 6–10 years old group. The average worm burden of ascariasis cases of the area was about eight ranging about four to 14 in different age groups and from one to 82 in the individual cases (Peng Weidong et al., unpublished data). The relationship between age and intensity of ascariasis in the area follows the pattern demonstrated in Myanmar and other regions rather than the pattern reported from Iran (see Crompton, 1989). In the case of the frequency distribution of numbers of worms per host, about 70% of infected villagers who accepted treatment harboured only 25% of worms expelled, while 30% of the treated cases harboured 75% of the worm population (Peng Weidong et al., unpublished data). Generally, the intensity of ascariasis in male and female villagers did not show a significant difference whether measured by egg or worm counts (Peng Weidong et al., 1996; Peng Weidong et al., unpublished data).

Table 10 Fluctuation in constitution of intensity of ascariasis cases in Manhu villages of Xinjian county, Jiangxi Province, China (Peng Weidong *et al.*, unpublished results)

Villages	June 1993	August 1993	October 1993	January 1994	April 1994	June 1994
Laochi						
Light infection	54(31.76)[a]	73(43.45)	76(41.53)	59(38.82)	41(35.04)	58(35.58)
Moderate infection	84(49.41)	71(42.26)	85(46.45)	63(41.45)	62(52.99)	89(54.60)
Heavy infection	32(18.82)	24(14.29)	22(12.02)	30(19.73)	14(11.97)	16(9.82)
Total	170(100)	168(100)	183(100)	152(100)	117(100)	163(100)
Panchi						
Light infection	71(38.38)	117(66.86)	90(47.37)	67(44.37)	67(52.34)	53(30.64)
Moderate infection	79(42.70)	53(30.29)	70(36.84)	67(44.37)	49(38.28)	90(52.02)
Heavy infection	35(18.92)	5(2.85)	30(15.79)	17(11.26)	12(9.38)	30(17.34)
Total	185(100)	175(100)	190(100)	151(100)	128(100)	173(100)

[a] The number of ascariasis cases followed by the constituent ratio (%)

8. OBSERVATIONS ON THE ORIGIN OF THE HUMAN–*ASCARIS* ASSOCIATION

Recently Peng Weidong *et al.* (1995) tentatively suggested that the now well-established host-parasite relationship between humans and *A. lumbricoides*, or one version of it, might have had its origins in relatively recent times in China. There are several strands of argument to this proposal, some being somewhat stronger and less fragile than others.

China, despite numerous reorganizations during the history of its people (Spence, 1990), has experienced a sustained civilization for several thousand years. Although there is great diversity within China, there is a single written language, mutually intelligible to every literate person, and a diet in which rice and pork predominate (Reader, 1988). The rural portion of China's population, roughly 928 million people, spend considerable time in close proximity to pigs on land that has been farmed by traditional methods since Neolithic times (An Zhimin, 1989). We may ask if there are any grounds for thinking that humans have acquired what we now recognize as *A. lumbricoides* from pigs? If this question is to stand examination, the host–parasite relationship we know today would have arisen during the last few thousand years. Can that proposal be supported, at least to a point that would justify further investigation?

The domestication of animals began with the wolf about 12 000 years ago and that of the wild boar *Sus scrofa* probably began about 7000 years ago in western Asia and in other places after that (Clutton-Brock, 1987). According to studies reviewed by Davis (1987), all domestic pigs examined to date possess 38 chromosomes; however, wild boar from Western Europe have 36 chromosomes while those in Asia and Far East have 38. Assuming that the modern distribution of boar karyotypes was similar in antiquity, the modern domestic pig probably originated somewhere between the Balkans and the Far East. Davis (1987) states that it is likely that the pig was independently domesticated in China, but that as yet we have little evidence from that part of the world. There is, however, archaeological evidence to suggest that in China the pig may be as ancient a domesticated animal as the dog (An Zhimin, 1989).

Since humans are primates, with close affinities to chimpanzees and gorillas (Diamond, 1992; Wood, 1996), it might at first be expected that there would be many infections in common between the three types of primate. With regard to helminth infections, however, qualitative information suggests that humans share surprisingly few helminth infections with other primate species (Coombs and Crompton, 1991). After an investigation of reports of helminth infections in nonhuman primates, Orihel (1970) considered that an infection such as *A. lumbricoides* in a nonhuman primate host could be referred to as an anthroponosis or disease of humans transmitted to animals. In an analysis of the social history of humans and diseases, Karlen (1995) wrote that during recent human evolution we have provided new ecological niches for microbes by tilling fields and domesticating animals. Karlen (1995) cites research to indicate that humans now share about 300 infections with species of domesticated animal and stressed that our helminths are largely acquired from livestock and pets. Apparently, we share 42 diseases with domesticated pigs (see Karlen, 1995). This interaction between infectivity and susceptibility perhaps began to occur in recent times following events when humans largely abandoned the life of being hunter–gatherers and became settled in farming communities.

For some time, scientists have recognized that pigs are rather like people in some respects. Modern pigs have anatomical, physiological, immunological, metabolic and nutritional similarities with humans (Pond and Haupt, 1978). Not surprisingly, more experimental work on human parasitic infections has been proposed with pigs seen as realistic experimental hosts. Perhaps in China, humans met what we now know as *A. lumbricoides*, or its ancestor, in the wild pigs which came to scavenge around their settlements.

Rapid speciation, driven by the speed of human technological development and the associated selective breeding of domesticated animals, may have given rise to the nematodes commonly described as *A. lumbricoides* and *A. suum*.

The proposition that the present human–*Ascaris* association originated in the region of the world now known as China could be investigated further by making a comparative survey of the current number of infections found in pigs and humans in the main centres of human population. We should expect strong correlations between the range and frequency of shared infections in China; this would not be in any way conclusive, but it would help to support the view that human–*Ascaris* associations could have started several thousand years ago in China. Recently, Wood (1996) warned that the more we know about human evolutionary history, the more complex it becomes. Exactly the same cautious approach is needed for the consideration of the evolutionary history of human–parasite relationships. The invasion of *Homo sapiens* by *Ascaris* may well have taken place on several occasions in various locations, but the prevailing infections between humans, *Ascaris* and pigs in China remain a tantalizing topic for research.

9. TRADITIONAL CHINESE MEDICINE FOR THE TREATMENT OF ASCARIASIS

Chemotherapy dependent on the use of anthelmintic drugs developed, tested and distributed by the research-based pharmaceutical industry has become the strategic foundation of the programmes for the control of ascariasis and other soil-transmitted helminthiases (WHO, 1987, 1995 a,b). Generic forms of these drugs have also become widely available as patents expired. There are warning signs to suggest that such drugs may begin to lose their effectiveness before long and may not be replaced. Increasing numbers of reports of the development of populations of nematode parasites resistant to modern anthelminthic drugs are now being published. Drug resistance is defined as the genetically transmitted loss of sensitivity in a helminth population which was previously sensitive to the appropriate therapeutic dose of the drug in question. Strongyle nematode parasites of grazing livestock and horses are now well known for their drug resistance (see Prichard *et al.*, 1980; Wescott, 1986). Although there is no unequivocally confirmed report of the presence of drug-resistant populations of *A. lumbricoides* in humans, the threat of its emergence must be taken seriously (WHO, 1996). Knowledge of the development of pesticide resistance in insect populations (Comins, 1977), if applied to the development drug resistance in parasitic nematodes, suggests that up to 100 or so generations are required for the problem to emerge. Drug resistance may be delayed by careful management of chemotherapeutic regimes, but can we expect to respond effectively if the problem is evident? The approach to drug resistance by the pharmaceutical industry to such nematodes as *A. lumbricoides* may be to withdraw from that aspect of their business because of the financial risk involved. The countries where ascariasis is endemic can least afford to buy drugs whose sales must cover the costs of development and realize an acceptable profit for the manufacture. Furthermore the research-based pharmaceutical industry has to contend with the losses that accrue as a result of the production and world-wide sales of counterfeit drugs (WHO, 1992). The counterfeiting problem does not encourage legitimate companies to invest in research and development of drugs for a needy and impoverished market.

Traditional Chinese medicine, in contrast, offers considerable hope for the millions of people needing treatment for ascariasis, especially where the traditional preparations are integrated with the practices of western medicine for the management of infected patients. A rich experience has been acquired in China for over 3000 years in the use of about 5000 herbal preparations (Wang Pei, 1983). How this pharmacopoeia of traditional drugs is to be used is not easily understood by western-trained health workers and requires insight into the knowledge compiled in the *Chinese Internal Classics,* a book in which the principles of traditional medicine are enunciated (Wang Pei, 1983). For the treatment of ascariasis, patients are prescribed decoctions of fruits and seeds with extracts prepared from *Aconitum carmichaeli, Angelica sinensis, Ascarum heterotropoides, Cinnamomum cassia, Coptis chinensis, Panax ginseng, Phellodendron amurense, Prunus mume, Zanthoxylum bungeanum* and *Zingiber officinale* being commonly recommended (Li Peisheng, 1987).

There is plenty of evidence to attest to the effectiveness of Chinese traditional medicine for the treatment of ascariasis. For example, a 99.5% cure rate was reported by Wang Zhixing (1982) for 778 cases of biliary ascariasis using one of the traditional decoctions and a cure rate of 99% was observed by Li Dengyu (1991) when another traditional decoction was applied. Use of these well-tried remedies in China and elsewhere is to be encouraged, given the range of problems that are beginning to threaten the continued use of synthetic anthelminthic drugs.

10. CONCLUSIONS

Ascariasis is considered to present a significant public health problem to the population in China. Infection with *Ascaris lumbricoides* exists in all administrative regions and is most prevalent in the south-east of the country. More than 500 million people or nearly half of the total population of the country are estimated to harbour this infection, with most of them living in rural areas. The infection is unevenly distributed both within the country and within regions. Many factors can be correlated with the prevalence of ascariasis including those of demographic, environmental and socio-economic origin. Overall, the epidemiology of ascariasis in China is similar to that described for the infection in most parts of the world. In addition, polyparasitism and ascariasis in pigs might be correlated with the prevalence of ascariasis in humans. There is a need to gain much more information about the intensity of the infection in humans in China and to investigate the role of pigs in the transmission or otherwise of the infection. A major contribution to understanding the possibility of cross-infection between pigs and humans would be the development of a reliable procedure for distinguishing eggs passed by humans from those passed by pigs. The enduring question of the identities of *A. lumbricoides* and *A. suum* also needs to be resolved. Planning for ascariasis control in China will need to take into account the full epidemiology, including intensity, since control strategies should aim to reduce intensity rather than eradicate the infection. Education will eventually follow when modern sanitation is installed throughout the country.

ACKNOWLEDGEMENTS

This work was made possible through financial support of the National Natural Science Foundation of China (NNSFC), the International Bureau of the NNSFC and of the WHO Centre for Soil-transmitted Helminthiases, Institute of Biomedical and Life Sciences, University of Glasgow, Scotland, UK. We thank Jane Grant for the preparation of Figure 1, K. Purser for technical help and Dr T. Anderson, Dr A. Morrison and Professor P. J. Whitefield for helpful suggestions.

REFERENCES*

An Zhimin (1989) Prehistoric agriculture in China. In: *Foraging and Farming* (D.R. Harris and G.C. Hillman, eds). pp. 643–649. London: Unwin Hyman.
Anderson, R.M. (1986) The population dynamics and epidemiology of intestinal nematode infections. *Transactions of the Royal Society of Tropical Medicine and Hygiene* **80**, 686–696.
Anderson, T.J.C. (1995) *Ascaris* infections in humans from North America: molecular evidence for cross-infection. *Parasitology* **110**, 215–219.
Anderson, T.J.C., Romero-Abal, M.E. and Jaenike, J. (1993). Genetic structure and epidemiology of *Asacaris* populations: patterns of host affiliation in Guatemala. *Parasitology* **107**, 319–334.
Atlas of Provinces of China (1992). China Map Publishing House (in Chinese).
Beazley, M. (1982). *The Atlas of Mankind*. London: Mitchell Beazley International.
Bundy, D.A.P. and Guyatt, H. 1996. Schools for health: focus on health, education and the school-age child. *Parasitology Today* **12**(Insert), 1–16.
Bunge, F.M. and Shinn, R.P. (eds.). (1981). *China*. 3rd edn. Washington, DC: United States Government.
Chan, M.S., Medley, G.F., Jamison, D. and Bundy, D.A.P. (1994). The evaluation of potential global morbidity attributable to intestinal nematode infections. *Parasitology* **9**, 373–387.
Chen, E.R. and Hsieh, H.C. (1989). Control of soil-transmitted nematode in Taiwan. In: *Collected Papers on the Control of Soil-transmitted Helminthiases*, Vol. IV, pp. 131–146. Tokyo: Asian Parasite Control Organization.
Chen Jizhang *et al.* (1994). Investigation and analysis of intestinal parasite infection in rural children in Hainan Province. *Chinese Journal of Parasitology and Parasitic Diseases, Special Issue for the Nationwide Survey of the Distribution of Human Parasites* **12**, 157–159 (in Chinese).
Chen Sili *et al.* (1993a). An analysis on intestinal parasites of children in Hubei province. *Chinese Journal of Zoonoses* **9** (Suppl.), 48–49 (in Chinese).
Chen Sili *et al.* (1993b). Survey of distribution of human parasites in Hubei province. *Chinese Journal of Zoonoses* **9**, (Suppl.), 8–12 (in Chinese).
Chen Xiqi *et al.* (1993). Preliminary report on distribution investigation of human parasites in Guangdong province. *Chinese Journal of Zoonoses* **9** (Suppl.), 18–21 (in Chinese).
Chen Youzhu *et al.* (1991). Survey of human parasite infections in the suburbs of Zhangzhou city and Longhai county. *Chinese Journal of Parasitology and Parasitic Diseases, Special Issue for the Nationwide Survey of the Distribution of Human Parasites*, 52–55 (in Chinese).
Chen Youzhu *et al.* (1993). Survey on the species of human intestinal helminth and their distribution in Fujian province. *Chinese Journal of Zoonoses* **9** (Suppl.), 60–64 (in Chinese).
Chen Zhaoyi *et al.* (1994). A survey of human parasite distribution in Guizhou Province. *Chinese Journal of Parasitology and Parasitic Diseases, Special Issue for the Nationwide Survey of the Distribution of Human Parasites*, **12**, 37–43 (in Chinese).

---

* Some papers with only Chinese titles were translated into English and the names of authors were transliterated into the Chinese phonetic alphabet.

Clutton-Brock, J. (1987). *A Natural History of Domesticated Animals.* Cambridge: Cambridge University Press.
Comins, H.N. (1977). The development of insecticide resistance in the presence of migration. *Journal of Theoretical Biology* **64**, 177–197.
Coombs, I. and Crompton, D.W.T. (1991). *A Guide to Human Helminths.* London and Philadelphia: Taylor & Francis Ltd.
Crompton, D.W.T. (1973). The sites occupied by some parasitic helminths in the alimentary tract of vertebrates. *Biological Reviews of the Cambridge Philosophical Society* **48**, 27–83.
Crompton, D.W.T. (1989). Prevalence of ascariasis. In: *Ascariasis and its Prevention and Control* (D.W.T. Crompton, M.C. Nesheim and Z.S. Pawlowski eds), pp. 45–69. London: Taylor & Francis Ltd.
Davis, S.J.M. (1987). *The Archaeology of Animals.* New Haven: Yale University Press.
Din Jian-Guo (1987). Supervision report on diseases of students in primary and middle schools of Yinchuan city, 1986. *Journal of Medicine of Ningxia* **9**, 77–78. (in Chinese).
Diamond, J. (1992). *The Third Chimpanzee.* New York: Harper Collins Publishers.
Dong Qiang *et al.* (1991). A survey of parasitic infection in Yining county, Xinjiang. *Chinese Journal of Parasitology and Parasitic Diseases, Special Issue for the Nationwide Survey of the Distribution of Human Parasites* 158. (in Chinese)
Fang Jixin *et al.* (1991). A survey of parasitic infections in Gangzhou prefecture. Jiangxi province. *Chinese Journal of Parasitology and Parasitic Diseases, special issue For the Nationwide Survey of the Distribution of Human Parasites* 140 (in Chinese).
Feng Bu-Wei *et al.* (1991). Survey of human intestinal parasite infections in Jinzhong prefecture, Shanxi province. *Chinese Journal of Parasitology and Parasitic Diseases, Special Issue for the Nationwide Survey of the Distribution of Human Parasites* 16–19 (in Chinese).
Feng Shixian *et al.* (1994). A report of human intestinal parasites in eastern plain of Henan Province. *Chinese Journal of Parasitology and Parasitic Diseases, special issue for the nationwide survey of the distribution of human parasites* **12**, 170–173 (in Chinese).
Fong Zhishan *et al.* (1989). Survey of *Ascaris* infection of primary school children in Qiaoxi district, Zhangjiakuo city. *Journal of Zhangjiakuo Medical College* **6**, 12–13 (in Chinese).
Guan Qin-Li, *et al.* (1991). Distribution of human parasite in four counties of Baoji prefecture of Shaanxi province. *Chinese Journal of Parasitology and Parasitic Diseases, special issue for the Nationwide Survey of the Distribution of Human Parasites* 85–88 (in Chinese).
Guo Wen-Min, *et al.* (1991). Distribution of human parasites in four counties of Linzhi prefecture, Lhasa, Xizang. *Chinese Journal of Parasitology and Parasitic Diseases, Special Issue for the Nationwide Survey of the Distribution of Human Parasites* 107–109 (in Chinese).
Guo Wen-Min *et al* (1994). Survey on distribution of human parasite in Xizang. *Chinese Journal of Parasitic Disease Control* **7**, 131–132 (in Chinese).
Guo Shou-De *et al.* (1991). Relationship between intestinal parasite infection and socio-economic factors in rural population of Jiamusi, Heilongjiang. *Chinese Journal of Parasitology and Parasitic Diseases, special issue for the Nationwide Survey of the Distribution of Human Parasites* 105–107 (in Chinese).

Guyatt, H.L. and Bundy, D.A.P. (1993). Estimation of intestinal nematode prevalence: influence of parasite mating patterns. *Parasitology* **107**, 99–106.
Han Jia-Jun *et al.* (1991). Human intestinal parasite infections and their family aggregation in Sichuan. *Chinese Journal of Parasitology and Parasitic Diseases, Special Issue for the Nationwide Survey of the Distribution of Human Parasites* 76–79 (in Chinese).
Han Jiajun *et al.* (1994). Epidemiological aspects of human parasitoses in Sichuan Province. *Chinese Journal of Parasitology and Parasitic Diseases, special issue for the nationwide survey of the distribution of human parasites* **12**, 8–16 (in Chinese).
Han Xiumin *et al.* (1994). Situation of human intestinal nematode infection in Qinghai Province. *Chinese Journal of Parasitology and Parasitic Diseases, Special Issue for the Nationwide Survey of the Distribution of Human Parasites* **12**, 141–144 (in Chinese).
Hang Panyu *et al.* (1994). Investigation on parasitic infections in people of Jiangsu Province. *Chinese Journal of Parasitology and Parasitic Diseases, Special Issue for the Nationwide Survey of the Distribution of Human Parasites* **12**, 93–96 (in Chinese).
Hu Shao-Yuan *et al.* (1991). Investigation of human intestinal parasitization in Guiyang city, Guizhou province. *Chinese Journal of Parasitology and Parasitic Diseases, Special Issue for the Nationwide Survey of the Distribution of Human Parasites* 72–75 (in Chinese).
Hu Wanchueng *et. al.* (1991). Analysis on relationship between ascariasis and natural factors. *Journal of Research and Control of Parasitic Diseases* **20**, 153–154 (in Chinese).
Huang Jian *et al.* (1994). Intestinal nematode infection in Children and juveniles in Guangxi. *Chinese Journal of Parasitology and Parasitic Diseases, Special Issue for the Nationwide Survey of the Distribution of Human Parasites* **12**, 160–162 (in Chinese).
Huang Jian *et. al.* (1995). A survey of human intestinal parasite infection in Guangxi Zhuang Autonomous Region. *Guangxi Preventive Medicine* **1**, 25–28 (in Chinese).
Huang Xuemin *et al.* (1993). An analysis on intestinal nematoda of children in Zhejiang province. *Chinese Journal of Zoonoses,* **9**, (Supplement), 50–53 (in Chinese).
Human parasitic distribution survey team of Henan. 1991. Survey of human parasite distribution in Henan province. *Chinese Journal of Parasitology and Parasitic Diseases, Special Issue for the Nationwide Survey on the Distribution of Human Parasites* 62–65 (in Chinese).
Jiangxi Office for Schistosomiasis and Endemic Disease Control. 1991. A survey of human parasite infections in Jiangxi province. *Chinese Journal of Parasitology and Parasitic Diseases, Special iIssue for the Nationwide Survey of the Distribution of Human Parasites* 40–42 (in Chinese).
Kang Qingde *et al.* (1994). An evaluation on survey of human parasite distribution in Helongjiang Province. *Chinese Journal of Parasitology and Parasitic Diseases, Special Issue for the Nationwide Survey of the Distribution of Human Parasites* **12**, 61–64 (in Chinese).
Karlen, A. (1995). *Plague's Progress.* London: Victor Gollancz.
Kofies, B.A.K. and Dipeolu, O.O. (1983). A study of human and porcine ascariasis in rural areas of Southwest Nigeria. *International Journal of Zoonoses* **10**, 66–70.
Lai, D. and Hsi, B.P. (1996). Soil-transmitted helminthiases in China: a spatial

statistical analysis. *Southeast Asian Journal of Tropical Medicine and Public Health* **27**, 754–759.

Lei Changqiu *et al.* (1993). Survey of human parasite infections in Zhejiang province. *Chinese Journal of Zoonoses* **9** (Supplement), 2–7 (in Chinese).

Li Baoshan *et al.* (1993). Survey of human parasite distribution in Xinjiang. *Chinese Journal of Zoonoses* **9**, (Suppl.), 13–16 (in Chinese).

Li Dengyu, (1991). 119 cases of biliary ascariasis treated with decoction of *Herba artemisiae capillaris, Rhizoma coptidis*. *Tianjin Journal of Traditional Chinese Medicine* **22**, 9–10 (in Chinese).

Li Peisheng, (1987). *Treatise on Exogenous Febrile Diseases*. Beijing: People's Health Press (in Chinese).

Li Shu-Ben *et al.* (1991). A survey of parasitic infections in Shijiazhuang region, Hebei province. *Chinese Journal of Parasitology and Parasitic Diseases, Special Issue for the Nationwide Survey of the Distribution of Human Parasites* 19–21 (in Chinese).

Li Yan-Bi *et al.* (1993). Epidemiological survey and study on human intestinal parasites in Hebei province. *Chinese Journal of Zoonoses* **9** (Suppl.), 22–26 (in Chinese).

Lin Jinxiang *et al.* (1993). Studies on the epidemiological law and distribution of human parasites in Fujian province. *Chinese Journal of Zoonoses* **9** (Supplement), 27–35 (in Chinese).

Liu Meizhen *et al.* (1993). Investigation of parasite infections of Children in Guangong province. *Chinese Journal of Zoonoses* **9** (Supplement), 75–77 (in Chinese).

Ma Cheng-Ji *et al.* (1991). Investigation of *Ascaris lumbricoides* infection in rural population of Ningxia. *Chinese Journal of Parasitology and Parasitic Diseases, Special Issue for the Nationwide Survey of the Distribution of Human Parasites* 98–100 (in Chinese).

Ma Chenji, Zhang Qiaolin and Fu Daren (1994). Human parasitic infection in children and juveniles in Ningxia. *Chinese Journal of Parasitology and Parasitic Diseases, Special Issue for the Nationwide Survey of the Distribution of Human Parasites* **12**, 82–84 (in Chinese).

Ma Wan-Hai *et al.* (1991). Investigation on the distribution of human parasites in the Yinmahe township, Jiutai county, Jilin province. *Chinese Journal of Parasitology and Parasitic Diseases, Special Issue for the Nationwide Survey of the Distribution of Human Parasites* 24–26 (in Chinese).

Ma Xing-Bao *et al.* (1991). Present status of intestinal parasite infections in Pudong new area of Shanghai. *Chinese Journal of Parasitology and Parasitic Diseases, Special Issue for the Nationwide Survey of the Distribution of Human Parasites* 9–11 (in Chinese).

Ma Xingbao *et al.* (1994). Influence of some factors on prevalence of intestinal parasites in Shanghai. *Journal of Parasitology and Parasitic Diseases, Special Issue for the Nationwide Survey of the Distribution of Human Parasites* **12**, 27–30 (in Chinese).

Mao Shou-Bai (1992). *Excerpts of papers of human parasitic diseases in China, 1949–1986*. Beijing: Public Health Publishing House (in Chinese).

Nesheim, M.C. (1989). Ascariasis and human nutrition. In: *Ascariasis and its Prevention and Control* (D.W.T. Crompton, M.C. Nesheim and Z.S. Pawlowski eds), pp. 101–107. London: Taylor & Francis Ltd.

Orihel, T.C. (1970). The helminth parasites of non-human primates and man. *Laboratory Animal Care* **20**, 395–401.

Peng Weidong, Zhou Xianmin and Crompton, D.W.T. (1995). Aspects of ascariasis in China. *Helminthologia* **32**, 97–100.
Peng Weidong *et al.* (1996). *Ascaris*, people and pigs in a rural community of Jiangxi Province, China. *Parasitology* **113**, 545–558.
Pond, W.G. and Haupt, K.A. (1978). *The Biology of Pig*. Ithaca and London: Cornell University Press.
Prichard, R.K., Hall, C.A., Kelly, J.D., Martin, I.C.A. and Donald, A.D. (1980). The problem of anthelmintic resistance in nematodes. *Australian Veterinary Journal* **56**, 239–251.
Reader, J. (1988). *Man on Earth*. New York: Harper & Row.
Shanxi Sanitary and Anti-epidemic Station (1991). Summary of the survey on distribution of human intestinal parasite in Shanxi province. *Chinese Journal of Parasitology and Parasitic Diseases, Special Issue for the Nationwide Survey on the Distribution of Human Parasites* 15–17 (in Chinese).
Spence, J.D. (1990). *The Search for Modern China*. New York and London: W.W. Norton & Company.
Tang Min *et al.* (1991). Survey of parasite infections in Changchun city, Jilin province. *Chinese Journal of Parasitology and Parasitic Diseases, Special Issue for the Nationwide Survey on the Distribution of Human Parasites* 129–130 (in Chinese).
Thein Hlaing (1985). *Ascaris lumbricoides* infection in Burma. In: *Ascariasis and its Public Health Significance* (D.W.T. Crompton, M.C. Nesheim and Z.S. Pawlowski eds), pp. 83–112. London and Philadelphia: Taylor & Francis.
Thein Hlaing (1993). Ascariasis and childhood malnutrition. *Parasitology* **107**, S125–S136.
Thein Hlaing, Toe, T. Saw, T. Kyin, M.L. and Lwin, M. (1991). A controlled chemotherapeutic intervention trial on the relationship between *Ascaris lumbricoides* infection and malnutrition in children. *Transactions of the Royal Society of Tropical Medicine and Hygiene* **85**, 523–528.
Tong Shu-Fen *et al.* (1991). Survey of parasitic infections in urban and rural population in Harbin. *Chinese Journal of Parasitology and Parasitic Diseases, Special Issue for the Nationwide Survey of the Distribution of Human Parasites* 26–29 (in Chinese).
Wang Guilin *et al.* (1991). Survey of parasite infections in Benxi prefecture, Liaoning province. *Chinese Journal of Parasitology and Parasitic Diseases, Special Issue for the Nationwide Survey on the Distribution of Human Parasites* 129 (in Chinese).
Wang Pei (1983). Traditional Chinese Medicine. In *Traditional Medicine and Health Care Coverage* (R.H. Bannerman, J. Burton and Ch'en Wen-Chieh eds). Geneva: World Health Organization.
Wang Qing *et al.* (1994). An investigation of human parasite distribution in Tianjin. *Chinese Journal of Parasitology and Parasitic Diseases, Special Issue for the Nationwide Survey of the Distribution of Human Parasites* **12**, 58–60 (in Chinese).
Wang Weizhou *et al.* (1994). Prevalence of *Ascaris* infection in population of Jiangxi Province. *Chinese Journal of Parasitology and Parasitic Diseases, Special Issue for the Nationwide Survey of the Distribution of Human Parasites* **12**, 155–156 (in Chinese).
Wang Zhao and Xu Shu-Hui. (1991). Survey of parasite distribution is an important and initiative task. *Chinese Journal of Parasitology and Parasitic Diseases, Special Issue for the Nationwide Survey of the Distribution of Human Parasites* 1–2 (in Chinese).

Wang Zixing (1982). A clinical summary of 778 cases of biliary ascariasis. *Hubei Journal of Traditional Chinese Medicine* **5**, 37 (in Chinese).

Wei Shaokuan. (1989). Survey on family aggregation of *Ascaris* and hookworm infection in the district inhabited by Mulao people. *Journal of Medical College of Youjiang Nationalities* **11**, 35–37. (in Chinese)

Wescott, R.B. (1986). Anthelminthics and drug resistance. In *Veterinary Clinics of North America* (ed. R.P. Herd). pp. 367–380. Philadephia: W.B. Saunders Co.

Wood, B. (1996). Human evolution. *BioEssays* **18**, 945–954.

WHO (1987). Prevention and Control of Intestinal Parasitic Infections. *Technical Report Series*, 749. Geneva: World Health Organization.

WHO (1992). *Counterfeit Drugs.* Report of a joint WHO/IFPMA Workshop WHO/DMP/CFD/92. Geneva: World Health Organization.

WHO (1995a). The Use of Essential Drugs. *Technical Report Series* 850. Geneva: World Health Organization.

WHO (1995b). *WHO Model Prescribing Information*, 2nd edn. Geneva: World Health Organization.

WHO (1996). Report on an Informal Consultation on the Use of Chemotherapy for the Control of Mobidity due to Soil-transmitted Nematodes. Geneva: World Health Organization.

Wu Chixu *et al.* (1994). Distribution of human parasitic infections in Hubei Province. *Chinese Journal of Parasitology and Parasitic Diseases, Special Issue for the Nationwide Survey of the Distribution of Parasites* **12**, 49–53 (in Chinese).

Wu Xianhong *et al.* (1994). Human parasite distribution in pasturage area of Qinghai Province. *Chinese Journal of Parasitology and Parasitic Diseases, Special Issue for the Nationwide Survey of the Distribution of Parasites* **12**, 194–198 (in Chinese).

Xu Feng-Hui *et al.* (1991). A survey of human parasite infections in Lincheng town, Lingao county, Hainan province. *Chinese Journal of Parasitology and Parasitic Diseases, Special Issue for the Nationwide Survey of the Distribution of Human Parasites* 113–115 (in Chinese).

Xu Fengshui *et al.* (1994). Investigation on human parasite distribution in Hainan Province. *Chinese Journal of Parasitology and Parasitic Diseases, Special Issue for the Nationwide Survey of the Distribution of Human Parasites* **12**, 22–25 (in Chinese).

Xu Funiu *et al.* (1991). Survey of human parasites in Anhui province. *Journal of Research and Control of Parasitic Diseases* **20**, 7–12 (in Chinese).

Xu Jing-Tian *et al.* (1991). Survey of human parasite infections in 10 cities of Liaoning province. *Chinese Journal of Parasitology and Parasitic Diseases, Special Issue for the Nationwide Survey of the Distribution of Human Parasites* 21–23 (in Chinese).

Xu Maolin and Xu Hongjun (1991). Survey of parasitic infections of population in Baicheng prefecture, Jilin province. *Chinese Journal of Parasitology and Parasitic Diseases, Special Issue for the Nationwide Survey of the Distribution of Parasites* 130–131 (in Chinese).

Xu Rong-Qi *et al.* (1991). Clinical manifestations of ascariasis in china. *Chinese Journal of Parasitology and Parasitic Diseases, Special Issue for the Nationwide Survey of the Distribution of Human Parasites* 115–121 (in Chinese).

Xu Shu-Hui (1993). Preface to the supplement of Chinese Journal of Zoonoses 1993, **9** (in Chinese).

Xu Yunhua and Tong Nuei (1991). Survey of parasite infections in Fuyuan county, Yunnan province. *Chinese Journal of Parasitology and Parasitic Diseases, Special*

Issue for the Nationwide Survey on the Distribution of Human Parasites 153–154 (in Chinese).
Yang Baogui et al. (1991). Survey of parasite infections in Mojiang country, Yunnan province. *Chinese Journal of Parasitology and Parasitic Diseases, Special Issue for the Nationwide Survey on the Distribution of Human Parasites* 152 (in Chinese).
Yang Guohua et al 1991. Survey of parasitic infections of population in Liaoyuan city, Jilin province. *Chinese Journal of Parasitology and Parasitic Diseases, Special Issue for the Nationwide Survey of the Distribution of Human Parasites* 130 (in Chinese).
Yang Jia-Lun et al. (1991). Survey of human parasite infections in Lisu and Leme minority areas in Nujiang autonomous prefecture of Yunnan Province. *Chinese Journal of Parasitology and Parasitic Diseases, Special Issue for the Nationwide Survey of the Distribution of Human Parasites* 82–85 (in Chinese).
Yang Jia-Lun et al. (1992). A survey of human intestinal parasite in Yunnan. *Chinese Journal of Parasitic Disease Control* **5**, 245–248 (in Chinese).
Yang Jia-Lun et al. (1994). A survey of epidemiological factors in hookworm, *Ascaris* and *Trichuris* infections. *Chinese Journal of Parasitology and Parasitic Diseases, special issue for the nationwide survey of the distribution of human parasites* **12**, 145–147 (in Chinese).
Yu Hongbao et al. (1991). Survey of parasite infections in Anshan prefecture, Liaoning province. *Chinese Journal of Parasitology and Parasitic Diseases, Special Issue for the Nationwide Survey of the Distribution of Human Parasites*, 128 (in Chinese).
Yu, S.H., Jiang, Z.X. and Xu, L.Q. (1989). The present status of soil-transmitted helminthiases in China. In: *Collected Papers on the Control of Soil-transmitted Helminthiases*, vol. IV, pp. 5–17. Tokyo: Asian Parasite Control Organization.
Yu Senhai (1992). Work report on the first nationwide survey of the distribution of human parasites in China. *Chinese Journal of Parasitic Diseases Control* **4**, (Suppl.), 59–63 (in Chinese).
Yu Senhai et. al. (1994). Report on the first nationwide survey on the distribution of human parasites in China. 1. Regional distribution of parasite species. *Chinese Journal of Parasitology and Parasitic Diseases* **12**, 241–247 (in Chinese).
Zhang Bei-Xiu and Chen Feng-Yi. (1991). Survey of parasite infection in Dalian prefecture, Liaoning province. *Chinese Journal of Parasitology and Parasitic Diseases, Special Issue for the Nationwide Survey of the Distribution of Human Parasites* 126–127 (in Chinese).
Zhang Bin et al. (1994). Distribution of human intestinal parasites in Inner Mongolia. *Chinese Journal of Parasitology and Parasitic Diseases, Special Issue for the Nationwide Survey of the Distribution of Human Parasites* **12**, 65–67 (in Chinese).
Zhang Guangyuan et al. (1994). Human parasite distribution in Shannxi Province. *Chinese Journal of Parasitology and Parasitic Diseases, Special Issue for the Nationwide Survey of the Distribution of Human Parasites* **12**, 68–73 (in Chinese).
Zhang Shou-Yi and Chen Ming (1991). Preliminary investigation on the distribution of human parasites in Gansu province. *Chinese Journal of Parasitology and Parasitic Diseases, Special Issue for the Nationwide Survey of the Distribution of Human Parasites* 88–90 (in Chinese).
Zhang Shouyi et al. (1994). First survey of human parasite distribution in Gansu Province. *Chinese Journal of Parasitology and Parasitic Diseases, Special Issue for the Nationwide Survey of the Distribution of Human Parasites* **12**, 54–57 (in Chinese).

Zhang Xiang-Jun *et al.* (1991). Investigation of intestinal parasite infections in Hunan province. *Chinese Journal of Parasitology and Parasitic Diseases, Special Issue for the Nationwide Survey on the Distribution of Human Parasites* 55–59 (in Chinese).

Zhang Xiangjun *et al.* (1994). An investigation on human parasite distribution in Hunan Province. *Chinese Journal of Parasitology and Parasitic Diseases, Special Issue for the Nationwide Survey of the Distribution of Human Parasites* **12**, 17–21 (in Chinese).

Zhang Xizhong *et al.* (1991). Survey of parasitic infections of population in Siping city, Jilin province. *Chinese Journal of Parasitology and Parasitic Diseases, Special Issue for the Nationwide Survey of the Distribution of Human Parasites* 131 (in Chinese).

Zhang Xuelin *et al.* (1991). Survey of parasitic infections of population in Gongzhuling city, Jilin province. *Chinese Journal of Parasitology and Parasitic Diseases, Special Issue for the Nationwide Survey of the Distribution of Human Parasites* 132 (in Chinese).

Zhang Zhongshan *et al.* (1991). Survey of parasitic infection of population in Tonghua city, Jilin province. *Chinese Journal of Parasitology and Parasitic Diseases, Special Issue for the Nationwide Survey of the Distribution of Parasites* 131–132 (in Chinese).

Zhao Gui-Su *et al.* (1991). Survey of human parasites in Daxing, Tongxian and Miyun counties, Beijing suburbs. *Chinese Journal of Parasitology and Parasitic Diseases, Special Issue for the Nationwide Survey of the Distribution of Human Parasites* 6–7 (in Chinese).

Zheng Chang-Qian *et al.* (1991). An investigation on the distribution of parasite infections in eleven counties in Jiangsu province. *Chinese Journal of Parasitology and Parasitic Diseases, Special Issue for the Nationwide Survey of the Distribution of Human Parasites* 29–31 (in Chinese).

Zhong Huilan (1986). *Tropical Medicine*, pp. 768–769. Beijing: Public Health Publishing House (in Chinese).

Zhou Daren, Li Yuesheng and Yang Xianming. (1994). Schistosomiasis control in China. *World Health Forum* **15**, 387–389.

Zhu Yu-Guang *et al.* (1991). Investigation of the species and distribution of human intestinal parasites in Shandong Province. *Chinese Journal of Parasitology and Parasitic Diseases, Special Issue for the Nationwide Survey of the Distribution of Human Parasites* 128–129 (in Chinese).

The Generation and Expression of Immunity to *Trichinella spiralis* in Laboratory Rodents

R. G. Bell

James A. Baker Institute for Animal Health, College of Veterinary Medicine, Cornell University, Ithaca, NY 14853, USA

1. Introduction 150
2. The Development of Mechanistic Theories of Rejection of *T. spiralis* in the Period up to 1970 150
3. Analysis of the Cellular and Humoral Immune Response to Infection since 1970 155
   3.1 Antibody responses 155
   3.2 T cells and their role in protection 162
   3.3 Immunosuppression 170
4. Variation in the Host Response to Infection 173
   4.1 Major histocompatibility complex-linked variation 174
   4.2 Non-major histocompatibility complex-linked variation 178
5. The Response of Granulocytic Cell Populations to Infection 180
   5.1 Eosinophils 180
   5.2 Mast cells 183
   5.3 Neutrophils 189
6. Assessment of Proposed Mechanisms of Protection 190
   6.1 Non-specific inflammation 190
   6.2 Allergic inflammation 195
   6.3 Intestinal mucus 195
   6.4 Intestinal epithelial cells 196
7. Synthesis 197
8. Conclusions 201
Acknowledgements 202
References 202

1. INTRODUCTION

The last comprehensive review of immunity against *Trichinella spiralis* was published in 1983 by Wakelin and Denham. Since then, numerous reviews have analyzed the genetics of responsiveness in mice, but no comprehensive review has appeared. A great deal has changed in the 14 years since 1983 and the accumulation of new information and insights with *Trichinella spiralis* merit comprehensive analysis. The primary object in assembling and reviewing the published literature is an analysis of the experimental rodent work and the theories of rejection that have developed from it. Where appropriate, new concepts are advanced which suggest different areas of emphasis for future experimental work. Because a great deal of current work has its origin in experiments conducted 40–60 years ago, the contributions of these workers is acknowledged in a historial assessment of work prior to 1970. The *T. spiralis* studies took place concurrently with work on many different nematodes, e.g., *Nippostrongylus brasiliensis*, *Heligmosomoides polygyrus*, *Trichuris muris* and *Strongyloides ratti*; some of this work is also mentioned where appropriate. Two emerging concepts in the study of intestinal nematode rejection include the view that there are multiple independent mechanisms of rejection and that the primary rodent species, mice and rats, may differ in their final effector mechanisms against individual parasite species. Until rejection processes are defined mechanistically for rats and mice both views remain speculative and are brought to the attention of readers here but not emphasized in the body of this review.

2. THE DEVELOPMENT OF MECHANISTIC THEORIES OF REJECTION OF *T. SPIRALIS* IN THE PERIOD UP TO 1970

The early years of research on trichinosis were characterized by the use of a variety of host species, including man, monkeys, pigs, rats and guinea pigs, in investigations that were directed largely at the biology of the parasite. The first attempt at passive transfer of protection with serum antibody was conducted in infected patients in New York in 1916 (Salzer, 1916), which led, almost immediately, to conflicting claims regarding the efficacy of passively transferred immune serum (Schwartz, 1917). A pattern that was to be repeated later with immune serum transfer in rodents. With this exception, early investigators were less concerned about the protective role of serum than they were about whether immunity existed at all. McCoy (1931) provided the first quantitative evidence for immunity in rodents against *T. spiralis* when he demonstrated that rats infected once were resistant (10 out of 11 survived) to an infection that killed nine out of 11 nonimmune controls. This paper demonstrated that immunity was intestinal, by finding reduced numbers of adult worms in the gut of immune rats and that larval counts in muscle could be used to measure immunity. McCoy (1931) also showed a pronounced loss of muscle larvae in rats given multiple infections but this has not been confirmed since. In 1932, McCoy showed that the primary infection was terminated by an active host response and that the period for which worms persisted in the gut was directly dose-dependent. Three years later, McCoy (1935) demonstrated effective immunization of rats (approximately 50% reduction in muscle larvae numbers) with killed larvae and larval powder. Larval immunization shortened the intestinal life-span of adult worms in immunized rats. McCoy (1932) considered that there was little evidence for immunity against the muscle invasive stages (i.e., newborn larvae), a belief that persisted well into the 1980s. McCoy thus demonstrated host responsiveness and laid down procedures, principles and areas of investigation that are still the foundation for current investigations of immunity to *T. spiralis*.

During the 1930s, the existence of immunity was established experimentally for several intestinal nematode species, and the first dose response curves and *T. spiralis* LD_{50} values were defined (Roth, 1939; Fischthal, 1943). The first quantitative analysis of passively transferred immunity against *T. spiralis* in mice successfully used immune rabbit serum as the source of antibody (Culbertson and Kaplan, 1938). Antibody was generally considered to be an important mediator of host immunity against nematodes (Taliafferro, 1940; Culbertson, 1942b) and this view continued to gain support well into the 1940s. However, there was also scepticism from individuals who felt that antibody titers in infected animals correlated poorly with protection (Chandler, 1939). Other concepts that were developed at this time included the view that intestinal immunity was local in nature, and that its effects were reversible upon transfer of damaged worms to a normal host (Chandler, 1939; Taliafferro, 1940). As pointed out by Culbertson (1942a) most of the authors who supported the concept of local intestinal immunity had failed in attempts to passively transfer protection with immune serum. McCoy (1940) in a landmark study described a rapid loss of trichinae fed to actively immune rats and attributed protection to increased mucus production and/or peristalsis rather than antibody. Despite this view, the consensus at that point was clearly that antibody had a role in expulsion of *T. spiralis* and *N. brasiliensis* infections (Taliafferro, 1940).

The most conspicuous new idea was Oliver-Gonzalez's (1940, 1941) radical proposal that antibody was directed separately against larval and adult worms. This 'dual antibody' hypothesis introduced the concept of stage-specificity to the analysis of nematode immunity. Oliver-Gonzalez (1941) protected rats with hyperimmune rabbit serum absorbed with larval powder leading him to conclude that protective antibodies were adult-specific, although both stages of the life cycle could elicit antibodies. This early demonstration of the antigenic uniqueness of adult and larval parasites was, like McCoy's data, virtually ignored until stage-specificity was rediscovered (Bell *et al.*, 1979; Philipp *et al.*, 1980).

However, by the late 1940s and early 1950s the ability of adult mice to express immunity upon reinfection demonstrated by Culbertson (1942a) was being questioned. In rats and guinea pigs, resistance to reinfection was expressed in the rapid loss of the challenge infection but in mice the challenge infection often persisted (Larsh and Fletcher, 1948). Rappaport and Wells (1951) attempted to resolve these differences by studying the effect of time since the primary infection and different challenge doses on expulsion. They found that immunity was variably expressed, being inversely related to both time and dose of challenge, as McCoy (1932) had shown in nonimmune rats. Rapaport and Wells (1951) and Despommier and Wostmann (1969) also recorded male and female stunting and reduced fecundity as prominent effects of immunity, both of which were shown to precede worm expulsion. Larsh *et al.* (1952) determined worm distribution in the small and large intestine in immune and normal mice, demonstrating that in both groups larvae preferentially colonized the upper small intestine and then moved down the small and into the large intestine as immunity developed. This intraintestinal movement of adults also preceded worm expulsion (Larsh *et al.*, 1952). It is interesting that despite his evidence for localized effects in the intestine — i.e., rejection from the upper portion of the small intestine first — the convention of treating rejection as a process that is only measured on the basis of total small intestine worm burden was continued by Larsh *et al.* (1952), as it has been by essentially all other workers (Bell, 1992).

The issue of local immunity in the gut raised by Chandler (1939), and revisited by Larsh *et al.* (1952), was specifically addressed by Larsh in 1953 when he compared a local injection of larvae (immunizing infection) into the cecum versus a natural infection which deposits larvae in the upper small intestine. Larsh (1953) concluded that there was no local immunity as mice immunized in the cecum rejected a natural (oral) challenge infection from the upper small intestine as fast as mice immunized in that site. This was supported by Zaiman *et al.* (1955a,b), working with parabiotic rats, who suggested that rejection was mediated by circulating factors that could be transferred between parabionts. Larsh and Race (1954) followed their earlier work with an examination of the histopathology of the upper small intestine during infection of normal and immune mice. The local inflammatory response they observed convinced Larsh and Race (1954) that secondary cellular (inflammatory) reactions were also important for worm expulsion, particularly in immune mice. The evidence for this was entirely histological and correlative, being dependent on the time frames at which the predominant early cellular influx of polymorphs and later mononuclear cells preceded rejection. This work was the first to suggest that inflammatory cells contribute substantially to rejection of a *T. spiralis* infection in primary and challenge infections. However, the idea was not new having been first advanced by Taliafferro and Sarles (1939) in *N. brasiliensis* (=*muris*) infections.

Several important new themes were developing in the 1950s. Treatment of human cases of trichinosis with adrenocorticotrophic hormone (ACTH) (Luongo *et al.*, 1951) led to an examination of the effects of corticosteroid treatment (Coker, 1956; Markell and Lewis, 1957) which diminished resistance. Experiments were also undertaken using irradiation to sterilize *T. spiralis* infectious larvae and hence restrict infections to the intestinal stages of the life cycle (Alicata and Burr, 1949; Gomberg and Gould, 1953; Gould *et al.*, 1955). Campbell (1955) demonstrated that tissue culture fluid containing the excretions and secretions of incubated muscle larvae were immunogenic, paving the way for future analysis. Larsh and Race (1954), Rappaport and Wells (1951) and Hendricks (1950), all working with mice, failed to find the rapid loss (rapid expulsion) of a challenge infection observed by McCoy (1940) in rats. Rapid expulsion thus disappeared from view until the late 1970s. Finally, Cox (1952), showed that a concurrent or prior infection with *Ancylostoma caninum* increased the resistance of mice to infection with *T. spiralis*. The effects of concurrent infections upon each other was to become a critical element in the development of theories on the nature of gut immunity to *T. spiralis*.

In 1963, Larsh published the first of several reviews of *T. spiralis* infections. These reviews have proved to be seminal in providing working hypotheses of mechanisms of intestinal immunity to *Trichinella* and they still dominate much current thinking about intestinal immunity to nematodes. In his review of the literature, conducted species by species, Larsh (1963) dealt first with rats and acknowledged the strong immunity present in this species but largely ignored McCoy's pioneering work of the 1930s and did not refer to the rapid loss of challenge larvae demonstrated by McCoy (1940). Larsh proposed that the transfer of immunity Zaiman *et al.* (1955 a,b) consistently demonstrated was due to the transfer of antigen from the infected to the uninfected parabiont. Larsh (1963) also addressed the concurrent infection experiments of Cox (1952) in which *T. spiralis* infections superimposed on a 10-day-old *A. caninum* infection were rejected earlier than usual, concluding that the earlier rejection of *T. spiralis* was due to a nonspecific intestinal inflammation produced by *A. caninum*. Most attention in this review was directed at immunity in mice and what

Larsh viewed as the pre-eminent role of inflammation. In his next review, Larsh (1967) proposed that immunity was due to a nonspecific intestinal inflammation, although he devoted considerable attention to emerging evidence for the presence of immediate hypersensitivity, recently shown in *T. spiralis*-infected mice and guinea pigs (Sharp and Olson, 1962; Briggs, 1963b; Briggs and DeGiusti, 1966). However, Larsh viewed the possible role of immediate hypersensitivity to be that of potential initiator of inflammation rather than the mediator itself. The 1967 review was important in two respects: it further diminished the role of antibody, and it re-emphasized the view that nonspecific gut inflammation was the critical event leading to worm expulsion. According to this theory, a delayed hypersensitivity (DH) reaction initiated an allergic inflammation that nonspecifically led to worm rejection by creating an unsuitable biochemical environment. Larsh buttressed his argument with the dual infection data mentioned above and the fact that corticosteroid and irradiation treatment had been shown to prolong adult worm residence time in the gut. Both agents eliminated acute inflammation and reduced circulating or fixed tissue lymphocyte number, yet had little effect on antibody levels. Simultaneously, the need for antibody was called into question and the role of several types of cells made more prominent. Successful transfer of immunity with cells from lymph nodes (Larsh *et al.*, 1964a) or peritoneal cells (Larsh *et al.*, 1964b), albeit with a low success rate, nevertheless established the view that DH was critical. Based largely on histopathology, the critical role in nonspecific inflammation was thought to be played by polymorphs. Larsh (1967) allowed the possibility that antibodies might be responsible for interfering with growth and the reproduction of adult worms.

Allergic reactions were gaining recognition as important host response elements during helminth infections. Andrews (1962) was careful to define immediate hypersensitivity as the 'allergic' response of interest. The older use of the term allergy meant hyperreactive and encompassed several types of acquired immunity including immediate hypersensitivity and DH. It was this original meaning of allergic that Larsh had in mind when he discussed 'allergic reactivity' (Larsh, 1967; Larsh and Race, 1975) but largely focused on DH. Briggs (1963b) however, showed active and passive mast cell degranulation in *T. spiralis*-infected mice. By 1964 homocytotropic reaginlike antibodies had been detected in *N. brasiliensis*-infected rats (Ogilvie 1964) and by 1968 in mice and rabbits infected with *T. spiralis* (Sadun *et al.*, 1968). Although attempts to induce protection with immediate hypersensitivity were unsuccessful (Arnold and Olson, 1966), support for the idea came from increased adult worm counts in mice treated with antihistamine or antiserotonin in the diet (Campbell *et al.* 1963).

A final thread that emerged in the 1960s was the observation that by using various anthelminthic agents it was possible to restrict the exposure of experimental animals to relatively defined stages of the life cycle (Campbell, 1965). This was a technical improvement over the use of irradiated larvae, which had earlier been shown by several investigators to produce immunity (Levin and Evans, 1942; Gould et al., 1955; Zaiman et al., 1955b), and led to the conclusion that larval and adult worms were immunogenic. While individual investigators were clearly considering stage-specific immunity none, with the exception of Oliver-Gonzalez (1941), had provided evidence for it. By the end of the 1960s a direct role for antibody in expulsion had been largely replaced by a proposed role for cells, as expounded by Larsh (1967). Although an indirect role for antibody through immediate hypersensitivity was attracting attention, both themes shared the view that expulsion was a nonspecific process whether mediated by cells (probably polymorphs) in Larsh's version, or by pharmacologic mediators through immediate hypersensitivity.

These early workers thus defined the procedures by which immunity is examined today and the major questions that still occupy researchers. At least two of them defined phenomena that were rejected by their peers but which are now well established: McCoy (1940) defined what is now called rapid expulsion and Oliver Gonzalez (1940, 1941) recognized stage-specificity. Larsh (1963, 1967) championed the view that inflammation and nonspecific elements could produce rejection. Furthermore, immediate hypersensitivity had been thrust into the foreground of discussion.

3. ANALYSIS OF THE CELLULAR AND HUMORAL RESPONSE TO INFECTION SINCE 1970

3.1. Antibody Responses

3.1.1. Serum Immunoglobulin Levels in Mice

A *T. spiralis* infection provides a dramatic stimulus to the whole immune system. This is reflected in serum immunoglobulin levels in which an initial reduction during the first 7–10 days (Crandall and Crandall, 1972; Almond and Parkhouse, 1986; Zackroff et al., 1989) is followed by substantial increases in serum levels. These increases are isotype-specific, with IgE generally by far the strongest at 50–100-fold and IgG1 next at 10–20-fold with smaller increases in IgM and IgA values (Crandall and Crandall, 1972; Almond and Parkhouse, 1986; Zakroff et al., 1989). Mouse strain specific variations in the degree and timing of these changes were evident in Almond and Parkhouse's 1986 study. Little overall change is evident in

IgG2b and IgG3 levels (Zakroff *et al.*, 1989; Almond and Parkhouse, 1986) during the first 2 months of infection.

This pattern of selective hypergammaglobulinemia (strongly elevated IgE and IgG1 levels) reflects the strong T_H2-specific activation stimulus provided by a *T. spiralis* infection. Noteworthy also is the fact that the strongest-responding mouse strain (NIH) in the Almond and Parkhouse (1986) study appeared to have a higher total IgG1 and IgA response than any other strain.

3.1.2. *The Specific Antibody Response to Infection*

(a) *In mice.* The first analyses of isotype distribution in specific antibody to *T. spiralis* originated in attempts to dissociate the contribution of IgE and IgG1 to immediate hypersensitivity. Sadun *et al.* (1968) first detected IgE antibodies 5 weeks after initial infection in mice using a 3-day latent period passive cutaneous anaphylaxis (PCA) reaction. These levels peaked 9 weeks after infection. PCA reactivity declined rapidly after 9 weeks and in some mice was undetectable 11 weeks after infection. Whole IgG antibodies, as measured by a soluble fluorescent antigen test, peaked at 5–6 weeks but remained at high levels thereafter. The persistent PCA reactivity present after 11 weeks was found to be due to short latent period (4 h) IgG1 which appeared with the same kinetics as the long (72 h) latent period antibody, IgE (Mota *et al.*, 1969). Reinfection resulted in an increase in both IgE and IgG1 antibodies whereas repeated infections led ultimately to the disappearance of IgE antibodies but the continued production of IgG1 (Mota *et al.*, 1969).

Rivera-Ortiz and Nussenzweig (1976) compared IgG1 and IgE production in various inbred mice after infection with 200 muscle larvae. They found substantial interstrain variation in IgE production (SJL and AKR, weak; ASW, LP, RF and Bub, intermediate; and DBA/1, A, A/He, C57L, DBA/2 and C57BL, strong) but considerably less interstrain variation in IgG1 levels, although SJL and AKR again responded poorly. IgE was first detected between 2 and 5 weeks after infection but Rivera-Ortiz and Nussenzweig (1976) did not find the precipitous reagin decline 9 weeks after infection reported by Mota *et al.* (1969). Subsequently, Gabriel and Justus (1979) demonstrated reaginic antibody as early as 8 days after infection using a more sensitive active cutaneous anaphylaxis (ACA) test and low doses of antigen to induce the ACA reaction.

Almond and Parkhouse (1986) measured isotype responses of strong rejection NIH mice and intermediate rejection C3H/He mice, to surface and secreted antigens of muscle larvae, adults and newborn larvae. IgM antisurface values for each life cycle stage increased marginally, but equally, for both mouse strains, whereas specific IgG1 responses to muscle larvae were higher in C3H than in NIH mice. While IgG1 responses to adult worms were higher for NIH than C3H mice, those to newborn larvae were equal. The same stage-specific pattern held for IgG2 antibody to muscle larvae and adults, but with this isotype, C3H had a stronger anti-newborn larvae response. The NIH IgA response to surface antigens exceeded that of C3H mice to all stages of the life cycle. In contrast, IgG1 and IgG2 responses to secreted antigens of muscle larvae and adults peaked between days 20 and 30 after infection and there were no strain-related differences in amount. However, the NIH and C3H IgG1 response to muscle larvae and adult worm surface antigens recognized distinct proteins which, the authors felt, could be related to protection (Almond and Parkhouse, 1986). Pond *et al.* (1989) measured antibody responses to muscle larvae antigens in AKR (resistant/intermediate responders) and B10.BR (weak responders). B10.BR mice mounted stronger IgG1, IgA and IgE responses and those of AKR mice were higher only in IgG2a production. A different perspective was given by Denkers *et al.* (1990) who noted a biphasic response to infection, with an early (day 13) peak of IgG1 and IgM antibody to a set of larval antigens (Group I) that differed from the targets of a second antibody peak at or after day 42 to Group II antigens.

Another approach was taken by Goyal and Wakelin (1993) who evaluated variations in host response to different *T. spiralis* isolates. They examined three isolates, one each from England, Spain and Poland in NIH and C57B1/10 (weak rejection) mice. These *Trichinella* isolates differed in the speed with which they were rejected from both mouse strains, with the English being fast, and the Polish the slowest by a few days. The IgM response was higher in C57B1/10 than NIH mice at days 12 and 22 but not day 40 after infection. While the total IgG response was comparable at most time points, in NIH mice the IgG1 isotype was greater at day 40 and IgG2a at day 12. Only IgE antibody produced by C57B1/10 mice showed a consistent pattern of increases over that of NIH mice. The distribution of specific antibody by isotype largely followed the pattern of increase observed in total Ig values. Based on these results, Goyal and Wakelin (1993) proposed that the length of exposure to the parasite determined the strength of the antibody response, as the C57B1/10 mouse retained adult worms for approximately twice the period that NIH mice did. Attractive as this hypothesis appears, since it equates degree of parasite exposure to the ensuing antibody response, it does not fully account for the results of Almond and Parkhouse (1986). For Almond and Parkhouse, the NIH strain antibody response exceeded C3H mice in three out of the four antibody isotypes examined, despite the slower capacity of C3H mice to reject adult worms.

Most of these studies were initiated either to describe the host response pattern or to search for strain-specific clues to significant resistance factors.

The only reasonable conclusion is that there is no obvious connection between the specific antibody response of any isotype in serum, quantitatively or temporally, and adult worm rejection. Conversely, the Rivera-Ortiz and Nussenzweig (1976) study showed an interesting inverse relationship between the strength of the serum IgE response and the number of muscle larvae establishing in that strain (DBA/1, highest IgE response, lowest number of larvae; SJL, lowest IgE response, highest number larvae). This pattern of response suggests a protective role for IgE that has not been followed up since the initial report. If protection is largely a consequence of what happens in the gut, and this is dependent on local factors operative at that site, then we can, with hindsight, propose that serum antibody values may not reflect much of what is happening in the gut. This appears to be the case and it is noteworthy that the one indication of a protective link to serum antibody in the Rivera-Ortiz and Nussenzweig (1976) study, was directed at the systemic component of the life cycle — the newborn larvae.

(b) *In rats*. There have been no studies of changes in total immunoglobulin levels in rats, all researchers have measured specific antibody levels. In rats, the strong and persistent specific IgG response and the equally strong but less persistent serum IgE response were first described by Ottesen *et al.* (1975). Their data have since been confirmed (Ahmad *et al.*, 1991a) in detail even though different rat strains were used (Lewis versus AO): total serum IgG and IgE antibody responses peak at around 4 weeks with IgE declining thereafter. Dessein *et al.* (1981) measured isotype-specific responses as part of their study of IgE-suppressed Wistar/Lewis rats. They found peak IgE, IgM, IgG2b and IgG2c levels at day 33, the last day they tested. IgG1 peaked at day 25 and IgG2a was the same at day 25 and day 33. The IgG values did not vary by more than a single (log2) dilution at their peak. Nine weeks after an initial infection of AO rats the serum IgG response was dominated by IgG1 with IgG2a following closely behind, then IgG2c, with IgG2b barely detectable (Appleton *et al.*, 1988). Subsequent studies in PVG/c rats (Peters *et al.*, 1997) focused on the early B cell response to infection in the spleen, mesenteric and cervical lymph nodes. The authors showed that while all isotypes increased at 8 and 16 days the response was dominated by IgG2c and a strong but volatile IgG2b response. High IgA levels were also present throughout the 50-day sampling period. At 50 days IgG2c was still dominant and IgG2a, IgA and IgM were about equal. The IgG2c response was found to be largely T-cell dependent by the use of athymic nude rats. Throughout the first 50 days the mesenteric node was the principal source of B cells producing antibody of the three tissues tested. The cellular response was broadly proportional to serum antibody levels. The rat studies examined antibody levels only to muscle larvae antigens, and there is an urgent need to define antibody responses to adults and newborn larvae.

(c) *Secretory antibody responses.* To date, almost all studies of rodents have examined serum antibody levels or B cell responses in lymph nodes or the spleen. In one of the few attempts to document responses in both serum and the gut, van Loveren *et al.* (1988) found that intestinal IgA peaked 10 days after infection and then declined. The significance of the gut lamina propria for IgE and IgA production was underlined recently by the observation that from 100–500 times as much IgE enters the gut lumen as the plasma on a daily basis during the intestinal stages of infection (Negrao-Correa *et al.*, 1996). Since the IgE that enters the gut is not removed from serum, although mechanisms for this exist (Ramaswamy *et al.*, 1994a), it must be produced entirely in the lamina propria. Serum levels do not reflect the amount of IgE produced locally and the specificity for adult worm antigens of gut IgE is distinct from that of serum IgE (D. Negrao-Correa and R.G. Bell, unpublished observations). Our data confirm those of van Loveren *et al.* (1988), in showing a strong secretory IgA response that peaks during the first two weeks of infection. As mentioned above, serum antibody responses have not provided insight into the kinetics or strength of adult worm expulsion in mice or rats. This is partly because most studies have examined the response to muscle larvae antigens, but it also reflects the fact that the most important responses apparently occur in the gut and are directed to specific adult antigens. There is a need for more specific data in this area.

3.1.3. *Is Antibody Involved in Protection?*

Antibody was originally considered to be the mediator of worm expulsion until it was supplanted by the rise of hypotheses involving inflammation (Larsh 1963, 1967; Larsh and Race, 1975). While direct experimental evidence for a protective role of antibody continued to accumulate (Wakelin and Lloyd, 1976; Love *et al.*, 1976; Crum *et al.*, 1977; Despommier *et al.*, 1977b; Dessein *et al.*, 1981; Jacqueline *et al.*, 1981), no recent reviews have accorded antibody a significant role in expulsive processes. A role for antibody simply fell from favour and disappeared from view. Despite this, a substantial body of data has, over the last 10 years, established a role for antibody in rapid expulsion and anti-newborn larvae immunity. There are also indications that antibody is important in adult worm expulsion. Essentially, all of this recent data derives from studies conducted in rats.

(a) *Rejection of muscle larvae.* In adult, as opposed to neonatal rats (Appleton and McGregor, 1987) antibody causes rapid expulsion only when it is transferred with T cells or when some form of intestinal stimulation that could activate local T cells takes place (Bell and McGregor, 1980a,b; Ahmad et al., 1990; Bell et al., 1992). In adult rats, protection resulting from rapid expulsion has been reliably transferred with purified serum IgE and the IgG fraction of immune serum, but not monoclonal IgG (Ahmad et al., 1991a). Protection with IgE required immune thoracic duct cells (CD4$^+$, OX22$^-$; Ahmad et al., 1991b) to be transferred before the serum (Ahmad et al., 1990). The protective cells (CD4$^+$, OX22$^-$) home to the intestine in large numbers but are short-lived (Wang et al., 1990a; R.G. Bell, unpublished data), although the effect of these cells in priming for rapid expulsion was remarkably long-lived as antibody could be transferred for up to 49 days after the T cells (Ahmad et al., 1990). Subsequently, it was shown that monoclonal IgG could transfer rapid expulsion provided that an infection with *H. polygrus* preceded the transfer of IgG (Bell et al., 1992). Thus, evidence that *H. polygyrus* but not CD4$^+$, OX22$^-$ cells would prime the gut differentiated the functional requirements for expression of IgG-mediated rapid expulsion from those required for IgE-mediated rapid expulsion. Interestingly, in neonatal rats, IgE is without effect, although immune serum or monoclonal IgG1, IgG2a and IgG2c are effective. There is no requirement for cell transfer and antibody can be directly injected into pups, fed to them or transferred in milk from the mother (Appleton and McGregor, 1987; Appleton et al., 1988). It seems probable that the neonatal Fcγ receptor may be involved in the neonatal version of rapid expulsion and this may account for the absence of a T cell requirement.

Insight into the potential role of CD4$^+$, OX22$^-$ T cells in the transfer of rapid expulsion was provided by experiments showing that a specific IgE-transport process could be induced in the intestine by transferred CD4$^+$ OX22$^-$ cells or by treatment with recombinant rat interleukin 4 (IL-4) (Ramaswamy et al., 1994a). IL-4 dependent activation of intestinal cell populations resulted in the translocation of plasma IgE into the intestinal lumen. The relevance of this process to an infection with *T. spiralis* was established by showing that 100–500 times as much IgE was transported to the gut as to serum in a primary infection (Negrao-Correa et al., 1996).

(b) *Adult worm rejection*. In rats, adult worm expulsion (approximately 40%) was reported by Love et al. (1976) after the transfer of immune serum alone, but when cells and serum were transferred together, expulsion was about 80% effective. Comparable results in mice were reported by Wakelin and Lloyd (1976) who showed that immune cells and immune serum that were ineffective after transfer alone, reduced adult worm numbers significantly when combined. Crum et al. (1977) and Wakelin and Wilson (1979a), both transferred protection, as assessed by reduced adult worm counts, to naive recipients with B cells alone. The latter authors concluded that the effector cells in their B cell population were contaminating T cells. It is possible that the contaminating T cells were necessary for B cell function, as the local release of T cell cytokines may be required for appropriate delivery of antibody (see below). Most of the above authors compared adult worm counts of control and cell/serum transfer groups six or more days after the challenge infection and found that adult worm numbers were already depressed. We cannot be sure, therefore, that the effect was not some form of rapid expulsion which reduced the initial worm take. Only the experiments of Despommier *et al.* (1977b) in which adult worm rejection occurred after 7 days showed a definitive anti-adult effect of B cells. These authors found no effect of transferred hyperimmune serum alone possibly because, in the absence of T cells, the antibody could not gain access to the gut.

A direct connection between gut IgE and adult *T. spiralis* rejection was suggested by data showing that adult-worm specific IgE is present in the gut lumen at the time of rejection and before antibody of this specificity appears in serum (D. Negrao-Correa and R.G. Bell, unpublished data). A novel rejection mechanism was also indicated by *in vitro* evidence using rat enterocyte cell lines (IEC-6, SLC-44) which are susceptible to invasion by adult worms (D. Negrao-Correa and R.G. Bell, unpublished data). When either cell line was incubated with IL-4 plus specific IgE isolated from the immune intestine, adult worm invasion was inhibited. Neither IL-4 treated cells alone, IL-4 treated cells plus a nonspecific myeloma IgE (1R162), serum IgE plus IL-4, nor intestinal IgE alone, could prevent invasion. These results provide a direct mechanism by which IgE and enterocytes might act to eliminate adult worms from the gut during a primary infection.

Unlike most other processes proposed as mediators for worm expulsion, there is little ambiguity about the mechanism of action of antibody. Its primary role is to bind to antigen and it can do this either free or when attached to cells. While the binding to antigen component is strategically recognized to be important, the process that leads from there to rejection is opaque. At present, there is a substantial body of data demonstrating that rapid expulsion is an antibody-mediated process which is dependent on $CD4^+$ cell-derived T_h2-type cytokines to provide the milieu in which the antibody can act (see Section 3.2.6). Similarly, there is direct evidence indicating a similar cytokine-dependent mechanism which facilitates the direct action of IgE to prevent adult worm invasion of enterocytes. Antibody can thus be proposed to effect both of the major expulsive intestinal responses.

3.2. T Cells and Their Role in Protection

3.2.1. *The Role of the Thymus*

Interest in the role of the thymus in immunity was pre-eminent in the basic immunology literature in the 1960s and this carried over into all areas of immunoparasitology. DiNetta *et al.* (1972) and Larsh *et al.* (1972) showed that antithymocyte serum treatment of mice delayed rejection of adult worms. However, this approach was superseded by the use of congenitally athymic 'nude' mice or thymectomized mice (Walls *et al.*, 1973; Ljungstrom and Ruitenberg, 1976; Ruitenberg and Elgersma, 1976; Ruitenberg *et al.*, 1977a,b, 1979; Gustowska *et al.*, 1980; Parmentier *et al.*, 1982; Vos *et al.*, 1983). The basic pattern was laid out by the first group of authors to use nude mice (Walls *et al.*, 1973). They found that adult worms were retained for longer periods (up to 38 days) in the gut, that the intestinal inflammatory response was delayed and weaker, eosinophilia developed poorly and that progressively larger numbers of larvae established in muscle. The inflammatory response around mature, encapsulated muscle larvae was weak or absent. In addition, the intestinal mast cell and globule leukocyte response was impaired in nude mice (Ruitenberg and Elgersma, 1976) and antibody responses were also lower in thymectomized animals. Ultimately, adult worms were shown to persist for up to 83 days in nude mice whereas littermate controls eliminated their worms 10 days after infection (Ruitenberg *et al.*, 1977b). In keeping with the poorly developed intestinal immune response, few pyroninophilic (blast) cells developed in intestinal tissues (Ruitenberg *et al.*, 1977a). Comparable pathologies were observed in nude rats: delayed rejection of adult worms, increased muscle larvae deposition in muscle, weak or absent antibody response (IgM, IgG and IgE) and weak mast cell and globule leukocyte responses (Vos *et al.*, 1983).

3.2.2. *Delayed Hypersensitivity*

Work in the mid to late 1960s established that DH was present in mice after immunization with muscle larvae antigen in Freunds complete adjuvant (Kim 1966; Kim *et al.*, 1967). Under these conditions, DH was evident in most mice and reactivity was transferrable with isolated cell populations (Kim *et al.*, 1967; Larsh *et al.*, 1969). In addition, Larsh *et al.* (1969, 1970) showed that recipients of spleen, lymph node or peritoneal cavity cells from immunized or infected mice were partially protected from the intestinal phase of the infection. Although Larsh *et al.* (1969, 1970) attributed these effects to the transer of DH, it is not evident from their experiments that this was so.

In actively infected mice Gabriel and Justus (1979) demonstrated the presence of DH by footpad swelling and showed that it appeared concurrently with rejection of adults in the primary infection. Grove *et al.* (1977a) failed to observe positive footpad swelling reactions 4 and 8 weeks after infection, but found strong reactivity from 14 to 40 weeks after infection. Since then, DH has not been pursued, although Pond *et al.* (1989, 1992) suggested that T_h1 responses were more important for worm rejection than T_h2.

3.2.3. *Site of Production of Activated T Cells*

Studies of athymic nude mice and their littermate controls indicated that strong activation of intestinal lymphocytes occurred and peaked (pyroninophilic cell count) on day 10. A powerful mesenteric node response characterized by increased cellularity (Manson-Smith *et al.*, 1979b) and large numbers of dividing cells, peaked on day 8 in NIH mice (Grencis and Wakelin, 1982). Manson-Smith *et al.* (1979b) noted strain differences in the speed of mesenteric node cell proliferation and in gut homing which correlated with the strength of rejection for each strain. Comparable, but later (peak day 10–20) increases in the number of *T. spiralis* reactive cells in the mesenteric node were noted in rats by Ottesen *et al.* (1975). However, the peak output of dividing, protective *T. spiralis*-reactive cells occurred between 3 and 4 days in the thoracic duct lymph of infected AO rats (Bell *et al.*, 1987) and these cells were protective on adoptive transfer. Korenaga *et al.* (1989) showed that protective $CD4^+$ $OX22^-$ cells migrated from the afferent lymphatics of the gut after removal of the mesenteric node. This work pinpointed the origin of the protective T cells to the intestine rather than the mesenteric node. Possible sites of development were assessed by Wang *et al.* (1990a), based on the appearance of dividing cells in various compartments of the gut. They found that there was an increase in the number of dividing cells in the epithelium and lamina propria as early as 12 h after infection. However, the number of dividing cells in the Peyer's patches did not increase until 96 h. These data suggest that reactive cells may not originate in Peyer's patches but arrive there after being activated in, and migrating from, the lamina propria. Attempts to transfer protection with either Peyer's patch or intraepithelial lymphocytes have not been successful (R.G. Bell, unpublished data) and the yield of lamina propria cells is so low as to preclude experimentation. The observations in rats are more consistent with the direct activation of epithelial and lamina propria lymphocytes than they are of mesenteric node or Peyer's patch cells. It is worth noting that the increase in dividing cells in the mesenteric node and Peyer's patches corresponds with the arrival of activated, antigen-specific T cells from the lamina propria.

3.2.4. *Migration and Tissue Distribution of Dividing Cell Subsets*

Homing of gut-derived dividing lymphocytes back to the gut has been recognized for many years (Griscelli *et al.*, 1969). The dividing cells in the mesenteric node produced during a *T. spiralis* infection enter intestinal tissues at the site of infection in much higher numbers than usual (Rose *et al.*, 1976). This effect was observed between 2 and 4 days after infection. Because cells from both infected and uninfected donors migrated equally efficiently, the change was evidently a property of local vessels rather than the cells themselves. Furthermore, the increased cell influx could not be ascribed to an increased delivery of cells as cardiac output to the gut was unchanged until day 5 after infection, when blast cell localization declined (Ottaway and Parrott, 1980; Ottaway *et al.*, 1980). These early studies identified T cells as the primary population entering the gut (Rose *et al.*, 1976). In a follow-up study, Manson-Smith *et al.* (1979b) used NIH and BALB/c mice, which differ both in the time of expulsion of adult worms (NIH fast; BALB/c intermediate) and in their site of residence in the small intestine (NIH anterior; BALB/c posterior) to demonstrate that the site of the primary infection produces local changes that increase cell migration into that site. Secondly, they showed that expulsion of worms begins 2–3 days after the increased local entry of blast cells takes place. A similar pattern of blast cell accumulation was found by Despommier *et al.* (1977b) who examined intestinal localization of thoracic duct immunoblasts from hyperimmune rats. In their experiments, most of the dividing cells were B cells indicating that B immunoblasts can localize as effectively as T immunoblasts in the infected gut (Despommier *et al.*, 1977b).

Further studies on rat $CD4^+$ $OX22^-$ and $OX22^+$ subsets of thoracic duct lymphocytes revealed distinct intraintestinal migration pathways (Wang *et al.*, 1990a). The $OX22^-$ subset migrated extensively through the lamina propria with cells distributed from the muscularis mucosae to the villous tips. In addition, some 15% of intestinally recovered $CD4^+$ $OX22^-$ cells entered the epithelium. In contrast, $CD4^+$ $OX22^+$ cells remained in a band extending from the base of the villi at the crypts down to the muscularis mucosae layer; relatively few enter the epithelium. Protective cells are dependent on the integrin VLA-4 for their initial movement from the gut and, subsequently, from the bloodstream to re-enter the gut (Bell and Issekutz, 1993). This work identified three sites at which VLA-4 appeared to be crucial to effect protection. They included: (1) initial activation/migration of cells from the gut to the draining afferent lymphatics; (2) extravasation from the blood to re-enter the gut; and (3) migration or effector function within the gut after cells have crossed the vascular endothelium (Bell and Issekutz, 1993). This sequence was defined in actively infected animals for step 1 and in recipients of adoptively transferred cells for steps 2 and 3. It is not established that the lymph → blood → intestine dissemination sequence of CD4⁺ OX22⁻ cells is essential for rejection to occur in actively infected rats. Activation and effector function of locally stimulated sessile CD4⁺ cells may be sufficient for rejection of the primary infection. However, the repeated observation, particularly in mice, has been that the adult *T. spiralis* infection is initially a focal event that is

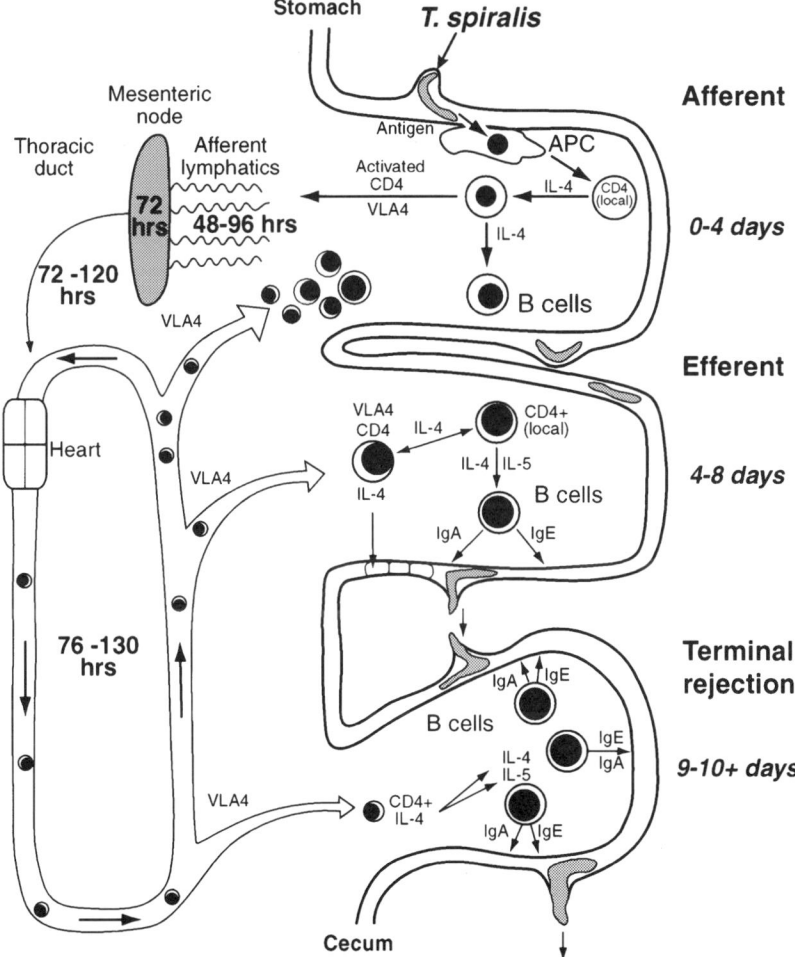

Figure 1 Schematic depiction of the sequence of events that accompany rejection of the primary infection in rats. The time course for the parasite is entered on the right along with the stage of immune reactivity that is elicited (afferent, efferent). Activation of immune cells and the deployment of effector B cells is depicted as being largely intraintestinal.

well illustrated by NIH and BALB/c mice (Manson-Smith et al., 1979b). Since the parasite can survive and reproduce successfully at many sites in the gut, dissemination of focally produced effector $CD4^+$ cells (as described above) may act to prevent re-establishment of adults downstream after their eviction from initially preferred sites. The connection between the local T cell response, the migration of activated T cells and worm rejection suggests the following relationships (based on NIH strain mice and several rat strains).

T cells are stimulated in the gut epithelium and lamina propria by 12 h after infection. A fraction of these locally activated cells migrate from the gut 48 h after infection through the mesenteric node to re-enter the gut preferentially at the site of infection, but also at all other sites. The key $OX22^-$ subset of $CD4^+$ cells do not produce IL-4 at 48 h but they begin to before 72 h. The process of cell proliferation and accumulation continues until about day 6 and worm movement down the gut begins at this time. At each successive downstream movement, residence time for the worms shortens because activated cells have already occupied these sites. By day 9 or 10, worm counts for the entire small intestine begin to drop as worms are now passing into the large intestine. This analysis suggests that worm rejection occurs earlier than is usually thought to be the case and that the dissemination of cells to nonparasitized sites is important in determining overall rejection time. This process is illustrated in Figure 1.

3.2.5. Cytokine Production

The significance of T cell derived products has been evident since the early studies of athymic mice and rats documented depressed or absent antibody, mast cell, eosinophil and goblet cell responses. An appreciation of the significance of the phenotypic polarization of T cells into the T_h1 and T_h2 subsets (Mossman and Coffman, 1989) with their distinct cytokine repertoires, has had a major impact on attempts to understand protective immunity and immunopathology with many infectious diseases, including helminth infections. Riedlinger et al. (1986) first characterized cytokine profile of $L3T4^+$ T cell lines capable of transferring protection to naive mice. They found that supernatants taken from *in vitro* cultures 24 hrs after stimulation with *T. spiralis* antigen contained IL-2, IL-3 and IFN-γ by bioassay. *In vivo*, the transferred cell lines induced both an intestinal mastocytosis and an enhanced eosinophilia. The presence of IL-5, evident from induction of eosinophilia by transferred cells, is consistent with secretion of T_h2 cytokines. The presence of IL-2, IFN-γ and IL-3 was confirmed in a kinetic study examining the response of mesenteric node cells collected from days 4–15 after infection and restimulated with *Trichinella* antigen *in vitro* (Grencis et al., 1987). A comparison of IL-2 production and IL-2R expression in inbred mice of different rejection phenotypes (Zhu and Bell, 1989) showed that the strong responder, NFS strain mice, produced more IL-2, had more receptors for IL-2 and responded better to exogenous IL-2 than either C3H (intermediate responder) or B10.BR (weak responder) mice. Interestingly, the B10.BR mice had better IL-2 responses and produced more IL-2 than C3H mice, suggesting that IL-2 production did not determine response phenotype. This was confirmed when worm rejection and IL-2 secretion were shown to segregate independently in individual offspring from NFS/B10.BR F1 mice backcrossed to B10.BR. No relationship between IL-1 or IL-3 secretion was evident in strength of rejection phenotype for individual mouse strains (Zhu and Bell, 1990). Further studies of cytokine production in mice infected with *T. spiralis* suggested that immune responses were compartmentalized with IL-5 producing cells dominating in the mesenteric lymph node but IFN-γ-producing cells dominating in spleen (Kelly *et al.*, 1991). This work was extended to an examination of the cytokine profile of mice of different rejection phenotypes, i.e., AKR (intermediate = 'resistant') and B10.BR (weak = 'susceptible') (Pond *et al.*, 1992). They found that both IFN-γ/IL-2 secreting cells and IL-4/IL-5 secreting cells could develop simultaneously but that in B10.BR mice IL-4/IL-5 secreting cells dominated whereas in AKR mice IL-2/IFN-γ secreting cells were predominant. This led these authors to propose that the resistant (strong rejection) phenotype was T_h1 dependent and the weak rejection phenotype was T_h2 determined. However, a contemporaneous study compared NIH (strong) and B10.G (weak) responder phenotype mice (Grencis *et al.*, 1991) and found a predominantly T_h2 response with essentially no difference between the two strains in cytokine profile. Equivalent profiles of high and low responder mice for IL-2 and IL-3 were also found by Crook and Wakelin (1994). While these results were in sharp contrast to those of Pond *et al.* (1992), they also differed from findings with other intestinal nematodes in which a strong dependence of immunity on the activation of T_h2 cells and IL-4 function was evident (Else and Grencis, 1991; Urban *et al.*, 1991). Recent evidence supports the view that the capacity to reject adult *T. spiralis* is IL-4 related (see Section 3.2.6). Perhaps the conflicting data in mice and the absence of distinct cytokine differences between strains of distinct phenotype reflects the fact that cell populations that include activated but nonprotective cells have usually been tested. Furthermore, differences in cytokine levels in the gut could be expected to be more important than mesenteric node cells and may be more consistent with a T_h2 profile.

In rats, there are fewer reagents for detecting and analyzing the cytokine profile, but the ability to separate cells based on the monoclonal antibody (mAb) OX22 has clearly defined *T. spiralis* reactive cell populations that have distinct cytokine profiles. This was first evident through *in vivo* experiments in which recipients of protective $CD4^+$ $OX22^-$ cells displayed a pronounced intestinal eosinophilia and increased antibody formation, whereas recipients of the reciprocal population, $CD4^+$ $OX22^+$ cells, elicited a strong mastocytosis, but no protection (Wang et al., 1990b). The separation of IL-3 production between these two subsets in rats contrasts with mouse studies where IL-3 has been found in protective but uncloned cell lines (Riedlinger et al., 1986). After in vitro stimulation, the $CD4^+$ $OX22^-$ cell secretes little or no γ-IFN or TNF-α, but the CD4 $OX22^+$ cells secreted substantial amounts of γ-IFN. Secretion of IL-4 was common to both cell types and mRNA levels for IL-2, 3, 4, 5 and 10 or γ-IFN, from both cell types did not differ (Ramaswamy et al., 1994a). Thus, while the two cell populations are clearly functionally different they appear to be in the process of differentiating as they enter thoracic duct lymph and apparently continue this process after they have re-entered the gut. Recent studies of afferent intestinal lymph in rats injected with $CD4^+$, $OX22^+$ or $OX22^-$ cells have shown that IL-4 production is shut off within 7.5 h in the former population, but continues past 24 h in the latter (Ramaswamy et al., 1996). Analysis of cytokine profile in the afferent lymph draining the infected gut has provided a continuous monitor of tissue cytokine levels and fluctuations (Ramaswamy et al., 1996). This has shown that IFN-γ is produced intermittently in the gut during the first 8 days of infection and that the presence of an ongoing strong IL-4 response does not exclusively commit local cells to a T_h2-type cytokine repertoire. While IFN-γ was produced in waves, IL-5 was present continuously from early in the infection. Most IFN-γ was synthesized in the gut until 57 h after infection and then the mesenteric node took over, although it produced less. From 55 h, IL-4 was detected in substantial amounts in both afferent and efferent lymph, with about twice as much in the latter. It is not clear whether the increased IL-4 in efferent lymph arises from production in the node or from the absence of consumption of IL-4 from afferent lymph in the node. It is probably no coincidence that the rise in IL-4 occurs simultaneously in afferent and efferent lymph at around 55 h. This is the precise time at which the first activated $CD4^+$ $OX22^-$ cells are carried from the gut in afferent lymph to the mesenteric node. These results contrast with those of Grencis et al. (1991) who were unable to find IFN-γ in NIH mice at days 8 and 13 after infection but are perhaps more comparable to those of Grencis et al. (1987) who also found IFN-γ production by mesenteric node cells. The NIH mouse rejects T. spiralis with similar kinetics to the AO strain of rat used by Ramaswamy et al. (1996) and the underlying kinetics of cytokine production might be expected to be similar. The in vivo pattern of cytokine production, particularly that from the gut, has not been defined for mice. In rats, cytokine production from the gut was not accurately reflected by apparent production from the mesenteric lymph node, as measured in efferent lymph of these rats. Overall, these results suggest that the predominant response is universally of a T_h2 type but the production of T_h1 cytokines continues possibly throughout the entire enteral period of infection.

3.2.6. The Role of T Cells in Protection

An essential role for thymus-derived cells was demonstrated in work on nude mice and rats where adult *T. spiralis* survived for much longer in the gut. Crum *et al.* (1977) showed that protection against adult worms was conferred on naive rats by T cells isolated from thoracic duct lymph. The T cells were nondividing, which probably reflected the use of hyperimmune rats (Crum *et al.*, 1977). Using mice, both Manson-Smith *et al.* (1979a) and Wakelin and Wilson (1979a) transferred protection with T cells. Protective cells were present in the mesenteric node on days 4 and 8 after infection (Grencis and Wakelin, 1982). In the thoracic duct lymph of rats however, they were present only for about 2 days, from days 3–5 after infection (Bell *et al.*, 1987). Protective T cells were short-lived and dividing whether from mice or rats (Wakelin *et al.*, 1982; Bell *et al.*, 1987) and $CD4^+$ (Grencis *et al.*, 1985; Korenaga *et al.*, 1989).

Results from several other laboratories working with different intestinal nematode infections of mice such as *N. brasiliensis*, *T. muris* and *H. polygyrus* have established an important role for T_h2 responses, and specifically IL-4 in protection (Urban *et al.*, 1991; Else *et al.*, 1994; Urban *et al.*, 1995). Furthermore, the results of studies with *T. spiralis* in rats have consistently shown that the protective $CD4^+$ $OX22^-$ cell population is more T_h2 than T_h1 in cytokine profile (Ramaswamy *et al.*, 1994a) and able to sustain IL-4 production *in vivo* after a challenge infection (Ramaswamy *et al.*, 1996). In the mouse–*T. spiralis* system, it is peculiar that no differences in cytokine production has been found among strains where pronounced differences in eosinophils (Wakelin and Donachie, 1983b; Lammas *et al.*, 1992) or gut mast cell numbers have been observed (Alizadeh and Wakelin, 1982). The simplest interpretation of these results is that the genes that determine strong and weak adult worm rejection are unrelated to cytokine profile. In this regard it is remarkable that almost all mouse and rat strains, whether strong or weak in rejection strength, display eosinophilia, mastocytosis and elevated IgE and IgG1 responses. These responses are also typical of the other intestinal nematodes for which host response patterns have been defined. All are hallmarks of T_h2 activation in mice, although mastocytosis in rats may be more T_h1 dependent (Wang *et al.*, 1990b). Another plausible explanation is that investigators have simply not sampled the critical cytokine-producing tissues (most likely the intestine) at the right time.

There is increasing evidence to suggest that IL-4 is important in adult worm rejection. In rats, IgE has been shown to be capable of transferring immunity (Ahmad et al., 1991a) and IgE is IL-4 dependent (Finkelman et al., 1986). Further, IL-4 has now been shown to stimulate the intestinal translocation of IgE (Ramaswamy et al., 1994b) in a process that is probably connected to the expression of IgE-dependent rapid expulsion. Treatment with OX-81, a monoclonal anti-rat IL-4 impairs rejection in the primary infection (R.G. Bell, unpublished observations). In BALB/c mice, treatment with anti-IL-4R mAbs enhances muscle larvae burden in the primary infection and treatment with anti-IL-4R mAb and anti-IL-4 mAb together prevents worm expulsion in immune BALB/c mice (J.F. Urban, personal communication). Recently, IL-4 KO mice have been shown to display delayed rejection of adult worms from the gut (J.A. Appleton, personal communication). Thus, in both rats and mice there is a growing body of direct and indirect evidence indicating a role for IL-4 in the *T. spiralis* rejection processes. However, there is no evidence that IL-4 or any other cytokine directly affects *T. spiralis* adult worms *in vivo*, but with increasing evidence for multiple mechanisms of protection, possible direct effects of some cytokines cannot be excluded. The usual model has T cells exerting an effect through cytokines which act on a second (and perhaps third) step in a sequence leading to worm expulsion. I would suggest that the first step is to activate and promote the growth of local B cells in the gut and the second is to stimulate non-B cells of the gut (such as, but not only, enterocytes) that may have several roles in protection.

3.3. Immunosuppression

In 1970, Svet-Moldavsky and colleagues ushered in a new approach to the study of *Trichinella* infections when they demonstrated that BALB/c mice infected with *T. spiralis* 23 days earlier were unable to reject skin grafts as effectively as uninfected controls. Svet-Moldavsky et al. (1970) proposed that *T. spiralis* secreted a molecule that suppressed the immune response, thus prolonging graft rejection time. Within a year, Faubert and Tanner (1971) showed that mice infected with *T. spiralis* for 30 days responded poorly to sheep red cells, as assessed by anti-sheep red cell antibody level. Following the suggestion of Svet-Moldavsky et al. (1970) Faubert and Tanner injected test mice with sera from *T. spiralis*-infected rabbits and mice in a search for a *T. spiralis*-derived circulating immunosuppressive factor. As controls, they included rabbit anti-mouse lymphocyte serum and normal rabbit serum injected groups. All groups, except the normal mouse serum recipient, showed a two- to three-fold reduction in hemagglutinin titer. Although Faubert and Tanner felt that their results supported the notion of a circulating *T. spiralis*-derived factor, they also entertained other hypotheses, notably antigenic competition. Barriga (1975) found that anti-SRC antibody levels were depressed in C57Bl/6 mice when immunized on day 10 of infection. In addition, Cypess *et al.* (1973) showed that a *T. spiralis* infection could impair antibody responses to a concurrent infection with Japanese B encephalitis virus (Cypess *et al.*, 1973).

Trichinella spiralis infection was not universally suppressive; for example, antibody levels to polyvinylpyrrolidone were normal, and graft-versus-host reactions were similarly unaffected (Barriga, 1975, 1978). Furthermore, the timing of antigen presentation was important — mice infected for 56 and 72 days mounted normal anti-SRC responses (Lubiniecki and Cypess, 1975) whereas antibody to Japanese B encephalitis virus was reduced when mice were infected 7–28 days before, but infections 42, 56 or 70 days before did not reduce the antibody response. Jones *et al.* (1976) found that antibody responses to SRC were unimpaired when the antigen was given 3, 7 or 14 days after infection, but strong impairment was seen at 20 days. Overall, the results suggested that antibody formation was most influenced from 14 days after infection to around 30 days after infection. That is, after adults have been eliminated from the intestine of all but a few strains of mice and after the great majority of newborn larvae have entered muscle in all strains.

Several attempts were made to demonstrate that parasite-derived substances were directly immunosuppressive. Faubert and Tanner (1974) showed that mice injected with serum from infected mice or treated with a saline extract of muscle larvae had levels of rosette-forming cells that were identical to actively infected mice. Barriga (1975) found that *T. spiralis* muscle larvae antigen injected i.p. for 7 days impaired anti-sheep red cell antibody responses to a greater extent than active infection. Faubert (1976) demonstrated that newborn larvae, but not muscle larvae or adults, could inhibit the development of anti-sheep red cell antibody by spleen cells *in vitro*. Barriga (1978) published the final paper in this area when he demonstrated delayed skin graft rejection in mice receiving extracts of muscle larvae.

Overall, the evidence that any stage of the life cycle produces immunosuppressive molecules must be regarded as inconclusive at best. No molecule was ever isolated, antigen doses did not reflect actual infection levels, and there were discrepancies in the results from different laboratories. The lack of a direct connection between the temporal sequence of immunosuppression as seen in active infection (i.e., from 14 to 30 days) and the use of extracts from muscle larvae argue against these being related phenomena. While serum transfers were used frequently, there was no direct evidence that serum from infected animals contained *T. spiralis* proteins in the amounts likely to have been achieved after direct *in vivo* injection of extracts. The case for a direct immunosuppressive effect of *T. spiralis* products can only be regarded as tenuous and runs counter to clear evidence of strong host responsiveness to the parasite.

An alternative view of the immunological phenomena associated with the infection was suggested by Jones *et al.* (1984) who showed that the addition of splenocytes from *T. spiralis* infected mice inhibited the anti-sheep RC response of normal splenocytes *in vitro*. This suppression was abolished if the infected splenocytes were treated with anti-Θ-immune serum before addition to the culture. The work thus implicated T cells in the infected mice as the source of 'suppressive' agents. While this theme was not pursued, it is interesting that one study of DH to BCG in *T. spiralis* infected mice showed that the peak DH response after infections with 200 larvae occurred at 28 days after infection (Blackwood and Molinari, 1981). This suggests that at 28 days T_h1 cytokines may be relatively dominant.

The immunosuppression period in *T. spiralis* experimentation followed closely a time where immunosuppression was a strong theme in the basic immunology literature. Essentially all of the studies examined antibody responses and, as shown in Sections 3.1.1 and 3.1.2, *T. spiralis* induces dramatic but selective elevations in certain immunoglobulins (IgG1, IgM, IgA and IgE) reflecting the strong T_h2 activation. The other IgG isotypes IgG2b and IgG3 may even fall in concentration. It appears probable that the suppressive effect noted by researchers, especially to a relatively benign, nonreproducing antigen such as sheep red cells, reflected the powerful shift to different isotypes produced by the *T. spiralis* infection. In no sense did this change reflect an inability of mice to mount an antibody response or to upregulate immunoglobulin production. An examination of anti-sheep red cell isotype profile and of anti *T. spiralis* antibody response concurrently might have helped delineate the true nature of the changes in antibody profile induced by the infection.

While interest in immunosuppressive *T. spiralis* molecules declined, the concept of immunosuppression lingered. Wassom *et al.* (1984a) found that the number of worms present in the small intestine 9 and 12 days after infection (for C3H/He mice) or 12 and 15 days after infection (C57Bl/6) increased proportionately to the dose of worms given, as had been shown earlier (McCoy 1932; Bell *et al.*, 1979, 1983). However, Wassom *et al.* (1984a) also found that the *in vitro* response to *T. spiralis* larval antigen was reduced in C3H/He mice at higher doses, leading to the suggestion that H-2, specifically $H-2^k$ linked genes, produced inhibition of immune responses with higher doses of worms. The delay in worm rejection associated with dose was also observed by Wakelin *et al.* (1985) and again attributed to selective, dose-dependent immunosuppression. In an attempt to resolve this issue, Bell and Liu (1988), detailed the relationship between worm dose and worm expulsion. They found that in all mouse strains, from the strongest responder to the weakest, and independently of genotype, the rate of rejection was slower as the dose of worms increased. Slower rejection with increasing worm dose was also evident in immune rats (Bell et al., 1979). However, if the total number of worms rejected in a given period of time (in this case 13 days) was measured, there was a linear increase in the number rejected that was proportional to dose up to a dose of 800 given (Bell and Liu, 1988). The principal reason for the apparent 'immunosuppression' was a delay in the start of rejection that was proportional to dose; this occurred in all strains of mouse and rat independently of MHC haplotype, rejection phenotype or immune status (Bell and Liu, 1988). It has been suggested that the delay in rejection is due to a requirement for an initial level of immune-mediated damage to occur before any worms are rejected (Bell and Liu, 1988). Obviously, as worm number rises, this will require more host mediators (= a stronger response) and hence more time to produce the mediators. Subsequent experiments by Kennedy et al. (1991) showed that the antibody response of NIH and CBA mice increased in range of components detected and total amount of antibody produced, up to a dose of 800–1000 muscle larvae. This observation supports the view that host immune responses increase proportionately with dose up to a dose of about 800–1000 infectious larvae (Bell and Liu, 1988).

Despite considerable work over a 15-year period, observations on immunosuppression remain phenomenological and the existence of any distinct form of immunosuppression has not been established. Rather, the data suggest that *T. spiralis* initiates a powerful, dose-dependent stimulation predominantly of T_h2-type CD4 cells for about the first month of infection. After this, as the muscle-stage matures, very limited data suggests a switch to a T_h1-dominated response. Most of the phenomena advanced as evidence of 'immunosuppression' can be explained based on the initial commitment by the host of substantial resources to dealing with the intestinal infection, and then of the development of a T_h2-type response.

4. VARIATION IN THE HOST RESPONSE TO INFECTION

Interest in the degree to which inbred animals differed in their responsiveness to *T. spiralis* began in the mid 1970s (Perrudet-Badoux et al., 1975; Rivera-Ortiz and Nussenzweig, 1976) and peaked in the 1980s. A focus on the major histocompatibility complex (MHC) during that period derived from the view of immunologists that the MHC played a central role in regulating the immune system and that insights into protective mechanisms and cellular interactions would accrue from the analysis of MHC variation. However, few specific insights into the overall immune response to *T.*

spiralis or its control have emerged from the genetic studies. The interesting basis for response variation due to genes outside the MHC is still a neglected area. While the last comprehensive review of the immunology of *T. spiralis* infections dates to 1983, there have been no less than six reviews directly addressing the genetics, particularly the role of the MHC in mice since then (Wakelin, 1985, 1988; Wassom *et al.*, 1987; Wassom, 1988; Robinson and David, 1989; Wassom and Kelly 1990) and a further four that address the genetics of resistance to *T. spiralis* in mice in a context that includes other parasites and their immunogenetics (Wakelin, 1986, 1989, 1992; Wassom, 1993). As there have been no new studies on the MHC since the most recent of these reviews, the MHC data will not be reviewed in great detail here and readers are referred to these papers.

Perrudet-Badoux *et al.* (1975) used the Biozzi mice selected for high and low antibody responsiveness to sheep red cells to examine resistance to *T. spiralis*. The high antibody line produced elevated levels of IgE and IgG1 but there was no difference in muscle larvae burden between the strains. However, in subsequent experiments, Perrudet-Badoux *et al.* (1978) showed that the high antibody line had a significantly lower muscle larvae burden than low-line mice after the primary infection, but did not control the challenge infection as well as low line, again as assessed by muscle larvae burden. Using conventional inbred lines, Rivera-Ortiz and Nussenzweig (1976) demonstrated that the establishment of muscle larvae was inversely proportional to the amount of IgE produced by each strain. The DBA/1 strain produced most IgE and harbored the fewest muscle larvae, whereas SJL mice produced least IgE and had the most muscle larvae. Comparable muscle larvae burdens were also found by Bell *et al.* (1985) and Wassom *et al.* (1983a) with most of the same strains. Rivera-Ortiz and Nussenzweig (1976) concluded that there was some association of IgE production with the MHC. However, these authors also concluded that there were no differences between the strains in worm rejection from the gut, a factor that could clearly influence the number of muscle larvae establishing. Subsequently, several laboratories showed reproducible differences in the rate and time of adult worm rejection between most of these strains (Wakelin, 1980; Bell *et al.*, 1982b, 1984a; Bell, 1988). Neither Rivera-Ortiz and Nussenzweig (1976) or Tanner (1978) who examined muscle larvae burden in H-2 disparate mice appreciated the H-2 linked differences that subsequently became apparent.

4.1 Major Histocompatibility Complex-linked Variation

Wassom *et al.* (1979) presented the first evidence for MHC-linked effects on resistance of mice to *T. spiralis* infections. Using muscle larvae burden to assess degree of resistance to a standard infection in B10 congenic mice they found a slightly larger than two-fold range of muscle larvae values (B10.Q and B10.S lowest and B10.P and B10.BR highest). Comparison of several inbred strains, including a group of five that were all H-2^k, showed a range of about three-fold overall (highest C3HeB/Fe, lowest BUB/Bn) and just under twofold within the H-2^k identical group. This evidence established the existence of genes both within and outside the MHC that influenced levels of resistance to *T. spiralis*. The authors interpreted these data to indicate that genes within the MHC were more important overall than the non-MHC-linked genes. In the same paper, Wassom *et al.* (1979) examined muscle larvae burden in congenic intra-H-2 recombinant mice and localized the principal determinant of low resistance to H-2^k genes (either I-A or I-B) mapping to the K end of the H-2 region. Wakelin (1980) analyzed the genetics of resistance by measuring the speed of rejection of adult worms from the gut of mice of distinct strains. He showed that H-2^q mice of different backgrounds were able to substantially reduce intestinal adult worm number by day 12 and that B10 mice took several more days (day 20) to achieve the same result. Wakelin (1980) also determined whether various treatements designed to inactivate suppressor cells, such as irradiation, cyclophosphamide treatment or thymectomy, would enhance the weak rejection of B10 mice. No treatment was able to modify the rejection pattern, leading Wakelin (1980) to state that there was no evidence for suppressor cell activity in the B10 mice. Wakelin (1980) concluded that genetic control of resistance was complex, with both H-2 linked and non-H-2 genes likely to be involved. He proposed that MHC genes may determine immunological responses to infection whereas non-MHC genes influenced the degree of inflammation that occurs. The role of non-MHC genes was further examined in experiments in which immune mesenteric node cells were transferred between H-2^q identical strong responders (NIH) and weak responder (B10.G) mice where the recipient-phenotype determined the outcome after challenge (Wakelin and Donachie, 1980). Similar findings regarding the strength of MHC and non-MHC genes were presented by Bell *et al.* (1982a,b) after examining both rapid expulsion and adult worm expulsion either in the primary infection or after challenge of mice given stage-specific immunization. Thus, a pattern was established in which MHC-related variation was most readily evident in congenic strains that were poor overall responders (e.g. C57Bl/6 or 10) due to the overriding effects of their non-MHC genes on adult worm expulsion. If muscle larvae burden rather than adult worm rejection was used as the primary criterion for assessing the strength of response, the relative strength of non-MHC genes is obscured (Wassom *et al.*, 1983a). This is probably due to host effects on either fecundity or resistance to newborn larvae implantation

(Bell et al., 1985) which can counteract the effects of weak worm rejection in establishing muscle larvae load.

In addition to demonstrating H-2 related differences in mesenteric node cell proliferation in vitro (H-2^s > H-2^k) Krco et al. (1982) were able to show that Ly1^+ cells (CD4^+) and Ia antigen recognition were important for proliferation after antigen stimulation. These results were extended when Krco et al. (1983) showed that cloned T cell lines were inhibited by I-E specific alloantisera but lost the I-A recognition pattern displayed by freshly isolated immune cells. Thus, both I-A and I-E class II molecules seemed to be involved in recognition of *T. spiralis* antigens by B10.K mice. These data were in accord with the view that one MHC-linked gene (Ts-1, Wassom et al., 1983b) lay in the I-A region. A second MHC-linked gene mapping to the D end of the H-2 complex was also proposed by Wassom et al. (1983b) based on comparisons of B10.S recombinants (s and d alleles) and B10.A related strains (b and d recombinants). One or more strains in each group e.g, B10.S (7R), typing as D^d and B10.A (18R) typing as D^d was significantly more susceptible to *T. spiralis*, as judged by muscle larvae burden, than were sister strains of apparently identical H-2 configuration (all D^d). Wassom et al. (1983b) postulated that a crossover must have occurred between the S and D loci of the H-2 region during strain development. While there was no known gene product or independent antibody to mark the locus, it was designated Ts-2 and defined by theoretical crossover maps giving the Ts-2 gene to four recombinant strains B10.S (7R), B10.M (11R), B10.A (18R), and B10.T (6R) (Wassom et al., 1983b). Wakelin and Donachie (1983a) reported that the B10.T (6R) ($K^q I^q D^d$) mice were slower to reject adult worms than were B10.G ($K^q I^q D^q$) suggesting that D^d affected worm rejection. Subsequently, the localization of the I-A gene, Ts-1, was reappraised when Wassom et al. (1987) demonstrated that mouse strains expressing I-E products were more susceptible to infection, as measured by muscle larvae burdens. The authors pointed out that the putative I-A-linked gene Ts-1 may represent an I-E defect (Wassom and Kelly, 1990). It is interesting that the susceptibility effect assigned to expresson of I-E is dominant when F_1 crosses are made between B10 mice (B10.BR × B10.Q − H-2k/q) and muscle larvae burden assessed but recessive, or largely so, when crosses are made between NFR and NFS mice and B10.BR (Bell et al., 1982b; Bell, 1988; Bell and Liu, 1988) and intestinal worm rejection is assessed. This suggests that the effect of Ts-2 may lie primarily in the extent to which newborn larvae successfully implant.

Since the principal known role of class II molecules is the presentation of antigen to CD4 cells leading to antibody formation by B cells, an effect of the putative I-A or I-E gene might be reflected in antibody repertoire. Indeed, Wassom et al. (1983b) proposed that the Ts-1 gene would function as a classical Ir gene controlling the response to a functional antigen. Subsequently, Wassom et al. (1984b) proposed that H-2 linked and non-linked genes interacted independently to influence the outcome of anti-adult responses, anti-fecundity responses and rapid expulsion. This proposal did not specify a site or a mechanism by which these gene products could interact.

Analysis of antibody responses to *T. spiralis* muscle larvae antigens showed differences between strong (NIH) and weak (C3H) strains in timing and amount of antibody produced (Jungery and Ogilvie, 1982) and isotype distribution (Almond and Parkhouse, 1986). However, attempts to identify unique recognition patterns for muscle larvae antigens between different inbred mouse strains showed that the overwhelming response was to three dominant antigens of 41, 46 and 55 kDa in all the strains (Kennedy et al., 1991). When I-A repertoire was specifically examined (Kennedy et al., 1991) differences were apparent in specificity of antigens detected, but usually only to minor antigens. Although comparisons between all B10 recombinants defining Ts-1 (Wassom et al., 1983b) were not done, sufficient were analyzed for the authors to conclude that there was no association between antibody repertoire and susceptibility or resistance to infection. This result raises questions about the function of the putative Ts-1 gene. Either one must propose that the antibody repertoire may be distinct from the T cell repertoire as Kennedy et al. (1991) did and then rejected, or one must propose as yet unknown functions for I-A and/or I-E genes. Earlier papers (Wassom et al., 1984a; Wakelin, 1988; Wassom and Kelly, 1990) had proposed a role for suppressor cells, pointing out (Wassom et al., 1987) that the I–J region recently displaced from the MHC had been (temporarily) relocated on chromosome 4, a position that was linked to the FV-1 locus that discriminated the AKR/J and the AKR-Fv-1 strains. However, the dose-dependent MHC-linked suppression invoked by Wassom et al. (1984a) and Wakelin (1988) to explain the susceptibility of $H-2^k$ mice to *T. spiralis* was shown to be a general effect that occurred in all inbred mouse and rat strains independently of genotype (Bell and Liu, 1988). This eliminated selective immunosuppression and we are presently left with no putative function for Ts-1. While some significance was attached to the fact that the genes for TNF-α and TNF-β are in the general region mapped for Ts-2 (Wassom and Kelly, 1990), no evidence for a particular of differential role of TNF in *T. spiralis* infections has yet appeared.

In seeking a role for the Ts-1 and Ts-2 genes we should bear in mind that these genes, and their positions, were defined by the same techniques that localized the I–J and I–B regions in the MHC, both of which are now defunct. For the I–J region there were many serologically defined products at the peak of their fame but no comparable markers have yet been found for Ts-1 or Ts-2. Of particular significance is a paper by Klein *et al.* (1982) which examined the chromosomal composition of 104 H-2 congenic lines. They found that large portions (sometimes all) of chromosome 17 telomeric to the H-2 region in the congenic strain had been inherited from the H-2 donor strain. In other words, identification of genes within H-2 recombinant mice using the assumption that the H-2 region is the only differential segment between these congenic lines is not tenable. Because many of these chromosomal segments are considerably larger than the H-2 component of interest, the chance that a given phenotype will result from gene(s) lying outside the H-2 complex is greater than the probability it will result from genes within the H-2. Without some other form of localization, through molecular or classical genetic studies, the location of Ts-1 and Ts-2 within the H-2 region on chromosome 17 should be considered tentative at this time. If these genes are not in the MHC, we should not be surprised if we do not find effects related to Class II or Class I antigen presentation.

4.2. Non-Major Histocompatibility Complex-linked Variation

While the contributions and site of action of MHC-linked genes appear to be less secure than they were in the late 1980s, the significance of the non-MHC genes has, if anything, consolidated. This has occurred despite the fact that no defined functions have yet been ascribed to any of these genes. The most obvious genetic effects are those that influence the timing of adult worm expulsion from the gut of mice during the primary infection. In the early work, some confusion resulted from the different classifications used by different groups to define the phenotypes that were observed. For example, Wakelin (1980) and Wakelin and Donachie (1980) using NIH and B10 congenic mice defined two phenotypes, rapid (NIH) and slow (B10), a terminology which they retained with several different inbred strains. In contrast, Bell *et al.* (1983) found consistent differences between NFR/N and NFS/N (related to NIH), C3H/He and B10 congenics and their F_1s leading to a classification of three phenotypes, strong, intermediate and weak. These phenotypes were subsequently demonstrated to be applicable to a variety of different strains (Bell *et al.*, 1984a) and over a wide range of doses of infectious larvae (Bell and Liu, 1988). Finally, a genetic proof of the validity of these designations was provided when backcrosses between the NFS (strong) and B10.BR (weak) produced a 1:2:1 segregation of offspring of high, intermediate and weak phenotype respectively, whereas C3H/B10.BR × B10.BR segregated in at 1:1 intermediate:low ratio (Bell, 1988). These data thus established the existence of two independently segregating genes that determined the basic phenotype of adult worm expulsion. These genes have not been further defined or named. It is within the framework established by the non-MHC genes (strong, intermediate or weak) that the putative MHC-linked genes operate. The effect of Ts-1 or Ts-2 is to move response type (particularly weak responders) to one or other end of the phenotypic range (e.g. weak/weak or strong/weak) defined by non-MHC genes. While the phenotypes defined as strong, intermediate and weak remained stable in relative position over a wide dose range (from 100 muscle larvae–1000 muscle larvae), they were impossible to distinguish at doses below 50 muscle larvae (Bell and Liu, 1988). Rats show as much strain variation in expulsion strength of adult worms from the gut as do mice (Bell, 1992) but there have been no genetic studies.

A third gene that has been proposed to influence resistance, based on muscle larvae counts, is thought to be linked to the FV-1 locus on chromosome 4 (Wassom et al., 1987). The FV-1 gene distinguishes the AKR/J and the AKR/FV-1^b strains. After a standard infection, the AKF-FV-1^b strain carries about three times as many muscle larvae as the AKR/J strain (Wassom et al., 1987). The original work interpreted this result in terms of the expression of I-J^k near FV-1 (Hayes et al., 1984) and the induction of I-J^k positive suppressor cells, but this hypothesis is no longer tenable. In the absence of further data all we can say is that a gene that may be on chromosome 4 influences muscle larvae burden in AKR mice. The effects in AKR mice on muscle larvae burden are similar to results obtained in several other strains of mice in experiments that examined resistance to newborn larvae (Bell et al., 1985). In those studies, a 'nonspecific' resistance to the establishment of newborn larvae in muscle was identified. This resistance varied between strains independently of the phenotype of adult worm rejection and clearly had a genetic basis, with C3H mice showing the least resistance and DBA/1 mice the most. No genetic experiments were carried out but the gene(s) could be related to the FV-1-linked locus on chromosome 4 identified by Wassom et al. (1987).

A fourth gene, designated Ihe-1 was proposed based on the segregation of rapid expulsion in mice (Bell et al., 1984c). However, with evidence now available indicating that mice cannot express rapid expulsion (Bell, 1992) it seems more likely that this gene may be one of the two genes that determine adult worm rejection phenotype, as discussed above. Genetic variation in resistance to newborn larvae is also clearly present but whether or not one of these genes is synonymous with Ts-3 remains open to question. The experiments that have been conducted to date have demonstrated genetically determined variation between mouse and rat strains in virtually every measure of host immunity that has been examined. However, it has proven extremely difficult to define the genes or assign them a specific immunological or physiological role. This is an area that still promises rich insights for the persistent investigator.

5. THE RESPONSE OF GRANULOCYTIC CELL POPULATIONS TO INFECTION

5.1. Eosinophils

The pronounced effect of a *T. spiralis* infection on the numbers of circulating eosinophils was recognized clinically and quantitated experimentally at the turn of the century (Brown, 1897; Opie, 1904). Opie (1904) provided temporal analyses of the blood pattern in guinea pigs and showed substantial increases in eosinophil numbers in the third week of infection. He also showed that the blood eosinophil count was influenced by the dose of worms given — it was larger and earlier with higher doses — and that eosinophils originated in bone marrow. This work was extended to rats (Beahm and Downs, 1939; Beahm and Jorgensen, 1941) and rabbits (Wantland, 1937) and showed that the blood eosinophil response of rats occurred earlier than in guinea pigs, with an increase detectable by the end of the first week and a peak occurring by the end of the second week. Eosinophilia was thought to be initiated by the migrating newborn larvae since it occurred after newborn larvae migration began (Opie, 1904; Bachman and Oliver-Gonzalez, 1936; Beahm and Downs, 1939). However, Zaiman and Villaverde (1964) demonstrated that an eosinophilia in the uninfected partner of a parabiotic pair of rats that had been surgically separated 4.5 days after infection, i.e., prior to the initiation of newborn larvae migration. The role of the infectious L1/adult worm was established by the studies of Basten *et al.*, (1970a,b) who produced an eosinophilia with intravenously injected L1 larvae and Pincus *et al.* (1986) who produced eosinophilia with an intestinal infection abbreviated by drug treatment before newborn larvae production began. The work of Basten *et al.* (1970a,b) established another important point by showing that the L1-induced eosinophilia could be transferred to uninfected rats with thoracic duct T cells from infected rats. This observation highlighted the role of host T cells and was an important step in the analysis of T cell products and their biological roles. More importantly, it demonstrated that eosinophlia was not produced by some specific eosinopoietic component of the *T. spiralis* parasite. This had been a popular theory among researchers until the striking observations of Basten *et al.* (1970a,b).

Pincus *et al.* (1986) reported a biphasic eosinophilia in inbred mice of the C3H/He, DBA/1 and B10.D2 strains with peaks at 5 or 8 days and again at 17–23 days. However, the early peak identified around 1 week after infection by Pincus *et al.* (1986) was not evident to all investigators using mice and comparable sampling points (see Wilkes and Goven, 1984; Lammas *et al.*, 1992). Furthermore, it has not been described in rats at all, where a gradual increase, starting by the end of the first week and peaking around the end of the second week or early in the third week seems more typical (Zaiman and Villaverde, 1964; Basten et al., 1970a,b; Ismail and Tanner, 1972a). Variation between strains of mice in their capacity to produce an eosinophilia was first described for *T. spiralis* by Wakelin and Donachie (1983b) who demonstrated that NIH mice mounted an earlier and stronger response than C57B1/10 strain mice.

5.1.1. *Role in Protection*

The first evidence of eosinophil-mediated protection (Grove et al., 1977b) showed that treatment of mice with a rabbit anti-eosinophil serum increased the numbers of muscle larvae. In a follow-up study Kazura and Grove (1978) observed antibody-dependent killing of newborn larvae *in vitro* by eosinophils. Optimum conditions resulted in up to 99% killing but required a cell ratio of 10 000:1 newborn larvae, 24-h exposure, and induced peritoneal exudates from mice that had been infected 4–6 weeks before cell harvest. While the data were convincing, there were nevertheless questions, for example, the requirement for such high cell:larvae ratios was disquieting, as was the use of 24-h exposure. Neither seemed likely to occur naturally *in vivo*. In addition, the optimum peritoneal exudate cell population was obtained after the period during which newborn larvae production would be expected to occur *in vivo*. Nevertheless, further analysis detailed the attachment process of cells to the cuticle (Kazura and Aikawa, 1980) binding of C3 to the cuticle, and analysis of the role of other cells, such as macrophages which enhanced killing (Mackenzie et al., 1980a). Since, in the absence of immune serum cells did not bind to the cuticle, either Fc receptor or complement receptor-mediated binding was presumably occurring. Interestingly, Mackenzie et al. (1980b) found that neutrophil degranulation on the cuticle of newborn larvae left a deposit that reduced nitro blue tetrazolium (nbt — an indicator of peroxidase enzymes) but eosinophils, which also degranulated, did not. One possible explanation was that eosinophils were killing newborn larvae with a non-peroxidase-based system. Wassom and Gleich (1979) had ealier indicated that major basic protein could be an effector. Peroxidases were, nevertheless, shown to be capable of killing newborn larvae by Buys et al. (1981) who isolated human eosinophil peroxidase and combined this with H_2O_2 and chloride *in vitro* at pH 5.5 to effect newborn larvae killing within 20 min of exposure. They attributed this effect to the formation of hypochlorous acid. Subsequently, myeloperoxidase derived from macrophages and neutrophils was shown to be more efficient at killing newborn larvae than eosinophil peroxidase, thus leaving open the question of which cell population may be more effective (Buys et al., 1984). Kazura and

Meshnick (1984) showed that adult worms and muscle larvae contained three to five times as much superoxide dismutase and five times as much glutathione peroxidase as newborn larvae, suggesting that the stage-specificity of eosinophil activity (only against newborn larvae) lies in the relative amounts of antioxidant possessed by each stage. An attempt to connect the distinct lines of evidence from *in vitro* and *in vivo* experiments was made by Lee (1991) who showed that eosinophils isolated from the intestine of infected rats could be induced to kill newborn larvae, provided that IL-5 was added to the culture medium.

The recognition that eosinophil growth and maturation was dependent on the cytokine IL-5 ultimately led to the use of an mAb to IL-5 to prevent the appearance of eosinophils at the source, the bone marrow. Herndon and Kayes (1992) used this approach to abrogate both eosinophilia and bone marrow eosinopoiesis during infection. Analysis of muscle larvae numbers showed no differences betwen anti IL-5 treated mice and untreated controls or controls treated with a control mAb. This was true in both primary and secondary infections. These results raise doubts about all the earlier work showing effects of anti-eosinophil sera on muscle larvae burden as well as the *in vitro* data indicating killing by eosinophils.

There are a variety of potential explanations for these discrepancies which should be viewed against a background of strong *in vivo* evidence for the existence of anti-newborn larvae immunity in both mice and rats (Bell *et al.*, 1985; Wang and Bell, 1987, 1988). Herndon and Kayes (1992) mention several: the possible activity of anti-eosinophil sera against other cell types, the possibility that eosinophils lysed/damaged as a result of antibody treatment impair some non-eosinophil component of the immune system used for anti-newborn larvae immunity. Two other elements that should be kept in mind are: (1) the evidence that cell types other than eosinophils can damage and kill newborn larvae, and (2) that Herndon and Kayes' (1992) experiments used complete infections and muscle larvae numbers as their assay. No attempt to measure anti-newborn larvae immunity directly was made although this may have been quite informative. The role of anti-newborn larvae immunity in a primary infection is relatively small compared with the effects of anti-adult responses (Wang and Bell, 1987). In secondary infections of both the strains used by Herndon and Kayes (1992) very few newborn larvae are produced as few adults survive long enough to produce them. These experiments were, therefore, not that definitive in determining whether eosinophils might be an important component of anti-newborn larvae responses. The question should remain open at this point.

To summarize, antibody-dependent killing of newborn larvae by eosinophils has been described *in vitro* under relatively contrived conditions. *In vivo*, immune destruction of newborn larvae has been described in mice and rats, including under conditions in which fecund adult worms were transplanted to the gut (Wang and Bell, 1987). Granulocytic cells and macrophages have been associated with *in vivo* killing but the role of eosinophils specifically has not been defined fully and there are conflicting data.

5.2. Mast Cells

Since the discovery of the dramatic mastocytosis that accompanies most intestinal nematode infections (Miller and Jarrett, 1971), enormous effort has been directed at defining a role for these cells in protection. Early work with *T. spiralis* infections established the thymus-dependency of the intestinal mast cell response and its kinetics (Ruitenberg and Elgersma, 1976; Ruitenberg *et al.*, 1979; Brown *et al.*, 1981). Mast cell numbers rise at the end of the first week (Day 7, Ruitenberg and Elgersma, 1976) or as late as days 10–12 after infection begins in rats and mice (Tronchin *et al.*, 1979). Peak intestinal mast cell levels usually occur late in the second or early in the third week and decline quickly to an elevated plateau thereafter, although the time at which they fall to baseline levels has not been defined. Variation in intestinal mast cell numbers and their kinetics was shown by Alizadeh and Wakelin (1982) to be genetically determined in mice with B10 congenics having a slow mast cell response in the primary infection (begins at day 12), and NIH and DBA mice having a rapid response beginning by day 8. These authors also showed that, while immune mesenteric lymph node cells could transfer an enhanced mastocytosis to infected mice, the ensuing pattern of response was that of the recipient phenotype, not that of the donor cells (Alizadeh and Wakelin, 1982). Comparable results were found when lethally irradiated F_1 mice of high mastocytosis phenotype (NIH) crossed with low mastocytosis mice (B10.G) were reconsituted with parental bone marrow cells. The mastocytosis phenotype of the resulting chimera was that of the recipient F_1 mice. Alizadeh and Wakelin (1982) suggested that for intestinal mast cells either the number of mast cell precursors or their rate of division or response to exogenous growth factors (stem cell factor/IL-3) are genetically determined. The authors excluded the possibility that different amounts of T-derived factors (e.g., IL-3) produce mouse strain-specific variation in mast cells, even though T cells are required for mast cell hyperplasia in this system.

5.2.1. *Role in Protection*

The role of mast cells and mast cell derived agents in rejection of intestinal nematode infections has been intensively studied but is still unresolved. One of the earliest models was the W/W^v mouse in which the stem cell factor (SCF) receptor, c-kit (a protooncogene tyrosine kinase), is aberrant. Among other effects, this mutation profoundly decreases numbers of mucosal mast cells (Kitamura *et al.*, 1978). W/W^v mice infected with *T. spiralis* have a poor intestinal mast cell response to infection and show a variably delayed rejection of the primary, e.g., by 6–10 days (Ha *et al.*, 1983; Alizadeh and Murrell, 1984) or only 1 or 2 days (Oku *et al.*, 1984). A delay in the rejection of a challenge infection was also evident in Alizadeh and Murrell's (1984) study. Both Alizadeh and Murrell, (1984) and Oku *et al.* (1984) were able to reconstitute the mast cell deficiency with bone marrow cells and restore rejection capacity. Treatment of normal mice with a monoclonal antibody to c-kit prevented mast cell hyperplasia and also reduced the capacity for adult worm rejection (Grencis *et al.*, 1993). In *T. spiralis*-infected mice, the slower rejection of adult parasites from c-kit dysfunctional mice has been a consistent and repeated observation. As might be expected after these effects from dysfunctional c-kit, interference with its ligand, SCF, also impairs worm rejection (Donaldson *et al.*, 1996). In the experiments of Donaldson *et al.* (1996) treatment with anti-SCF antibody increased intestinal worm burden at day 10, but by day 16 worm burdens were comparable to those of control mice.

Of the cytokines that can influence mast cell numbers in addition to SCF, IL-3 is the most important. Using injections of rIL-3 in C3H/He mice prior to infection (day − 5 to 1 day after), Korenaga *et al.* (1996a) showed reduced intestinal worm numbers at day 5 and day 10 but not at day 14 after infection. The most pronounced effect was seen on day 5 when treated mice had 47% fewer worms, whereas on day 10 there were only 38% fewer worms than in untreated controls. In their experiments, Korenaga *et al.* (1996a) did not examine worm burden earlier than day 5 but these results suggest that it may have been the initial take of worms in the first 24 hours after infection that was critically altered. Treatment of mice with anti-IL-4 or anti-IL-5 did not modify the capacity of IL-3-treated mice to reject adult *T. spiralis*, implying specificity for IL-3.

Measurement of intestinal mast cell proteases (IMCP) made during the period of rejection in strong and weak responder mice (Woodbury *et al.*, 1984; Tuohy *et al.*, 1990) showed mast cell degranulation. These authors measured IMCP in serum and gut homogenates of NIH, SWR and B10.G mice and found a direct correlation between the amount of IMCP released and rejection strength between strains of mice. Thus, observations from several laboratories and with a number of experimental systems make a compelling, albeit circumstantial case for a role of mast cells in rejection of *T. spiralis*. Are there other possible explanations for these associations?

Experiments with W/W^v mice hinge on the role of c-kit. This surface receptor is a member of the platelet-derived-growth-factor/M-CSF family of receptors and it is present and functional on a variety of cell types. It is known to be involved in hematopoiesis, melanogenesis and gametogenesis (Besmer, 1991) and all of these components are deficient in W/W^v mice. In addition, c-kit is expressed on bone marrow stem cells that give rise to lymphopoietic activity of both thymus and B cells (de Vries *et al.*, 1992; Yasunaga *et al.*, 1995). Dendritic cell precursors are another important cell type that are influenced by c-kit (Young *et al.*, 1995). It is noteworthy that no attempts to detail the immune response to *T. spiralis* of infected W/W^v mice have been made and all interpretations on the site of defect have ignored this possibility in favor of mast cells. Reconstitution of W/W^v mice with normal bone marrow (Ha *et al.*, 1983; Oku *et al.*, 1984) provides no help in defining the site of function as the many immune and granulocyte-specific effects noted above would also be compensated. The only specific treatment that would be informative with W/W^v mice would be reconstitution with pure populations of intestinal mast cells. This has not been done. Treatment with antibody to c-kit or SCF (Grencis *et al.*, 1993; Donaldson *et al.*, 1996) suffers the same drawbacks, although to a lesser extent, since the immune and granulocytic defects are not long-standing.

In addition to the defects listed above for SCF/c-kit interactions, there are others that are relevant to the rejection of nematodes from the gut. One is the apparent role of SCF/c-kit in regulating the cholera toxin-induced secretory response of the gut. W/W^v mice have an aberrant intestinal fluid secretory response to cholera toxin (Klimpel *et al.*, 1995). Similarly, elements of the gut neuronal system also express c-kit. The interstitial cells of Cajal are believed to regulate the rhythmic contractions of the gut (Torihashi *et al.*, 1997) and are absent from W/W^v mice. Mast cell–neuronal interactions have been proposed as important mediators of intestinal responsiveness (Castro, 1989). For *T. spiralis* infections, intestinal myoelectric activity and fluid secretion in the small intestine have been related to intestinal rejection processes (Castro *et al.*, 1979; Palmer *et al.*, 1984). These observations do not prove that intestinal epithelial cells or the neuronal network are critically involved in the rejection of *T. spiralis*. In fact, it is noteworthy that the cholera toxin secretory defect of W/W^v mice cannot be reversed by reconstitution with normal bone marrow, unlike *T. spiralis* rejection. Nevertheless, these examples illustrate the pleiotropic nature of c-kit and its broad functional distribution. The degree of effect can be quantitated roughly: the longest delay in rejection in W/W^v was reported as 6 days (Ha *et al.*, 1983), which increases time to rejection by about 42%. However, the data for T-cell deficient mice show that in nude mice some intestinal worms are still present 83 days after infection (Ruitenberg *et al.*, 1977b) or 492% longer than in euthymic controls (using 14 days as the average time to rejection). Thus, the effect of the c-kit deficiency is about 10% of that observed in full T-cell-deficiency. It is important to also note that the difference of 6 days in rejection between c-kit normal and c-kit dysfunctional mice is less than the difference between strong responders (NFS or NIH mice) and weak responders (B10 and congeners) which varies from 6 days to more than 10 days depending on dose (Bell *et al.*, 1982b; Bell, 1988; Bell and Liu, 1988). The latter strains are, in fact, fully mast cell competent, suggesting that genes that do not affect mast cells may be quantitatively as important in protection as the c-kit-related effects. Overall, without more direct evidence and the use of different procedures, the weaker, but not completely absent, rejection of adult *T. spiralis* from W/W^v mice cannot be attributed to mast cells.

Many of the same comments made above for SCF/c-kit also apply to IL-3. While not believed to be as pleiotropic as SCF/c-kit, it is premature to consider the function of IL-3 as being restricted to mast cells, particularly in the gut where so many distinct and potentially reactive cell types are present. Korenaga *et al.* (1996b) demonstrated that IL-3 enhanced IgE responses after infection in W/W^v mice indicating that mucosal mast cells were not required for this effect. Since IgE is known to be protective, it is possible that the effect of IL-3 treatment noted by Korenaga *et al.* (1996a) is IgE-mediated.

What about the association of worm rejection with high levels of released mast cell mediators such as rat mast cell protease II, IMCP and leukotrienes (Woodbury *et al.*, 1984; Moqbel *et al.*, 1987; Tuohy *et al.*, 1990)? These experiments show that mast cells are active close to the time of worm expulsion. Without more direct evidence for a role of the proteases themselves or of the other pharmacologically active agents released by mast cells, the experiments are not definitive, simply correlating these events in time. Furthermore, if we take the data of Manson-Smith *et al.* (1979b) to indicate that rejection occurs locally within the small intestine before an overall loss of worms is evident, then the timing of IMCP release is not as closely related to the primary rejection process. Finally, the experiments of Wang *et al.*, (1990b) clearly dissociated the mast cell inductive properties of CD4$^+$ OX22$^+$ cells, but not CD4$^+$ OX22$^-$ cells from protection.

Mast cell function can also be analyzed by providing selected agents (histamine, serotonin (5HT) and prostaglandins (PG), etc.) directly to parasitized animals and determining whether or not they produce rejection. Briggs (1963a) injected 1 mg 5-hydroxytryptamine (5-HT, serotonin) intravenously 90 min after infection and, 7 days later, assessed worm burdens in four groups given infections with 200, 250, 300, and 350 larvae. Reduced worm burdens were found only in mice receiving 200 larvae, subsequent experiments with higher doses of 5-HT did not increase resistance. In contrast, Dutoit *et al.* (1979) injected PG, histamine and 5-HT separately into infected mice and rats. Single intraduodenal injections were given on day 3 after infection and mice were killed on day 7 for adult worm counts. Histamine (8 mg) and 5-HT (4 mg) did not influence worm count at day 7, nor did 50 µg of $PGA_{1\&2}$, $B_{1\ or\ 2}$ or $F_1\alpha$. However, PGE_1 and PGE_2 both increased worm burden at doses of 50 µg. This result was repeated in a further experiment using later times of injection (day 6) of PGE_2 and lower doses (25 µg) also in CBA mice. In Wistar rats, using the same procedure, PGE_2 inhibited rejection at a dose of 250 µg and enhanced rejection at a dose of 500 µg. The timing of injection suggests that Briggs (1963a) was examining events associated with larval implantation/penetration in the epithelial layer whereas Dutoit et al. (1979) were examining events related to the adult phase of the infection. Neither set of experiments could be considered conclusive but, overall, 5-HT and histamine exerted little effect on worm rejection. Conversely, PGE_2 prolonged worm residence time in most cases (Dutoit et al., 1979).

In a more direct test of the role of PGE_2 Zhang and Castro (1990) injected 50 µg kg^{-1} of PGE_2 subcutaneously 30 min before infection of nonimmune rats. One principal effect of PGE_2 on the gut is the induction of fluid secretion. While this was achieved, no effect of PGE_2 infusion on worm implantation was observed. Zhang and Castro (1990) also infused serotonin directly into the carotid artery beginning 5 min before infection and continuing for a further 30 mins, at which time rats were killed for worm counts. The number of worms implanting was reduced by about 50%. It is difficult to compare these results directly with Briggs (1963a) as there were significant differences in subject species (mice and rats) and experimental procedure with Briggs injecting 4–16 times the dose used by Zhang and Castro (1990). It is unfortunate that Zhang and Castro's (1990) single experiment has not been followed up as the possible role of 5-HT in implantation or rapid expulsion is not resolved by these data which show both enhanced and diminished rejection. Stewart et al. (1985) found that histamine administration to CD-1 swiss mice reduced fecundity in vivo and had a direct effect on fecundity when administered in vitro. There have been no other studies of the direct effects of mast cell products on worm physiology or function, although they are an essential adjunct to the in vivo studies.

An alternative approach is to selectivity inhibit one or more of the pharmacologically active components of mast cells. This method was used by Campbell et al. (1963) with antiserotonin and antihistamine agents. Drugs were added to food and provided ad libitum to mice infected with 875 or 575 larvae in two experiments. In both experiments, large numbers of mice died for undetermined reasons. All drugs, the antihistamine (chlorpheniramine maleate), and antiserotonin (1-benzyl-2-methyl-5-methoxytryptamine), or a drug with both properties (cyproheptadine) were effective in prolonging adult worm residence time. A similar protocol was used by Parmentier et al. (1987) who injected the antiserotonins methysergide or ketanserin subcutaneously (s.c.) 2 × daily into mice infected with 300 *T. spiralis* larvae. This treatment dramatically inhibited mast cell, eosinophil and goblet cell increases in the gut but failed to modify worm expulsion. Ismail and Tanner (1972b) injected methysergide and promethazine (antihistamine) and found some increases in muscle larvae burden. Bell et al. (1982c) isolated the effector arm of rapid expulsion in rats to examine the role of a broader range of potential mediators. In their experiments, most drugs were injected 30 min prior to challenge of immune rats and worm counts were conducted 8 h after challenge. Neither antihistamines, antiserotonins, antiprostaglandins, smooth muscle stimulants nor relaxants were able to influence rejection. Some inhibition of rejection was observed after treatment with the decomplementing agent cobra venom factor and once with promethazine, but the only consistent effects were observed with irradiation and corticosteroid treatments of immune rats (Bell et al., 1982c). The two positive results (promethazine and cobra venom factor) were considered to be statistical artifacts of the tightly grouped positive controls rather than indications of the biological role of histamine or complement.

Zhang and Castro (1990) also examined the role of various inhibitors on rapid expulsion. In this case, the authors had developed a variety of assays for modified epithelial cell function. These included Cl^- exchange and fluid secretion into the gut lumen, both of which could be inhibited by 5-HT and histamine inhibitors *in vitro* (Russell, 1986; Castro et al., 1987). To inhibit fluid secretion *in vivo* Zhang and Castro (1990) used the cyclooxygenase inhibitor indomethacin, the H1 inhibitor diphenhydramine and L-651,392 a lipoxygenase inhibitor. Neither indomethacin nor L-651,392 inhibited fluid secretion by themselves, but together they inhibited worm-induced fluid secretion by about 25%. Diphenhydramine reduced fluid secretion by over 50% and, when combined with indomethacin, by over 80%. No treatment prevented or even inhibited worm rejection, leading the authors to conclude that fluid secretion was not the mechanism of worm rejection in rapid expulsion. However, when the 5-HT antagonists ketanserin and MDL-72222 were used together there was an increase in the intestinal worm burden of immune rats. This increase was to levels of less than half of the worm burden of nonimmune controls, suggesting that rapid expulsion was only partly affected.

These results have left the potential role of mast cells, or their principal recognized mediators, rather in limbo. Only 5-HT has been shown to have any effect (and that only partial) both from infusion of the molecule or through the use of inhibitors. It would be of considerable interest to determine the source as well as the mechanism of action of serotinin. Although all of the above studies have considered serotonin in the context of mast cells, it is present in large amounts in secretory form in enterochromaffin cells and is a recognized neurotransmitter in mammals and nematodes (Horvitz et al., 1982). In the absence of other direct evidence, it is premature to consider that a role for 5-HT, which is itself not established, is equivalent to a role for mast cells. Even more importantly, when there have been effects of pharmacologic mediators on rejection, a determination needs to be made of the site of action. Does the agent influence some non-mast cell mediated process, such as, delivery/activation of cells or antibody?

The mast cell studies based on cellular effects of growth factors and cytokines (SCF/c-kit and IL-3) have consistently demonstrated an impairment of rejection when there is a defect (W/W^v) or faster rejection for specific treatment (IL-3). These experiments lack definitive proof that connects the pleiotropic and multifunctional factors to the proposed mediator, the mast cells. However, attempts to relate mast cell degranulation products directly to worm rejection, by either inhibition or administration studies, have produced little consistent or convincing proof of their direct or even indirect involvement in worm rejection. Considering the effort that has gone into attempts to produce or inhibit rejection with mast cell products, it is reasonable to conclude that mast cell products *per se* are largely ineffective until more direct evidence is produced. This conclusion leaves open the possibility that mast cells might be indirectly involved in worm rejection. This has been suggested before (the leak–lesion hypothesis) and would be compatible with the results of growth-factor type experiments.

5.3. Neutrophils

Of all the granulocytic cells, neutrophils have attracted the least attention and few studies have bothered to quantitate or follow them. The florid and conspicuous increases of mast cells and eosinophils, and their selectivity for helminth infections partly accounts for this. Early workers (van Someren, 1938) recorded a neutrophilia in rats with multiple peaks at 2, 6 and 14 days and a long elevated plateau from about 32 days of infection. Increases were also evident in Coker's (1956) study of the effects of cortisone on *T. spiralis* infections of mice. However, Larsh and Race (1954) and Larsh (1967) were the first to relate neutrophil numbers to intestinal pathology when they recorded an early phase, from 4 to 6 days, with mild polymorphonuclear cell invasion, followed by an acute phase that peaked at day 8, with a considerable accumulation of polymorphonuclear lymphocytes. In 1954, Larsh and Race did not confer much significance on the presence of neutrophils but later, Larsh (1967) looked on them principally as the visible evidence of local inflammation. Subsequently, Larsh and Race (1975) suggested that many of the 'neutrophils' they had identified earlier were inappropriately stained eosinophils.

Analysis of the interaction of granulocytes or macrophages with newborn larvae led Mackenzie et al. (1980a) to suggest that only neutrophils were able to reduce nitroblue tetrazolium dye in the reaction/degranulation process on the cuticular surface. A mechanism for neutrophil accumulation in the intestine of *T. spiralis*-infected mice and rats is suggested by the recent demonstration that mast cells, and specifically TNF-α, is required for recruitment and localization of neutrophils at sites of bacterial invasion (Malaviya et al., 1996). However, possible roles of neutrophils in any of the forms of immunity that are effective against *T. spiralis* are entirely speculative.

6. ASSESSMENT OF PROPOSED MECHANISMS OF PROTECTION

6.1. Non-specific Inflammation

6.1.1. *Definition*

As outlined earlier, inflammation was first proposed as a mechanism of immunity by Larsh (1963). This evolved to allergic inflammation (Larsh and Race, 1975). There is obviously considerable overlap in these topics and neither has been defined in an exclusive or mechanistic fashion with respect to *T. spiralis* infections. The terminology is doubly confusing as the T_h1 response which leads to macrophage activation, neutrophil accumulation and DH is usually termed inflammatory (Abbas et al., 1996). The use of the same term, inflammation, for a T_h2 dominated response, whose signal markers include eosinophilia and mastocytosis and whose cardinal feature, downregulation of DH is termed anti-inflammatory (Abbas et al., 1996), is very confusing and misleading. The continuing use of inflammation in the literature without an 'allergic' modifier justifies a redefinition of terms and may help to focus attention on precise functional differences between them. The following is proposed: nonspecific inflammation is the theory that nematode rejection from the intestine can be effected by the products of bone marrow-derived inflammatory cells, such as neutrophils, acting alone; allergic inflammation might be restricted to mast cell (perhaps including eosinophil)-dependent or mediated effects on the parasite.

6.1.2. *Observations*

The original idea of inflammation developed from histologic studies of the gut in *T. spiralis*-infected mice (Larsh and Race, 1954). Among the other observations that have contributed to the nonspecific inflammation hypothesis, the effect of dual or concurrent infections in enhancing rejection have been critical. For example, Howard *et al.* (1978) showed that a *Hymenolepis microstoma* infection superimposed on a *T. spiralis* infection at or prior to the rejection of *T. spiralis* led to stunting or poor infection with the *H. microstoma*. Several other reports confirmed the general principle that the rejection phase of one intestinal nematode could lead to enhanced rejection of a second intestinal helminth (e.g., Cox, 1952 with *A. caninum* and *T. spiralis*; Behnke *et al.* (1977); Howard *et al.* (1978) with *T. spiralis* and *Hymenolepis diminuta*; Louch (1962) with *T. spiralis* and *N. brasiliensis*). However, this is not an invariant principle as a prior infection with *Nematospiroides dubius* (now *Heligmosomoides polygyrus*) delays rejection of *T. spiralis* (Behnke *et al.*, 1978) and with other combinations, e.g., *Strongyloides ratti* and *T. spiralis* (Moqbel and Wakelin, 1979) demonstrable cross immunity exists. When there is cross immunity, the degree or existence of a nonspecific component in rejection is impossible to assess. Nonspecific inflammation obtained support from two further sources. First, Wakelin and Wilson (1979b) showed that adult worms collected 8 days after infection (i.e., as rejection begins) could be transferred into normal mice, where they would survive for many days before rejection. Wakelin and Wilson (1979b) then showed that if 8-day-old worms were transplanted into mice that were rejecting adult *T. spiralis* from an earlier infection, the transplanted worms were rejected with the same kinetics as the worms from the original infection. These results were interpreted to indicate that, at the time of rejection, the inflamed intestine had become hostile to any worms, regardless of age. Second, Wakelin and Wilson (1977b) demonstrated that bone marrow cells were required to restore the capacity to reject adult *T. spiralis* in sublethally irradiated mice. Comparable results had been found in a rat–*N. brasiliensis* model (Dineen and Kelly, 1973) and so a general principle of a requirement for a bone marrow population to effect rejection was established. Both Dineen and Kelly (1973) and Wakelin and Wilson (1977b) speculated that the bone marrow component was myeloid in nature rather than lymphoid. Similar evidence for a role of non-T cells was provided by experiments showing that the transfer of immune $CD4^+$ T cells between histocompatible strong (NIH) and weak (B10.G) worm expulsion phenotype mice produced immunity with the kinetics of the recipient and not the donor cells. Thus, transfer of B10.G T cells to NIH mice conferred fast rejection but transfer of NIH T cells to B10.G mice conferred slow rejection (Wakelin and Donachie, 1980). This result was reinforced when the same authors (Wakelin and Donachie, 1981) demonstrated that radiation chimeras, prepared between the same two mouse strains had the *T. spiralis* responder phenotype of the reconstituting bone marrow cells (e.g. B10.G in B10.G bone marrow →NIH/B10.G chimeras). Again, the authors interpreted these results to mean that the phenotype of rejection was determined by bone marrow, not the T cells of the two strains.

A final argument that has been used to support the case for a nonspecific final effector is the apparent lack of damage suffered by adult worms collected during expulsion. This is illustrated in statements such as 'T-cell-mediated inflammation in the small intestine renders the environment unsuitable for worm survival and the worms leave of their own accord' (Wassom and Kelly, 1990). This view is partly based on the longterm survival of adult worms collected during expulsion after their transfer to a naive recipient.

6.1.3. Concurrent Infections, What They Tell Us

When analyzing the above experiments, the question is whether or not the data demonstrate a role for granulocytic inflammation or whether there are other possible explanations. The difficulty in analyzing the concurrent infection results lies in the complexity of the systems involved. For example, it has proven difficult to dissect the mechanism of rejection for a single nematode species, partly because of the complexity of the two interacting systems: a vertebrate host and a large multicellular parasite. To expect that the addition of a further complex multicellular parasite to this basic interaction will clarify underlying processes is a tall order. Several questions remain from these experiments. The first concerns the possibility of specific immunological interplay, perhaps as a result of cross-reacting antigens. Many cross-reacting antigens have already been identified in nematodes, e.g., phosphorylcholine is present on *T. spiralis* (Lim and Choy, 1986) and many other nematodes (Maizels *et al.*, 1987); *Ascaris*, *Toxocara* and *Trichinella* all cross react (Sharp and Olson, 1962) as do many other unrelated nematodes. In general, researchers have focused on finding species-specific antigens rather than cross-reactive antigens which frequently present obstacles to their research. Furthermore, analyzing cross-reactivity by determining whether animals previously infected with parasite A display immunity to parasite B is not a reliable test. Previous work with *T. spiralis* pre-adult immunization in mice showed that it was largely ineffective in providing homologous protection (Bell *et al.*, 1982b; Wakelin *et al.*, 1986) even though the antigens were known to be immunogenic and were, by multiple exposure, demonstrably protective. Similar results were found by Robinson *et al.* (1995) who transferred protection with serum from immunized mice, although the mice themselves were not protected. These examples suggest that the immunological reactions and cross-reactions that might be possible when an animal is mounting a full-fledged intestinal rejection response are likely to be unpredictable and encompass several possibilities in addition to the proposed one of inflammation.

A second possibility is that during concurrent infections, the first infection enhances the normal immune response to the second infection so that it starts several days earlier. This is particularly likely with *T. spiralis* as each life-cycle stage is highly immunogenic and elicits a strong host response. An example of this is that mice and rats will both reject a superinfection with *T. spiralis* muscle larvae before adult worm rejection begins (Bell, 1992). In some strong responder F_1 hybrids (DBA/C3H and B10.BR/NFS), rejection of a larval superinfection was measurable a mere 2 days after the primary infection began. This was well prior to the start of adult worm rejection, and hence before the advent of inflammation. Concurrent infection studies suggest that the most effective method to demonstrate enhanced rejection is when the second parasite is given 4–10 days before rejection of the first parasite. Thus, there is ample time for a specific immune response to occur to the second parasite. Enhanced reactivity to the superimposed infection does not appear to have been considered as a process. An additional concern is that it has not been demonstrated that rejection of the second parasite is caused by the same processes that reject the initial parasite. In other words, the 'inflammation' (or other factor) could be a consequence of the rejection process of the first parasite rather than the cause of it. If the processes are not functionally identical then no conclusions can be reached about their specificity. Finally, there is no 'smoking gun' in which a product of the inflammatory process or cell population can be shown to affect the parasite *in vitro* or *in vivo*. This is a crucial step if we are to move beyond the phenomenological to analysis.

6.1.4. *The Role of Bone Marrow Cells*

Another element of the inflammation hypothesis relates to the transfer of activated T cells to mice of defined rejection phenotype. The observation (described above) is that recipients respond without changing their phenotype, even though the transferred T cells adoptively transfer immunity as outlined above. This has been interpreted to show that the rate-limiting step is determined by the bone marrow of the recipient and hence that inflammation will follow the phenotype of the recipient. What these experiments indicate is that the phenotype of rejection is not determined by the donor T cells alone. They leave open a number of possible sites at which this might be controlled. In view of the increasing evidence for a role of antibody in rejection mechanisms (Appleton and McGregor, 1987, 1988; Ahmad *et al.*, 1990; Bell *et al.*, 1992; D. Negrao-Correa and colleagues, unpublished data), the possibility that B cells, number, specificity, deployment or rate of expansion, are involved, must be considered. These cells are bone marrow derived and conform to the only established criterion of bone marrow dependency. A role for intraepithelial lymphocytes, also bone marrow derived, could also be considered.

Effects at the level of T cells have generally been excluded but even this may be premature. A characteristic of the protective T cells of mice or rats is that they are dividing and short-lived (Grencis and Wakelin, 1982; Wang et al., 1990a). In fact, the thoracic duct T cells of rats that transfer protection are undetectable in the recipient gut by 48 h post-transfer (R.G. Bell, unpublished data), well before parasite rejection occurs. It is very unlikely that these cells clonally proliferate to become effector cells; more likely is the possibility that they have critical roles to play in the continued expansion of indigenous reactive cell populations (both T cells and non-T cells) in the gut. The rate of expansion of the resident T cells, after stimulation by donor cells, may determine the phenotype of fast or slow immunity independently of any granulocytic cell population. The radiation chimera experiments have also been interpreted to favour a bone-marrow origin of strain phenotype differences (Wakelin and Donachie, 1981). Because these mice were subjected to lethal irradiation, which eliminates T, B and granulocyte/macrophage precursors, all of these cell systems would be reconstituted from the donor bone-marrow inoculum. Under these conditions no persuasive evidence exists that allows one to favour one differentiated cell population (T, B or granulocyte) or the other in eventual rejection processes.

6.1.2. *Are Adults 'Undamaged' at Expulsion?*

The usual prelude to adult worm rejection is stunting and reduction of fecundity. Stunting has been noted since the 1930s and fecundity, in particular, drops in virtually all strains of mice prior to adult worm expulsion (Despommier et al., 1977a; Wakelin and Wilson, 1977a; Wakelin and Donachie, 1980; Bell et al., 1982b). Since we know that in nude mice adult worm survival is protracted and muscle larvae burden achieves very high numbers (Ruitenberg et al., 1977a) it seems likely that effects on fecundity are due to a direct host response on the parasite as originally proposed by Despommier et al. (1977a). While the parasite antigens that are the targets, or the host mechanism that leads to either stunting or fecundity reduction are not defined, their existence surely constitutes evidence that the worm suffers damage while maintaining its niche. That such damage occurs was clearly evident in Love et al.'s 1976 study which documented considerable ultrastructural damage suffered by adult *T. spiralis* prior to rejection. Furthermore, adult *T. spiralis* progressively move away from the most intense point of host attack at their initial site of residence (Larsh et al, 1952; Larsh and Race, 1954; Kennedy, 1980). They may continue to reside in sites in the large intestine for several days after they have been dislodged from the small intestine (Bell, 1992). This pattern of effects is more compatible with the view that the worms are subjected to a multisystem attack which impairs reproduction, growth and function and eventually dislodges the worm from the epithelium.

There is little hard or direct evidence that points to inflammation or inflammatory cells as the terminal effector of rejection. Although this hypothesis has often been advanced since Larsh (1963) first proposed it, possible alternative explanations are conspicuously absent and the hypothesis has not led to identified mediators.

6.2. Allergic Inflammation

Larsh and Race (1975) first used the term 'allergic inflammation', in a manner that followed the original view of 'allergic' as immunologically hyperreactive rather than the current meaning associated with immediate hypersensitivity. Larsh and Race (1975) specifically designated their view of the histopathology as an initial DH followed by a nonspecific inflammatory response. The latter component was viewed as the 'allergic inflammation' and while various cell types were observed at sites of intestinal infection (neutrophils, eosinophils, macrophages and lymphocytes), no particular type was identified as causal to rejection.

Wakelin (1993) analyzed this paper and pointed out that Larsh and Race (1975) probably did not observe mast cells in their studies because of inappropriate fixation. Whether or not an awareness of mast cells and their local infiltration into the intestine would have changed Larsh and Race's (1975) views is arguable but irrelevant. It would appear more logical to discuss cell-mediated/delayed hypersensitivity under T-cell-mediated responses. To resolve this issue it is proposed that allergic inflammation be restricted to mean 'worm expulsion mediated by the secreted products of mast cells'. Such a definition would exclude the postulated role of mast cells in the 'leak lesion' hypothesis, whereby their degranulation facilitates the entry of protective antibody (Barth et al., 1966). The evidence for a direct role of mast cells in protection has been discussed in Sections 5.2 and 5.2.1.

6.3. Intestinal Mucus

The first evidence that mucus might have a part to play in helminth expulsion was provided by Lee and Ogilvie (1980, 1982) who demonstrated 'trapping' of larvae in intestinal mucus during rapid expulsion. However, Bell et al. (1984b) showed that rapid expulsion could take place in the absence of significant mucus trapping and that mucus trapping required muscle larvae development in actively infected rats. Antibody was shown to be critical for rapid expulsion in neonatal rats (Appleton and McGregor, 1987) and this was accompanied by mucus trapping as in adult rats (Carlisle *et al.*, 1990). Specific antibody could be added to mucus collected from normal rats to produce larval trapping (Carlisle *et al.*, 1991a) but, larvae trapped in mucus could escape and penetrate the epithelium. Carlisle *et al.* (1991b) concluded that mucus trapping was not the cause of rapid expulsion although it might have a role at times. More important was the presence of antibody in the circulation which can expel larvae even if it is injected after larvae have penetrated the epithelium. Thus, it is likely that the principal effector is antibody acting at the level of the epithelial cell/basement membrane which either damages the parasite or impairs entry, forcing its return to the lumen. Once back in the lumen, mucus trapping, in conjunction with antibody at the epithelial layer, act together to prevent recolonization of the epithelium.

6.4. Intestinal Epithelial Cells

Interest in a possible role of epithelial cells in the protective immune response to *T. spiralis* developed from observations of modified epithelial cell structure and function during a challenge infection. These changes included reduced wheat germ lectin binding to the epithelial brush border membrane in infected rats (Harari and Castro, 1983) reduced hexose transport (Hessel *et al.*, 1982), increased potential difference across epithelial tight junctions (Russell and Castro, 1985), and chloride ion secretion (Harari *et al.*, 1987), which is linked to a change from fluid uptake to net fluid secretion by enterocytes. The functional changes in epithelial cells described in these studies have been linked to the immune system through immune serum transfer (Harari *et al.*, 1987). Comparable changes in epithelial function can be observed *in vitro* and are also dependent on the presence of immune serum. The susceptibility of immune serum to heat treatment at 56°C suggested an underlying immediate hypersensitivity as IgE is degraded at this temperature (Harari *et al.*, 1987). In this model, antigen induces the degranulation of mucosal mast cells to produce a reaction by the epithelial cells, which results in changes in net ion flow and water secretion. This is an appealing hypothesis because it brings together mucosal mast cells, a conspicuous component of the rodent response to infection with *T. spiralis*, and enterocytes, which comprise the principal niche of the intestinal infectious larvae and adults (Wright 1979). While no aggressive host–defensive molecule in enterocytes has been described, a relatively small change in one of many normal physiological parameters might render the enterocyte unsuitable for occupancy by the invading larvae. Furthermore, there is good evidence that worms are not damaged by rapid expulsion (McCoy, 1940; Hessel *et al.*, 1982) and that mast cell degranulation is temporally associated with rapid expulsion (Tuohy *et al.*, 1990). While these observations are all consistent with an effect of enterocyte modification on worm penetration, the case is currently largely correlative and circumstantial. Many of these issues have been discussed under the section on mast cells. More recently, a direct role of enterocytes in IgE transport to the gut has been defined in *T. spiralis*-infected rats, which raises the possibility that enterocytes may be able to interact directly with IgE to prevent adult worm penetration (Negrao-Correa *et al.*, 1996; Ramaswamy *et al.*, 1996).

7. SYNTHESIS

In attempting to draw together the tangled and often contradictory lines of evidence produced over the last 50 years, a degree of selectivity is essential. Also important is an awareness that rats and mice may have significant differences in the protective mechanisms they display. In a further simplifying move it is suggested that there are just three basic host responses whose consequences are: rejection of infectious larvae (rapid expulsion in rats; associative expulsion in mice), rejection of adult worms (adult worm rejection or self-cure) and anti-newborn larvae immunity. It is evident that swine can also kill muscle larvae but evidence for this response has only appeared once in rodents. Other responses, such as antifecundity, lack experimental definition and are, I believe, more reasonably considered to be a part of adult worm expulsion until shown to be distinct.

Over the years most attention has been directed to an analysis of adult worm rejection. All experimental systems have demonstrated that this is a $CD4^+$ T cell-dependent process and, increasingly, that IL-4, probably secreted by the $CD4^+$ cells, is an important, although possibly not an essential ingredient. The steps leading from IL-4 (and other cytokine) secretion to expulsion are less defined. We favour the view that local antibody, particularly IgE and perhaps IgA, are key components. Both isotypes have strong local intestinal responses during the primary infection (D. Negrao-Correa and R.G. Bell, unpublished data) and antibody of the IgE isotype is carried into the intestinal lumen and can apparently block penetration of epithelial cells by adult worms. It is envisaged that IgA could function similarly. A role for IgE in rapid expulsion has already been established. In addition, it is probable that serum IgG will function if some local reaction were to facilitate its entry into the gut. The mast cell mediated 'leak-lesion' hypothesis seems a strong candidate but may not be the only one. These mechanisms are summarized in Figure 2A.

At least two distinct processes underlie rapid expulsion in adult rats and there may be a third. In the first, larvae-specific IgE interacts with IL-4 activated enterocytes to produce high concentrations of IgE at the enterocyte–lumen interface. This IgE can bind to larval glycoproteins that are important in invasion or occupancy of the enteral niche and either prevent invasion or dislodge the parasite. Local intestinal secretion of IL-4

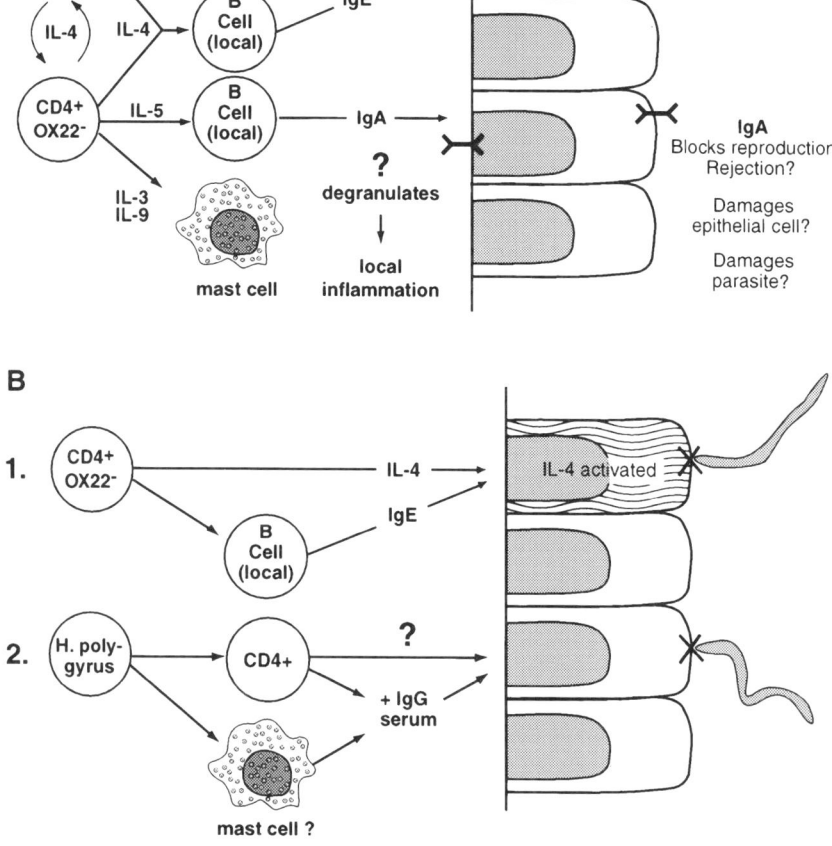

Figure 2 The major participants in the processes leading to adult worm expulsion (A) and rapid expulsion (B). The critical role of $CD4^+$ $OX22^-$ in stimulating B cell growth and in promoting enterocyte transport of Ig is shown.

appears essential for this mechanism to be expressed in adult rats. It is suggested that this response occurs during the primary infection and for a period of a few weeks afterwards. This response may also account for 'associative expulsion' which appears to be larvae-specific. The second form of rapid expulsion involves serum antibodies of various IgG isotypes. It seems likely, since $CD4^+$ $OX22^-$ cells will not prime the intestine for IgG activity, that it does not involve IL-4. However, because IgG is ineffective unless rats have been given a second stimulus (in our hands *H. polygyrus*), we believe a second component is necessary. This has not been further defined but could derive from any of a wide variety of cells: mast cells, macrophages, T_h1 cells or enterocytes, etc. IgG antibodies appear in functional amounts after muscle larvae maturation, usually about 4 weeks after the primary infection and they persist for very long periods. The undefined second component is presumably important in delivering antibody rapidly and in sufficient concentration to the site of larval invasion in the intestine. Because of its kinetics, and because there are suggestions that mast cells have a role, it seems plausible that a local mast cell degranulation might release functional levels of IgG across the basement membrane. This is essentially an identical process to the 'leak-lesion' process envisaged by Barth *et al.* (1966). We suggest that IgG molecules function in the same manner as IgE to block larval penetration and occupancy of the epithelial niche. Further, it is suggested that the 25-fold higher requirements for IgG (≥ 5 mg) than IgE (≤ 200 μg) in transfer experiments results from the fact that only a small proportion of the IgG that is infused intravenously 'leaks' into the critical environment, whereas IgE is selectively transported to the lumen. Functional concentrations of IgG at the site of attack may be comparable to those of IgE, with the difference being an efficiency of delivery. This proposal would accommodate many of the earlier difficulties experienced with immune serum transfer. That is, some form of local stimulation is required for both IgG and IgE isotypes to be effective; thus, antibody transfer would only be effective when recipients had deliberately or inadvertently received an appropriate stimulus (Figure 2B).

We further propose that the effect of antibody is dose-dependent and local in nature because its action must be facilitated by the local release of cytokines from activated T cells. Low levels of antibody impair adult worm nutrition and reproduction, whereas higher levels prevent worms from moving from cell to cell impairing their capacity to maintain their position by re-entering enterocytes (Figure 3). Worms lost from one position in the gut can successfully re-establish lower down if the immune response disseminated to that site is not yet strong enough to prevent it. In weak responder mice or rats, re-establishment is readily accomplished for a time by adult worms and prolongs their overall residence period in the gut. For strong responder mice and most rat strains, the downstream compartment already has a response effective enough to prevent re-establishment of displaced worms. This view would link strength of response to the rate at which an initial antibody response develops and to the rate at which it disseminates. For both phases there is a requirement for T cells which have two primary functions: (1) the stimulation of an appropriate B cell response, and (2) modification of local cell populations such as the enterocyte and/or mast cell so that they are able to interact with antibody or move antibody to its effective site. Worms are ejected from their niche because they can no longer maintain themselves there, having suffered nonpermanent structural and biochemical damage prior to rejection. Damage is evident in stunting, reduced fecundity, movement within the intestine and through ultrastructural lesions in the worm.

Anti-newborn larvae immunity appears to be antibody-mediated and may also be dependent on myeloid cells. While there is a great deal of evidence indicating eosinophils have activity *in vitro*, there is little *in vivo* data to support this. However, if the only *in vivo* experiments that indicate a site of effect in the intestine (Wang and Bell, 1987) can be taken at face value, there is ample opportunity for eosinophils, neutrophils or even macrophages to engage newborn larvae in the lamina propria. This is the least explored facet of immunity to *T. spiralis* and one that deserves more analysis.

This synthesis proposes that antibody is the major effector of immunity to *T. spiralis* at all points of host attack. Other effectors are not excluded,

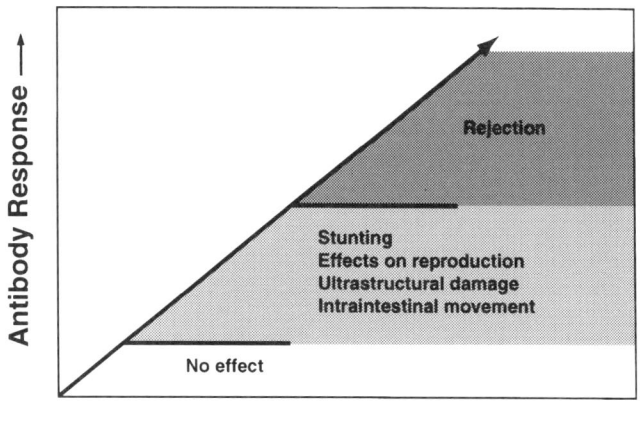

Effects on the parasite

Figure 3 The scale of response to a graded series of detrimental effects on the parasite which culminate in expulsion. Expulsion is viewed as an active process that is dependent on the amount of effectors present relative to the number of worms.

although the evidence for individual agents needs to be strengthened before they should be accepted. It is suggested that the parasite must work to maintain itself in all its niches and that the molecules that carry out functions such as host cell recognition, host cell penetration, nutrient transport, mate recognition, etc., are all subject to inactivation by antibody. The degree of inactivation (host response strength) determines the outcome, partial inactivation leads to adverse effects on the parasite which have been well described in the literature, more complete inactivation evicts the parasite from its niche as its self maintenance systems fail (Figure 3).

8. CONCLUSION

Immunoparasitology is a hybrid field with enormous debts to its two parent fields, parasitology and immunology. In this review I have tried to give credit to the many early investigators who brought the insights of parasitology to the investigation of host responsiveness. These investigators established many of the techniques and concepts that are still used or which guide experimentation today. Arbitrarily, I selected 1970 as the turning point between the early investigators and their more recent counterparts largely on the basis of Larsh's prior reviews and the fact that immunology had an enormous impact from 1970. The 1970 and after group is marked by the external influence of immunology in its thinking and particularly, experimentation. While using the concepts largely of the parasitologists the post-1970s group incorporated immunological techniques and insights from a distance as a second cousin of immunology. At some point in the early 1990s immunoparasitology gained independence and recognition as it was realized that helminthic infections could help explain the basic immunological paradigm then dominating research endeavor — the T_h1 and T_h2 model of CD4 cell function.

This seemed an opportune point at which to examine the strengths and weaknesses of past approaches in order to develop new and more direct means of investigating host responsiveness in the future. A key, although unargued element of this view is that nematode parasites are ancient adversaries of the immune system which has evolved specific strategies for dealing with them that owe little to other infectious agents, such as bacteria and viruses. There is little novelty in host adaptive mechanisms of defence and these have probably all been defined — antibody and activated cells of several lineages. The novelty is how they are brought together for different parasitic organisms. The next 20 years of investigation will surely define the novel nematode parasite-specific mechanisms of protection that exist. Our efforts to date borrowed mechanisms and processes from basic immunology and applied them to immunoparasitology with a somewhat mixed outcome. This reinforces the notion that it is only when we approach infectious agents, particularly complex multicellular ones, on their own terms, by understanding their biology, that real insights into the host response develop. It is with this anticipation that the models outlined in this review are proposed. Antibody is resurrected from a period of neglect and given the most important role in protection, proponents of different mechanisms will, I hope, define molecular and cellular mediators and mechanisms of action. The future promises an intricate and exciting delineation of parasite biology and host responsiveness.

ACKNOWLEDGEMENTS

Thanks are due to the following people: Judy Appleton, K. Ramaswamy and Catherine O'Neill for reading and critiquing the manuscript, Joe Urban and Judy Appleton for providing me with some of their unpublished work, Anita Hesser for secretarial work and Barbara Tefft for producing the figures. The work was supported by National Institutes of Health grant number AI17484.

REFERENCES

Abbas, A.K., Murphy, K.M. and Sher, A. (1996). Functional diversity of helper T lymphocytes. *Nature* **383**, 787–793.
Ahmad, A., Wang, C.H., Korenaga, M., Bell, R.G. and Adams, L.S. (1990). Synergistic interaction between immune serum and thoracic duct cells in the adoptive transfer of rapid expulsion of *Trichinella spiralis* in adult rats. *Experimental Parasitology* **71**, 90–99.
Ahmad, A., Wang, C.H. and Bell, R.G. (1991a). A role for IgE in intestinal immunity: Expression of rapid expulsion of *Trichinella spiralis* in rats transfused with IgE and thoracic duct lymphocytes. *Journal of Immunology* **146**, 3563–3570.
Ahmad, A., Bell, R.G., Wang, C.H. and Sacuto, F. (1991b). Characterization of the thoracic duct T helper cells that co-mediate, with antibody, the rapid expulsion of *Trichinella spiralis* in adult rats. *Parasite Immunology* **13**, 147–159.
Alicata, J.E. and Burr, G.D. (1949). Preliminary observations on the biological effects of radiation on the life cycle of *Trichinella spiralis*. *Science* **109**, 595–596.
Alizadeh, H. and Murrell, K.D. (1984). The intestinal mast cell response to *Trichinella spiralis* infection in mast cell-deficient W/W^v mice. *Journal of Parasitology* **70**, 767–773.
Alizadeh, H. and Wakelin, D. (1982). Genetic factors controlling the intestinal mast cell response in mice infected with *Trichinella spiralis*. *Clinical and Experimental Immunology* **49**, 331–337.

Almond, N.M. and Parkhouse, R.M.E. (1986). Immunoglobulin class specific responses to biochemically defined antigens of *Trichinella spiralis*. *Parasite Immunology* **8**, 391–406.
Andrews, J.M. (1962). Parasitism and allergy. *Journal of Parasitology* **48**, 3–12.
Appleton, J.A. and McGregor, D.D. (1987). Characterization of the immune mediator of rapid expulsion of *Trichinella spiralis* in suckling rats. *Immunology* **62**, 477–484.
Appleton, J.A., Schain, L.R. and McGregor, D.D. (1988). Rapid expulsion of *Trichinella spiralis* in suckling rats: Mediation by monoclonal antibodies. *Immunology* **65**, 487–492.
Arnold, C.S. and Olson, L.J. (1966). Persistence of adult *Trichinella spiralis* in Pertussis-treated mice. *Journal of Parasitology* **52**, 1036–1037.
Bachman, G.W. and Oliver-González, J. (1936). Immunization in rats against *Trichinella spiralis*. *Proceedings of the Society for Experimental Biology and Medicine* **35**, 215–217.
Barriga, O.O. (1975). Selective immunodepression in mice by *Trichinella spiralis* extracts and infections. *Cellular Immunology* **17**, 306–309.
Barriga, O.O. (1978). Depression of cell-mediated immunity following inoculation of *Trichinella spiralis* extract in the mouse. *Immunology* **34**, 167–173.
Barth, E.E.E., Jarrett, W.F.H. and Urquhart, G.M. (1966). Studies on the mechanism of the self-cure reaction in rats infected with *Nipponstrongylus brasiliensis*. *Immunology* **10**, 459–464.
Basten, A. and Beeson, P.B. (1970). Mechanism of eosinophilia. II. Role of the lymphocyte. *Journal of Experimental Medicine* **131**, 1288–1305.
Basten, A., Boyer, M.H. and Beeson, P.B. (1970). Mechanism of eosinophilia. I. Factors affecting the eosinophil response of rats to *Trichinella spiralis*. *Journal of Experimental Medicine* **131**, 1271–1286.
Beahm, E.H. and Downs, C.M. (1939). Differential blood picture and total cell count on normal and trichina infected albino rats. *Journal of Parasitology* **25**, 405–411.
Beahm, E.H. and Jorgensen, N.M. (1941). Some effects of experimental trichinosis in the dog. *Proceedings of the Society for Experimental Biology and Medicine* **47**, 294–299.
Behnke, J.M., Bland, P.W. and Wakelin, D. (1977). Effect of the expulsion phase of *Trichinella spiralis* on *Hymenolepis diminuta* infection in mice. *Parasitology* **75**, 79–88.
Behnke, J.M., Wakelin, D. and Wilson, M.M. (1978). *Trichinella spiralis*: delayed rejection in mice concurrently infected with *Nematospiroides dubius*. *Experimental Parasitology* **46**, 121–130.
Bell, R.G. (1988). Genetic analysis of expulsion of adult *Trichinella spiralis* in NFS, C3H/He and B10.BR mice. *Experimental Parasitology* **66**, 57–65.
Bell, R.G. (1992). *Trichinella spiralis*: evidence that mice do not express rapid expulsion. *Experimental Parasitology* **74**, 417–430.
Bell, R.G. and Issekutz, T. (1993). Expression of a protective intestinal immune response can be inhibited at three distinct sites by treatment with anti-α4 integrin. *Journal of Immunology* **151**, 4790–4802.
Bell, R.G. and Liu, W.-M. (1988). *Trichinella spiralis*: Quantitative relationships between intestinal worm burden, worm rejection, and the measurement of intestinal immunity in inbred mice. *Experimental Parasitology* **66**, 44–56.
Bell, R.G. and McGregor, D.D. (1980a). Requirement for two discrete stimuli for

induction of the intestinal rapid expulsion response against *Trichinella spiralis* in the rat. *Infection and Immunity* **29**, 186–193.

Bell, R.G. and McGregor, D.D. (1980b). Rapid expulsion of *Trichinella spiralis*: Co-induction using antigenic extracts of larvae and intestinal stimulation with an unrelated parasite. *Infection and Immunity* **29**, 194–199.

Bell, R.G., McGregor, D.D. and Despommier, D.D. (1979). *Trichinella spiralis*: Mediation of the intestinal component of protective immunity in the rat by multiple, phase-specific, antiparasitic responses. *Experimental Parasitology* **47**, 140–157.

Bell, R.G., McGregor, D.D. and Adams, L.S. (1982a). *Trichinella spiralis*: Characterization and strain distribution of rapid expulsion in inbred mice. *Experimental Parasitology* **53**, 301–314.

Bell, R.G., McGregor, D.D. and Adams, L.S. (1982b). *Trichinella spiralis*: Genetic basis for differential expression of phase-specific intestinal immunity in inbred mice. *Experimental Parasitology* **53**, 315–325.

Bell, R.G. McGregor, D.D. and Adams, L.S. (1982c). Studies on the inhibition of rapid expulsion of *Trichinella spiralis* in rats. *International Archives of Allergy and Applied Immunity* **69**, 73–80.

Bell, R.G. McGregor, D.D., Woan, M.C. and Adams, L.S. (1983). *Trichinella spiralis*: Selective intestinal immune deviation. *Experimental Parasitology* **56**, 129–142.

Bell, R.G., Adams, L.S. and Ogden, R.W. (1984a). *Trichinella spiralis*: Genetics of worm expulsion in inbred and F_1 mice infected with different worm doses. *Experimental Parasitology* **58**, 345–355.

Bell, R.G., Adams, L.S. and Ogden, R.W. (1984b). Intestinal mucus trapping in the rapid expulsion of *Trichinella spiralis* by rats: Induction and expression analyzed by quantitative worm recovery. *Infection and Immunity* **45**, 267–272.

Bell, R.G., Adams, L.S. and Ogden, R.W. (1984c). A single gene determines rapid expulsion of *Trichinella spiralis* in mice. *Infection and Immunity* **45**, 273–275.

Bell, R.G., Wang, C.H. and Ogden, R.W. (1985). *Trichinella spiralis*: Non-specific resistance and immunity to newborn larvae in inbred mice. *Experimental Parasitology* **60**, 101–110.

Bell, R.G., Korenaga, M. and Wang, C.H. (1987). Characterization of a cell population in thoracic duct lymph that adoptively transfers rejection of adult *Trichinella spiralis* to normal rats. *Immunology* **61**, 221–227.

Bell, R.G., Appleton, J.A., Negrao-Correa, D.A. and Adams, L.S. (1992). Rapid expulsion of *Trichinella spiralis* in adult rats mediated by monoclonal antibodies of distinct IgG isotypes. *Immunology* **75**, 520–527.

Besmer, P. (1991). The kit ligand encoded at the murine Steel locus: a pleiotropic growth and differentiation factor. *Current Opinion in Cell Biology* **3**, 939–946.

Blackwood, L.L. and Molinari, J.A. (1981). Dose dependence of *Trichinella spiralis*-induced immunopotentiation. *International Archives of Allergy and Applied Immunology* **66**, 55–58.

Briggs, N.T. (1963a). Hypersensitivity in murine trichinosis: Some responses of *Trichinella*-infected mice to antigen and 5-hydroxytryptophan. *Annals of the New York Academy of Sciences* **113**, 456–466.

Briggs, N.T. (1963b). Immunological injury of mast cells in mice actively and passively sensitized to antigens from *Trichinella spiralis*. *Journal of Infectious Diseases* **113**, 22–32.

Briggs, N.T. and DeGiusti, D.L. (1966). Generalized allergic reactions in *Trichinella*-infected mice: The temporal association of host immunity and sensitivity to exogenous antigens. *American Journal of Tropical Medicine and Hygiene* **15**, 919–929.

Brown, P.J., Bruce, R.G., Manson-Smith, D.F. and Parrott, D.M.V. (1981). Intestinal mast cell response in thymectomised and normal mice infected with *Trichinella spiralis*. *Veterinary Immunology and Immunopathology* **2**, 189–198.

Brown, T.R. (1897). Studies on Trichinosis. *Bulletin of the Johns Hopkins Hospital* **8**, 79–81.

Buys, J., Wever, R., van Stigt, R. and Ruitenberg, E.J. (1981). The killing of newborn larvae of *Trichinella spiralis* by eosinophil peroxidase *in vitro*. *European Journal of Immunology* **11**, 843–845.

Buys, J., Wever, R. and Ruitenberg, E.J. (1984). Myeloperoxidase is more efficient than eosinophil peroxidase in the *in vitro* killing of newborn larvae of *Trichinella spiralis*. *Immunology* **51**, 601–607.

Campbell, C.H. (1955). The antigenic role of the excretions and secretions of *Trichinella spiralis* in the production of immunity in mice. *Journal of Parasitology* **41**, 483–491.

Campbell, W.C. (1965). Immunizing effect of enteral and enteral-parental infections of *Trichinella spiralis* in mice. *Journal of Parasitology* **51**, 185–194.

Campbell, W.C., Hartman, R.K. and Cuckler, A.C. (1963). Effect of certain antihistamine and antiserotonin agents upon experimental trichinosis in mice. *Experimental Parasitology* **14**, 23–28.

Carlisle, M.S., McGregor, D.D. and Appleton, J.A. (1990). The role of mucus in antibody-mediated rapid expulsion of *Trichinella spiralis* in suckling rats. *Immunology* **70**, 126–132.

Carlisle, M.S., McGregor, D.D. and Appleton, J.A. (1991a). Intestinal mucus entrapment of *Trichinella spiralis* larvae induced by specific antibodies. *Immunology* **74**, 546–551.

Carlisle, M.S., McGregor, D.D. and Appleton, J.A. (1991b). The role of the antibody Fc region in rapid expulsion of *Trichinella spiralis* in suckling rats. *Immunology* **74**, 552–558.

Castro, G.A. (1989). Immunophysiology of enteric parasitism. *Parasitology Today* **5**, 11–19.

Castro, G.A., Hessel, J.J. and Whalen, G. (1979). Altered intestinal fluid movement in response to *Trichinella spiralis* in immunized rats. *Parasite Immunology* **1**, 259–266.

Castro, G.A., Harari, Y. and Russell, D. (1987). Mediators of anaphylaxis-induced ion transport changes in small intestine. *American Journal of Physiology* **253**, G540–G548.

Chandler, A.C. (1939). The nature and mechanism of immunity in various intestinal nematode infections. *American Journal of Tropical Medicine* **19**, 309–317.

Coker, C.M. (1956). Some effects of cortisone in mice with acquired immunity to *Trichinella spiralis*. *Journal of Infectious Diseases* **98**, 39–44.

Cox, H.W. (1952). The effect of concurrent infection with the dog hookworm, *Ancylostoma caninum*, on the natural and acquired resistance of mice to *Trichinella spiralis*. *Journal of the Elisha Mitchell Scientific Society* **68**, 222–235.

Crandall, R.B. and Crandall, C.A. (1972). *Trichinella spiralis*: Immunologic response to infection in mice. *Experimental Parasitology* **31**, 378–398.

Crook, K. and Wakelin, D. (1994). Induction of T lymphocyte subsets and levels of interleukin-2 and interleukin-3 after infection with *Trichinella spiralis* are similar in mice of high- and low-responder phenotypes. *International Journal for Parasitology* **24**, 119–126.

Crum, E.D., Despommier, D.D. and McGregor, D.D. (1977). Immunity to *Trichinella spiralis*. I. Transfer of resistance by two classes of lymphocytes. *Immunology* **33**, 787–795.

Culbertson, J.T. (1942a). Active immunity in mice against *Trichinella spiralis*. *Journal of Parasitology* **28**, 197–206.

Culbertson, J.T. (1942b). Passive transfer of immunity to *Trichinella spiralis* in the rat. *Journal of Parasitology* **28**, 203–206.

Culbertson, J.T. and Kaplan, S.S. (1938). A study upon passive immunity in experimental trichiniasis. *Parasitology* **30**, 156–166.

Cypess, R.H., Lubiniecki, A.S. and Hammon, W.M. (1973). Immunosuppression and increased susceptibility to Japanese B encephalitis virus in *Trichinella spiralis*-infected mice. *Proceedings of the Society of Experimental and Biological Medicine* **143**, 469–473.

Denkers, E.Y., Wassom, D.L. and Hayes, C.E. (1990). Characterization of *Trichinella spiralis* antigens sharing an immunodominant, carbohydrate-associated determinant distinct from phosphorylcholine. *Molecular and Biochemical Parasitology* **41**, 241–250.

Despommier, D.D. and Wostmann, B.S. (1969). *Trichinella spiralis*: Immune elimination in mice. *Experimental Parasitology* **24**, 243–250.

Despommier, D.D., Campbell, W.C. and Blair, L.S. (1977a). The *in vivo* and *in vitro* analysis of immunity to *Trichinella spiralis* in mice and rats. *Parasitology* **74**, 109–119.

Despommier, D.D., McGregor, D.D., Crum, E.D. and Carter, P.B. (1977b). Immunity to *Trichinella spiralis*. II. Expression of immunity against adult worms. *Immunology* **33**, 797–805.

Dessein, A.J., Parker, W.L., James, S.L. and David, J.R. (1981). IgE antibody and resistance to infection. I. Selective suppression of the IgE antibody response in rats diminishes the resistance and the eosinophil response to *Trichinella spiralis* infection. *Journal of Experimental Medicine* **153**, 423–436.

de Vries, P., Brasel, K.A., McKenna, H.J., Williams, D.E. and Watson, J.D. (1992). Thymus reconstitution by *c-kit*-expressing hematopoietic stem cells purified from adult mouse bone marrow. *Journal of Experimental Medicine* **176**, 1503–1509.

Dineen, J.K. and Kelly, J.D. (1973). Expulsion of *Nippostrongylus brasiliensis* from the intestine of rats: the role of a cellular component derived from bone marrow. *International Archives of Allergy and Applied Immunology* **45**, 759–766.

DiNetta, J., Katz, F. and Campbell, W.C. (1972). Effect of heterologous anti-lymphocyte serum on the spontaneous cure of *Trichinella spiralis* infections in mice. *Journal of Parasitology* **58**, 636–637.

Donaldson, L.E., Schmitt, E., Huntley, J.F., Newlands, G.F.J. and Grencis, R.K. (1996). A critical role for stem cell factor in host protective immunity to an intestinal helminth. *International Immunology* **8**: 559–567.

Dutoit, E., Tronchin, G., Vernes, A. and Biguet, J. (1979). The influence of prostaglandins and vasoactive amines on the intestinal phase of experimental trichinellosis. *Annales de Parasitologie* **54**, 465–474.

Else, K.J. and Grencis, R.K. (1991). Cellular immune responses to the murine nematode parasite *Trichuris muris*. I. Differential cytokine production during acute or chronic infection. *Immunology* **72**, 508–513.

Else, K.J., Finkelman, F.D., Maliszewski, C.R. and Grencis, R.K. (1994). Cytokine-mediated regulation of chronic intestinal helminth infection. *Journal of Experimental Medicine* **179**, 347–351.

Faubert, G.M. (1976). Depression of the plaque-forming cells to sheep red blood cells by the newborn larvae of *Trichinella spiralis*. *Immunology* **30**, 485–489.

Faubert, G.M. and Tanner, C.E. (1971). *Trichinella spiralis*: Inhibition of sheep hemagglutinins in mice. *Experimental Parasitology* **30**, 120–123.

Faubert, G.M. and Tanner, C.E. (1974). The suppression of sheep rosette-forming cells and the inability of mouse bone marrow cells to reconstitute competence after infection with the nematode *Trichinella spiralis*. *Immunology* **27**, 501–505.

Finkelman, F.D., Katona, I.M., Urban, J.F., Jr., Snapper, C.M., Ohara, J. and Paul, W.E. (1986). Suppression of *in vivo* polyclonal IgE responses by monoclonal antibody to the lymphokine B-cell stimulatory factor 1. *Proceedings of the National Academy of Sciences of the USA* **83**, 9675–9678.

Fischthal, J.H. (1943). Number of larvae and time required to produce active immunity in rats against *Trichinella spiralis*. *Journal of Parasitology* **29**, 123–126.

Gabriel, B.W. and Justus, D.E. (1979). Quantitation of immediate and delayed hypersensitivity responses in *Trichinella*-infected mice. Correlation with worm expulsion. *International Archives of Allergy and Applied Immunology* **60**, 275–285.

Gomberg, H.J. and Gould, S.E. (1953). Effect of irradiation with Cobalt-60 on *Trichinella* larvae. *Science* **118**, 75–77.

Gould, S.E., Gomberg, H.J., Bethell, F.H., Villella, J.B. and Hertz, C.S. (1955). Studies on *Trichinella spiralis*. IV. Effect of feeding irradiated *Trichinella* larvae on production of immunity to reinfection. *American Journal of Pathology* **31**, 949–953.

Goyal, P.K. and Wakelin, D. (1993). Influence of variation in host strain and parasite isolate on inflammatory and antibody responses to *Trichinella spiralis* in mice. *Parasitology* **106**, 371–378.

Grencis, R.K. and Wakelin, D. (1982). Short lived, dividing cells mediate adoptive transfer of immunity to *Trichinella spiralis* in mice. I. Availability of cells in primary and secondary infections in relation to cellular changes in the mesenteric lymph node. *Immunology* **46**, 443–450.

Grencis, R.K., Riedlinger, J. and Wakelin, D. (1985). L3T4-positive T lymphoblasts are responsible for transfer of immunity to *Trichinella spiralis* in mice. *Immunology* **56**, 213–218.

Grencis, R.K., Riedlinger, J. and Wakelin, D. (1987). Lymphokine production by T cells generated during infection with *Trichinella spiralis*. *International Archives of Allergy and Applied Immunology* **83**, 92–95.

Grencis, R.K., Hültner, L. and Else, K.J. (1991). Host protective immunity to *Trichinella spiralis* in mice: activation of Th cell subsets and lymphokine secretion in mice expressing different response phenotypes. *Immunology* **74**, 329–332.

Grencis, R.K., Else, K.J., Huntley, J.F. and Nishikawa, S.I. (1993). The *in vivo* role of stem cell factor (c-kit ligand) on mastocytosis and host protective immunity to the intestinal nematode *Trichinella spiralis* in mice. *Parasite Immunology* **15**, 55–59.

Griscelli, C., Vassalli, P. and McCluskey, R.T. (1969). The distribution of large dividing lymph node cells in syngeneic recipient rats after intravenous injection. *Journal of Experimental Medicine* **130**, 1427–1451.

Grove, D.I., Hamburger, J. and Warren, K.S. (1977a). Kinetics of immunological responses, resistance to reinfection, and pathological reactions to infection with *Trichinella spiralis*. *Journal of Infectious Diseases* **136**, 562–570.

Grove, D.I., Mahmoud, A.A.F. and Warren, K.S. (1977b). Eosinophils and resistance to *Trichinella spiralis*. *Journal of Experimental Medicine* **145**, 755–759.

Gustowska, L., Ruitenberg, E.J. and Elgersma, A. (1980). Cellular reactions in tongue and gut in murine trichinellosis and their thymus-dependence. *Parasite Immunology* **2**, 133–154.

Ha, T.-Y., Reed, N.D. and Crowle, P.K. (1983). Delayed expulsion of adult *Trichinella spiralis* by mast cell-deficient W/W^v mice. *Infection and Immunity* **41**, 445–447.

Harari, Y. and Castro, G.A. (1983). Sialic acid deficiency in lectin-resistant intestinal brush border membranes from rats following the intestinal phase of trichinellosis. *Molecular and Biochemical Parasitology* **9**, 73–81.

Harari, Y., Russell, D.A. and Castro, G.A. (1987). Anaphylaxis-mediated epithelial Cl$^-$ secretion and parasite rejection in rat intestine. *Journal of Immunology* **138**, 1250–1255.

Hayes, C.E., Klyczek, K.K., Krum, D.D., Whitcomb, R.M., Hullett, D.A. and Cantor, H. (1984). Chromosome 4 Jt gene controls murine T cell surface I-J expression. *Science* **223**, 559–561.

Hendricks, J.R. (1950). The relationship between precipitin titer and number of *Trichinella spiralis* in the intestinal tract of mice following test infections. *Journal of Immunology* **64**, 173–177.

Herndon, F.J. and Kayes, S.G. (1992). Depletion of eosinophils by anti-IL-5 monoclonal antibody treatment of mice infected with *Trichinella spiralis* does not alter parasite burden or immunologic resistance to reinfection. *Journal of Immunology* **149**, 3642–3647.

Hessel, J., Ramaswamy, K. and Castro, G.A. (1982). Reduced hexose transport by enterocytes associated with rapid, noninjurious rejection of *Trichinella spiralis* from immune rats. *Journal of Parasitology* **68**, 202–207.

Horvitz, H.R., Chalfie, M., Trent, C., Sulston, J.E. and Evans, P.D. (1982). Serotonin and octopamine in the nematode *Caenorhabditis elegans*. *Science* **216**, 1012–1014.

Howard, R.J., Christie, P.R., Wakelin, D., Wilson, M.M. and Behnke, J.M. (1978). The effect of concurrent infection with *Trichinella spiralis* on *Hymenolepis microstoma* in mice. *Parasitology* **77**, 273–279.

Ismail, M.M. and Tanner, C.E. (1972a). *Trichinella spiralis*: Peripheral blood, intestinal, and bone-marrow eosinophilia in rats and its relationship to the inoculating dose of larvae, antibody response and parasitism. *Experimental Parasitology* **31**, 262–272.

Ismail, M.M. and Tanner, C.E. (1972b). *Trichinella spiralis*: The effect of antiserotonin and antihistamine reagents on the eosinophilic response in rats. *Experimental Parasitology* **31**, 273–283.

Jacqueline, E., Crinquette, J., Bout, D., Barrois, J. and Vernes, A. (1981). *Trichinella spiralis* in rats: *in vivo* effects of the bile and *in vitro* action of secretory IgA from bile. *Annales de Parasitologie* **56**, 395–400.

Jones, J.F., Crandall, C.A. and Crandall, R.B. (1976). T-dependent suppression of the primary antibody response to sheep erythrocytes in mice infected with *Trichinella spiralis*. *Cellular Immunology* **27**, 102–110.

Jones, J.F., Crandall, C.A. and Crandall, R.G. (1984). *In vivo* and *in vitro* responses to sheep erythrocytes by lymph node cells from mice with trichinellosis. *Clinical and Experimental Immunology* **57**, 301–306.

Jungery, M. and Ogilvie, B.M. (1982). Antibody response to stage-specific *Trichinella spiralis* surface antigens in strong and weak responder mouse strains. *Journal of Immunology* **129**, 839–843.

Kazura, J.W. and Grove, D.I. (1978). Stage-specific antibody-dependent eosinophil-mediated destruction of *Trichinella spiralis*. *Nature* **274**, 588–589.

Kazura, J.W. and Aikawa, M. (1980). Host defense mechanisms against *Trichinella spiralis* infection in the mouse: Eosinophil-mediated destruction of newborn larvae *in vitro*. *Journal of Immunology* **124**, 355–361.

Kazura, J.W. and Meshnick, S.R. (1984). Scavenger enzymes and resistance to oxygen mediated damage in *Trichinella spiralis*. *Molecular and Biochemical Parasitology* **10**, 1–10.

Kelly, E.Z.B., Cruz, E.S., Hauda, K.M. and Wassom, D.L. (1991). IFN-γ- and IL-5-producing cells compartmentalize to different lymphoid organs in *Trichinella spiralis*-infected mice. *Journal of Immunology* **147**, 306–311.

Kennedy, M.W. (1980). Effects of the host immune response on the longevity, fecundity and position in the intestine of *Trichinella spiralis* in mice. *Parasitology* **80**, 49–60.

Kennedy, M.W., Wassom, D.L., McIntosh, A.E. and Thomas, J.C. (1991). H-2 (I-A) control of the antibody repertoire to secreted antigens of *Trichinella spiralis* in infection and its relevance to resistance and susceptibility. *Immunology* **73**, 36–43.

Kim, C.W. (1966). Delayed hypersensitivity to larval antigens of *Trichinella spiralis*. *Journal of Infectious Diseases* **116**, 208–214.

Kim, C.W., Savel, H. and Hamilton, L.D. (1967). Delayed hypersensitivity to *Trichinella spiralis*. I. Transfer of delayed hypersensitivity by lymph node cells. *Journal of Immunology* **99**, 1150–1155.

Kitamura, Y., Go, S. and Hatanaka, K. (1978). Decrease of mast cells in W/W^v mice and their increase by bone marrow transplantation. *Blood* **52**, 447–452.

Klein, D., Tewarson, S., Figueroa, F. and Klein, J. (1982). The minimal length of the differential segment in *H-2* congenic lines. *Immunogenetics* **16**, 319–328.

Klimpel, G.R., Chopra, A.K., Langley, K.E., Wypych, J., Annable, C.A., Kaiserlian, D., Ernst, P.B. and Peterson, J.W. (1995). A role for stem cell factor and c-kit in the murine intestinal tract secretory response to cholera toxin. *Journal of Experimental Medicine* **182**, 1931–1942.

Korenaga, M., Wang, C.H., Bell, R.G., Zhu, D. and Ahmad, A. (1989). Intestinal immunity to *Trichinella spiralis* is a property of OX8$^-$ OX22$^-$ T helper cells that are generated in the intestine. *Immunology* **66**, 588–594.

Korenaga, M., Watanabe, N., Abe, T. and Hashiguchi, Y. (1996a). Acceleration of IgE responses by treatment with recombinant interleukin-3 prior to infection with *Trichinella spiralis* in mice. *Immunology* **87**, 642–646.

Korenaga, M., Abe, T. and Hashiguchi, Y. (1996b). Injection of recombinant interleukin 3 hastens worm expulsion in mice infected with *Trichinella spiralis*. *Parasitology Research* **82**, 108–113.

Krco, C.J., David, C.S. and Wassom, D.L. (1982). Characterization of an *in vitro* proliferation response to solubilized *Trichinella spiralis* antigens: role of Ia antigens and Ly-1$^+$ T cells. *Cellular Immunology* **68**, 359–367.

Krco, C.J., Wassom, D.L., Abramson, E.J. and David, C.S. (1983). Cloned T cells recognize *Trichinella spiralis* antigen in association with an E_b^κ E_α^κ restriction element. *Immunogenetics* **18**, 435–444.

Lammas, D.A., Wakelin, D., Mitchell, L.A., Tuohy, M., Else, K.J. and Grencis, R.K. (1992). Genetic influences upon eosinophilia and resistance in mice infected with *Trichinella spiralis*. *Parasitology* **105**, 117–124.

Larsh, J.E. (1953). Studies in old mice to test the hypotheses of local and general immunity to *Trichinella spiralis*. *Journal of Infectious Diseases* **93**, 282–293.

Larsh, J.E., Jr. (1963). Experimental Trichiniasis. *Advances in Parasitology* **1**, 213–286.

Larsh, J.E., Jr. (1967). The present understanding of the mechanism of immunity to *Trichinella spiralis*. *American Journal of Tropical Medicine and Hygiene* **16**, 123–132.

Larsh, J.E., Jr. and Fletcher, O.K., Jr. (1948). Further studies in mice on the effect of alcohol on acquired immunity to *Trichinella spiralis*. *Journal of the Elisha Mitchell Scientific Society* **64**, 196–203.

Larsh, J.E. Jr. and Race, G.J. (1954). A histopathologic study of the anterior small intestine of immunized and non-immunized mice infected with *Trichinella spiralis*. *Journal of Infectious Diseases* **94**, 262–272.

Larsh, J.E., Jr. and Race, G.J. (1975). Allergic inflammation as a hypothesis for the expulsion of worms from tissues: A review. *Experimental Parasitology* **37**, 251–266.

Larsh, J.E., Jr., Gilchrist, H.B. and Greenberg, B.G. (1952). A study of the distribution and longevity of adult *Trichinella spiralis* in immunized and non-immunized mice. *Journal of the Elisha Mitchell Scientific Society* **68**, 1–11.

Larsh, J.E., Jr., Goulson, H.T. and Weatherly, N.F. (1964a). Studies on delayed (cellular) hypersensitivity in mice infected with *Trichinella spiralis*. I. Transfer of lymph node cells. *Journal of the Mitchell Society* **80**, 133–135.

Larsh, J.E., Jr., Goulson, H.T. and Weatherly, N.F. (1964b). Studies on delayed (cellular) hypersensitivity in mice infected with *Trichinella spiralis*. II. Transfer of peritoneal exudate cells. *Journal of Parasitology* **50**, 496–498.

Larsh, J.E., Jr., Goulson, H.T., Weatherly, N.F. and Chaffee, E.F. (1969). Studies on delayed (cellular) hypersensitivity in mice infected with *Trichinella spiralis*. IV. Artificial sensitization of donors. *Journal of Parasitology* **55**, 726–729.

Larsh, J.E., Jr., Goulson, H.T., Weatherly, N.F. and Chaffee, E.F. (1970). Studies on delayed (cellular) hypersensitivity in mice infected with *Trichinella spiralis*. V. Tests in recipients injected with donor spleen cells 1, 3, 7, 14, or 21 days before infection. *Journal of Parasitology* **56**, 978–981.

Larsh, J.E., Jr., Weatherly, N.F., Goulson, H.T. and Chaffee, E.F. (1972). Studies on delayed (cellular) hypersensitivity in mice infected with *Trichinella spiralis*. VII. The effect of ATS injections on the numbers of adult worms recovered after challenge. *Journal of Parasitology* **58**, 1052–1060.

Lee, G.B. and Ogilvie, B.M. (1980). The mucus layer in intestinal nematode infections. In: *Mucosal Immune System in Health and Disease (Proceedings of the 81st Ross Conference on Pediatric Research)* (P.L. Ogra and J. Bienenstock, eds), pp. 175–187. Columbus: Ross Laboratories.

Lee, G.B. and Ogilvie, B.M. (1982). The intestinal mucus layer in *Trichinella spiralis*-infected rats. In: *Recent Advances in Mucosal Immunity* (W. Strober, L.A. Hanson and K.W. Sell, eds), p. 319–329. New York: Raven Press.

Lee, T.D.G. (1991). Helminthotoxic responses of intestinal eosinophils to *Trichinella spiralis* newborn larvae. *Infection and Immunity* **59**, 4405–4411.

Levin, A.J. and Evans, T.C. (1942). The use of roentgen radiation in locating an origin of host resistance to *Trichinella spiralis* infections. *Journal of Parasitology* **28**, 477–483.

Lim, P.-L. and Choy, W.-F. (1986). Monoclonal IgM/A hybrid antibodies: artifacts due to anti-idiotype (T15) antibodies in commercial anti-α sera. *Molecular Immunology* **23**, 909–916.

Ljungström, I. and Ruitenberg, E.J. (1976). A comparative study of the immunohistological and serological response of intact and T cell-deprived mice to *Trichinella spiralis*. *Clinical and Experimental Immunology* **24**, 146–156.

Louch, C.D. (1962). Increased resistance to *Trichinella spiralis* in the laboratory rat following infection with *Nippostrongylus muris*. *Journal of Parasitology* **48**, 24–26.

Love, R.J., Ogilvie, B.M. and McLaren, D.J. (1976). The immune mechanism which expels the intestinal stage of *Trichinella spiralis* from rats. *Immunology* **30**, 7–15.

Lubiniecki, A.S. and Cypess, R.H. (1975). Immunological sequelae of *Trichinella spiralis* infection in mice: effect on the antibody responses to sheep erythrocytes and Japanese B encephalitis virus. *Infection and Immunity* **11**, 1306–1311.

Luongo, M.A., Reid, D.H. and Weiss, W.W. (1951). The effect of ACTH in Trichinosis. A clinical and experimental study. *New England Journal of Medicine* **245**, 757–760.

Mackenzie, C.D., Jungery, M., Taylor, P.M. and Ogilvie, B.M. (1980a). The *in vitro* interaction of eosinophils, neutrophils, macrophages and mast cells with nematode surfaces in the presence of complement or antibodies. *Journal of Pathology* **133**, 161–175.

Mackenzie, C.D., Jungery, M., Taylor, P.M. and Ogilvie, B.M. (1980b). Activation of complement, the induction of antibodies to the surface of nematodes and the effect of these factors and cells on worm survival *in vitro*. *European Journal of Immunology* **10**, 594–601.

Maizels, R.M., Burke, J. and Denham, D.A. (1987). Phosphorylcholine-bearing antigens in filarial nematode parasites: analysis of somatic extracts, *in-vitro* secretions and infection sera from *Brugia malayi* and *B. pahangi*. *Parasite Immunology* **9**, 49–66.

Malaviya, R., Ikeda, T., Ross, E. and Abraham, S.N. (1996). Mast cell modulation of neutrophil influx and bacterial clearance at sites of infection through TNF-α. *Nature* **381**, 77–80.

Manson-Smith, D.F., Bruce, R.G. and Parrott, D.M.V. (1979a). Villous atrophy and expulsion of intestinal *Trichinella spiralis* are mediated by T cells. *Cellular Immunology* **47**, 285–292.

Manson-Smith, D.F., Bruce, R.G., Rose, M.L. and Parrott, D.M.V. (1979b). Migration of lymphoblasts to the small intestine. III. Strain differences and relationship to distribution and duration of *Trichinella spiralis* infection. *Clinical and Experimental Immunology* **38**, 475–482.

Markell, E.K. and Lewis, W.P. (1957). Effect of cortisone treatment on immunity to subsequent reinfection with *Trichinella* in the rat. *American Journal of Tropical Medicine and Hygiene* **6**, 553–561.

McCoy, O.R. (1931). Immunity of rats to reinfection with *Trichinella spiralis*. *American Journal of Hygiene* **14**, 484–494.

McCoy, O.R. (1932). Size of infection as an influence on the persistence of adult Trichinae in rats. *Science* **75**, 364–365.

McCoy, O.R. (1935). Artificial immunization of rats against *Trichinella spiralis*. *American Journal of Hygiene* **21**, 200–213.

McCoy, O.R. (1940). Rapid loss of *Trichinella* larvae fed to immune rats and its bearing on the mechanism of immunity. *American Journal of Hygiene* **32**, 105–116.

Miller, H.R.P. and Jarrett, W.F.H. (1971). Immune reactions in mucous membranes. I. Intestinal mast cell response during helminth expulsion in the rat. *Immunology* **20**, 277–288.

Moqbel, R. and Wakelin, D. (1979). *Trichinella spiralis* and *Strongyloides ratti*: immune interaction in adult rats. *Experimental Parasitology* **47**, 65–72.

Moqbel, R., Wakelin, D., MacDonald, A.J., King, S.J., Grencis, R.K. and Kay, A.B. (1987). Release of leukotrienes during rapid expulsion of *Trichinella spiralis* from immune rats. *Immunology* **60**, 425–430.

Mosmann, T.R. and Coffman, R.L. (1989). Th1 and Th2 cells: Different patterns of lymphokine secretion lead to different functional properties. *Annual Review of Immunology* **7**, 145–173.

Mota, I., Sadun, E.H., Bradshaw, R.M. and Gore, R.W. (1969). The immunological response of mice infected with *Trichinella spiralis*: biological and physio-chemical distinction of two homocytotropic antibodies. *Immunology* **16**, 71–81.

Negrao-Correa, D., Adams, L.S. and Bell, R.G. (1996). Intestinal transport and catabolism of IgE. *Journal of Immunology* **157**, 4037–4044.

Ogilvie, B.M. (1964). Reagin-like antibodies in animals immune to helminth parasites. *Nature* **204**, 91–92.

Oku, Y., Itayama, H. and Kamiya, M. (1984). Expulsion of *Trichinella spiralis* from the intestine of W/W^v mice reconstituted with haematopoietic and lymphopoietic cells and origin of mucosal mast cells. *Immunology* **53**, 337–344.

Oliver-Gonzalez, J. (1940). The *in vitro* action of immune serum on the larvae and adults of *Trichinella spiralis*. *Journal of Infectious Diseases* **67**, 292–300.

Oliver-Gonzalez, J. (1941). The dual antibody basis of acquired immunity in Trichinosis. *Journal of Infectious Diseases* **69**, 254–270.

Opie, E.L. (1904). An experimental study of the relation of cells with eosinophile granulation to infection with an animal parasite (*Trichinella spiralis*). *American Journal of the Medical Sciences* **127**, 477–493.

Ottaway, C.A. and Parrott, D.M.V. (1980). Regional blood flow and the localization of lymphoblasts in the small intestine of the mouse. I. Examination of normal small intestine. *Immunology* **41**, 955–961.

Ottaway, C.A., Manson-Smith, D.F., Bruce, R.G. and Parrott, D.M.V. (1980). Regional blood flow and the localization of lymphoblasts in the small intestine of the mouse. II. The effect of a primary enteric infection with *Trichinella spiralis*. *Immunology* **41**, 963–971.

Ottesen, E.A., Smith, T.K. and Kirkpatrick, C.H. (1975). Immune response to *Trichinella spiralis* in the rat. I. Development of cellular and humoral responses during chronic infection. *International Archives of Allergy and Applied Immunology* **49**, 396–410.

Palmer, J.M., Weisbrodt, N.W. and Castro, G.A. (1984). *Trichinella spiralis*: intestinal myoelectric activity during enteric infection in the rat. *Experimental Parasitology* **57**, 132–141.

Parmentier, H.K., Ruitenberg, E.J. and Elgersma, A. (1982). Thymus dependence of the adoptive transfer of intestinal mastocytopoiesis in *Trichinella spiralis*-infected mice. *International Archives of Allergy and Applied Immunology* **68**, 260–267.

Parmentier, H.K. de Vries, C., Ruitenberg, E.J. and van Loveren, H. (1987). Involvement of serotonin in intestinal mastocytopoiesis and inflammation during a *Trichinella spiralis* infection in mice. *International Archives of Allergy and Applied Immunology* **83**, 31–38.

Perrudet-Badoux, A., Binaghi, R.A. and Biozzi, G. (1975). *Trichinella* infestation in mice genetically selected for high and low antibody production. *Immunology* **29**, 387–390.

Perrudet-Badoux, A., Binaghi, R.A. and Boussac-Aron, Y. (1978). *Trichinella spiralis* infection in mice. Mechanism of the resistance in animals genetically selected for high and low antibody production. *Immunology* **35**, 519–522.

Philipp, M., Parkhouse, R.M.E. and Ogilvie, B.M. (1980). Changing proteins on the surface of a parasitic nematode. *Nature* **287**, 538–540.

Pincus, S.H., Cammarata, P.V., Delima, M. and Despommier, D. (1986). Eosinophilia in murine trichinellosis. *Journal of Parasitology* **72**, 321–325.

Pond, L., Wassom, D.L. and Hayes, C.E. (1989). Evidence for differential induction of helper T cell subsets during *Trichinella spiralis* infection. *Journal of Immunology* **143**, 4232–4237.

Pond, L., Wassom, D.L. and Hayes, C.E. (1992). Influence of resistant and sus-

ceptible genotype, IL-1, and lymphoid organ on *Trichinella spiralis*-induced cytokine secretion. *Journal of Immunology* **149**, 957–965.
Ramaswamy, K., Goodman, R.E. and Bell, R.G. (1994a). Cytokine profile of protective anti-*Trichinella spiralis* CD4$^+$ OX22$^-$ and non-protective CD4$^+$ OX22$^+$ thoracic duct cells in rats: Secretion of IL-4 alone does not determine protective capacity. *Parasite Immunology* **16**, 435–445.
Ramaswamy, K., Hakimi, J. and Bell, R.G. (1994b). Evidence for an interleukin 4-inducible, immunoglobulin E uptake and transport mechanism in the intestine. *Journal of Experimental Medicine* **180**, 1793–1803.
Ramaswamy, K., Negrao-Correa, D. and Bell, R. (1996). Local intestinal immune responses to infections with *Trichinella spiralis*. Real-time, continuous assay of cytokines in the intestinal (afferent) and efferent thoracic duct lymph of rats. *Journal of Immunology* **156**, 4328–4337.
Rappaport, I. and Wells, H.S. (1951). Studies in Trichinosis. I. Immunity to reinfection in mice following a single light infection. *Journal of Infectious Diseases* **88**, 248–253.
Riedlinger, J., Grencis, R.K. and Wakelin, D. (1986). Antigen-specific T-cell lines transfer protective immunity against *Trichinella spiralis in vivo*. *Immunology* **58**, 57–61.
Rivera-Ortiz, C.-I. and Nussenzweig, R. (1976). *Trichinella spiralis*: anaphylactic antibody formation and susceptibility in strains of inbred mice. *Experimental Parasitology* **39**, 7–17.
Robinson, K., Bellaby, T. and Wakelin, D. (1995). Immunity to *Trichinella spiralis* transferred by serum from vaccinated mice not protected by immunization. *Parasite Immunology* **17**, 85–90.
Robinson, M. and David, C.S. (1989). The genetics of the immune response to *Trichinella spiralis* antigens in the mouse. In: *Immunobiology of Proteins and Peptides. V. Vaccines* (M.Z. Atassi, ed.), pp. 329–340. Plenum Press.
Rose, M.L., Parrott, D.M.V. and Bruce, R.G. (1976). Migration of lymphoblasts to the small intestine. I. Effect of *Trichinella spiralis* infection on the migration of mesenteric lymphoblasts and mesenteric T lymphoblasts in syngeneic mice. *Immunology* **31**, 723–730.
Roth, H. (1939). Experimental studies on the course of Trichina infection in guinea pigs. II. Natural susceptibility of the guinea pig to experimental trichina infection. *American Journal of Hygiene* **29**, 89–104.
Ruitenberg, E.J. and Elgersma, A. (1976). Absence of intestinal mast cell response in congenitally athymic mice during *Trichinella spiralis* infection. *Nature* **264**, 258–260.
Ruitenberg, E.J., Elgersma, A., Kruizinga, W. and Leenstra, F. (1977a). *Trichinella spiralis* infection in congenitally athymic (nude) mice. Parasitological, serological and haematological studies with observations on intestinal pathology. *Immunology* **33**, 581–587.
Ruitenberg, E.J., Leenstra, F. and Elgersma, A. (1977b). Thymus dependence and independence of intestinal pathology in a *Trichinella spiralis* infection: a study in congenitally athymic (nude) mice. *British Journal of Experimental Pathology* **58**, 311–314.
Ruitenberg, E.J., Elgersma, A. and Kruizinga, W. (1979). Intestinal mast cells and globule leucocytes: Role of the thymus on their presence and proliferation during a *Trichinella spiralis* infection in the rat. *International Archives of Allergy and Applied Immunology* **60**, 302–309.

Russell, D.A. (1986). Mast cells in the regulation of intestinal electrolyte transport. *American Journal of Physiology* **251**, G253-G262.

Russell, D.A. and Castro, G.A. (1985). Anaphylactic-like reaction of small intestinal epithelium in parasitized guinea-pigs. *Immunology* **54**, 573–579.

Sadun, E.H., Mota, I. and Gore, R.W. (1968). Demonstration of homocytotropic reagin-like antibodies in mice and rabbits infected with *Trichinella spiralis*. *Journal of Parasitology* **54**, 814–821.

Salzer, B.F. (1916). A study of an epidemic of fourteen cases of trichinosis with cures by serum therapy. *Journal of the American Medical Association* **67**, 579–580.

Schwartz, B. (1917). Serum therapy for Trichinosis. *Journal of the American Medical Association* **69**, 884–886.

Sharp, A.D. and Olson, L.J. (1962). Hypersensitivity responses to *Toxocara-*, *Ascaris-*, and *Trichinella*-infected guinea pigs to homologous and heterologous challenge. *Journal of Parasitology* **48**, 362–367.

Stewart, G.L., Kramar, G.W., Charniga, L. and Kramar, M. (1985). The effects of histamine and an antihistamine on *Trichinella spiralis* and on trichinous enteritis in the host. *International Journal for Parasitology* **4**, 327–332.

Svet-Moldavsky, G.J., Shaghijan, G.S., Chernyakhovskaya, I.Y., Mkheidze, D.M., Litovchenko, T.A., Ozeretskovskaya, N.N. and Kadaghidze, Z.G. (1970). Inhibition of skin allograft rejection in *Trichinella*-infected mice. *Transplantation* **9**, 69–70.

Taliafferro, W.H. (1940). The mechanism of acquired immunity in infections with parasitic worms. *Physiological Reviews* **20**, 469–492.

Taliafferro, W.H. and Sarles, M.P. (1939). The cellular reactions in the skin, lungs and intestine of normal and immune rats after infection with *Nippostrongylus muris*. *Journal of Infectious Diseases* **64**, 157–192.

Tanner, C.E. (1978). The susceptibility of *Trichinella spiralis* of inbred lines of mice differing at the H-2 histocompatibility locus. *Journal of Parasitology* **64**, 956–957.

Torihashi, S., Ward, S.M. and Sanders, K.M. (1997). Development of c-Kit-positive cells and the onset of electrical rhythmicity in murine small intestine. *Gastroenterology* **112**, 144–155.

Tronchin, G., Dutoit, E., Vernes, A. and Biguet, J. (1979). Oral immunization of mice with metabolic antigens of *Trichinella spiralis* larvae: Effects on the kinetics of intestinal cell response including mast cells and polymorphonuclear eosinophils. *Journal of Parasitology* **65**, 685–691.

Tuohy, M., Lammas, D.A., Wakelin, D., Huntley, J.F., Newlands, G.F.J. and Miller, H.R.P. (1990). Functional correlations between mucosal mast cell activity and immunity to *Trichinella spiralis* in high and low responder mice. *Parasite Immunology* **12**, 675–685.

Urban, J.F., Jr., Katona, I.M., Paul, W.E. and Finkelman, F.D. (1991). Interleukin 4 is important in protective immunity to a gastrointestinal nematode infection in mice. *Proceedings of the National Academy of Sciences, USA*, **88**, 5513–5517.

Urban, J.F., Jr., Maliszewski, C.R., Madden, K.B., Katona, I.M. and Finkelman, F.D. (1995). IL-4 treatment can cure established gastrointestinal nematode infections in immunocompetent and immunodeficient mice. *The Journal of Immunology* **154**, 4675–4684.

van Loveren, H., Osterhaus, A.D.M.E., Nagel, J., Schuurman, H.J. and Vos, J.G. (1988). Detection of IgA antibodies and quantification of IgA-antibody-producing cells specific for ovalbumin or *Trichinella spiralis* in the rat. *Scandinavian Journal of Immunology* **28**, 377–381.

van Someren, V.D. (1938). Eosinophilia and the differential blood count in Trichinosis of the rat. *Journal of Helminthology* **14**, 83–92.
Vos, J.G., Ruitenberg, E.J., Basten, N.V., Buys, J., Elgersma, A. and Kruizinga, W. (1983). The athymic nude rat. IV. Immunocytochemical study to detect T-cells, and immunological and histopathological reactions against *Trichinella spiralis*. *Parasite Immunology* **5**, 195–215.
Wakelin, D. (1980). Genetic control of immunity to parasites. Infection with *Trichinella spiralis* in inbred and congenic mice showing rapid and slow responses to infection. *Parasite Immunology* **2**, 85–98.
Wakelin, D. (1985). Genetic control of immunity to helminth infections. *Parasitology Today* **1**, 17–23.
Wakelin, D. (1986). Genetic and other constraints on resistance to infection with gastrointestinal nematodes. *Transactions of the Royal Society of Tropical Medicine and Hygiene* **80**, 742–747.
Wakelin, D. (1988). Helminth Infections. In: *Genetics of Resistance to Bacterial and Parasitic Infection* (D. Wakelin and J.M. Blackwell, eds), pp. 153–224. London: Taylor and Francis.
Wakelin, D. (1989). Nature and nurture: overcoming constraints on immunity. *Parasitology* **99**, S21–S35.
Wakelin, D. (1992). Immunogenetic and evolutionary influences on the host-parasite relationship. *Developmental and Comparative Immunology* **16**, 345–353.
Wakelin, D. (1993). Allergic inflammation as a hypothesis for the expulsion of worms from tissues. *Parasitology Today* **9**, 115–116.
Wakelin, D. and Denham, D.A. (1983). The immune response: In: Trichinella *and* Trichinosis (W.C. Campbell, ed.), pp. 265–308. New York: Plenum Press.
Wakelin, D. and Donachie, A.M. (1980). Genetic control of immunity to parasites: adoptive transfer of immunity between inbred strains of mice characterized by rapid and slow immune expulsion of *Trichinella spiralis*. *Parasite Immunology* **2**, 249–260.
Wakelin, D. and Donachie, A.M. (1981). Genetic control of immunity to *Trichinella spiralis*. Donor bone marrow cells determine responses to infection in mouse radiation chimaeras. *Immunology* **43**, 787–792.
Wakelin, D. and Donachie, A.M. (1983a). Genetic control of immunity to *Trichinella spiralis*: influence of H-2-linked genes on immunity to the intestinal phase of infection. *Immunology* **48**, 343–350.
Wakelin, D. and Donachie, A.M. (1983b). Genetic control of eosinophilia. Mouse strain variation in response to antigens of parasite origin. *Clinical and Experimental Immunology* **51**, 239–246.
Wakelin, D. and Lloyd, M. (1976). Accelerated expulsion of adult *Trichinella spiralis* in mice given lymphoid cells and serum from infected donors. *Parasitology* **72**, 307–315.
Wakelin, D. and Wilson, M.M. (1977a). Transfer of immunity to *Trichinella spiralis* in the mouse with mesenteric lymph node cells: time of appearance of effective cells in donors and expression of immunity in recipients. *Parasitology* **74**, 215–224.
Wakelin, D. and Wilson, M.M. (1977b). Evidence for the involvement of a bone marrow-derived cell population in the immune expulsion of *Trichinella spiralis*. *Parasitology* **74**, 225–235.
Wakelin, D., and Wilson, M.M. (1979a). T and B cells in the transfer of immunity against *Trichinella spiralis* in mice. *Immunology* **37**, 103–109.
Wakelin, D., and Wilson, M.M. (1979b). *Trichinella spiralis*: Immunity and inflam-

mation in the expulsion of transplanted adult worms from mice. *Experimental Parasitology* **48**, 305–312.

Wakelin, D., Donachie, A.M. and Grencis, R.K. (1985). Genetic control of immunity to *Trichinella spiralis* in mice: capacity of cells from slow responder mice to transfer immunity in syngeneic and F_1 hybrid recipients. *Immunology* **56**, 203–211.

Wakelin, D., Grencis, R.K. and Donachie, A.M. (1982). Short lived, dividing cells mediate adoptive transfer of immunity to *Trichinella spiralis* in mice. II. *In vivo* characteristics of the cells. *Immunology* **46**, 451–457.

Wakelin, D., Mitchell, L.A., Donachie, A.M. and Grencis, R.K. (1986). Genetic control of immunity to *Trichinella spiralis* in mice. Response of rapid- and slow-responder strains to immunization with parasite antigens. *Parasite Immunology* **8**, 159–170.

Walls, R.G., Carter, R.L., Leuchars, E. and Davies, A.J.S. (1973). The immunopathology of Trichiniasis in T-cell deficient mice. *Clinical and Experimental Immunology* **13**, 231–242.

Wang, C.H. and Bell, R.G. (1987). *Trichinella spiralis*: Intestinal expression of systemic stage-specific immunity to newborn larvae. *Parasite Immunology* **9**, 465–475.

Wang, C.H. and Bell, R.G. (1988). Antibody-mediated *in-vivo* cytotoxicity to *Trichinella spiralis* newborn larvae in immune rats. *Parasite Immunology* **10**, 293–308.

Wang, C.H., Korenaga, M., Sacuto, F.R., Ahmad, A. and Bell, R.G. (1990a). Intraintestinal migration to the epithelium of protective, dividing, anti-*T. spiralis* $CD4^+$ $OX22^-$ cells requires MHC Class II compatibility. *Journal of Immunology* **145**, 1021–1028.

Wang, C.H., Korenaga, M., Greenwood, A. and Bell, R.G. (1990b). T helper subset function in the gut of rats: Differential stimulation of eosinophils, mucosal mast cells and antibody forming cells by $OX8^-$ $OX22^+$ and $OX8^-$ $OX22^-$. *Immunology* **71**, 166–175.

Wantland, W.W. (1937). Blood studies on normal and trichinized white rabbits. *Journal of Laboratory and Clinical Medicine* **23**, 32–38.

Wassom, D.L. (1988). Genetic control of immunity to parasite infection: Studies of *Trichinella*-infected mice. In: *The Biology of Parasitism* (Englund, P.T. and Sher, A., eds), pp. 329–346, New York: Alan R. Liss, Inc.

Wassom, D.L. (1993). Immunoecological succession in host-parasite communities. *Journal of Parasitology* **79**, 483–487.

Wassom, D.L. and Gleich, G.J. (1979). Damage to *Trichinella spiralis* newborn larvae by eosinophil major basic protein. *American Journal of Tropical Medicine and Hygiene* **28**, 860–863.

Wassom, D.L. and Kelly, E.A.B. (1990). The role of the major histocompatibility complex in resistance to parasite infections. *Critical Reviews in Immunology* **10**, 31–52.

Wassom, D.L., David, C.S. and Gleich, G.J. (1979). Genes within the major histocompatibility complex influence susceptibility to *Trichinella spiralis* in the mouse. *Immunogenetics* **9**, 491–496.

Wassom, D.L., Brooks, B.O., Cypess, R.H. and David, C.S. (1983a). A survey of susceptibility to infection with *Trichinella spiralis* of inbred mouse strains sharing common H-2 alleles but different genetic backgrounds. *Journal of Parasitology* **69**, 1033–1037.

Wassom, D.L., Brooks, B.O., Babish, J.G. and David, C.S. (1983b). A gene mapping between the *S* and *D* regions of the *H-2* complex influences resistance to *Trichinella spiralis* infections of mice. *Journal of Immunogenetics* **10**, 371–378.

Wassom, D.L., Dougherty, D.A., Krco, C.J. and David, C.S. (1984a). H-2-controlled, dose-dependent suppression of the response that expels adult *Trichinella spiralis* from the small intestine of mice. *Immunology* **53**, 811–818.

Wassom, D.L., Wakelin, D., Brooks, B.O., Krco, C.J. and David, C.S. (1984b). Genetic control of immunity to *Trichinella spiralis* infections of mice. Hypothesis to explain the role of H-2 genes in primary and challenge infections. *Immunology* **51**, 625–631.

Wassom, D.L., Krco, C.J. and David, C.S. (1987). I-E expression and susceptibility to parasite infection. *Immunology Today* **8**, 39–43.

Wilkes, S.D. and Goven, A.J. (1984). Tissue eosinophil numbers and phospholipase B activity in mice infected with *Trichinella spiralis*. *International Journal for Parasitology* **14**, 479–482.

Woodbury, R.G., Miller, H.R.P., Huntley, J.F., Newlands, G.F.J., Palliser, A.C. and Wakelin, D. (1984). Mucosal mast cells are functionally active during spontaneous expulsion of intestinal nematode infections in rats. *Nature* **312**, 450–452.

Wright, K.A. (1979). *Trichinella spiralis*: An intracellular parasite in the intestinal phase. *Journal of Parasitology* **65**, 441–445.

Yasunaga, M., Wang, F.-h., Kunisada, T., Nishikawa, S. and Nishikawa, S.I. (1995). Cell cycle control of c-kit^+ IL-7R$^+$ B precursor cells by two distinct signals derived from IL-7 receptor and c-kit in a fully defined medium. *Journal of Experimental Medicine* **182**, 315–323.

Young, J.W., Szabolcs, P. and Moore, M.A.S. (1995). Identification of dendritic cell colony-forming units among normal human CD34$^+$ bone marrow progenitors that are expanded by c-kit-ligand and yield pure dendritic cell colonies in the presence of granulocyte/macrophage colony-stimulating factor and tumor necrosis factor α. *Journal of Experimental Medicine* **182**, 1111–1119.

Zaiman, H. and Villaverde, H. (1964). Studies on the eosinophilic response of parabiotic rats infected with *Trichinella spiralis*. *Experimental Parasitology* **15**, 14–31.

Zaiman, H., Storey, J.M. and Headley, N.C. (1955a). Studies on the nature of immunity to *Trichinella spiralis* in parabiotic rats. VIII. The duration of the immune response in the 'uninfected' parabiotic rat following infection of one twin with *Trichinella spiralis*. *American Journal of Hygiene* **61**, 15–23.

Zaiman, H., Storey, J.M., Rubel, J. and Headley, N.C. (1955b). Studies on the nature of immunity to *Trichinella spiralis* in parabiotic rats. VII. The immune response of the 'uninfected' twin one month after its mate received an immunizing dose of irradiated (X-ray) larvae. *American Journal of Hygiene* **61**, 5–14.

Zakroff, S.G.H., Beck, L., Platzer, E.G. and Spiegelberg, H.L. (1989). The IgE and IgG subclass responses of mice to four helminth parasites. *Cellular Immunology* **119**, 193–201.

Zhang, S. and Castro, G.A. (1990). Involvement of Type I hypersensitivity in rapid rejection of *Trichinella spiralis* from adult rats. *International Archives of Allergy and Applied Immunology* **93**, 272–279.

Zhu, D. and Bell, R.G. (1989). IL-2 production, IL-2 receptor expression, and IL-2 responsiveness of spleen and mesenteric lymph node cells from inbred mice infected with *Trichinella spiralis*. *Journal of Immunology* **142**, 3262–3267.

Zhu, D. and Bell. R.G. (1990). Genetic analysis of the relationship between interleukin production and worm rejection in *Trichinella spiralis*-infected inbred mice. *Journal of Parasitology* **76**, 703–710.

Population Biology of Parasitic Nematodes: Applications of Genetic Markers

Timothy J.C. Anderson[1], Michael S. Blouin[2] and Robin N. Beech[3]

[1] Wellcome Trust Centre for Epidemiology of Infectious Disease, Department of Zoology, South Parks Rd., Oxford OX1 3PS, UK
[2] Department of Zoology, Cordley Hall 3029, Oregon State University, Corvallis, OR 97331–2714, USA
[3] Institute for Parasitology, 21111 Lakeshore Rd, Ste-Anne de Bellevue, Quebec H9X 3V9, Canada

1. Introduction 220
2. How Variable are Nematode Parasites? 221
   2.1. Mitochondrial DNA 221
   2.2. Nuclear genome 229
3. Population Structure — How is Genetic Variation Arranged in Populations? . . 237
   3.1. Geographical differentiation and gene flow 237
   3.2. Nematode life history traits and population structure 240
   3.3. Differences among loci in population structure 241
   3.4. Inferring natural selection from population structure 242
   3.5. Microspatial population structure and transmission patterns 242
   3.6. Heterozygote excess and mating patterns 244
4. Sibling Species, Host Affiliation, and Hybridization 245
   4.1. Sibling species at an early stage of divergence 247
   4.2. Sibling species at a later stage of divergence 250
   4.3. Morphological and genetic divergence 253
   4.4. Ubiquity of sibling species 255
   4.5. Inferring population history 258
5. Nematode Life-histories 260
   5.1. Sexual exchange in *Strongyloides* 260
   5.2. Identification of larval stages 262
6. Genetic Markers and Drug Resistance 262
   6.1. Genes involved in resistance 263
   6.2. Genetics of resistance 265

    6.3. Population biology 268
7. Other Applications of Genetic Markers 270
    7.1. Intrahost dynamics 270
    7.2. Nematode mating patterns and sociobiology 270
    7.3 Antigen evolution and immunology............................. 272
Acknowledgements 273
References... 273

1. INTRODUCTION

How can we measure patterns of movement and geneflow in nematode parasites? How can we identify sibling species and establish patterns of host affiliation and cross infection? How can we measure levels of drug resistance within populations and the fitness costs of drug resistance genes? These are all questions which are best answered using genetic markers. In this review we describe genetic approaches to answering a variety of questions in the population biology of parasitic helminths. We concentrate on nematode parasites, although we also choose occasional examples from other helminth groups. Our coverage of material will be somewhat biased, representing our respective interests. For example, most of the examples will be drawn from the literature on animal parasitic nematodes and much less from the extensive literature on plant parasitic nematodes. Furthermore, we restrict ourselves in only looking at patterns of variation at the generic level and below; those interested in phylogenetic relationships between different nematode groups are referred to work by Vanfleteren *et al.* (1994). Throughout, we aim to outline not only what has been done, but what could be done in future work.

    We have arranged this chapter as follows. In the following section, we describe levels of heterozygosity revealed by allozymes and patterns of variation in a variety of different sequence types. In the third section, we describe how this variation is distributed in parasite populations, and how we can use these patterns to ask questions about various aspects of nematode biology and epidemiology. In the fourth, we move to the interspecific level and describe how genetic approaches have clarified our ideas on nematode speciation, and revealed a wealth of sibling species, while in the fifth, examples of the use of markers for clarifying life cycles are discussed. The sixth section deals with genes responsible for drug resistance. Finally, the last section contains a mixed bag of other miscellaneous uses for genetic markers. Much of this section is speculative, and describes fields in which we feel genetic markers should see greater usage in the future.

2. HOW VARIABLE ARE NEMATODE PARASITES?

The various molecular tools used to detect and measure biochemical and molecular variation are broadly the same for nematodes as for any other group of animals or plants, and are reviewed elsewhere (Hillis and Moritz, 1990; Avise, 1994; Schierwater et al., 1994). In this section we describe the patterns of variation revealed by a variety of commonly used techniques. We compare patterns observed in nematodes with those seen in other groups of organisms, and evaluate patterns of evolution of different genes and genomes in nematode parasites.

2.1. Mitochondrial DNA

2.1.1. Sequence Diversity

Within-population sequence diversity in mitochondrial (mt)DNA varies considerably among nematode species (Table 1). No mtDNA variation was found in *Howardula aoronymphium* populations (Jaenike, 1996), and relatively little was found in *Strongyloides ratti* (Fisher, 1997), or in the plant parasites, *Meloidogyne* spp. (Hugall et al., 1994, 1997). In fact the patterns seen in these species are typical of patterns of mtDNA variation in free living organisms (Avise, 1994). In contrast, *Ascaris* spp. and trichostrongylid parasites of ruminants show very high levels of mtDNA diversity within populations — up to 10 times that typically observed in vertebrates (Blouin et al., 1992, 1995; Anderson et al., 1993, 1995b; Anderson and Jaenike, 1997). What might cause these disparate patterns? We explore three possible explanations for the elevated diversity seen in some nematode populations:

(a) *Admixture of populations.* Recent admixture of different geographical population could give rise to high levels of within-population diversity. This is almost certainly a contributory factor for *Ascaris*. Three divergent clusters of mtDNA sequences differing by up to 6% in sequence are found in *Ascaris* populations infecting both humans and pigs (see Figure 6). Furthermore, identical mtDNA genotypes (at 164 restriction sites surveyed) are found in worms collected from different continents. This indirect evidence for recent global mixing of populations is easily explicable when one considers the global movement of humans and domestic animals over the last few thousand years.

However, admixture of populations does not appear to explain the high mtDNA diversity observed in trichostrongylid populations. In a mixed population one would see distinct clusters of haplotypes, as seen in *Ascaris* (see Figure 6), with large genetic distances between clusters. In

Table 1 Nucleotide diversity in mtDNA and introns of nuclear genes.

Nematode sp.	Host (intermediate)	Genes	Nucleotide diversity	Method	Reference
Mitochondrial DNA					
Meloidogyne hapla	Various plant spp.	ND3, 16s rRNA	0.0128	Sequence	Hugall et al. (1997)
Meloidogyne spp. (except hapla)	Various plant spp.	ND3, 16s rRNA	0.0012	Sequence	Hugall et al. (1997)
Haemonchus contortus	Sheep	ND4	0.026	Sequence	Blouin et al. (1995)
Haemonchus placei	Cattle	ND4	0.019	Sequence	Blouin et al. (1995)
Teladorsagia circumcincta	Sheep	ND4	0.024	Sequence	Blouin et al. (1995)
Mazamastrongylus odocoilei	Deer	ND4	0.028	Sequence	Blouin et al. (1995)
Ostertagia ostertagi	Cattle	ND4	0.027	Sequence	Blouin et al. (1995)
Ostertagia ostertagi	Cattle	complete MtDNA	0.018	RFLP	Blouin et al. (1992)
Ascaris lumbricoides	Humans	lgRNA to ATPase 6^b	0.016	RFLP	Anderson et al. (1993)
Ascaris suum	Pigs	lgRNA to ATPase 6^b	0.009	RFLP	Anderson et al. (1993)
Howardula aoronymphium A	Drosophila spp.	Nd4, LNC, COI, COII, l-rRNA	0	RFLP	Jaenike (1996)
Howardula aoronymphium B	Drosophila spp.	Nd4, LNC, COI, COII, l-rRNA	0	RFLP	Jaenike (1996)
Strongyloides ratti	Rat (Berkshire and Germany)	Cyt b (304bp)	<0.001	Sequence	Fisher (1997)
Nuclear genes					
Haemonchus contortus	Sheep	β-Tubulin Isotype 1 intron	0.094	RFLP	Beech et al. (1994)
Haemonchus contortus	Sheep	β-Tubulin isotype 2 intron	0.091	RFLP	Beech et al. (1994)
Ascaris spp.	Human/pig	Hemoglobin intron (A–C)	0.002^a	RFLP	Anderson and Jaenike (1997)
Ascaris spp.	Human/pig	G4, G9, G12^c	$0.003–0.008^a$	RFLP	Anderson and Jaenike (1997)
Ascaris spp.	Human/pig	Myo D intron	0.004^a	RFLP	Anderson and Jaenike (1997)

^{a} Mean diversity of parasites (six genes per population).
^{b} See Figure 2.
^{c} Introns in anonymous single copy genes (see Anderson and Jaenike, 1997)

the trichostrongylid phenograms we see ladder-like tree topologies, with more or less continuous variation in distance among haplotypes (e.g. Figure 1). The mixed population hypothesis can also be tested more formally using Tajima's D statistic (Tajima, 1989; Rand and Kann, 1996), which compares the average number of substitutions between haplotypes with the total number of segregating sites. Significantly positive values of Tajima's D would suggest mixed populations or balancing selection. All four trichostrongylid species studied by Blouin *et al.* (1995) had negative values of D for the ND4 gene, although none were statistically significantly different from zero. (So if anything, the pattern is opposite to that expected in a mixture.)

(b) *Large effective population size.* The remaining two explanations follow from the fact that under mutation-drift equilibrium the sequence diversity in a population is a simple function of $N_e\mu$, where N_e is the effective population size and μ is the mutation rate (Kimura, 1983). High levels of diversity could result if either N_e or mutation rates are elevated. Consistent with this explanation is the fact that although *Ascaris* and the trichostrongylids have similar life cycles (one-host, obligately sexual), *Ascaris* species have much smaller census population sizes and lower nucleotide diversity than the trichostrongylids. Effective size may also explain the lack of mtDNA variation in the other species cited above. *Howardula* go though yearly bottlenecks, overwintering in small populations of hibernating mycophagous *Drosophila* (Jaenike, 1992). *Strongyloides* sp. may use either asexual (homogonic) or sexual (heterogonic) development, and the asexual pathway frequently predominates (Viney *et al.*, 1992; Fisher, 1997). Consequently, the per generation variance in number of gametes contributed to the next generation may be high. In such a situation effective size might be just a tiny fraction of census size (Hartl and Clark, 1989).

(c) *Rapid rate of mtDNA evolution.* Mitochondrial DNA evolves at different speeds in different groups of organisms (Rand *et al.*, 1994), and unfortunately there is currently no calibration of the nematode mtDNA clock. Nevertheless, two sources of evidence suggest that the rate of mtDNA molecular evolution is rapid in nematodes. (1) Phylogenetic trees showing relationships among metazoan taxa have been constructed using the mitochondrial rRNA genes (Okimoto *et al.*, 1994) and cytochrome oxidase II (COII) gene (Hoeh *et al.*, 1996). In both studies the longest or next-to-longest branches lead to nematodes, demonstrating an accelerated rate of mtDNA evolution in nematodes relative to other phyla (Figure 2). (2) Comparison of branch lengths in trees of cospeciating parasites and their hosts can be used to compare relative rates of sequence evolution (Hafner *et al.*, 1994; Moran *et al.*, 1995). Machado (personal communication) sequenced the complete mtDNA COII gene from several species of

Panamanian fig-pollinating wasps (*Pegoscapus* spp., *Tetrapus* spp.) and from their nematode parasites (*Parasitodiplogaster* spp.). Comparison of the gene phylogenies showed concordant branching patterns, which suggested cospeciation between the pollinators and their species-specific parasitic nematodes. They compared the branch lengths of strictly cospeciating pairs showing that the COII gene has been evolving two times faster in the nematodes. However, part of the difference could be explained by the lower codon bias in the nematodes (see Section 2.1.2).

One hypothesis worth testing is that nematode parasites of endotherms have higher substitution rates than nematodes of ectotherms. Martin and Palumbi (1993) showed that mtDNA evolves more rapidly in endotherms than ectotherms, and suggested that the higher metabolic rate of endotherms results in production of excess free radicals and consequent oxidative damage to mtDNA. Because nematodes of mammals are maintained at constant high temperatures, it could be argued that they too will have more rapid metabolic rates and greater levels of mutation than nematodes of ectotherms. We currently lack sufficient data to test this hypothesis, although it is interesting that mtDNA diversity is higher in *Ascaris* and the trichostrongylids than in the insect parasite *Howardula* or the plant parasite, *Meloidogyne* (Hugall *et al.*, 1994, 1997).

2.1.2. Substitution Bias

Nematode mtDNA tends to be extremely A+T rich, with typical values between 75 and 80% (Thomas and Wilson, 1991; Okimoto *et al.*, 1992; Powers *et al.*, 1993; Hyman and Azevedo, 1996; Hugall *et al.*, 1997). From patterns of base composition in these genomes it has been generally concluded that this bias results from an underlying mutational pressure (Thomas *et al.*, 1991; Powers *et al.*, 1993; Jermiin *et al.*, 1995; Hyman and Azevedo, 1996; Hugall *et al.*, 1997). M. S. Blouin (unpublished data) recently analysed substitutions occurring at fourfold degenerate sites of ND4 gene in trees of individuals from each of four species of trichostrongylid (40–50 individuals per species). This analysis verified that the A+T

Figure 1 (opposite) UPGMA trees showing the relationships between mtDNA sequences from *Teladorsagia circumcincta*. The locations from which parasites were collected are shown on the map and are marked at the branch tips. Values adjacent to the abbreviated location names are nucleotide diversities (π) within each population. These values are an order of magnitude greater than those typically observed in other taxa. Values of N_{ST} (a sequence based measure of population differentiation) are close to zero, indicating low differentiation among populations, a result consistent with high levels of gene flow. Reproduced from Blouin *et al.* (1995), with permission.© Genetics Society of America.

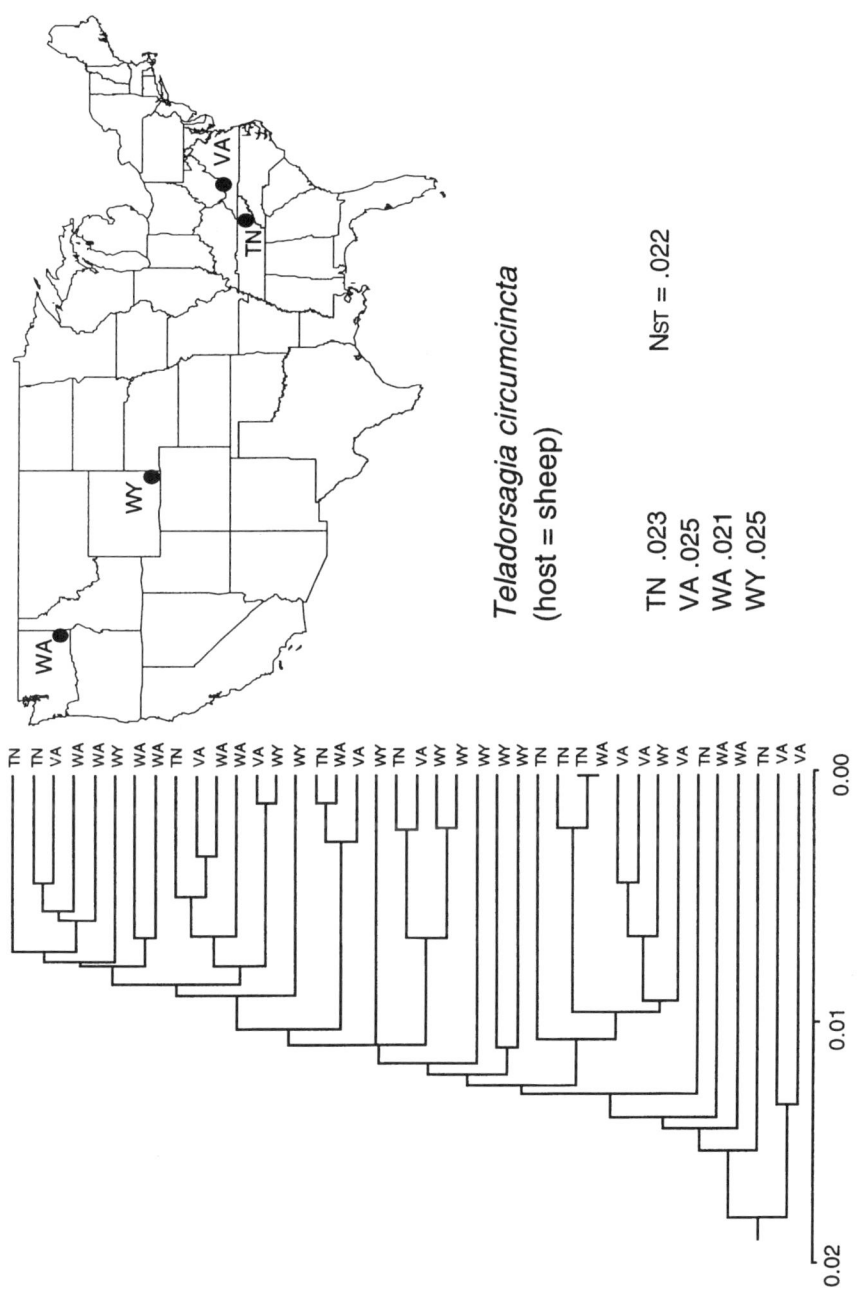

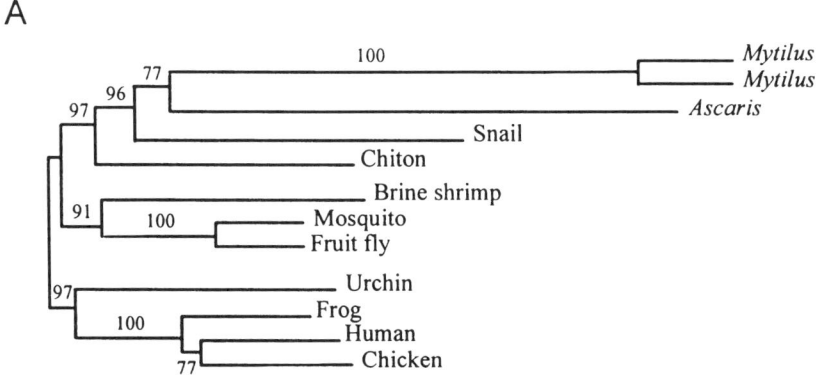

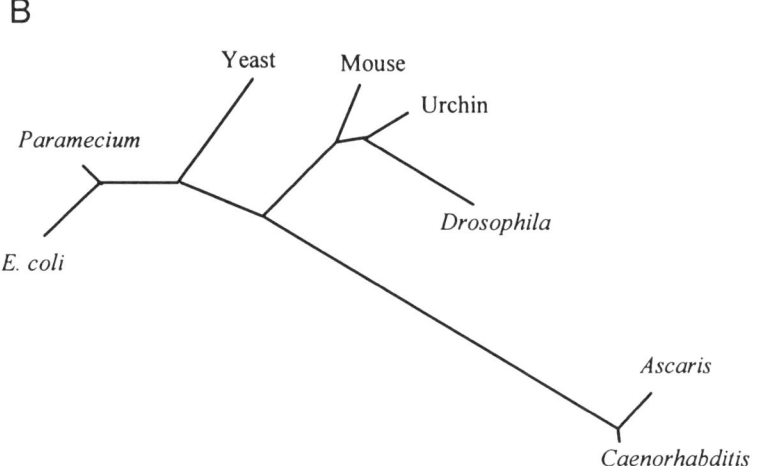

Figure 2 Phylogenetic trees of mtDNA genes in a variety of taxa. Branches leading to nematodes are long relative to other taxa, consistent with a faster rate of sequence evolution in nematodes. (A) Neighbour joining consensus tree based on nucleotide differences at first and second positions in the COI and COIII genes. The numbers indicate bootstrap support for each node. Redrawn from Hoeh *et al.* (1996), with permission. (B) Maximum likelihood phylogeny of the large subunit rRNA from metazoan mtDNA and yeast and bacterial homologues. For details of tree construction see Okimoto *et al.* (1994). Redrawn from Okimoto *et al.* (1994), with permission of Springer-Verlag GmbH & Co. KG.

bias indeed results from a strong underlying substitutional bias towards As and Ts (e.g., the probability of an A→C substitution is much lower than that of a C→A, and so on, after accounting for starting base frequencies). Thomas and Wilson (1991) inferred similar substitution patterns from tree analysis of substitutions at the COI gene among a few strains of *Caenorhabditis elegans*. Interestingly, this mutational bias and the fast rate of mtDNA evolution discussed above may both be caused by the same phenomenon, an elevated rate of oxidative DNA damage in the mitochondria (Martin, 1995; Hugall *et al.*, 1997). A more important issue is that although the high rate of substitution makes mtDNA ideal for low-level phylogenetics applications, failure to correct for this severe substitution bias might lead to serious phylogeny reconstruction errors using either distance or parsimony-based methods (Lockhart *et al.*, 1994; Jermiin *et al.*, 1995).

2.1.4. Structural Evolution

Whereas the physical arrangement of genes in mtDNA is highly conserved in some taxa such as vertebrates, gene order in nematodes may be much more labile. Recent data on gene order in nematodes was summarized by Hyman and Azevedo (1996). Complete or nearly complete transcriptional maps are available for three species in the class secernentia, *C. elegans*, *Ascaris suum* and *Meloidogyne javanica*, and for one species in the class adenophorea, *Romanomermis culicivorax* (Table 2). Gene order is nearly identical in *A. suum* and *C. elegans*, differing only in the position of an A+T-rich putative control region and a few tRNAs. Gene order is very different in both *Meloidogyne* in the adenophorean *Romanomermis*. It is currently hard to tell whether these major differences mean that nematode mtDNA is particularly liable to rearrangement, or just that we are comparing ancient lineages. Recombination (see Section 2.1.5) may also promote rapid structural evolution.

Repetitive DNA may be a common feature of nematode mtDNA. *Romanomermis* mtDNA contains variable numbers of a 3.0 kb repeat that is found in tandem and dispersed copies throughout the molecule (Hyman and Azevedo, 1996). This repeat carries two of the functional genes, ND3 and ND6. Whether the redundancy is adaptive is unclear, since gene copy number is not correlated with transcription levels (Hyman and Azevedo, 1996). Small tandem repeats (8–102 bp) are found in the *Meloidogyne* mtDNA, located between coding genes in a region of unknown function (Okimoto *et al.*, 1991). They appear to be variable in repeat number and repeat sequence, both within and among species, and have been used as diagnostic markers of *Meloidogyne* species (Okimoto *et al.*, 1991; Hyman and Whipple, 1996). Duplication or deletion of repeats are frequent and

Table 2 Mitochondrial DNA gene orders in four species of nematode for which a full or partial map is available. Genes are simply listed in order, arbitrarily starting with ND5. For relative sizes of each gene and positions of known tRNAs, see the original references.

Ascaris suum (14.3 kb) Entire sequence published in Okimoto et al. (1992).
ND5 ND6 ND4L s-rRNA AT[a] ND1 ATPase6 ND2 Cytb COIII ND4 LNC[b] COI COII l-rRNA ND3

Caenorhabditis elegans (13.8 kb). Entire sequence published in Okimoto et al. (1992).
ND5 AT[a] ND6 ND4L s-rRNA ND1 ATPase6 ND2 Cytb COIII ND4 LNC[b] COI COII l-rRNA ND3

Meloidogyne javanica (20.5 kb) Partial sequence and physical map published in Okimoto et al. (1991).
ND5 COI s-rRNA ND1 ND2 COIII ND6 ND4L COII ORF[c] l-rRNA ND3 Cytb ND4 102R 8R 63R[d] ATPase6

Romanomermis culicivorax, lineage 3B4 (26 kb) Partial sequence and partial physical map published in Hyman and Azevedo (1996) (positions of ND4L, s-rRNA, ND2 unknown).
ND5 l-rRNA COI   ND3   ND6[e] ND1   [f]   COII COIII ND4   ND6 ND3   ND6 ND3 Cytb ATPase6
                      ←   →   →

[a] AT in *C. elegans* and *A. suum* refers to an AT-rich, putative origin of replication.
[b] LNC (long noncoding region) is an approximately 100 bp region of unknown function. The positions of ND4, LNC and COI are the same in the trichostrongylids as in *Caenorhabditis* and *Ascaris* (Blouin et al., 1995).
[c] ORF: unidentified open reading frame
[d] 02R and 63R: region of unknown function that contains tandem repeats of 102 bp, 8 bp and 63 bp units.
[e] 3.0 kb repeat containing the ND3 and ND6 genes. Arrows indicate orientation.
[f] Part of the same repeat.

result in 80% of *Meloidogyne*, being heteroplasmic for repeat copy number (B.C. Hyman, personal communication). Copy number also appears to evolve rapidly in *Romanomermis*, and changes have been recorded during routine laboratory propagation (Hyman and Slater, 1990).

Finally, several other noncoding regions of unknown function are found in these mtDNA genomes. For example, in *C. elegans*, *A. suum*, and the trichostrongylids, there is a noncoding region of around 100 bp between the ND4 and COI gene (Okimoto *et al.*, 1992; M.S. Blouin, unpublished data). This regions evolves oddly. Intraspecific comparisons reveal high rates of substitution, rearrangement and insertion/deletion events in its 5' half, and a relatively conserved 3' half. Between congeners or confamilials the entire region is extremely variable, and between families there is no homology at all (M.S. Blouin, unpublished data). Okimoto *et al.* (1992) speculated that it might be an origin of replication, but its function remains a mystery.

2.1.5. *Recombination*

One of the assumptions made when using animal mtDNA is that recombination does not occur and that all the mtDNA genes are effectively linked (Avise, 1994). Most rules have exceptions: Lunt and Hyman (1997) describe strong evidence for recombination in some Meloidogyne mtDNAs. Future work will reveal whether this occurs commonly, and whether the mtDNA gospel must be revised for nematodes.

2.2. Nuclear Genome

2.2.1. *Allozyme Variation*

Multilocus enzyme electrophoresis (MLEE) has been widely used since the 1960s to investigate patterns of genetic variation in a wide range of organisms, including nematodes. As such allozymes provide a good source of data to compare patterns of variation in the nuclear genome of nematodes with that seen in other phyla. Table 3 lists results from allozyme studies of nematodes, updated since Nadler's (1990) review.

Levels of heterozygosity per locus (H_e), the number of alleles per locus (A), and the proportion of loci which are variable (P) are very similar to that seen in other free-living organisms (Table 3; Figure 3). However, there is a large range in values observed in different groups of nematodes. Notably, ascarids with direct life cycles show much lower values of heterozygosity than those with indirect life cycle. What are the reasons for these differences? Bullini *et al.* (1986) first suggested that variation was selectively

Table 3. Allozyme variation in parasitic nematodes: heterozygosity (H_e), proportion of polymorphic loci (P) and alleles per locus (A). N_g indicates the mean number of genes examined for each loci, N_l shows the number of loci examined and $P_{0.95}$ and $P_{0.99}$ indicate the proportion of polymorphic loci using the 95% and 99% criteria, respectively.

Nematode family	Species	Life cycle	N_g	N_l	H_e	$P_{0.95}$ ($P_{0.99}$)	A	Reference
Ascaridoidea	*Ascaris lumbricoides*	D	92	24	0.02	0.25	1.38	Bullini *et al.* (1986)
Ascaridoidea	*Ascaris suum*	D	445	24	0.03	0.17	1.29	Bullini *et al.* (1986)
Ascaridoidea	*Parascaris univalens*	D	447	28	0.03	0.11	1.18	Bullini *et al.* (1986)
Ascaridoidea	*Parascaris equorum*	D	131	28	0.02	0.07	1.11	Bullini *et al.* (1986)
Ascaridoidea	*Neoascaris vitulorum*	D	253	18	0.04	0.11	1.11	Bullini *et al.* (1986)
Ascaridoidea	*Toxocara canis*	D	157	18	0.1	0.33	1.5	Bullini *et al.* (1986)
Ascaridoidea	*Toxocara cati*	D	44	18	0.05	0.17	1.22	Bullini *et al.* (1986)
Ascaridoidea	*Toxascaris leonina*	D	46	18	0.02	0.11	1.11	Bullini *et al.* (1986)
Ascaridoidea	*Baylascaris transfuga*	D	10	18	0.05	0.17	1.17	Bullini *et al.* (1986)
Ascaridoidea	*Anisakis simplex* A	I	272	22	0.12	0.64	2.32	Bullini *et al.* (1986)
Ascaridoidea	*Anisakis simplex* B	I	251	22	0.21	0.64	2.6	Bullini *et al.* (1986)
Ascaridoidea	*Anisakis physeteris*	I	84	22	0.11	0.5	1.95	Bullini *et al.* (1986)
Ascaridoidea	*Anisakis cystophorae*	I	12	21	0.12	0.24	1.33	Bullini *et al.* (1986)
Ascaridoidea	*Contracaecum osculatum* A	I	78	21	0.12	0.48	2.14	Bullini *et al.* (1986)
Ascaridoidea	*Contracaecum osculatum* B	I	169	21	0.10	0.62	2.71	Bullini *et al.* (1986)
Ascaridoidea	*Contracaecum osculatum* A	I	>100	24	0.19	0.541 (0.750)	2.8	Orecchia *et al.* (1994)
Ascaridoidea	*Contracaecum osculatum* B	I	>100	24	0.19	0.520 (0.708)	3.0	Orecchia *et al.* (1994)
Ascaridoidea	*Contracaecum osculatum* C	I	>100	24	0.15	0.604 (0.792)	2.3	Orecchia *et al.* (1994)
Ascaridoidea	*Contracaecum osculatum* D	I	>100	24	0.20	0.709 (0.833)	3.3	Orecchia *et al.* (1994)
Ascaridoidea	*Contracaecum osculatum* E	I	>100	24	0.28	0.667 (0.854)	3.6	Orecchia *et al.* (1994)
Ascaridoidea	*Contracaecum rudolfii* A	I	87	21	0.17	0.57	2.24	Bullini *et al.* (1986)
Ascaridoidea	*Contracaecum rudolfii* B	I	65	21	0.21	0.62	2.33	Bullini *et al.* (1986)
Ascaridoidea	*Contracaecum* spp. I	I	27	11	0.14	0.54	2.55	Vrijenhoek (1978)
Ascaridoidea	*Contracaecum* spp. I	I	24	11	0.19	0.54	2.00	Vrijenhoek (1978)
Ascaridoidea	*Pseudoterranova decipiens* A	I	?	16	0.08	0.22 (0.32)	1.4	Paggi *et al.* (1991)

Superfamily	Species						Reference
Ascaridoidea	Pseudoterranova decipiens B	I	?	16	0.05	0.22 (0.34)	1.4	Paggi et al. (1991)
Ascaridoidea	Pseudoterranova decipiens C	I	?	16	0.02	0.10 (0.24)	1.3	Paggi et al. (1991)
Ascaridoidea	Ascaris suum	D	50	29	0.066	0.207	1.31	Leslie et al. (1982)
Ascaridoidea	Ascaris suum	D	50	29	0.053	0.167	1.28	Leslie et al. (1982)
Ascaridoidea	Toxoxara cati	D	41	18	0.137	— (0.39)	1.94	Nadler (1986)
Ascaridoidea	Toxoxara canis	D	59	18	0.135	— (0.33)	1.44	Nadler (1986)
Ascaridoidea	Parascaris equorum	D	54	18	0.085	— (0.22)	1.44	Nadler (1986)
Filaroidea	Onchocerca volvulus	I	439	7	0.17	0.57	1.9	Flockhart et al. (1986)
Filaroidea	Onchocerca volvulus (Mali)	I	80	19	0.16	0.69	1.8	Cianchi et al. (1985)
Filaroidea	Onchocerca volvulus (Ivory Coast)	I	40	19	0.20	0.5	1.6	Cianchi et al. (1985)
Filaroidea	Onchocerca volvulus (Mali)	I	30	19	0.14	0.5	1.6	Cianchi et al. (1985)
Filaroidea	Onchocerca gibsoni	I	21	23	0.024^a	0.087	1.09	Andrews et al. (1989)
Filaroidea	Onchocerca lienalis	I	20	23	0.049^a	0.043	1.04	Andrews et al. (1989)
Filaroidea	Onchocerca gutterosa	I	34	22	0.010^a	0.091	1.18	Andrews et al. (1989)
Filaroidea	Dirofilaria immitis	I	31	17	0.017	0.059	1.1	Cancrini et al. (1989)
Filaroidea	Dirofilaria repens	I	39	17	0.029	0.059	1.1	Cancrini et al. (1989)
Physalopteridea	Physaloptera sp.	I	13	30	—	0.200	1.27	Norman and Chilton (1994)
Strongyloidea	Hypodontus macropi A	D	22	28	0.10^a	0.39	1.43	Chilton et al. (1992)
Strongyloidea	Hypodontus macropi B	D	22	28	0.07^a	0.25	1.25	Chilton et al. (1992)
Strongyloidea	Hypodontus macropi C	D	42	28	0.09^a	0.29	1.36	Chilton et al. (1992)
Strongyloidea	Hypodontus macropi E	D	24	28	0.15^a	0.46	1.68	Chilton et al. (1992)
Strongyloidea	Hypodontus macropi G	D	20	28	0.05^a	0.18	1.25	Chilton et al. (1992)
Strongyloidea	Macropostrongyloides baylisi	D	12	27	—	0.259	1.44	Beveridge et al. (1993)
Strongyloidea	Paramacropostrongylus typicus	D	15	37	—	0.270	1.30	Chilton et al. (1993a)
Trichostrongyloidea	Teladorsagia circumcincta	D	45	13	0.096	0.38	1.38	Gasnier and Cabaret, (1996)
Trichostrongyloidea	Echinocephalus overstreeti	I	14	28	0.089	0.36	1.39	Andrews et al. (1988)

^{a} values calculated from published data assuming Hardy–Weinberg equilibrium.

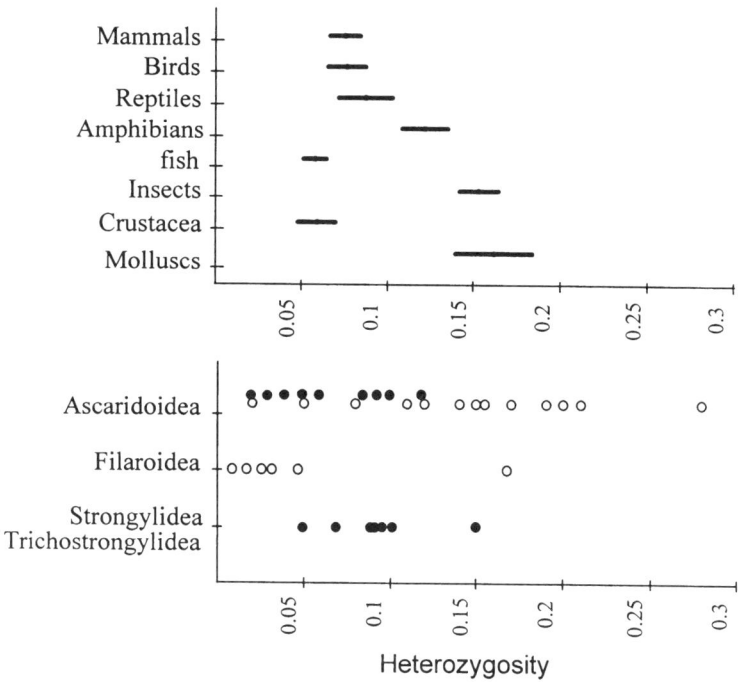

Figure 3 Allozyme diversity in nematodes compared with other phyla. Upper panel: mean (and SE) level of heterozygosity (H_e) in vertebrate and invertebrate phyla. Data from Ward *et al.* (1992). Lower panel: heterozygosity in four nematode families. Open circles represent parasite species utilizing indirect life cycles, while filled circles represent those with direct life cycles. Each point represents one species. For species in which H_e has been measured multiple times, average values were used. While ascarids with indirect life cycles have higher heterozygosity than ascarids with direct lifecycles, filarids (indirect) have lower mean heterozygosity than strongylids/trichostrongylids (direct).

maintained in indirect life cycle ascarids, perhaps owing to the greater heterogeneity in their environment. However, other features of ascarid populations, such differences in effective population sizes or recent expansion from historically smaller populations could explain the observed patterns without invoking selection. The selection hypothesis is interesting, but the comparative data supporting it is weak because indirect and direct lifecycle ascarids form monophyletic groups.

To further examine Bullini *et al.*'s (1986) hypothesis, we have summarized levels of variation from indirect and direct life cycle nematodes from a variety of nematode families. (Figure 3). The nematodes for which data are available are distributed among five families, Ascaridoidea, Filaroidea,

Trichostrongyloidea, Strongyloidea and Physalopteridea. Visual inspection of the data, reveals that among the Filaroidea, which have indirect life cycles, *Onchocerca volvulus* has extremely high levels of variation, while *Onchocerca* spp. infecting animals have much lower levels of heterozygosity. The direct life cycle strongylids and trichostrongylids show levels of variation overlapping both direct and indirect life cycle ascarids. The limited data from nematode families other than ascarididae do not provide strong support for or against Bullini *et al.*'s (1986) selection hypothesis. For a more rigorous comparative test further electrophoretic data should be collected from nematode sister taxa containing both monoxenous and heteroxenous representatives.

If the allozyme variation observed is mainly selectively neutral, then one would predict a positive relationship between long-term effective population size and levels of allozyme variation. However, quantitative estimation of nematode effective population size (Ne) is fraught with difficulty (Nadler *et al.*, 1995), and so one can only guess about the relative size of N_e in different species. The best method of differentiating between selection and neutral arguments for allozyme variation at individual loci will probably come from sequence analysis of enzyme-encoding genes. The low levels of allozyme variation in direct life cycle ascarids could result from (1) comparatively low N_e; (2) strong purifying selection preventing amino acid change; or (3) periodic selective sweeps in which one variant allele spreads to fixation and in doing so, removes variation at linked sites. Likewise, high levels of variation in indirect life-cycle Ascarids could result from (1) comparatively high N_e; (2) relaxed selection against amino acid change; or (3) positive selection for aminoacid variation, as proposed. The evidence for these competing hypotheses could be evaluated using comparative sequence data from allozyme encoding loci. Such data could be analysed by examining patterns of nonsynonymous and synonymous substitution within and between species (McDonald and Kreitman, 1991) and within different genes (Hudson *et al.*, 1987).

No such data are currently available. However, it is interesting that while *Ascaris* spp. show low levels of allozyme variation, appreciable nucleotide diversity is found in both introns and mitochondrial DNA (Anderson *et al.*, 1993; Anderson and Jaenike, 1997; see Sections 2.1.1 and 2.2.2. Patterns of nucleotide diversity within genes of parasitic nematodes are reviewed in the next section.

2.2.2. *Nuclear Sequence Diversity*

In section 2.1.1. we discussed evidence that mtDNA evolves unusually quickly in nematodes. Are both mtDNA and nuclear DNA evolution equally accelerated, or is the phenomenon restricted to mtDNA? In trees of metazoan phyla constructed using nuclear 18s rRNA sequences, the longest branches again led to nematodes (Philippe et al., 1994). Fitch et al. (1995) found that distances among rhabditid genera based on nuclear 18S rDNA sequences are eight times greater than distances among classes of tetrapods. Similarly, Ferris (1994) reports only 61% identity between the 5.8s rDNA of C. elegans and plant cyst nematodes, which is greater than the distance between echinoderms and amphibians. Although these differences are no doubt partly due to the fact that nematodes are morphologically conservative, it is again consistent with more rapid rDNA evolution in nematodes. Thomas and Wilson (1991) compared sequence variation within and between species of *Caenorhabditis* spp. at the mitochondrial COII gene and nuclear calmodulin gene. Their comparison of substitutions at silent sites in intra- and inter-specific pairings indicate that the mitochondrial gene is evolving at least 10–12 times faster than the nuclear gene. This rate difference is typical of that seen in other taxa, and so suggests that the mtDNA and nuclear genomes are equally accelerated.

Two groups have measured sequence diversity in single copy nuclear genes (scnDNA) of parasitic nematodes. Beech et al. (1994) measured nucleotide diversity within the introns of two β-tubulin isotypes from *Haemonchus contortus*, while Anderson and Jaenike (1997) measured sequence variation in a variety of introns in *Ascaris* (Table 1). Both studies used four-cutter digestion of amplified intron sequences. *Haemonchus contortus* β-tubulin introns were exceedingly variable with nucleotide diversities (π) of 0.094 and 0.091, respectively. For comparison, introns from 41 *Drosophila* genes had diversities ranging from 0 to 0.034 (Moriyama and Powell, 1996), and variation observed within other animals species also fall within this range (Palumbi and Baker, 1994; Slade et al., 1994). The five *Ascaris* introns studied also showed conventional patterns of variation, with π ranging from 0.002 to 0.008. Why are the *Haemonchus* β-tubulin introns so much more variable than introns in *Ascaris* or other taxa? The simplest explanation is large effective size, which also explains the high mtDNA diversities in *Haemonchus* (see Section 2.1).

Nevertheless, these β-tubulin introns seem excessively polymorphic, even after accounting for effective size, for the following reasons. For *H. contortus* we have two separate estimates of within-population diversity, one based on nuclear introns (Beech et al., 1994) and one based on mtDNA (Blouin et al., 1995). Under a model of drift-mutation equilibrium, the within-population sequence diversity, π, is a simple function of sequence mutation rate, μ, and population effective size, N_e. For mitochondrial genes, $\pi = N_e\mu$, while for nuclear genes, $\pi = 4N_e\mu$ (because the N_e of nuclear genes is four times greater than that for mitochondrial genes if one assumes strict maternal inheritance and a 1:1 sex ratio (Birky et al., 1989), which is a reasonable assumption for species such as *H. contortus*). The mtDNA and intron diversity estimates are both from agricultural populations in which values of N_e are likely to be similar. Consequently, we estimate the relative mutation rates of introns and mtDNA to be 0.023/N and 0.024/N, respectively. Because we expect nuclear genes to evolve about 10 times slower than mtDNA (see paragraph above), the mutation rate estimate for the β-tublin introns therefore seems high. More comparisons of nuclear and mitochondrial genes in the same nematode populations are required before we can generalize any more about the relative rates of evolution in the two types of DNA.

2.2.3. Ribosomal DNA Spacers

While many regions of ribosomal genes are extremely conserved, the internal transcribed spacers, ITS1 and ITS2, show extensive variation (Hillis and Dixon, 1991). Furthermore, primers in the conserved regions can be used to amplify DNA from many nematode species. ITS sequences and restriction fragment length polymorphisms (RFLPs) are finding widespread usage, in particular for distinguishing closely related nematode species. Work on nematode spacers has revealed a number of interesting features.

Ribosomal DNA sequences are normally encoded in long arrays of tandem repeats. Concerted evolution tends to homogenize rDNA sequences both within individuals and within populations (Arnheim, 1983). In nematodes, in contrast, multiple different sequences of the rDNA ITS are frequently found within individual worms (Back et al., 1984a; Anderson, 1995; Hoste et al., 1995; Brattey and Davidson, 1996). As a consequence of this, RFLP patterns often have multiple bands, and ambiguous bases are often found following direct sequencing of polymerase chain reaction (PCR) products.

Concerted evolution may fail to homogenize rDNA sequences within nematodes for two reasons. In *Ascaris*, in which rDNA sequence organization has been well characterized (Back et al., 1984b) rDNA arrays bearing different length variants are found in separate clusters on the same autosome. Furthermore, in *Ascaris* a site-specific transposable element (R4) (Burke et al., 1995) is inserted in some rDNA sequences. This large (7 kb) insertion may reduce levels of repeat loss and gain within *Ascaris* rDNA arrays by reducing homology and pairing between adjacent repeats. R4 type elements are also present in other nematode species (Burke et al., 1995). It would be interesting to compare 'within individual' ITS sequence heterogeneity in nematode species with and without rDNA insertion sequences to determine whether these insertions do indeed influence concerted evolution in repeat arrays.

However, despite the existence of intra-individual variation, the intraspecific variation is generally low relative to interspecific variation (Gasser and Hoste, 1995). As a result, ITS has been used extensively for distinguishing between closely related species (see Sections 4.2 and 4.3).

2.2.4. *Microsatellites, Random Amplified Polymorphic DNAs (RAPDs) and Repetitive DNA*

Microsatellite markers have not been widely used in nematode parasites. We know of only three works using these markers. Fisher and Viney (1996) cloned and sequenced a variety of microsatellites from *Strongyloides ratti*. However, attempts to amplify and score length variation using conventional techniques resulted in smears, despite prolonged attempts to optimize PCR conditions. They suggested that the *S. ratti* microsatellite primer sites were present in repetitive regions. Other workers have had more success. Zarlenga *et al.* (1996) repetition isolated and characterized microsatellite ($(TG)_n$ and $(TGC)_n$) variation in domain IV of the large subunit rDNA of *Trichinella*, and have found informative variation between individual isolates and species. Hoekstra *et al.* (1997b) have recently characterized allele frequencies at 13 $(CA)_n$ and $(GA)_n$ microsatellites in a variety of *H. contortus* populations. Over half the loci examined were polymorphic with up to six alleles per locus.

RAPD variation, resulting from amplification of genomic DNA with short random oligonucleotides (Hadrys *et al.*, 1992) has been widely used by plant nematologists, but rather less so by those interested in animal parasites (but see Nadler *et al.*, 1995). We discuss patterns revealed by RAPDs further in Section 3.1.2. This technique has many advantages in that no information is required to design primers, information is obtained randomly from many parts of the genome, and large numbers of markers can be scored. Although RAPDs are useful in linkage mapping projects, they are more difficult to apply in surveys of natural populations: the genetic basis of RAPD variation is usually unknown, there are serious problems with band reproducibility, and the fact that RAPD bands are dominant markers makes data analysis problematic (Lynch and Milligan, 1994; Grosberg *et al.*, 1996). A further potential problem with using RAPDs on small nematodes is that gut contents and symbionts may be the origin of some bands. A number of authors have recently described how RAPD-generated bands can be converted to produce codominant information (Paran and Michelmore, 1993; Burt *et al.*, 1996). We hope nematode population geneticists will adopt these, or similar techniques.

Repetive DNA, and transposable elements in particular, are potentially useful tools for DNA fingerprinting of individual worms. However, this approach is feasible only if individual worms are large enough to prepare sufficient genomic DNA from southern blots. Few transposable elements are known from nematodes other than *C. elegans*. However, in *Ascaris* two different transposable elements have been found. One of these, R4, has been mentioned briefly in Section 2.2.3. The second, known as TAS (transposable element of Ascaris) are found within the germline, but many are lost during chromosome diminution and are not found in somatic tissue. Hybridization of segments of TAS elements with southern blots of *Ascaris* DNA reveals extensive variation between individuals worms (Felder *et al.*, 1994). These, and other elements may prove to be useful markers for further work on parasite population genetics.

3. POPULATION STRUCTURE — HOW IS GENETIC VARIATION ARRANGED IN POPULATIONS?

One can define population genetic structure (PGS) in a species as the distribution of genetic variation among individuals sampled over different spatial scales. Here, we review patterns of PGS in animal and plant parasitic nematodes, and consider factors which may influence those observed patterns. We then describe comparative data on PGS revealed by different loci, and how heterogeneity in PGS may be used to infer selection. Finally, we review examples in which patterns of variation on a microgeographical scale can be used to make inferences about patterns of transmission and about mating systems.

Two types of data have been used to investigate the PGS of parasitic nematode populations. Most commonly, allele frequency data from allozyme surveys have been used, although equivalent data from RAPD markers (Nadler *et al.*, 1995) or single copy nuclear genes (Fisher, 1997; Anderson and Jaenike, 1997) have also been employed. In these data, the relationships between different alleles at each locus is unknown. Sequence information provides additional information on the relationships between alleles. This type of data has also been employed to investigate the 'phylogeography' (Avise, 1994) of some nematode species — that is, the distribution of sequence diversity within populations (Blouin *et al.*, 1992, 1995; Anderson *et al.*, 1993; Anderson and Jaenike., 1997). Use of both allele frequency and sequence data is illustrated in the following paragraphs.

3.1. Geographical Differentiation and Gene Flow

3.1.1. Animal Parasitic Nematodes

The PGS of a variety of species of animal parasitic nematodes have now been investigated. Summary statistics describing the amounts of genetic diversity partitioned within and between geographically separated populations is shown in Table 4. A striking feature of this table is that most parasite species studied show very little geographical population structure, a pattern consistent with high levels of gene flow. For example, in the four species of trichostrongylids of domestic animals studied by Blouin *et al.* (1992, 1995) 96–99% of nucleotide diversity is found within populations. Similarly, in marine ascarid species (Paggi *et al.*, 1991; Orecchia *et al.*, 1994) 94–99% of allozyme variation is found within populations, and 99% of variation in *S. ratti* was observed within locations. In contrast, the deer parasite *Mazamastrongylus odocoili* shows strong subdividision, with 31% of variation partitioned among subpopulations. Allozyme surveys of two other nematode species, *Onchocerca volvulus* in Africa, and *Teladorsagia circumcincta* in French goats, have also revealed high levels of subdivision (Cianchi *et al.*, 1985; Flockhart *et al.*, 1986; Gasnier and Cabaret, 1996). However, the *O. volvulus* populations analysed included both 'forest' and 'savannah' pathotypes, while some of the *T. circumcincta* populations contained mixed populations of different cryptic species (see Section 4.3.). The elevated values observed in these cases probably reflect intrinsic mating barriers rather than levels of gene flow between populations.

The genetic similarity between most populations indicate high levels of gene flow. For example, Paggi *et al.* (1991) estimate that N_e for member of the *Pseudoterranova decipiens* complex range from 4 to 10. That is, for every generation, 4–10 parasites contribute genes to a different population from that in which they were born. Are these patterns of population structure, with high levels of gene flow linking populations, typical for nematode parasites? It is difficult to say because the species studied are not very representative, the majority being parasites of humans, domestic animals, or commensals. Further investigation of the PGS of nematodes of 'undisturbed' species of animals and plants would be particularly interesting.

3.1.2. *Plant Parasitic Nematodes*

Most molecular population genetics work on plant parasitic nematodes has focused on identification of species and pathotypes. However, some workers have stressed the need to investigate patterns of genetic variation at a variety of levels (among genera, among species and among populations) in order to make rational taxonomic and management decisions (Ferris, 1994; Hyman, 1996; Hyman and Whipple, 1996). RAPDs have frequently been used, but in most cases RAPD profiles have been compared between single isolates from each locale, or between samples of pooled individuals. Data of this nature are difficult to interpret unless levels of intrapopulation variation are also known. The presence of additional bands in some populations may indicate that all nematodes in that population carry the

Table 4 Measurements of geographical population structure for parasitic nematodes.

Nematode sp.	Sampling scale	Technique	F_{ST}	Reference
Ascaris lumbricoides	World wide	scnDNA RFLP	0.180^b	Anderson and Jaenike (1997)
Ascaris lumbricoides	Guatemala only	scnDNA RFLP	0.013^b	Anderson and Jaenike (1997)
Ascaris suum	World wide	scnDNA RFLP	0.212^b	Anderson and Jaenike (1997)
Ascaris suum	Guatemala only	scnDNA RFLP	0.048^b	Anderson and Jaenike (1997)
Ascaris suum	N. American Midwest	MLEE/RAPDs	0.078/0.062	Nadler et al. (1995)
Ascaris suum	Iowa, New Jersey	MLEE	0.031	Leslie et al. (1982)
Contracaecum osculatum D	Antarctic	MLEE	0.013	Orrechia et al. (1994)
Contracaecum osculatum E	Antarctic	MLEE	0.010	Orrechia et al. (1994)
Haemonchus contortus	N. America	MtDNA Sequence	0.01^a	Blouin et al. (1995)
Haemonchus placei	N. America	MtDNA Sequence	0.04^a	Blouin et al. (1995)
Mazamastrongylus odocoileus	N. America	MtDNA Sequence	0.31^a	Blouin et al. (1995)
Meloidogyne spp.	Queensland, Australia	MtDNA RFLP	0.130^b	Hugall et al. (1994)
Onchocerca volvulus	Mali, Ivory Coast, Zaire	MLEE	0.155^b	Cianchi et al. (1985)
Onchocerca volvulus	Liberia, Ivory Coast, Burkina Faso, Sudan	MLEE	0.148	Flockhart et al. (1986)
Ostertagia ostertagi	N. America	MtDNA RFLP	0.012^a	Blouin et al. (1992)
Pseudoterranova decipiens A	N.E. Atlantic	MLEE	0.059	Paggi et al. (1991)
Pseudoterranova decipiens B	N.E. Atlantic/Atlantic Canada	MLEE	0.055	Paggi et al. (1991)
Pseudoterranova decipiens C	N.E. Atlantic/Atlantic Canada	MLEE	0.021	Paggi et al. (1991)
Strongyloides ratti	United Kingdom	scnDNA RFLPs	0.014	Fisher (1997)
Teladorsagia circumcincta	France — sheep	MLEE	0.02^b	Gasnier and Cabaret (1996)
Teladorsagia circumcincta	France — goats	MLEE	0.189^b	Gasnier and Cabaret (1996)
Teladorsagia circumcincta	N. America	MtDNA Sequence	0.02^a	Blouin et al. (1995)

^{a} Analogous measures of population structure, such as N_{ST} (Lynch and Crease, 1990), ^{b} G_{ST} (Nei, 1973) and ^{c} θ (Wier and Cockerham, 1984) were calculated instead of F_{ST} (Wright, 1951) in some cases.

sequence variant underlying the production of this fragment, or alternatively, that only some individuals within that population bear this sequence variant. As such, RAPD data from populations carry no information on frequencies of alleles within populations, which is essential for measurement of population subdivision.

Nevertheless, available data suggest that many plant nematode populations show substantial geographical subdivision. Caswell-Chen *et al.* (1992) measured RAPD variation in six populations of *Heterodera schachtii* from California. They found that populations isolated from sites 5 km apart showed only 45% similarity in band profiles. Furthermore, RAPD variation was also measured in individual nematodes from a single population, and shown to be low relative to interpopulation differences observed. In comparison, low levels of interpopulation variation were observed among *Globodera pallida* (83–100% similarity among isolates) and *Globodera rostochiensis* (87–97% similarity among isolates) populations from the Netherlands (Folkertsma *et al.*, 1994), or among populations of *Radophilus similis* from Sri Lanka (>90% of bands shared between isolates) (Hahn *et al.*, 1994). In *Globodera*, the variation was strongly structured, with nearby locations having distinctive RAPD profiles. Similarly, in *Radophilus*, Kaplan *et al.* (1996) showed that specific primers for one polymorphic RAPD fragment amplified from some populations of *Radophilus* sp. from Florida, Central America and Puerto Rico, but not from others, suggesting strong genetic structure within this species. Similarly, sequence variation in mtDNA also reveals strong subdivision among parthenogenetic *Meloidogyne* populations from Australia (Hugall *et al.*, 1994).

3.2. Nematode Life History Traits and Population Structure

Parasite species differ substantially in PGS (Table 4). Which life history or life cycle characteristics of different species predispose them to different patterns? PGS will be mainly a function of effective sizes of populations and the rates of migration among them, so any life cycle traits that affect those two variables will affect PGS (Nadler, 1995). In the only formal test to date, Blouin *et al.* (1995) investigated the hypothesis that the high rate of gene flow observed among trichostrongylid parasites of livestock results from the high rate at which livestock are moved by humans. These trichostrongylids all have a simple, one-host life cycle, and minimal powers of dispersal on their own, so host movement should be the major force controlling gene flow in these species. Because wild ruminants are moved about much less by humans than domestic ruminants, the parasites of wild ruminants should show more population differentiation. As predicted, a trichostrongylid parasite of wild deer in the USA, *Mazamastrongylus odocoilei*, showed strong population subdivision and isolation by distance, while trichostrongylids of cattle (*Ostertagia ostertagi* and *Haemonchus placei*) and of sheep (*T. circumcincta* and *H. contortus*) showed very little differentiation among populations (Table 4; Figure 1). Recent data on *O. ostertagi* extend the comparison to Australia. Approximately 2–3% of the total mtDNA variation among *O. ostertagi* individuals is distributed among populations within Australia, and 4% is distributed between the United States and Australia (C. Constantine, unpublished data). Thus, movement of cattle appears to have spread their parasites to such an extent that allele frequencies are homogenized across continents. Long-distance dispersal of parasites resulting from the movement of livestock by humans may be much more common than suspected, and has obvious implications for the spread of anthelmintic resistance. Note that movement of soil and plants by agriculture may similarly be enhancing gene flow in soil-dwelling and plant-parasitic nematodes.

3.3. Differences Among Loci in Population Structure

Different genetic markers may provide qualitatively different patterns of genetic structure within parasite populations. Because mtDNA is haploid and transmitted maternally, the effective size (N_e) for mtDNA is smaller than for diploid nuclear loci in the same population (Birky *et al.*, 1989). Consequently, mtDNA should, on average, show more variation among populations than nuclear loci such as allozymes, microsatellites or introns, owing to its greater rate of genetic drift. One consequence of this is that mtDNA is a more sensitive indicator of patterns of microheterogenity and transmission of parasites (see Section 3.5). Comparisons between mtDNA and nuclear loci can also be used to test hypotheses of gender biased dispersal (Karl *et al.*, 1992b; Melnick and Hoelzer, 1992). Repetitive genes, such as rDNA tend to become homogenized within species more rapidly than other nuclear genes, as a result of concerted evolution (Arnheim, 1983; see Section 2.2.3). Consequently, rDNA and other repetitive sequences tend to be particularly useful for differentiating between closely related species (see Section 4.3). These differences in the behaviour of different markers are illustrated by comparative data on mtDNA scnDNA and rDNA polymorphism in *Ascaris* (Anderson and Jaenike, 1997). When variation in 12 *Ascaris* populations from both humans and pigs from different countries was partitioned among hosts (humans and pigs) and among geographical populations, rDNA ITS revealed a strong discontinuity between populations infecting different hosts. In contrast, geography explained most of the heterogeneity observed in the distribution of mtDNA haplotypes (see Figure 6). Six scnDNA markers showed intermediate patterns with both host and geography explaining similar amounts of variation.

3.4. Inferring Natural Selection from Population Structure

When using genetic data to infer patterns of population structure, the assumption is made that loci used are selectively neutral. In some cases, one or more loci may show patterns of PGS which are very different from other loci. For example, consider the situation in which most loci show a panmictic distribution, while one reveals subdivision. A likely explanation is that selection maintains allele frequency differences between populations at this locus in the face of high levels of gene flow (Lewontin and Krakauer, 1973; McDonald, 1994). Similarly, lack of population genetic structure at some loci may indicate balancing selection at those loci (Karl and Avise, 1992a). This kind of comparison may also be used to make inferences about patterns of selection on phenotypic traits, if they are known to have a genetic basis (Taylor *et al.*, 1995). For example, *O. ostertagi* in the north and the south of the USA undergo developmental arrest at different times of the year, and this difference has been shown to have a genetic basis (Frank *et al.*, 1988). MtDNA RFLP data indicate high gene flow between northern and southern populations, so the life history difference must be maintained by strong natural selection in the face of gene flow (Blouin *et al.*, 1992). A similar approach may prove useful for inferring selection on loci thought to be involved in drug resistance (see Section 6), or on loci involved in local adaptation to hosts.

3.5. Microspatial Population Structure and Transmission Patterns

Patterns of differentiation over very small spatial scales in parasite populations can be used to make inferences about modes of transmission. Thus, for example, over-representation of some alleles within parasite populations from a human family might suggest that the household is the focus of transmission. Nevertheless, genetic markers have been used rather infrequently for investigating fine-scale genetic structuring in macroparasite populations.

Three groups have investigated patterns of microspatial genetic variation within *Ascaris* populations. Ibrahim *et al.* (1994) used allozymes to investigate population genetics of *Ascaris lumbricoides* burdens from 20 children in a heavily infected region of Bangladesh, while Nadler *et al.* (1995) characterized both allozymes and RAPD variation in 96 *A. suum* obtained from seven pigs from five locations in the North American Midwest.

Anderson et al. (1995b) investigated the distribution of 41 different mtDNA haplotypes among worms collected from 65 individuals (both humans and pigs) in 35 households from three Guatemalan villages. The results from these three studies have some similarities, but also interesting differences. Nadler et al. (1995) observed heterozygote deficiencies within the worm burdens of individual pigs, which was reflected by elevated values of inbreeding (F_{IS}). In Guatemalan populations, mtDNA haplotypes were clustered nonrandomly within worm burdens in both humans and pigs. In comparison, Ibrahim et al. (1994) found no evidence for deviations from panmixia among human *Ascaris* from Bangladesh in samples of comparable size to those used in Nadler et al.'s (1995) study.

High F_{IS} and F_{IT} values and clustering of mtDNA haplotypes within individual worm burdens might result if related infective stages (eggs) are clustered in space. This could be tested directly by spatial sampling and PCR-based genotyping of eggs collected from soil (see Gasser et al., 1993 and Anderson et al., 1995a for technical details). Spatial clustering of infective eggs may result from the overdispersion of parasites among hosts. In this situation the majority of hosts will contain only a few parasites and, consequently, eggs excreted by individual hosts will be related (Anderson et al., 1995b). Why were heterozygote deficiences not observed in Bangladesh? One possibility is that parasite intensities were much higher in Bangladesh than in North America or Guatemala. When mean worm burdens are high, the eggs excreted by individual hosts will be less closely related than when mean parasite burdens are low. Thus, even if spatial patterns of transmission are similar in regions of low and high intensity, patterns of population subdivision revealed by genetic markers may be very different. Population genetic models investigating the relationship between parasite intensity, overdispersion and allelic distributions within populations would be a valuable aid to empirical work in this field.

There is considerable interest in whether transmission of *Ascaris* and other intestinal helminths of humans is focused within the household unit. Clustering of like genotypes within households would provide strong evidence for household-based transmission. Within a lightly infected village in Guatemala, one particular mtDNA haplotype was found within all the members of one household (13 out of 21 worms surveyed), but in only two out of 14 parasites from the other households in the village, suggesting familial transmission (Anderson et al., 1995b). However, in the heavily infected village in the central highlands, from which parasites in multiple heavily infected families were surveyed, no significant clustering was observed. Once again, the large size of parasite burdens and low levels of gene flow between parasites from different households may result in genetic homogeneity among parasites from different households. Even if the household is the predominant source of infection, gene flow among households may obscure any genetic trace of this.

The discussion above suggests that genetic markers may be rather insensitive tools for investigating transmission in nematodes. One way to improve the sensitivity of this technique would be to investigate the genetic structure of recolonizing parasite populations following mass chemotherapy. In such a nonequilibrium situation, invasions of particular parasite alleles within a household may provide much clearer indications of parasite transmission patterns. This approach may not work for *Ascaris*, where eggs are long lived but for other human helminths, such as hookworm, pinworms and *Trichuris*, which have short-lived infective stages, this approach may be particularly useful.

There is one caveat concerning the use of genetic markers for investigating transmission. Host immune selection against particular parasite genotypes could also result in clustering of genotypes within hosts or households if loci under selection are in linkage disequilibrium with marker loci. Experimental infections using different marked parasite genotypes (see Section 7.1) are likely to provide the most persuasive evidence for interactions between parasite genotype and host genetic background and/or immune status.

Molecular markers may also be useful for investigating microspatial genetic structuring of plant parasite nematodes. Lasserre *et al.* (1996) sampled *Heterodera avenae*, an obligately sexual cyst nematode, from plants within a single field and scored variation at two isozyme loci. Over the entire field they observed Hardy–Weinberg equilibrium (HWE), while samples from individual potted plants showed deviation from HWE. The authors speculate that annual ploughing of the field mixes genotypes at each generation, even though there is nonrandom mating at the scale of individual plants.

3.6. Heterozygote Excess and Mating Patterns

In addition to reflecting transmission patterns, microspatial genetic structure may reflect nematode mating systems. Fisher (1997) scored restriction site variation at three loci in *S. ratti* collected from rats in four regions in England. An excess of heterozygotes at a biallelic actin locus was observed in 22 out of 24 rats sampled from Berkshire (Figure 4). Selection for heterozygotes might explain this result, but heterozygote excess was not observed in other populations. A second explanation is based on the peculiar life history of *Strongyloides* sp., which facultatively utilize either a homogonic (asexual life cycle) or heterogonic sexual cycle (see Section 5.1). The pattern observed in the Berkshire population might result

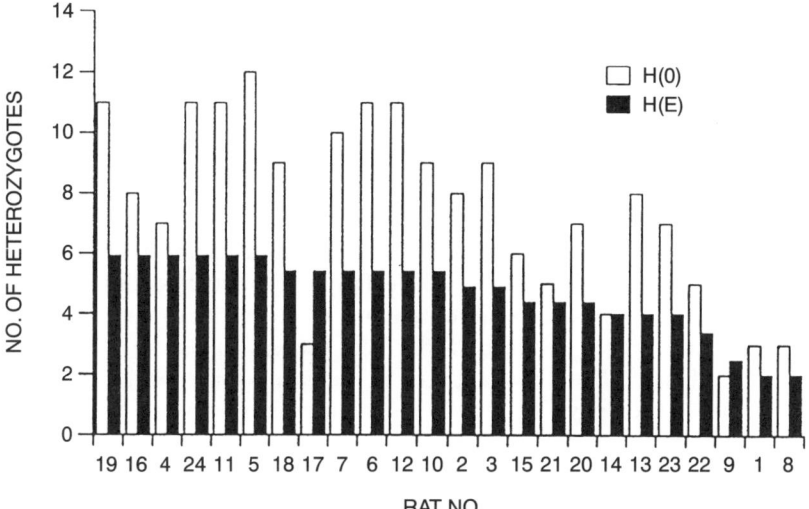

Figure 4 Excess heterozygosity at the Actin locus in L3 larvae of *Strongyloides ratti* from Berkshire, UK. Each pair of bars show observed and expected heterozygosity in Larvae from individuals rats. In 24 out of 27 cases observed values were greater than expected and in nine cases, significantly so. Reproduced with permission from Fisher (1997).

because the heterogonic route of development is rare in this location, resulting in little recombination. By chance, clones of *S. ratti* that are heterozygous at the actin locus may have risen to high frequencies in this population. Consistent with this hypothesis of limited recombination is the observation that significant linkage disequilibrium was observed in two out of three pairwise comparisons among three loci studied when samples from different regions of England were pooled. Note that this disequilibrium is unlikely to have resulted from the pooling of populations because very little differentiation was observed among *S. ratti* populations at individual loci (Fisher, 1997; see Table 4).

4. SIBLING SPECIES, HOST AFFILIATION, AND HYBRIDIZATION

Genetic markers are invaluable tools for differentiating between morphologically identical sibling species, for detecting hybridization, for identifying eggs or larval forms and for quantifying patterns of host affiliation and specificity. In this section we describe examples of the use of genetic markers for each of these purposes. We will be concerned mainly with 'sibling' or 'cryptic' species for which there are no readily discernible morphological characters allowing species discrimination. Such cryptic species appear to be ubiquitous in nematodes. Unravelling the components of nematode species complexes is therefore of primary importance for understanding parasite epidemiology and designing control programmes for species of medical, veterinary or agricultural importance. Similarly, such knowledge is fundamental to work on nematode community ecology and evolutionary biology.

When one species splits into two, the genomes of the two daughter species diverge by genetic drift. Allele frequencies drift to different frequencies in the two daughter populations, which eventually become fixed for alternative alleles. If we consider this process in terms of sequence variation, then at first the daughter populations share phylogenetically similar alleles (polyphyly) at any locus, then with time, some lineages are lost, resulting in paraphyly. Finally, the two populations reach a state in which phylogenetically distinct sequences are found in each daughter species (reciprocal monophyly). This process has been elegantly described by Avise (1994), and is illustrated in Figure 5. This process of divergence has important implications for the way in which we use genetic data to infer the presence of nematode sibling species and to interpret data from taxonomic work. When sibling species are at an early stage of divergence, all loci may be in a state of polyphyly or paraphyly. For these taxa, no single locus may

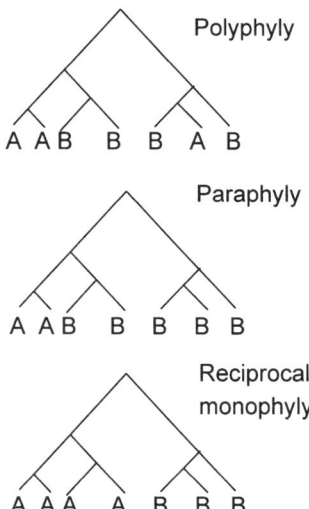

Figure 5 Phylogenies of alleles following splitting of parent species into two daughter species (A and B). Adapted from Avise (1994).

be used as a diagnostic marker for differentiating between populations, although information from multiple markers may be used. For parasite species at a later stage since separation, many loci may have attained a state of reciprocal monophyly. In these species, many loci will show fixed, diagnostic differences between species. For these species, single marker loci can be used to differentiate species.

4.1. Sibling species at an early stage of divergence

As described above, parasite populations at an early stage of divergence may frequently share phylogenetically similar alleles at all loci (polyphyly or paraphyly). Such taxa fall in a grey area in taxonomy on the border of species, and are frequently put into poorly defined categories such as host races or strains. Regardless of their taxonomic status, the presence of strong intrinsic barriers to gene flow within many parasite populations may have important implications for epidemiology and control of nematode populations. Furthermore, detailed investigation of species at an early stage of divergence can provide valuable clues to mechanisms of speciation in nematodes.

4.1.1. Ascaris *Infection of Humans and Pigs*

In many regions of the world, *Ascaris* spp. are found in sympatry in both human and pig populations — those in pigs are named *A. suum* and those in humans *A. lumbricoides,* although there are no morphological features which can be used to differentiate between them. Host-associated parasite populations showing minimal levels of genetic differentiation have also been found for *Ascaris* (but see Nascetti *et al.*, 1979 and Bullini *et al.*, 1981). Allozymes revealed no fixed differences between pig- and human-associated parasite populations from villages in Guatemala (Anderson *et al.*, 1993). However differences in allele frequencies were found at two polymorphic loci (EST-1, and MPI), and an additional MPI allele was found in parasites from pigs. Characterization of variation in mtDNA and intron 6 nuclear genes revealed further allele frequency differences between sympatric populations, but all markers examined were polyphyletic or paraphyletic with respect to host species. This is exemplified by the mtDNA tree in Figure 6: each of the three divergent mtDNA haplotypes are found in parasites from both humans and pigs. Of all the markers examined a single restriction site in the rDNA ITS was almost fixed in parasites from humans but found at very low frequencies in parasites from pigs. In this respect the distribution of rDNA ITS repeats is similar to the distribution OV-150 repeat types in 'forest' and 'savannah' forms of *O. volvulus* described below. To investigate whether cross-infection and hybridization are occurring in *Ascaris*, phenetic relationships among multilocus genotypes of individual worms were investigated in two villages in Guatemala. Trees based on the number of shared alleles failed to reveal a single case of cross-infection among parasites from pigs and parasites from humans from two villages (Figure 7).

However, cross-infection did account for many human *Ascaris* infections in regions of North America nonendemic for *A. lumbricoides*. ITS spacers of the type characteristic of parasites from pigs from worldwide locations were also found in all nine worms examined from North American cases (Figure 7). Further dissection of the mating barriers in this system are now possible because *Ascaris* crosses can be staged in domestic pigs (Jungersen *et al.*, 1996).

4.1.2. Repetitive Probes, Ocular Pathology and Onchocerciasis.

Onchocerca volvulus infected patients may differ in ocular pathology, with the 'savannah' form causing blinding onchocerciasis, and the 'forest' form causing a much milder form of the disease. Allozyme surveys of adult parasites (Cianchi *et al.*, 1985; Flockhart *et al.*, 1986) revealed minor differences between parasites from different countries, but failed to differentiate between different pathotypes of the parasite. More recently, molecular approaches have been used to find markers linked to ocular pathogenesis (Meredith *et al.*, 1991; Zimmerman *et al.*, 1992, 1993). These authors amplified a 150 bp repetitive sequence from worms (adults from nodules, microfilaria from skin snips and larvae from *Simulium* vectors). The amplification products were then hybridized consecutively with two probe sequences (pFS and pSS-1BT). Hybridization patterns with these probes were strongly correlated with *Onchocerca* pathotypes, and could therefore be used as molecular markers of pathogenesis. However, it is

Figure 6 (opposite) Neighbor joining tree showing the relationships between *Ascaris* mtDNA haplotypes from world-wide locations. MtDNA sequences from parasites infecting humans and pigs are polyphyletic, and the existence of identical haplotypes on different continents suggests recent mixing of parasite populations. The host (human or pig) and location from which parasites were collected are shown opposite the branch tips. Location abbreviations are as follows: GUA, Guatemala; BAN, Bangladesh; MAD, Madagascar; SWI, Switzerland; SCO, Scotland; PHI, Philippines; PERU, Peru. The map indicates frequencies of worms bearing the three major haplotypes in different locations (A = black, B = white, C = stippled). Modified from Anderson and Jaenike (1997), with permission of Cambridge University Press. Additional data on haplotype frequencies in sympatric Chinese populations were kindly provided by W. Peng *et al.* (unpublished data).

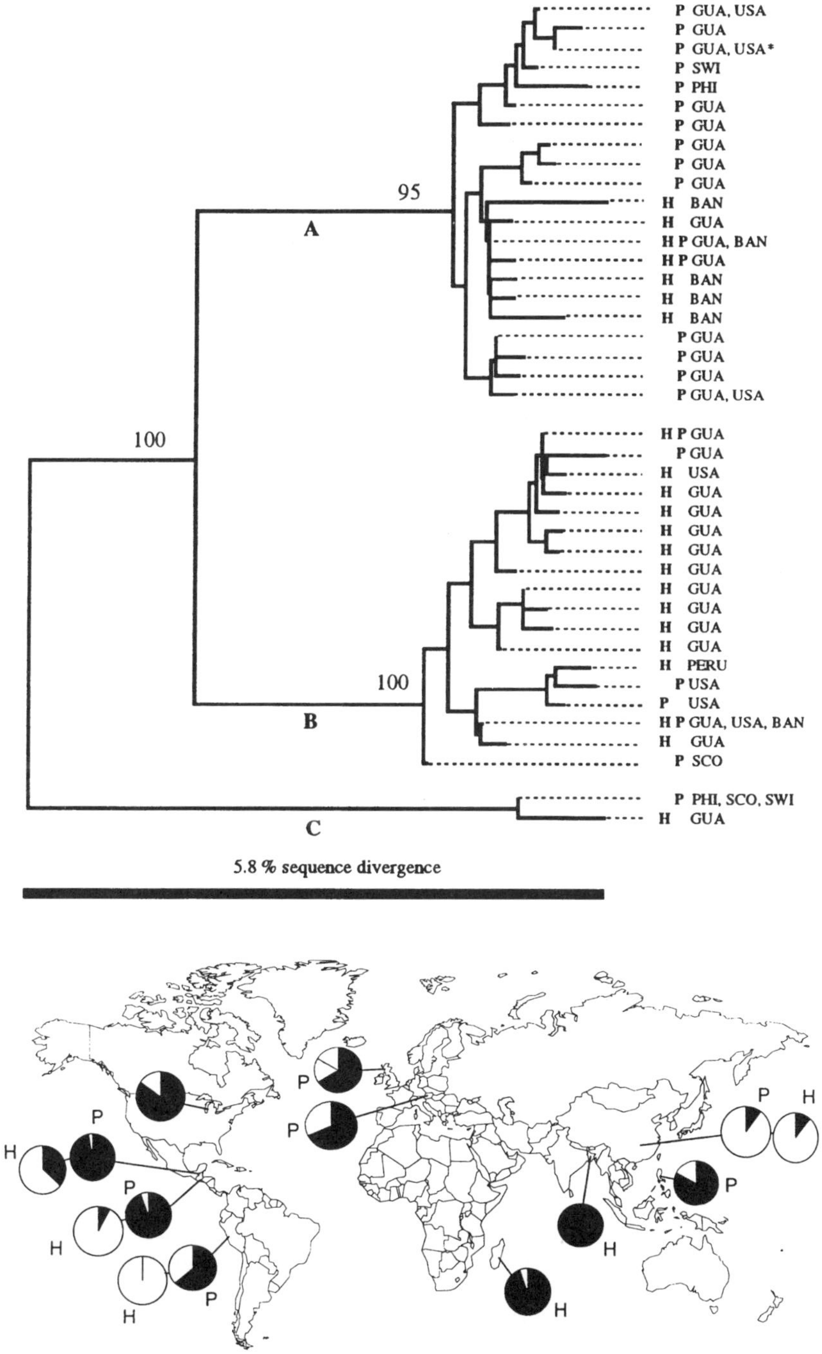

POPULATION BIOLOGY OF NEMATODES

```
                                            ┌─── P  GUA, USA
                                         ┌──┤
                                         │  └─── P  GUA
                                         │  ┌─── P  GUA, USA*
                                         ├──┤
                                         │  ├─── P  SWI
                                         │  └─── P  PHI
                                         ├─── P  GUA
                                         ├─── P  GUA
                                         │  ┌─── P  GUA
                                         ├──┤
                                         │  ├─── P  GUA
                                    95   │  └─── P  GUA
                              ┌─────A────┤    ┌─── H  BAN
                              │          │    ├─── H  GUA
                              │          ├────┤ H P GUA, BAN
                              │          │    ├─── H P GUA
                              │          │    ├─── H  BAN
                              │          │    ├─── H  BAN
                              │          │    └─── H  BAN
                              │          │  ┌─── P  GUA
                              │          └──┤ ├─── P  GUA
                              │             └─┤─── P  GUA
                              │               └─── P  GUA, USA
                        100   │             ┌─── H P GUA
                   ┌──────────┤           ┌─┤─── P  GUA
                   │          │           │ ├─── H  USA
                   │          │           │ ├─── H  GUA
                   │          │           │ ├─── H  GUA
                   │          │           ├─┤─── H  GUA
                   │          │           │ ├─── H  GUA
                   │          │           │ └─── H  GUA
                   │          │    100    │ ┌─── H  GUA
                   │          └─────B─────┤ ├─── H  GUA
                   │                      │ ├─── H  GUA
                   │                      │ └─── H  GUA
                   │                      │ ┌─── H  PERU
                   │                      │ ├─── P  USA
                   │                      └─┤─── P  USA
                   │                        ├─── H P GUA, USA, BAN
                   │                        ├─── H  GUA
                   │                        └─── P  SCO
                   │              ┌─── P  PHI, SCO, SWI
                   └──────C───────┤
                                  └─── H  GUA

        ├──────────────────┤
        5.8 % sequence divergence
```

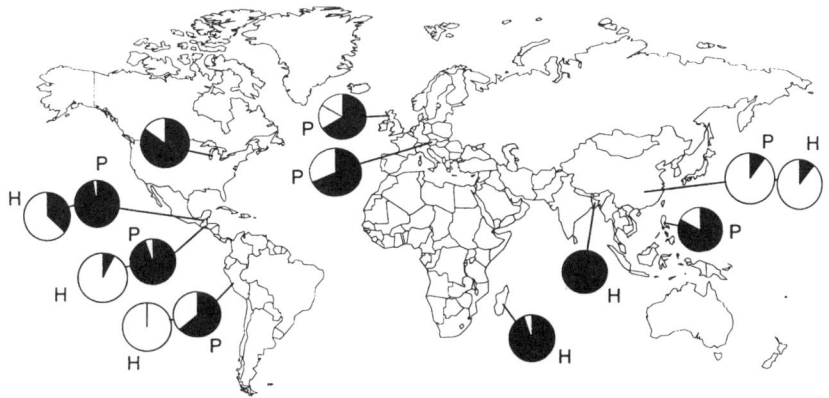

important to note that both major repeat types were found in some nematodes, while others contained one of the two and no hybridization was observed rarely. Thus, repeat arrays were not homogenized for a particular sequence type within worms associated with particular symptoms. This may be due to the limited time since divergence or to introgression of the repeat loci, following hybridization between two worm populations.

In some regions of Sierra Leone, with habitat midway between savannah and forest, the patterns of disease epidemiology and pathology and of probe hybridization results suggest that both 'savannah' and 'forest' *O. volvulus* may be circulating in sympatry, or that this region represents a zone of hybridization between these two pathotypes (Toe *et al.*, 1994). In this region Toe *et al.* (1997) have characterized both *O. volvulus* (using repeat probes) and cryptic *Simulium* species (using mtDNA sequence variation). They found no association between vector species and repeat sequences found in *O. volvulus*, suggesting that parasite–vector complexes do not play an important role in the epidemiology of this disease in these regions. However, because the repeat markers are unlikely to be directly linked with parasite pathogenesis, these markers may be rather poor markers of disease in areas where the two parasite populations are potentially hybridizing (Toe *et al.*, 1997). More detailed work on *O. volvulus* population genetics in these intermediate regions would be particularly informative. Such studies could usefully use multiple mendelian markers (microsatellite markers would be one option) to characterize individual parasites from across the forest/savannah transition zone, and measure how much hybridization and introgression occurs between the two parasite populations.

4.2. Sibling Species at a Later Stage of Divergence

An impressive series of allozyme surveys on Ascarid nematodes have been carried out by Bullini and colleagues (Cianchi *et al.*, 1985; Bullini *et al.*,

Figure 7 (opposite) Analysis of host specificity in *Ascaris*. (A) Human *Ascaris* infections may result from cross-infection from pigs. PCR amplified rDNA ITS sequences from North America and Guatemala following digestion with *HaeIII*. Parasites from humans from North America show similar digestion patterns to those from pigs from both North America and Guatemala. Reproduced from Anderson (1995) with permission of Cambridge University Press. (B) Cross-infection is rare or absent in Guatemalan villages. UPGMA phenograms showing the relationships between individual *Ascaris* from two villages in Guatemala. Pairwise distances between worms were calculated from the number of shared alleles at six nuclear loci. Redrawn from Anderson and Jaenike (1997), with permission of Cambridge University Press.

POPULATION BIOLOGY OF NEMATODES 251

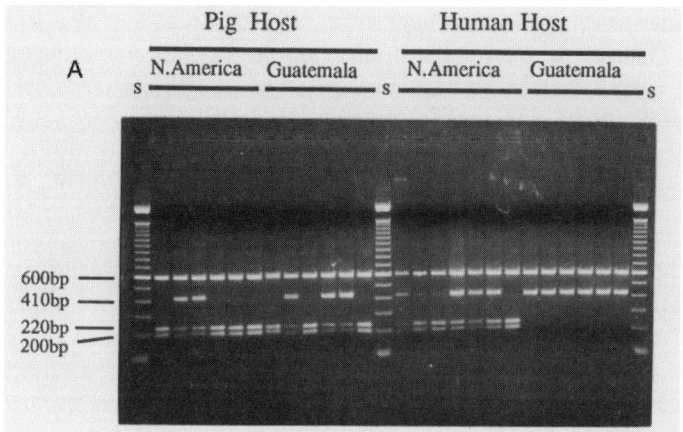

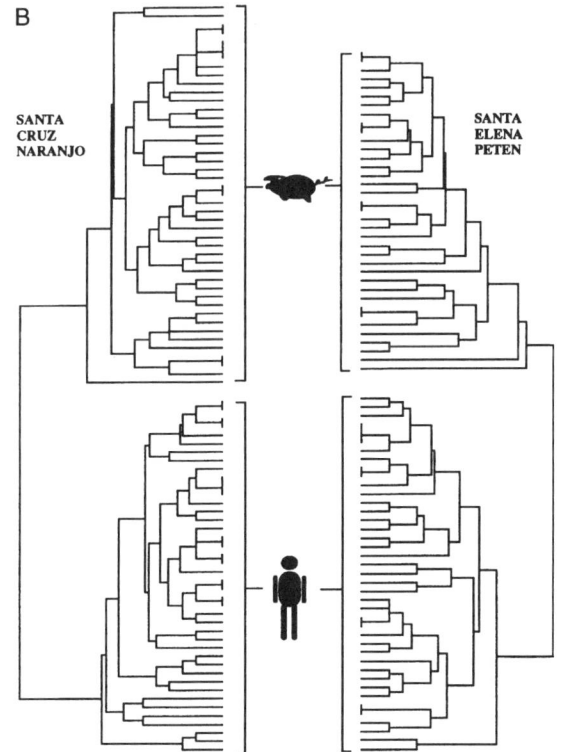

1986; Cancrini *et al.*, 1989; Paggi *et al.*, 1991; Orecchia *et al.*, 1994; see Table 3). Their work provide textbook examples of the use of allozymes for detecting cryptic species. Nematodes of the genus *Pseudoterranova* utilise fish definitive hosts and seal intermediates. Paggi *et al.* (1991) measured allozyme variation at 16 loci in over 1000 parasites of both fish and seal populations from locations in the North Atlantic and Norwegian and Barents seas. At three sites (Norway, Scotland and south-west Iceland) they found that population samples showed striking deficiency or complete absence of heterozygote individuals at multiple loci, suggesting that the samples contained more than one parasite population. When samples were subdivided into two species, designated A and B, according to allelic state at *Np*, *Est-2*, *Mdh-1*, *6Pgdh* and *Pgm*, each of the populations conformed with Hardy–Weinberg equilibrium. Similar approaches have revealed host affiliations and geographic ranges of sibling species of *Contracaecum* and *Anisakis* spp. (Orrechia *et al.*, 1986, 1994).

Vrijenhoek (1978) described a situation similar to that described above among larval parasites (*Contracaecum* spp.) infecting topminnows (*Poeciliopsis* spp.) and Cichlids (*Cyclastoma baeni*) in ponds and streams in north-west Mexico. In this case, electrophoretic examination of 11 enzyme loci revealed a complete absence of heterozygote individuals at Mdh-2 and Gpi. Three models were suggested to explain these patterns: (1) breeding systems involving parthenogenesis or high levels of inbreeding; (2) selection against heterozygotes; and (3) two sympatric noninterbreeding species. A number of lines support the third explanation. When the population sample was divided according to Mdh-2 and Gpi genotype, significant Hardy–Weinberg equilibrium was restored in each of the two subpopulations. Furthermore, there was strong linkage between enzyme loci, which disappeared on subdividing the population.

Interestingly, F_1 hybrids have been found in three studies of ascarids. Paggi *et al.* (1991) found that one in 275 parasites in regions where *Pseudoterranova decipiens* A and B co-occur was an F_1 hybrid, had heterozygous patterns at all five discriminatory loci. Similarly, in Vrijenhoek's (1978) study one out of 52 parasites examined was heterozygous at both discriminatory loci. Bullini *et al.* (1978) have also observed a hybrid among mixed populations of *Parascaris equorum* and *univalens* within a horse. The fact that backcross and recombinant genotypes have not been observed in any of these situations suggests that hybrid sterility prevents geneflow and introgression between these genetically well differentiated congeners. The levels of allozyme divergence between sibling species indicate that the parental species for each of the hybrids shared common ancestors between 2 and 4 million years ago (Paggi *et al.*, 1991).

When species occur in sympatry, as in the examples discussed above, single fixed differences or merely allele frequency differences between populations (see Section 4.1) are sufficient to demonstrate mating barriers and to detect sibling species. When populations are allopatric, fixed differences may evolve as a result of geographical separation, even when there are no intrinsic mating barriers. When dealing with allopatric comparisons, more care is required for interpretation of genetic data. Allozymes have been extensively used to investigate the relationships between nematodes infecting Australian animals (see Table 1; Chilton *et al.*, 1992, 1993a,b; Beveridge *et al.*, 1993, 1995). When making allopatric comparisons, these authors have assigned specific status if more than 15% of loci show fixed differences between allopatric populations. This 'cut off' was proposed by Richardson *et al.* (1986), on the basis of patterns of allozyme divergence in other animal species, and is sensibly conservative. Using these criteria, allopatric host-associated sibling species have been detected in the nematodes of the genus *Physaloptera* (Norman and Chilton, 1994), *Paramacropostrongylus* (Chilton *et al.*, 1993a), *Macropostrongylus* (Beveridge *et al.*, 1993), and *Hypodontus* (Chilton *et al.*, 1992), among others.

In recent years molecular techniques, in particular sequencing or PCR–RFLP of the rDNA ITS and RAPDs, have increasingly been used for comparisons of closely related nematode species. These PCR-based techniques are particularly useful because extremely small nematodes, nematode eggs, and alcohol or formalin preserved samples can be used. However, there are a some problems with interpretation of data from these techniques. While allozyme divergence can be interpreted in terms of time since separation and development of mating barriers (Coyne and Orr, 1989), there is no useful calibration of rDNA spacer data. It would be extremely useful to measure both allozyme and rDNA divergence for a variety of species pairs. Similarly, for RAPD data we currently have no idea of the number of shared fragments to expect in geographically separated populations within the same species and in well-differentiated biological species.

4.3. Morphological and Genetic Divergence

Levels of genetic differentiation have now been measured in many pairs or complexes of cryptic species. The bulk of the data comes from allozyme studies. In this case divergence between taxa can be measured in terms of the proportion of loci at which there are fixed genetic differences, or by standard measures of genetic distance. More recently, the ribosomal spacers ITS1 and ITS2 have been measured in a variety of cryptic species.

The data do not reveal a clear correlation between morphological and genetic divergence. Morphologically well-defined species in the genus *Trichostrongylus* differed by between 1.6–7.6%, in the ITS2 spacer (Hoste *et al.*, 1995). In comparison, morphologically indistinguishable members of the *Hypodontus macropi* species complex infecting kangaroos and wallabies, differ in the ITS2 spacer region by 25–28.3% (Chilton *et al.*, 1995). Allozyme measures of divergence for this complex are also high, with 15–63% of loci showing fixed differences between members of this complex (Chilton *et al.*, 1992). Other nematode sibling species from marsupials also show considerable genetic divergence: *Rugopharynx omega* and *R. sigma* have fixed differences at 45% of loci (Chilton *et al.*, 1993b). Ibrahim *et al.* (1997) describe an interesting plant parasitic nematode example of decoupling of morphological and genetic divergence. Morphologically undifferentiated isolates of the False Root Knot nematode (*Nacobbus aberrans*) differed by 5.1% in the conserved 5.8S rDNA, which is more than twice the divergence observed among well-differentiated genera *Globodera* and *Heterodera* of cyst nematodes.

Many studies have described cryptic genetic variation in morphologically identical species. The opposite situation, morphologically variable species that prove to be genetically indistinguishable, is less common. The classic example occurs in trichostrongylids. In males of each of the trichostrongylid species there exist 'major' and 'minor' morphs that are found in the same population. These are discrete polymorphisms in which the numerically abundant morph has one suite of characters (e.g. long, thinner spicules) and the rare morph has another set (e.g. short, stout spicules). For years the morphs have been described as separate species. For example, in *Ostertagia* we have *O. ostertagi/O. lyrata* and *O. dikmansi/O. mossi*. In *Teladorsagia* we have *T. circumcincta/T. trifurcata*, and so on. Recent molecular analyses support the hypothesis that variants are, in fact, members of the same species (for *Teladorsagia*; allozyme studies — Andrews and Beveridge, (1990); ITS-1 DNA sequence — Stevenson *et al.*, (1996); for the genus *Ostertagia*, mtDNA COI and rDNA ITS sequence data — D. Zarlenga, unpublished data). That this remarkable polymorphism is repeated in each genus suggests that the ability to produce the discrete morphs is an ancestral trait in trichstrongylids, and raises the following questions: are morphs environmentally induced, or genetically based, what is the underlying developmental control of this major morphological shift? These results also emphasize how labile morphological traits can be in nematodes, with obvious implications for morphology-based phylogeny reconstruction.

The situation described in trichostrongylids is complicated by the fact that there is strong evidence for the existence of mixed populations of reproductively isolated *T. circumcincta* population in Goats. Gasnier and Caberet (1996) provide three lines of evidence to suggest that *T. circumcincta* populations in French goats comprise two separate populations. In parasites from many goat farms, an additional allele (sR) at MDH-1 was found which was not observed in parasite populations from sheep farms. Furthermore, significant heterozygote deficits were seen in populations carrying high levels of this allele, which suggests the existence of mixed populations. They also provide experimental evidence: when populations bearing the sR allele were passaged through sheep for three generations, the sR allele was eliminated from the population. The simplest explanation for both experimental and field data is that two populations of *T. circumcinta* are co-transmitted in goat herds in France, while only one of these infects sheep. An alternative explanation, that the sR allele is selected against in parasites infecting sheep, could also be possible. However, in this case, heterozygote deficits would not be expected in goat populations only, so the two population hypothesis is most likely. The same authors have also observed strong heterozygote deficits in *Trichostrongylus colubriformis* populations. Further analysis of patterns of variation will be required to determine the relationships within this important nematode group.

4.4. Ubiquity of Sibling Species

A surprising result arising from work with genetic markers is that sympatric cryptic species are often found infecting the same host species and may even be found within the same individual host. Perhaps the most dramatic examples was described by Aho *et al.* (1992) for acanthocephalan parasites. In this case, allozymes revealed that four cryptic species coexisted within a turtle population. Examples from the nematode literature are listed in Table 5. These documented examples are likely to represent the tip of iceberg. The existence of multiple cryptic species raises a number of issues.

Conventional ecological dogma suggests that one species should occupy one niche, yet coexisting cryptic species of nematodes appear to be in direct contravention of this rule. We list three possible explanations for the observed patterns. (1) The documented cases could be the result of admixture of species that have evolved in allopatry and as such they might represent an unstable transient situation, in which one species may be in the process of replacing the other. This explanation is unlikely to account for the acanthocephalan example or the seal/ascarid systems, but could explain the examples involving humans or domestic animals. (2) Coexistence may be a stable situation, and could be explained by the model proposed by Atkinson and Shorrocks (1984), for organisms showing aggregated distributions on 'patchy' resources. In this model aggregation among patches reduces the level of competition between competing species thereby allowing stable coexistence. It would be of great interest to use genetic

Table 5 Cases in which genetic markers have revealed the co-occurrence of cryptic species within hosts.

Nematode spp.	Host	Technique	Reference
Contracaecum spp.	Topminnows/cichlid	MLEE	Vrijenhoek (1978)
Contracaecum osculatum A and B	American plaice	MLEE	Brattey (1995)
Contracaecum osculatum D and E	Weddell seals	MLEE	Orrechia et al. (1994)
Howardula aoronymphium A and B	*Drosophila falleni*	MtDNA	Jaenike (1996)
Onchocerca volvulus 'savannah' and 'forest'	Man	Repetitive sequence	Toe et al. (1994)
Parascaris equorum/univalens	Horse	MLEE	Bullini et al. (1978)
Pseudoterranova decipiens A and B	Seal/Fish	MLEE	Paggi et al. (1991)
Rugopharynx mawsonae/delta	Wallaby (*Macropus dorsalis*)	MLEE	Beveridge et al. (1995)
Teladorsagia spp.	Goat	MLEE	Gasnier and Cabaret (1996)

markers to investigate the distribution of different cryptic species within a host population to determine whether the distributions are independent of each other or positively correlated, because independence of distributions minimizes the level of interspecific competition (Shorrocks et al., 1990). (3) One intriguing possibility is that mating barriers between some cryptic species are the result of infection with *Wolbachia* sp. bacteria, which are known to be involved in reproductive incompatibility and speciation in insect populations (Werren, 1997b). Sironi et al. (1995) has recently described a *Wolbachia*-like bacterial symbiont in *Dirofilaria immitis*. It will be exciting to see how widespread these bacteria are within nematode populations, and whether they are involved in reproductive manipulations of nematodes similar to those demonstrated in insects and crustaceans (Werren, 1997a).

There are estimated to be 500 000 species of nematodes, of which 15 000 have been described (Poinar, 1983). These estimates are based on morphologically identifiable species. If parasitic nematodes characteristically exist as 'complexes' of morphologically identical cryptic species, then even these figures may be gross underestimates. To get an idea of the magnitude of underestimation consider the data on marine ascarids in Table 3. Allozyme work revealed the existence of 14 species where only six were originally suspected. While the existence of cryptic species is a problem for diversity estimates in all organisms, it is perhaps even more of a problem for nematode parasites, because nematodes have very conserved body plans and fewer useful morphological characters than organisms with defined endo- or exo-skeletons. Furthermore, many nematode cryptic species show high levels of genetic differentiation, indicating that morphological evolution may progress fairly slowly in this phyla (see Section 4.3.). This would further reduce the discriminatory power of traditional morphology-based systematics for nematodes.

The existence of cryptic species has important implications for human helminth epidemiology and community ecology. Where possible community ecologists would do well to follow Aho et al.'s (1992) example and use genetic markers to verify the identification of at least a subset of sampled parasite populations. A cautionary tale concerning the problems caused by cryptic species is furnished by Jaenike (1993; 1996). Populations of *Howardula aoronymphium* nematodes apparently lost the ability to infect one of the drosophilid host species during maintenance in the laboratory, leading to the conclusion that a rapid change in host range had occurred (Jaenike, 1993). Subsequent mtDNA analysis revealed the existence of two *Howardula* sibling species, one with a broad host range and one specializing in one *Drosophila* species. The apparent loss in host range was probably a result of the host generalist species becoming extinct in the laboratory culture (Jaenike, 1996). Those interested in the epidemiology of human intestinal helminths might learn much from routine surveys of genetic variation in presumed monospecific parasite populations. It would be mildly embarassing, but scarcely surprising, if common human parasites such as *Trichuris trichiura*, *Necator americanus* and *Ancylostoma duodenale* actually prove to be complexes of sibling species.

4.5. Inferring Population History

Phylogenetic information from closely related species can be used to make inferences about the population history of parasitic nematodes. We present two examples to illustrate this.

4.5.1. Onchocerca volvulus

This parasite is found in large regions of Africa, and in isolated foci in Central and South America. Within Africa there are two forms of the parasite (see Section 4.1.2), known as the savannah and forest forms which differ in patterns of pathology, vector usage, and in hybridization patterns with a repetitive probe. To explore the hypothesis that the South American foci resulted from recent importation of African parasites with the slave trade, Zimmerman *et al.* (1994) compared the sequences of multiple repetitive elements from parasites from both the Old World and the New World. They found reduced variation in the South American parasites, suggesting that the disease in South America, was the derived population. Furthermore, the sequences found in the South American parasites were very similar to those found in the African savannah parasites (Figure 8). This strongly suggests that Onchocerciasis is a recent disease in South America resulting from movement of people from West Africa.

4.5.2. Meloidogyne *Spp.*

Four species of *Meloidogyne*, *M. incognita*, *M. arenaria*, *M. javanica* and *M. hapla*, are important pests of a wide variety of important crops. Isolates of the first three species reproduce by mitotic parthenogenesis, while *M. hapla* reproduces by either meiotic parthenogenesis or sexually. Castagnone-Serreno *et al.* (1993) investigated the relationships between 18 different *Meloidogyne* isolates by hybridization of restricted genomic DNA with three repetitive probes. Each of the three probes revealed similar patterns — the composite tree combining all three data sets is shown in Figure 9 and, reassuringly, it corresponds closely with the tree based on allozyme data (Esbenshade and Triantaphyllou, 1987) and mtDNA (Hugall

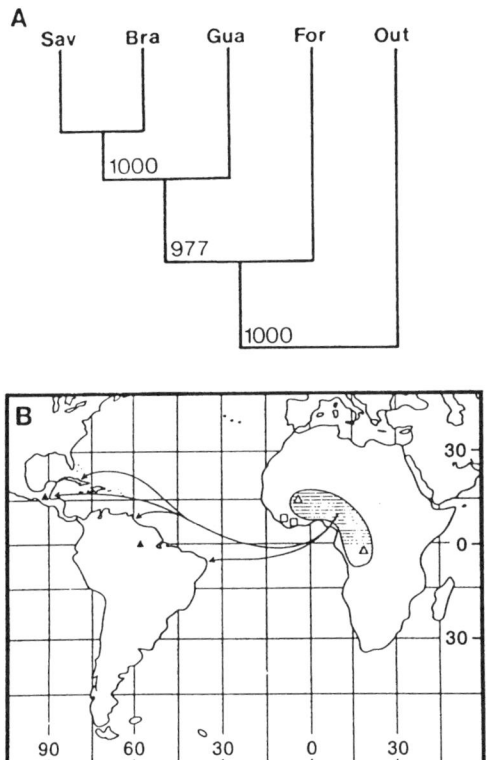

Figure 8 Inferring population history from phylogenetic information. Summary of the relationships between New World and African *Onchocerca volvulus*, and its relationships to the trans-Atlantic slave trade. (A) Strict consensus cladogram summarizing the relationships between the O-150 repeat family. (B) Summary of the historical record documenting the major sources of African slaves imported into the New World. The hatched region indicates the area of origin of slaves, while the arrows indicate final destinations in the Caribbean and Central and South America. The sources of parasite samples from which O-150 repeats were sequenced are indicated by the following symbols: □ = rainforest isolates (Liberia and Ivory Coast); ∆ = savannah isolates (Mali and Zaire); and ▲ = New World isolates (Guatemala and Brazil). Reproduced with permission from Zimmerman *et al.* (1994).

et al., 1994). Assuming similar rates of molecular evolution, the data suggest that *M. hapla* populations, which may utilize sexual as well as meiotic and mitotic parthenogenesis, are ancestral to the other populations which reproduce exclusively by mitotic parthenogenesis.

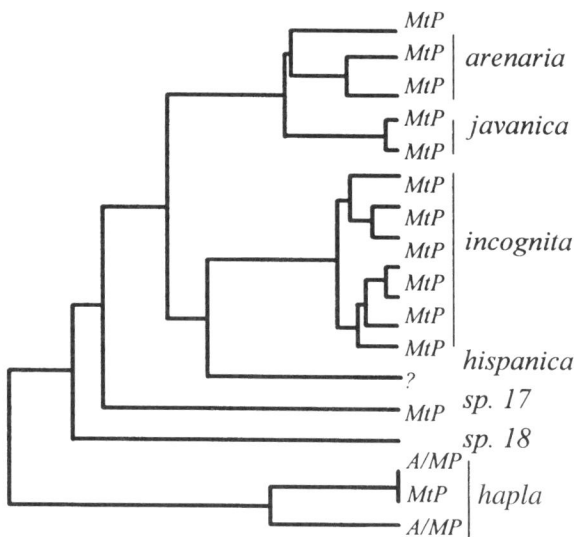

Figure 9 Phylogenetic relationships among *Meloidogyne* species with different modes of reproduction, suggesting that asexual cyst nematode lineages arose from sexual ancestors (MtP = mitotic parthengenesis; MP = meiotic parthenogenesis; A = amphimixis). Redrawn from Castagnone-Sereno *et al.* (1993), with permission of Blackwell Science Ltd.

5. NEMATODE LIFE-HISTORIES

5.1. Sexual exchange in *Strongyloides*

Most parasite nematodes of vertebrates use conventional amphimictic reproduction. Strongyloides, however, has a strange lifecycle involving alternative pathways of homogonic development or heterogonic development. Although both males and females copulate during the free-living generation, it was believed for a long time that no sperm was transferred (pseudogamy) and that offspring were genetically identical to the mother. Viney *et al.* (1993) tested this hypothesis by tracing the genotypes of females and their progeny at a single actin locus with a polymorphic *Mnl*1 site. The genotypes of offspring clearly demonstrated that inheritance of marker loci was biparental and that conventional sexual reproduction, rather than pseudogamy occurred (Figure 10). A similar analysis of reproduction in the parthenogenetic parasitic female revealed that parthenogenesis was functionally mitotic rather than meiotic — in other words offspring were genotypically identical to the mother (Viney, 1994).

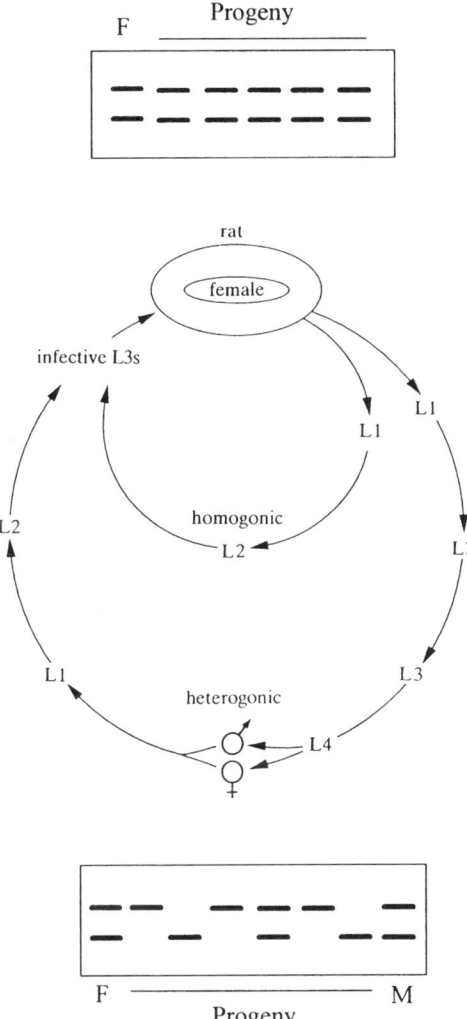

Figure 10 Unravelling the life cycle of *Strongyloides ratti* using genetic markers. Upper panel: Schematic of gel, showing analysis of inheritance patterns of a polymorphic restriction site in the actin locus from a parthenogenetic female and her progeny, indicating that the parthenogenesis is functionally mitotic. Lower panel: similar analysis of inheritance in progeny of controlled crosses in the free-living heterogonic males and females, showing mendelian segregation rather than pseudogamy. The life cycle figure was reproduced with permission from Viney *et al.* (1993), and The Royal Society.

5.2. Identification of Larval Stages

Genetic markers have also been used to identify correctly the larval stages of adult nematodes, for parasites with complex life cycles. We present two examples here. Adults of the parasitic nematode *Echinocephalus overstreeti* parasitize the shark *Heterodontus portusjacksoni*. However, larval stages of a variety of *Echinocephalus* spp. are indistinguishable on the basis of morphology and are found in marine molluscs. Andrews *et al.* (1988) investigated the identity of putative larvae of *E. overstreeti* from queen and king scallops (*Chlamys bifrons*, *Pecten albus*) by comparing electrophorectic patterns in larval and adult parasites. No fixed differences were found at 17 loci while in another 11 polymorphic loci common alleles were found in both larvae and adults, confirming that larvae and adults were part of a common life cycle. A very similar approach has been used by Orrechia *et al.* (1986) to identify the larval stages of sibling species of *Anisakis simplex* (A and B). Both *A. simplex* larvae and adults fell into two groups with fixed differences at three out of 22 enzyme loci. The distributions of the two sibling species were allopatric with *A. simplex* A in the Mediterranean sea, and *A. simplex* B in the north-east Atlantic Ocean.

6. GENETIC MARKERS AND DRUG RESISTANCE

Anthelmintic control of nematode parasites is often taken for granted until it becomes apparent that the drug removes fewer parasites than expected, at which point phenotypic resistance is diagnosed (Prichard, 1990). In this section we review the changes that are thought to occur in a parasite population as resistance develops, and the role that DNA-based technology can play in identifying these changes, monitoring the spread of resistance, and measuring parameters necessary for predictive modelling of resistance evolution.

When a new anthelmintic drug is introduced, alleles encoding resistance to this drug will be rare or absent. Selection will increase the frequency of resistance alleles in the population until drug efficacy is affected. The time from introduction of a new drug to the appearance of observable phenotypic resistance is influenced by several factors (Comins, 1977a,b; May, 1993). We discuss the number and type of genes which can affect anthelmintic resistance in Section 6.1. The frequency of resistance alleles in the population prior to drug use plays a crucial role in determining the rate of development of resistance. Closely coupled with this is the dominance relationship between resistance and susceptibility alleles. Section 6.2

focuses on these genetic aspects. The population biology of the parasite also influences the outcome of an anthelmintic control strategy. Section 6.3 discusses aspects of parasite biology relevant to the development of resistance such as population subdivision and migration rates, the presence of refugia and the association of deleterious fitness effects with resistance.

6.1. Genes Involved in Resistance

6.1.1. *How Many Genes?*

Single genes with a large effect on resistance will increase the progress made each generation toward resistance. If many genes, each with a small effect, are involved then a more complex change in the parasite population is necessary and a slower evolution of resistance is expected. The number of genes involved in conferring resistance can be difficult to determine. Genetic studies which involve crosses between susceptible and resistant parasite strains tend to reveal effects mediated by genes with large individual effects, and are relatively insensitive. In the case of benzimidazole (BZ) resistance, at least two genetic loci are involved in expression of full resistance (le Jambre *et al.*, 1979; Herlich *et al.*, 1981). Molecular analysis of the underlying mechanism of BZ resistance determined that an amino acid substitution of the β-tubulin protein is responsible (Roos *et al.*, 1990; Kwa *et al.*, 1993a,b, 1994, 1995) and that two β-tubulin loci are involved in *Haemonchus contortus* (Geary *et al.*, 1992). PCR-based restriction analysis of β-tubulin alleles in individual parasites has shown the association of a single allele at each of the two loci with resistance (Beech *et al.*, 1994). Recently, this same technique has implicated a third gene unrelated to tubulin in BZ resistance (W. Blackhall, personal communication).

Genetic analysis of levamisole resistance has identified only a single major gene, or tightly linked gene cluster, responsible for resistance, which is inherited differently in *H. contortus* and *T. colubriformis* (Martin and McKenzie, 1990; Dobson *et al.*, 1996). The gene encoding the drug-binding target, an acetyl choline receptor, has been cloned from *H. contortus* (Hoekstra *et al.*, 1997a), but examination of restriction enzyme pattern differences by Southern hybridization has provided no evidence that alleles at this locus are involved in selection during exposure of the parasite to levamisole. Unfortunately, the DNA examined was from a mixture of different individuals, and so only the most common allele types would be seen and estimates of allele frequencies could not be made.

Investigation of ivermectin resistance in the free living nematode, *C. elegans*, has shown that the drug binds to a glutamate-gated chloride channel, which is composed of at least two subunits (Cully *et al.*, 1993,

1994; Arena *et al.*, 1995). However, these receptor genes are unaltered in ivermectin-resistant mutants of *C. elegans* (JM Schaeffer, personal communication). Genetic crosses of ivermectin-resistant and -susceptible *H. contortus* have shown that resistance is mediated by a single gene, or gene complex (Dobson *et al.*, 1996). At the molecular level, there appear to be at least three distinct loci at which selection occurs on exposure to ivermectins (M.Xu, unpublished data; W. Blackhall, personal communication). Further work is needed to clarify the roles of these genes in the expression of resistance.

6.1.2. *Which Genes?*

Genetic analysis of parasite populations following selection provides a powerful technique for identifying genes involved in drug resistance. This approach involves measurement of the allele frequencies at candidate genes in parasite populations before and after selection. Changes in the frequency and diversity of alleles may be used to infer which genes are selected. This approach is particularly powerful if multiple selected lines are available. If the same alleles rise to high frequency in replicate populations, this strongly suggests that selection, rather than drift is involved in the changes observed (Roos *et al.*, 1990; Beech *et al.*, 1994; Kwa *et al.*, 1994). This approach has been used to find markers linked to resistance to benzimidazoles (Beech *et al.*, 1994; Kwa *et al.*, 1994) and ivermectin (W. Blackhall, personal communication). If no candidate loci are known then characterization of multiple markers spread over the genome in selected lines of parasites may suggest which regions of the genome are involved.

The changes that occur in parasites as they become resistant to anthelmintics can fall into several different categories. Alleles in which the target site for drug binding is modified can lead to resistance by reducing affinity of the anthelmintic for the target molecule. This has been demonstrated in the case of BZ resistance in *H. contortus*. Susceptible parasites express β-tubulin which bind the drug and are then prevented from polymerizing (Lubega and Prichard, 1991a,b,c; Lubega *et al.*, 1993, 1994). Resistant parasites express β-tubulin with a tyrosine substituted for a phenylalanine at position 200, which effectively prevents drug binding and allows the tubulin to polymerize at elevated drug concentrations (Lubega *et al.*, 1991a, 1994; Kwa *et al.*, 1994).

Modifications may also occur in proteins that can reduce the effective drug concentration in the parasite. The p-glycoprotein responsible for multidrug resistance in tumour cell lines is one potential candidate. Mice defective for the *mdr1a* gene, which encodes a p-glycoprotein, are unusually sensitive to ivermectin (Schinkel *et al.*, 1994). In the parasite, alleles with an increased drug efflux activity could confer resistance without modification of the drug target. This increased activity might be due to altered affinity for the drug, or increased expression of the protein. In order to determine if this gene plays a role in resistance, the association between specific p-glycoprotein alleles and resistance should be determined. Expression levels of the mRNA between susceptible and resistant parasites should also be examined because changes in the regulation of gene expression may be responsible for any changes.

When drug action is mediated through membrane receptor proteins, such as with levamisole and ivermectin, changes in the function of the receptor without an effect on drug binding could mediate resistance. Mutagenesis of *C. elegans* failed to identify any mutants of the glutamate-gated chloride channel resistant to ivermectin (J.M. Schaeffer, personal communication). However, the conditions used generate loss of function mutants with a complete absence of gene product. If the gene in question encodes an essential function, such mutants would be lethal. The spectrum of variation in natural populations may give rise to more subtle variants which might modify the drug effects while still maintaining normal gene function. These receptor genes should therefore not be ruled out as potential mechanisms of resistance without first examining the changes that occur in field strains of the parasite.

In the case of ivermectin resistance, it has been proposed that enzymatic modification of the macrocyclic lactone ring of the drug could abrogate drug efficacy (Coles, 1989). This logic can also be extended to other classes of anthelmintic. Changes in activity of the metabolic pathways by which anthelmintics are modified should therefore be examined as potential sites for change leading to drug resistance.

6.2. Genetics of Resistance

6.2.1. *Frequency of Resistance Alleles*

A major contributing factor to the rate at which resistance develops is the initial frequency of resistance alleles in the target population (May, 1993) (Figure 11). The ability to predict the dynamics of drug resistance evolution depends on measurement of allele frequency and not simply the ability to detect such alleles. One important aspect of using genetic tools to dissect the changes associated with developing resistance is that the DNA of individual parasites should be examined. When DNA is prepared from many individuals, the association between particular alleles is obscured. In fact, because analysis of such DNA reveals only the most common alleles present, the composition of the parasite gene pool can be impossible to determine (Roos *et al.*, 1990; Grant, 1994; Grant and Whitington, 1994;

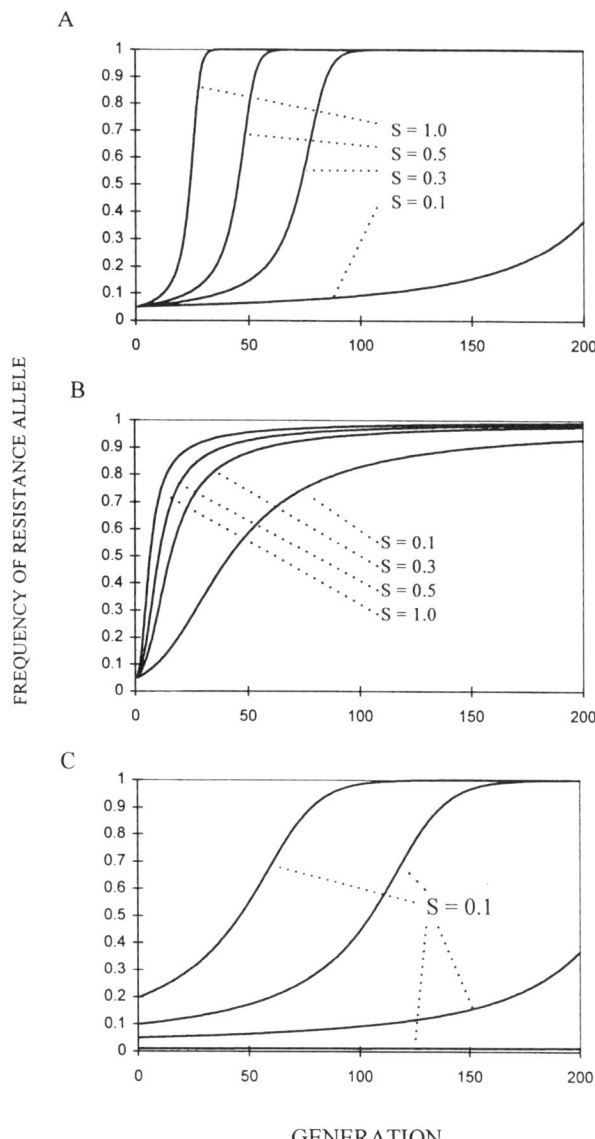

Figure 11 Graphs illustrating the importance of initial allele frequencies, selection pressure, and dominance on the rate of spread of alleles encoding drug resistance. The graphs show the trajectory of (A) recessive and (B) dominant resistance alleles under four different selection coefficients (S = 0.1–1.0). Selection coefficients, in this case, are related to the degree of drug coverage. The effect of different starting frequencies of the resistance allele is shown in (C). In this case the selection coefficient is fixed at 0.1. Adapted from Hartl and Clark (1989).

Lubega et al., 1994). PCR-based techniques allow rapid and accurate determination of allele frequencies and genotype distributions (Beech et al., 1994; Kwa et al., 1994).

Both Roos et al. (1990) and Lubega et al. (1994) have used bulk hybridization of DNA from *H. contortus* populations to investigate the selective changes in the two β-tubulin genes following drug selection. However, they could not distinguish whether the changes observed were due to changes in diversity at these loci or to rearrangements. Use of bulk DNA preparations from populations of nematodes may also account for the lack of detectable change at the acetylcholine receptor following levamisole resistance (Hoekstra et al., 1997). Analysis of individual parasite genotypes shows that prior to the imposition of drug selection the BZ resistance allele at each β-tubulin locus is present at quite high frequency (Beech et al., 1994). At the isotype-1 locus the initial frequency of the resistance allele was 46% and at the isotype-2 locus the frequency was 12%. Recent data suggests that the frequency of alleles associated with ivermectin resistance in unselected lines may be as high as 10–20% (W. Blackhall, personal communication). It should be remembered, however, that the resistance allele may represent only a subset of those identified by restriction enzyme analysis and therefore the frequency could be much lower. Rather than being extremely rare, these resistance alleles appear to be at surprisingly high frequencies. In the future, frequency distributions of resistance alleles for the three main classes of anthelmintic will allow more accurate modelling early in the development of resistance.

Work on anthelmintic resistance to date has been restricted to veterinary parasites, while resistance in medically important nematode parasites has received little attention (Geerts et al., 1997). We feel that such studies should be a priority in view of the high frequencies of resistance alleles observed in untreated populations of veterinary parasites (Beech et al., 1994) (W. Blackhall, personal communication). Large population-based chemotherapy programmes, such as the African Programme for Onchocerciasis Control (Remme, 1995) and school or community-based programmes against intestinal helminths (Bundy et al., 1997) impose strong selection for drug resistance on human parasites. For example, in many school-based chemotherapy programmes more than 50% of the parasite population may be exposed to anthelmintics. Characterization of β-tubulin isotypes and investigation of the frequencies of alleles bearing mutations (e.g., Phe → Tyr at position 200) in populations of human intestinal nematodes would be of considerable interest. If such mutations were detected at frequencies greater than say 5%, drug treatment programmes could be modified to slow the progress of BZ resistance.

6.2.2. Dominance

Closely associated with the effects of initial allele frequency is the role that dominance plays in the evolutionary response to drug selection (May, 1993; Dobson et al., 1996). When resistance alleles are rare, a majority are found as heterozygotes. If resistance alleles are dominant, heterozygotes will be resistant and the allele frequencies will increase rapidly. If resistance alleles are recessive, then heterozygotes are susceptible to the drug and only homozygotes are expected to survive treatment. Thus, dominant alleles are expected to give rise to a much faster initial rate of resistance development than codominant or recessive alleles (Dobson et al., 1996). However, while recessive resistance alleles increase in frequency slowly to begin with, fixation within populations will occur rapidly once they become common (see Figure 11).

Measurement of the degree of dominance of alleles responsible for resistance has been made primarily in genetic crosses. Alleles which confer BZ resistance appear to be recessive (le Jambre et al., 1979; Herlich et al., 1981), while alleles which confer levamisole resistance are sex-linked recessive in *T. colubriformis* but autosomal recessive in *H. contortus* (Martin and McKenzie, 1990; Dobson et al., 1996). For ivermectin resistance there appears to be a single gene or a gene cluster with primarily dominant effects (Dobson et al., 1996).

How can we use molecular markers to measure the relative dominance of alleles that confer drug resistance? If one has a candidate gene, and alleles thought to be associated with resistance, then determining the genotype of individuals prior to, and among the survivors of, drug treatment will give an estimation of the dominance of each allele. Deficits of homozygotes for the resistance allele among surviving parasites would indicate that these alleles are recessive, while loss of both homozygotes and heterozygotes indicates dominance. This experimental approach would also confirm, or refute, a commonly held belief regarding the effects of different drug doses on parasite survival. At low drug doses, more parasites survive treatment. The assumption is that at such doses, resistance alleles are effectively dominant. With high drug doses, more parasites are killed and these are presumed to be the heterozygotes because the resistance alleles become recessive under these conditions.

6.3. Population Biology

6.3.1. Population Subdivision and Migration

Application of an anthelmintic through treatment of infected host individuals can only act on parasites within the host. Many stages of the parasite exist outside the host and parasites in this component of the environment, or refugia, are unaffected by drug selection. Refugia have the effect of reducing the rate of selection for resistance because each generation of untreated individuals from refugia contribute to successive generations. Strategies that use the potential of refugia to minimize the development of resistance have been proposed (Comins, 1977a, 1979).

An extension to the concept of refugia is the immigration of parasites from geographic areas not treated with anthelmintic. There are two sides to the role that migration plays in the development of resistance. As immigration from untreated areas increases, the rate at which resistance arises decreases owing to the dilution effect described above. Conversely, resistance alleles which increase in frequency under drug treatment can emigrate and thus increase the frequency of resistance alleles elsewhere. As subdivisions of the parasite population under treatment become more isolated, resistance will develop at an accelerated rate, but the spread of this resistance will be reduced. The characterization of refugia and migration patterns, and their contribution to the development of resistance using molecular techniques are essentially those that have been dealt with in Section 3.

6.3.2. *Are Resistance Alleles Deleterious?*

An important parameter determining the spread and maintenance of resistance alleles within nematode populations is the fitness of these alleles in the absence of drug selection. The concept that resistance alleles are rare because of their deleterious effect in susceptible populations is inconsistent with the characteristics of alleles associated with resistance that have been identified so far. If resistance alleles were deleterious but recurrent mutation continually produced such alleles then one would expect resistance alleles to have many different linked restriction pattern haplotypes. In BZ resistance, the same allele, as identified by linked genetic polymorphism appears to be responsible for resistance in a wide variety of resistant lines (Roos *et al.*, 1990; Beech *et al.*, 1994; Kwa *et al.*, 1994). This implies that resistance arose once, and subsequently spread as a neutral allele, rather than being the result of recurrent mutation. This interpretation is supported by a quantitative genetic analysis of the association between resistance and fitness traits which exist in a population prior to selection (Chehresa *et al.*, 1997a) and following the development of resistance (Chehresa *et al.*, 1997b).

7. OTHER APPLICATIONS OF GENETIC MARKERS

In this final section we discuss a variety of fields of nematode biology where we feel genetic tools should have a major impact in future years.

7.1. Intrahost Dynamics

We know a considerable amount about the distributions of helminths among hosts, but very little about the age structure, mating patterns and competitive interactions among parasites within hosts. Population biologists treat infrapopulations as 'black boxes', resulting from a balance between worm recruitment and death. In the natural situation, hosts are repeatedly exposed to infective stages. To replicate the natural situation experimental trickle infection regimes are used (Keymer and Hiorns, 1986; Eriksen et al., 1992; Roepstorff et al., 1996). Knowledge of the age structure of parasites in such experiments would add considerably to our understanding of within-host dynamics and immunity to nematodes. Trickle infections using cohorts of genetically marked parasites would allow us to measure: (1) the age structure of parasite populations; (2) the relationship between worm size, age and reproductive success; (3) relationship between worm age and position in the gut; and (4) length and distribution of prepatent periods and larval arrest. In particular, such an approach would clarify the relative roles of worm expulsion and arrest/destruction of migrating larvae in determining burdens of adult parasites in the gut. This approach is currently being employed to investigate intrahost population dynamics of trickle infections with *Ascaris suum* in pigs at the Danish Centre for Experimental Parasitology in Copenhagen (G. Jungersen, personal communication). In this case natural mitochondrial variants are being employed to distinguish different parasite cohorts (Figure 12). Similar approaches could easily be adapted to other important veterinary parasites and to nematodes used in mouse model systems. The extensive mtDNA variation described from many nematode parasites (Blouin et al., 1992, 1995) provide a rich source of maternal markers for experimental epidemiology and should be exploited to the full.

7.2. Nematode Mating Patterns and Sociobiology

Nematode mating patterns have received considerable theoretical attention (May and Woolhouse, 1993; Haukisalmi et al., 1996) but there is almost no empirical data. Mating patterns could be monitored indirectly

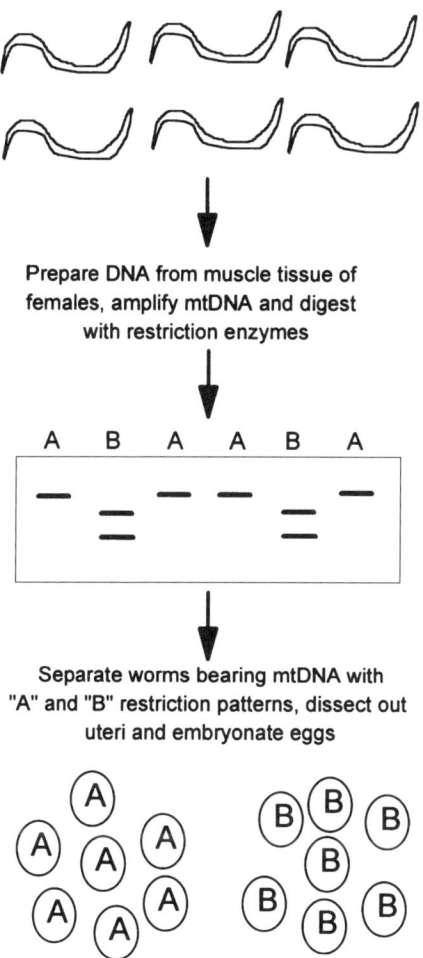

Figure 12 Generation of populations of nematodes bearing unique mitochondrial markers for use in experimental trickle infections.

using multiple genetic markers. Such an approach would allow us to investigate whether the offspring of a single female are sired by multiple male worms, the distributions of male and female mating success, and whether male size is a sexually selected trait that is reflected in greater mating success. Recent applications of genetic markers to free-living organisms have revealed many surprising facets of social structure and sexual selection (Amos *et al.*, 1993; Birkhead and Møller, 1995). However, a major problem when working with free-living organisms is that immigration and emigration make it difficult to identify all potential parents in any population. In nematode populations, which are confined to the host, this problem is largely eliminated. Mating patterns could be investigated most simply by collecting all adult worms from the gut, and obtaining progeny by dissecting fertilized eggs from adult female worms, and embryonating eggs or culturing larvae. Both parental and progeny genotypes could then be established using genetic markers with multiple alleles. Microsatellite markers would be ideal for this purpose. Triantaphylou and Esbenshade (1990) have used a similar approach to investigate mixing patterns for plant parasitic nematodes. They used isozyme markers to demonstrate that progeny from a single female may have multiple fathers in *Heterodera*. It would be useful to extend this approach to field populations.

7.3. Antigen Evolution and Immunology

Theoretical work on parasite host co-evolution suggests that there should be 'arms races' between host and parasite: parasites will be selected to avoid immune attack, while hosts will be selected to resist infection (Anderson and May, 1982; Austin and Anderson, 1996). Molecular studies of microparasites provides strong support for this view. Surface antigens show both high levels of allelic variation and nonsynonymous substitution (Bonhoeffer and Nowak, 1994; Hughes and Hughes, 1995; Smith *et al.*, 1995), suggesting that they are under diversifying selection. Surprisingly, there is no equivalent data published on variation in nematode antigens, despite the fact that many such antigens are known (Maizels *et al.*, 1993), and some are being developed as vaccine candidates (Wakelin, 1995; Jasmer *et al.*, 1996). Sequences of multiple alleles of antigens known to be involved in generating protective immunity would be of great practical importance, but would also provide fundamental insights into the nature of the host–parasite interaction. T. J. C. Anderson and M. W. Kennedy (unpublished data) have sequenced a short section of the DvA-1 polyprotein antigen of the cattle lung worm (*Dictyocaulus viviparus*) from a single allele of 10 individual worms. Up to 4.3% variation was observed between alleles. Furthermore, substitutions in nonsynonymous sites outnumbered those in synonymous sites ($D_{NS}/D_N > 1$), although insignificantly so. Further sequencing will reveal whether the high $D_{NS}:D_N$ ratios reflect an absence of functional constraints on this molecule or the action of diversifying selection. Other data suggest that there may be minimal variation in some nematode antigens. Comparison of the glutathione peroxidase (gp29) sequences from *Brugia malayi* and *Brugia pahangi* revealed only a few nonsynonymous changes between alleles in coding regions (Zvelebil *et al.*, 1993).

ACKNOWLEDGEMENTS

We thank Brad Hyman, Carlos Machado, Mat Fisher, Weidong Peng, Ruurdtje Hoekstra, Bill Blackhall, Ming Xu, Clare Constantine, Andrew Read and Gregers Jungersen for access to unpublished data and work in press, and to those who gave permission to reproduce published figures.

REFERENCES

Aho, J.M., Mulvey, M., Jacobson, K.C. and Esch, G.W. (1992). Genetic differentiation among congeneric acanthocephalans in the yellow-bellied slider turtle. *Journal Of Parasitology* **78**, 974–981.

Amos, W., Twiss, S., Pomeroy, P.P. and Anderson, S.S. (1993). Male mating success and paternity in the grey seal, *Halichoerus grypus*: a study using DNA fingerprinting. *Proceedings Of The Royal Society Of London Series B Biological Sciences* **252**, 199–207.

Anderson, R.M. and May, R.M. (1982). Coevolution of hosts and parasites. *Parasitology* **85**, 411–426.

Anderson, T.J.C. (1995). *Ascaris* infections in humans from North America: Molecular evidence for cross-infection. *Parasitology* **110**, 215–219.

Anderson, T.J.C. and Jaenike, J. (1997) Host specificity, evolutionary relationships and macrogeographic differentiation among *Ascaris* populations from humans and pigs. *Parasitology* **155**, 325–342.

Anderson, T.J.C., Romero Abal, M.E. and Jaenike, J. (1993). Genetic structure and epidemiology of *Ascaris* populations: Patterns of host affiliation in Guatemala. *Parasitology* **107**, 319–334.

Anderson, T.J.C., Komuniecki, R., Komuniecki, P.R. and Jaenike, J. (1995a). Are mitochondria inherited paternally in Ascaris? *International Journal for Parasitology* **25**, 1001–1004.

Anderson, T.J.C., Romero Abal, M.E. and Jaenike, J. (1995b). Mitochondrial DNA and *Ascaris* microepidemiology: The composition of parasite populations from individual hosts, families and villages. *Parasitology* **110**, 221–229.

Andrews, R.H. and Beveridge, I. (1990). Apparent absence of genetic differences among species of *Teladorsagia* (Nematoda: Trichostrongylidae). *Journal Of Helminthology* **64**, 290–294.

Andrews, R.H., Beveridge, I., Adams, M. and Baverstock, P.R. (1988). Identification of life cycles stages of the nematode *Echinocephalus overstreeti* by allozyme electrophoresis. *Journal Of Helminthology* **62**, 153–157.

Andrews, R.H., Beveridge, I., Adams, M. and Baverstock, P.R. (1989). Genetic characterization of three species of *Onchocerca* at 23 enzyme loci. *Journal Of Helminthology* **63**, 87–92.

Arena, J.P., Liu, K.K., Paress, P.S., Frazier, E.G., Cully, D.F., Mrozik, H. and Schsaeffer, J.M. (1995). The mechanism of action of avermectins in *Caenorhabditis elegans*: Correlation between activation of glutamate-sensitive chloride current, membrane binding, and biological activity. *Journal of Parasitology* **81**, 286–294.

Arnheim, N. (1983). Concerted evolution of multigene families. In: *Evolution of*

Genes and Proteins (M. Nei, and R.K. Koehn, eds), pp. 1–331. Sunderland, MA: Sinauer Associates Inc.

Atkinson, W.D. and Shorrocks, B. (1984). Aggregation of larval Diptera over discrete and ephemeral breeding sites: The implications for coexistence. *American Naturalist* **124**, 336–351.

Austin, D.J. and Anderson, R.M. (1996). Immunodominance, competition and evolution in immunological responses to helminth parasite antigens. *Parasitology* **113**, 157–172.

Avise, J.C. (1994). *Molecular Markers, Natural History and Evolution*, pp. 1–511. New York & London: Chapman & Hall.

Back, E., Felder, H., Muller, F. and Tobler, H. (1984a). Chromosomal arrangement of the two main rDNA size classes of *Ascaris lumbricoides*. *Nucleic Acids Research* **12**, 1333–1347.

Back, E., van Meir, E., Muller, F., Schaller, D., Neuhaus, H., Aeby, P. and Tobler, H. (1984b). Intervening sequences in the ribosomal RNA genes of Ascaris lumbricoides: DNA sequences at junctions and genomic organization. *Embo Journal* **3**, 2523–2530.

Beech, R.N., Prichard, R.K. and Scott, M.E. (1994). Genetic variability of the beta-tubulin genes in benzimidazole-susceptible and -resistant strains of *Haemonchus contortus*. *Genetics* **138**, 103–110.

Beveridge, I., Chilton, N.B. and Andrews, R.H. (1993). Sibling species within *Macropostrongyloides baylisi* (Nematoda: Strongyloidea) from macropodid marsupials. *International Journal for Parasitology* **23**, 21–33.

Beveridge, I., Chilton, N.B. and Andrews, R.H. (1995). Relationships within the *Rugopharynx delta* species complex (Nematoda: Strongyloidea) from Australian marsupials inferred from allozyme electrophoretic data. *Systematic Parasitology* **32**, 149–156.

Birkhead, T.R. and Møller, A.P. (1995). Extra-pair copulation and extra-pair paternity in birds. *Animal Behaviour* **49**, 843–848.

Birky, C.W.J., Fuerst, P. and Maruyama, T. (1989). Organelle gene diversity under migration, mutation, and drift: Equilibrium expectations, approach to equilibrium, effects of heteroplasmic cells, and comparison to nuclear genes. *Genetics* **121**, 613–628.

Blouin, M.S., Dame, J.B., Tarrant, C.A. and Courtney, C.H. (1992). Unusual population genetics of a parasitic nematode; mtDNA variation within and among populations. *Evolution* **46**, 470–476.

Blouin, M.S., Yowell, C.A., Courtney, C.H. and Dame, J.B. (1995). Host movement and the genetic structure of populations of parasitic nematodes. *Genetics* **141**, 1007–1014.

Bonhoeffer, S. and Nowak, M.A. (1994). Intra-host versus inter-host selection: Viral strategies of immune function impairment. *Proceedings of the National Academy of Sciences of the United States of America* **91**, 8062–8066.

Brattey, J. (1995). Identification of larval *Contracaecum osculatum* s.l. and *Phocascaris* sp. (Nematoda: Ascaridoidea) from marine fishes by allozyme electrophoresis and discriminant function analysis of morphometric data. *Canadian Journal of Fisheries and Aquatic Sciences* **52**, 116–128.

Brattey, J. and Davidson, W.S. (1996). Genetic variation within *Pseudoterranova decipiens* (Nematoda: Ascaridoidea) from Canadian Atlantic marine fishes and seals: Characterization by RFLP analysis of genomic DNA. *Canadian Journal of Fisheries and Aquatic Sciences* **53**, 333–341.

Bullini, L., Nascetti, G., Ciafre, S., Rumore, F. and Biocca, E. (1978). Ricerche cariologiche ed electroforetiche se *Parascaris univalens* e *Parascaris equorum*. *Accademia Nazionale dei Lincei, Rendiconti Classe Scienze Fisiche Matamatiche e Naturali, Serie VIII* **65**, 151–156.

Bullini, L., Nascetti, G. and Grappelli, C. (1981). Nuovi dati sulla divergenza e sulla variabilita genetica delle specie gemelle *Ascaris lumbricoides – A. suum* e *Parascaris univalens – P. equorum*. *Parassitologia* **23**, 139–142.

Bullini, L., Nascetti, G., Paggi, L., Orecchia, P., Mattiucci, S. and Berland, S. (1986). Genetic variation of ascaridoid worms with different life cycles. *Evolution* **40**, 437–440.

Bundy, D.A.P., Hall, A., Adjei, S., Kihamia, C., Gopaldas, T. and Khoi, H.H. (1997) Better health, nutrition and education for the school-aged child. *Transactions of the Royal Society of Hygiene and Tropical Medicine* **91**, 1–2.

Burke, W.D., Mueller, F. and Eickbush, T.H. (1995). R4, a non-LTR retrotransposon specific to the large subunit rRNA genes of nematodes. *Nucleic Acids Research* **23**, 4628–4634.

Burt, A., Carter, D.A., Koenig, G.L., White, T.J. and Taylor, J.W. (1996). Molecular markers reveal cryptic sex in the human pathogen Coccidioides immitis. *Proceedings of the National Academy of Sciences of the United States of America* **93**, 770–773.

Cancrini, G., Mattiuci, S., D'Amelio, S., Genchi, C. and Coluzzi, M. (1989). Genetic characterization of *Dirofilaria repens* and *Dirofilaria immitis* by electrophoretic analysis of gene-enzyme systems. *Parassitologia* **31**, 189–196.

Castagnone Sereno, P., Piotte, C., Uijthof, J., Abad, P., Wajnberg, E., Vanlerberghe Masutti, F., Bongiovanni, M. and Dalmasso, A. (1993). Phylogenetic relationships between amphimictic and parthenogenetic nematodes of the genus *Meloidogyne* as inferred from repetitive DNA analysis. *Heredity* **70**, 195–204.

Caswell Chen, E.P., Williamson, V.M. and Wu, F.F. (1992). Random amplified polymorphic DNA analysis of *Heterodera cruciferae* and *Heterodera schachtii* populations. *Journal Of Nematology* **24**, 343–351.

Chehresa, A., Beech, R.N. and Scott, M.E. (1997a). Changes in life history traits in *Heligosomoides polygyrus bakeri* associated with development of drug resistance. *International Journal for Parasitology*.

Chehresa, A., Beech, R.N. and Scott, M.E. (1997b). Life history variation among lines isolated from a laboratory population of *Heligosomoides polygyrus bakeri*. *International Journal for Parasitology* **27**, 541–551

Chilton, N.B., Beveridge, I. and Andrews, R.H. (1992). Detection by allozyme electrophoresis of cryptic species of *Hypodontus macropi* (Nematoda: Strongyloidea) from macropodid marsupials. *International Journal For Parasitology* **22**, 271–279.

Chilton, N.B., Beveridge, I. and Andrews, R.H. (1993a). Electrophoretic and morphological analysis of *Paramacropostrongylus typicus* (Nematoda: Strongyloidea), with the description of a new species, *Paramacropostrongylus iugalis* new species, from the eastern grey kangaroo *Macropus giganteus*. *Systematic Parasitology* **24**, 35–44.

Chilton, N.B., Beveridge, I. and Andrews, R.H. (1993b). Electrophoretic comparison of *Rugopharynx longiburasaris* Kung and *R. omega* Beveridge (Nematoda: Strongyloidea), with the description of *R. sigma* n. sp. from pademelons, *Thylogale* spp. (Marsupialia: Macropodidae). *Systematic Parasitology* **26**, 159–169.

Chilton, N.B., Gasser, R.B. and Beveridge, I. (1995). Differences in a ribosomal DNA sequence of morphologically indistinguishable species within the *Hypodontus macropi* complex (Nematoda: Strongyloidea). *International Journal for Parasitology* **25**, 647–651.

Cianchi, R., Karam, M., Henry, M.C., Villani, F., Kumlien, S. and Bullini, L. (1985). Preliminary data on the genetic differentiation of *Onchocerca volvulus* in Africa (Nematoda: Filarioidea). *Acta Tropica* **42**, 341–352.

Coles, G.C. (1989). The molecular biology of drug resistance in parasitic helminths. In: *Comparative Biochemistry of Parasitic Helminths.* (E. Bennet, C. Behm and C. Bryant, eds) pp. 125–144. London: Chapman and Hall.

Comins, H. (1977a). The development of insecticide resistance in the presence of migration. *Journal of Theoretical Biology* **64**,

Comins, H. (1977b). The management of pesticide resistance. *Journal of Theoretical Biology* **65**, 399–420.

Comins, H. (1979). Analytic methods for the management of pesticide resistance. *Journal of Theoretical Biology* **77**, 171–188.

Coyne, J.A. and Orr, H.A. (1989). Patterns of speciation in *Drosophila. Evolution* **43**, 362–381.

Cully, D.F., Paress, P.S., Liu, K.K. and Arena, J.P. (1993). Characterization and attempts to expression clone and avermectin and glutamate-sensitive chloride channel from *Caenorhabditis elegans. Journal of Cellular Biochemistry* **17C**, 115.

Cully, D.F., Vassilatis, D.K., Liu, K.K., Paress, P.S., Van Der Ploeg, L.H.T., Schaeffer, J.M. and Arena, J.P. (1994). Cloning of an avermectin-sensitive glutamate-gated chloride channel from *Caenorhabditis elegans. Nature* **371**, 707–711.

Dobson, R., Lejambre, L. and Gill, J. (1996). Management of anthelmintic resistance – inheritance of resistance and selection with persistent drugs. *International Journal for Parasitology* **26**, 993–1000.

Eriksen, L., Nansen, P., Roepstorff, A., Lind, P. and Nilsson, O. (1992). Response to repeated inoculations with *Ascaris* suum eggs in pigs during the fattening period: I. Studies on worm population kinetics. *Parasitology Research* **78**, 241–246.

Esbenshade, P.R. and Triantaphyllou, A.C. (1987). Enzymatic relationships and evolution in the genus *Meloidogyne* (Nematoda: Tylenchida). *Journal Of Nematology* **19**, 8–18.

Felder, H., Herzceg, A., De Chastony, Y., Aeby, P., Tobler, H. and Muller, F. (1994). Tas, a retrotransposon from the parasitic nematode *Ascaris lumbricoides. Gene* **149**, 219–225.

Ferris, V.R. (1994). The future of nematode systematics. *Fundamental and Applied Nematology* **17**, 97–101.

Fisher, M.C. (1997). Population genetics of the parasitic nematode *Strongyloides ratti.* Ph.D thesis, University of Edinburgh, UK.

Fisher, M.C. and Viney, M.E. (1996). Microsatellites of the parasitic nematode *Strongyloides ratti. Molecular and Biochemical Parasitology* **80**, 221–224.

Fitch, D.H.A., Bugaj Gaweda, B. and Emmons, S.W. (1995). 18S Ribosomal RNA gene phylogeny for some Rhabditidae related to Caenorhabditis. *Molecular Biology and Evolution* **12**, 346–358.

Flockhart, H.A., Cibulskis, R.E., Karam, M. and Albiez, E.J. (1986). *Onchocerca volvulus*: enzyme polymorphism in relation to the differentiation of forest and savannah strains of this parasite. *Transactions of the Royal Society of Tropical Medicine and Hygiene* **80**, 285–292.

Folkertsma, R.T., Rouppe van der Voort, J.N.A.M., van Gent Pelzer, M.P.E., De Groot, K.E., van den Bos, W.J., Schots, A., Bakker, J. and Gommers, F.J. (1994). Inter- and intraspecific variation between populations of *Globodera rostochiensis* and *G. pallida* revealed by random amplified polymorphic DNA. *Phytopathology* **84**, 807–811.

Frank, G.R., Herd, R.P., Marbury, K.S., Williams, J.C. and Willis, E.R. (1988).

Additional investigations of hypobiosis of *Ostertagia ostertagi* after transfer between northern and southern USA. *International Journal for Parasitology* **18**, 171–178.

Gasnier, N. and Cabaret, J. (1996). Evidence for the existence of a sheep and a goat line of *Teladorsagia circumcincta* (Nematoda). *Parasitology Research* **82**, 546–550.

Gasser, R.B. and Hoste, H. (1995). Genetic markers for closely-related parasitic nematodes. *Molecular and Cellular Probes* **9**, 315–320.

Gasser, R.B., Chilton, N.B., Hoste, H. and Beveridge, I. (1993). Rapid sequencing of rDNA from single worms and eggs of parasitic helminths. *Nucleic Acids Research* **21**, 2525–2526.

Geary, T.G., Nulf, S.C., Favreau, M.A., Tang, L., Prichard, R.K., Hatzenbuhler, N.T., Shea, M.H., Alexander, S.J. and Klein, R.D. (1992). Three beta-tubulin cDNAs from the parasitic nematode *Haemonchus contortus*. *Molecular and Biochemical Parasitology* **50**, 295–306.

Geerts, S., Coles, G.C. and Gryseels, B. (1997). Anthelmintic resistance in human helminths: learning from problems with worm control in livestock. *Parasitology Today* **13**, 149–151.

Grant, W. (1994). Genetic variation in parasitic nematodes and its implications. *International Journal for Parasitology* **24**, 821–830.

Grant, W.N. and Whitington, G.E. (1994). Extensive DNA polymorphism within and between two strains of *Trichostrongylus colubriformis*. *International Journal for Parasitology* **24**, 719–725.

Grosberg, R.K., Levitan, D.R. and Cameron, B.B. (1996). Characterization of genetic structure and genealogies using RAPD-PCR markers: a random primer for the novice and nervous. In: *Molecular Zoology: Advances, Strategies and Protocols* (J.D. Ferraris and S.R. Palumbi, eds), pp. 65–100. New York: Wiley-Liss.

Hadrys, H., Balick, M. and Schierwater, B. (1992). Applications of random amplified polymorphic DNA (RAPD) in molecular ecology. *Molecular Ecology* **1**, 55–63.

Hafner, M.S., Sudman, P.D., Villablanca, F.X., Spradling, T.A., Demastes, J.W. and Nadler, S.A. (1994). Disparate rates of molecular evolution in cospeciating hosts and parasites. *Science* **265**, 1087–1090.

Hahn, M.L., Burrows, P.R., Gnanapragasam, N.C., Bridge, J., Vines, N.J. and Wright, D.J. (1994). Molecular diversity amongst *Radopholus similis* populations from Sri Lanka detected by RAPD analysis. *Fundamental and Applied Nematology* **17**, 275–281.

Hartl, D.L. and Clark, A.G. (1989). *Principles of Population Genetics*. Sunderland, MA: Sinauer Associates, Inc.

Haukisalmi, V., Henttonen, H. and Vikman, P. (1996). Variability of sex ratio, mating probability and egg production in an intestinal nematode in its fluctuating host population. *International Journal for Parasitology* **26**, 755–763.

Herlich, H., Rew, R.S. and Colglazier, M.L. (1981). Inheritance of cambendazole resistance in *Haemonchus contortus*. *American Journal of Veterinary Research* **42**, 1342–1344.

Hillis, D.M. and Dixon, M.T. (1991). Ribosomal DNA: molecular evolution and phylogenetic inference. *Quarterly Review of Biology* **66**, 411–453.

Hillis, D.M. and Moritz, C. (1990). *Molecular Systematics*. Sunderland, MA: Sinauer Associates, Inc.

Hoeh, W.R., Stewart, D.T., Sutherland, B.W. and Zouros, E. (1996). Cytochrome c oxidase sequences comparisons suggest an unusually high rate of mitochondrial DNA evolution in *Mytilus* (Mollusca: Bivalvia). *Molecular Biology and Evolution* **13**, 418–421.

Hoekstra, R., Visser, A., Wiley, L., Weiss, A. and Sangster, N. (1997a). Characterization of an acetylcholine receptor gene of *Haemonchus contortus* in relation to levamisole resistance. *Molecular and Biochemical Parasitology* **84**, 179–187.

Hoekstra, R., Criado-Fornelio, A., Fakkeddij, J., Bergman, J. and Roos, M. (1997). Microsatellites of the parasitic nematode *Haemonchus contortus*: polymorphism and linkage with a direct repeat. *Molecular and Biochemical Parasitology* **89**, 97–107.

Hoste, H., Chilton, N.B., Gasser, R.B. and Beveridge, I. (1995). Differences in the second internal transcribed spacer (ribosomal DNA) between five species of *Trichostrongylus* (Nematoda: Trichostrongylidae). *International Journal for Parasitology* **25**, 75–80.

Hudson, R.R., Kreitman, M. and Aguade, M. (1987). A test of neutral molecular evolution based on nucleotide data. *Genetics* **116**, 153–160.

Hugall, A., Moritz, C., Stanton, J. and Wolstenholme, D.R. (1994). Low, but strongly structured mitochondrial DNA diversity in root knot nematodes (*Meloidogyne*). *Genetics* **136**, 903–912.

Hugall, A., Stanton, J. and Moritz, C. (1997). Evolution of the AT-rich mitochondrial DNA of the Root Knot Nematode *Meloidogyne hapla*. *Molecular Biology and Evolution* **14**, 40–48.

Hughes, M.K. and Hughes, A.L. (1995). Natural selection on *Plasmodium* surface proteins. *Molecular and Biochemical Parasitology* **71**, 99–113.

Hyman, B.C. (1996). Molecular systematics and population biology of phytonematodes: some unifying principles. *Fundamental and Applied Nematology* **19**, 309–313.

Hyman, B.C. and Azevedo, J.L.B. (1996). Similar evolutionary patterning among repeated and single copy nematode mitochondrial genes. *Molecular Biology and Evolution* **13**, 221–232.

Hyman, B.C. and Slater, T.M. (1990). Recent appearance and molecular characterization of mitochondrial DNA deletions within a defined nematode pedigree. *Genetics* **124**, 845–854.

Hyman, B.C. and Whipple, L.E. (1996). Application of mitochondrial DNA polymorphism to *Meloidogyne* molecular population biology. *Journal of Nematology* **28**, 268–276.

Ibrahim, A.P., Conway, D.J., Hall, A. and Bundy, D.A.P. (1994). Enzyme polymorphisms in *Ascaris lumbricoides* in Bangladesh. *Transactions of the Royal Society of Tropical Medicine and Hygiene* **88**, 600–603.

Ibrahim, S.K., Baldwin, J.G., Roberts, P.A. and Hyman, B.C. (1997). Genetic variation in *Nacobbus aberrans*: an approach towards taxonomic resolution. *Journal of Nematology* **29**, 241–249.

Jaenike, J. (1992). Mycophagous *Drosophila* and their nematode parasites. *American Naturalist* **139**, 893–906.

Jaenike, J. (1993). Rapid evolution of host specificity in a parasitic nematode. *Evolutionary Ecology* **7**, 103–108.

Jaenike, J. (1996). Rapid evolution of parasitic nematodes: not. *Evolutionary Ecology* **10**, 565.

Jasmer, D.P., Perryman, L.E. and McGuire, T.M. (1996). *Haemonchus contortus* GA1 antigens: Related, phospholipase C-sensitive, apical gut membrane proteins encoded as a polyprotein and released from the nematode during infection. *Proceedings of the National Academy of Sciences of the United States of America* **93**, 8642–8647.

Jermiin, L.S., Graur, D. and Crozier, R.H. (1995). Evidence from analyses of intergenic regions for strand-specific directional mutation pressure in Metazoan Mitochondrial DNA. *Molecular Biology and Evolution* **12**, 558–563.

Jungersen, G., Eriksen, L., Nielsen, C.G., Roepstorff, A. and Nansen, P. (1996). Experimental transfer of *Ascaris suum* from donor pigs to helminth naive pigs. *Journal of Parasitology 1996* **82**, 752–756.

Kaplan, D.T., Vanderspool, M.C., Garrett, C., Chang, S. and Opperman, C.H. (1996). Molecular polymorphisms associated with host range in the highly conserved genomes of burrowing nematodes, *Radopholus* spp. *Molecular Plant–Microbe Interactions* **9**, 32–38.

Karl, S.A. and Avise, J.C. (1992a). Balancing selection at allozyme loci in oysters: Implications from nuclear RFLPs. *Science* **256**, 100–102.

Karl, S.A., Bowen, B.W. and Avise, J.C. (1992b). Global population genetic structure and male-mediated gene flow in the green turtle (Chelonia mydas): RFLP analyses of anonymous nuclear loci. *Genetics* **131**, 163–173.

Keymer, A.E. and Hiorns, R.W. (1986). *Heligmosomoides polygyrus* (Nematoda): The dynamics of primary and repeated infection in outbred mice. *Proceedings of the Royal Society of London Series B Biological Sciences* **229**, 47–68.

Kimura, M. (1983). *The Neutral Theory of Molecular Evolution.* Cambridge: Cambridge University Press.

Kwa, M.S.G., Kooyman, F.N.J., Boersema, J.H. and Roos, M.H. (1993a). Effect of selection for benzimidazole resistance in *Haemonchus contortus* on beta-tubulin isotype 1 and isotype-2 genes. *Biochemical and Biophysical Research Communications* **191**, 413–419.

Kwa, M.S.G., Veenstra, J.G. and Roos, M.H. (1993b). Molecular characterisation of beta-tubulin genes present in benzimidazole-resistant populations of *Haemonchus contortus*. *Molecular and Biochemical Parasitology* **60**, 133–144.

Kwa, M.S.G., Veenstra, J.G. and Roos, M.H. (1994). Benzimidazole resistance in *Haemonchus contortus* is correlated with a conserved mutation at amino acid 200 in beta-tubulin isotype 1. *Molecular and Biochemical Parasitology* **63**, 299–303.

Kwa, M.S.G., Veenstra, J.G., van Dijk, M. and Roos, M.H. (1995). Beta-tubulin genes from the parasitic nematode *Haemonchus contortus* modulate drug resistance in *Caenorhabditis elegans*. *Journal of Molecular Biology* **246**, 500–510.

Lasserre, F., Gigault, F., Gauthier, J.P., Henry, J.P., Sandmeier, M. and Rivoal, R. (1996). Genetic variation in natural populations of the cereal cyst nematode (*Heterodera avenae*) submitted to resistant and susceptible cultivars of cereals. *Theoretical and Applied Genetics* **93**, 1–8.

le Jambre, L.F., Royal, W.M. and Martin, P.J. (1979). The inheritance of thiabendazole resistance in *Haemonchus contortus*. *Parasitology* **78**, 107–119.

Leslie, J.F., Cain, G.D., Meffe, G.F. and Vrijenhoek, R.C. (1982). Enzyme polymorphism in *Ascaris suum* (Nematoda). *Journal of Parasitology* **68**, 576–587.

Lewontin, R.C. and Krakauer, J. (1973). Distribution of gene frequency as a test of the theory of the selective neutrality of polymorphisms. *Genetics* **74**, 175–195.

Lockhart, P.J., Steel, M.A., Hendy, M.D. and Penny, D. (1994). Recovering evolutionary trees under a more realistic model of sequence evolution. *Molecular Biology and Evolution* **11**, 605–612.

Lubega, G.W. and Prichard, R.K. (1991a). Beta-tubulin and benzimidazole resistance in the sheep nematode *Haemonchus contortus*. *Molecular and Biochemical Parasitology* **47**, 129–138.

Lubega, G.W. and Prichard, R.K. (1991b). Interaction of benzimidazole anthelmintics with *Haemonchus contortus* tubulin: binding affinity and anthelmintic efficacy. *Experimental Parasitology* **73**, 203–213.

Lubega, G.W. and Prichard, R.K. (1991c). Specific interaction of benzimidazole anthelmintics with tubulin from developing stages of thiabendazole-susceptible and -resistant *Haemonchus contortus*. *Biochemical Pharmacology* **41**, 93–102.

Lubega, G.W., Geary, T.G., Klein, R.D. and Prichard, R.K. (1993). Expression of cloned beta-tubulin genes of *Haemonchus contortus* in *Escherichia coli*: Interaction of recombinant beta-tubulin with native tubulin and mebendazole. *Molecular and Biochemical Parasitology* **62**, 281–292.

Lubega, G.W., Klein, R.D., Geary, T.G. and Prichard, R.K. (1994). *Haemonchus contortus*: The role of two beta-tubulin gene subfamilies in the resistance to benzimidazole anthelmintics. *Biochemical Pharmacology* **47**, 1705–1715.

Lunt, D.H. and Hyman, B.C. (1997). End products of animal mitochondrial DNA recombination. *Nature* **387**, 247.

Lynch, M. and Crease, T.J. (1990). The analysis of population survey data on DNA sequence variation. *Molecular Biology and Evolution* **7**, 377–394.

Lynch, M. and Milligan, B.G. (1994). Analysis of population genetic structure with RAPD markers. *Molecular Ecology* **3**, 91–99.

Maizels, R.M., Bundy, D.A.P., Selkirk, M.E., Smith, D.F. and Anderson, R.M. (1993). Immunological modulation and evasion by helminth parasites in human populations. *Nature* **365**, 797–805.

Martin, A.P. (1995). Metabolic rate and directional nucleotide substitution in animal mitochondrial DNA. *Molecular Biology and Evolution* **12**, 1124–1131.

Martin, A.P. and Palumbi, S.R. (1993). Body size, metabolic rate, generation time, and the molecular clock. *Proceedings of the National Academy of Sciences of the United States of America* **90**, 4087–4091.

Martin, P. and McKenzie, J. (1990). Levamisol resistance in *Trichostrongylus colubriformis*: a sex linked character. *International Journal for Parasitology* **20**, 867–872.

May, R.M. (1993). Resisting resistance. *Nature* **361**, 593–594.

May, R.M. and Woolhouse, M.E.J. (1993). Biased sex ratios and parasite mating probabilities. *Parasitology* **107**, 287–295.

McDonald, J.H. (1994). Detecting natural selection by comparing geographical variation in protein and DNA sequences. In: *Non Neutral Evolution: Theories and Molecular Data* (B. Golding, eds.) pp. 88–100. New York: Chapman and Hall.

McDonald, J.H. and Kreitman, M. (1991). Adaptive protein evolution at the Adh locus in *Drosophila*. *Nature* **351**, 652–654.

Melnick, D.J. and Hoelzer, G.A. (1992). Differences in male and female dispersal lead to contrasting distributions of nuclear and mitochondrial DNA variation. *International Journal of Primatology* **13**, 379–393.

Meredith, S.E.O., Lando, G., Gbakima, A.A., Zimmerman, P.A. and Unnasch, T.R. (1991). *Onchocerca volvulus*: Application of the polymerase chain reaction to identification and strain differentiation of the parasite. *Experimental Parasitology* **73**, 335–344.

Moran, N.A., von Dohlen, C.D. and Baumann, P. (1995). Faster evolutionary rates in endosymbiotic bacteria than in cospeciating insect hosts. *Journal of Molecular Evolution* **41**, 727–731.

Moriyama, E.N. and Powell, J.R. (1996). Intraspecific nuclear DNA variation in *Drosophila*. *Molecular Biology and Evolution* **13**, 261–277.

Nadler, S.A. (1986). Biochemical polymorphism in *Parascaris equorum*, *Toxocara canis* and *Toxocara cati*. *Molecular and Biochemical Parasitology* **18**, 45–54.

Nadler, S.A. (1990). Molecular approaches to studying helminth population genetics and phylogeny. *International Journal for Parasitology* **20**, 11–29.

Nadler, S.A. (1995). Microevolution and the genetic structure of parasite populations. *Journal of Parasitology* **81**, 395–403.
Nadler, S.A., Lindquist, R.L. and Near, T.J. (1995). Genetic structure of midwestern *Ascaris suum* populations: A comparison of isoenzyme and RAPD markers. *Journal of Parasitology* **81**, 385–394.
Nascetti, G., Grappelli, C. and Bullini, L. (1979). Ricerche sul differenziamento genetico di *Ascaris lumbricoides* e *Ascaris suum*. *Atti Della Accademia Nazionale Dei Lincei Rendiconti Classe Di Scienze Fisiche Matematiche E Naturali* **67**, 457–465.
Nei, M. (1973). Analysis of gene diversity in subdivided populations. *Proceedings of the National Academy of Sciences of the United States of America* **70**, 3321–3323
Norman, R.J.D.B. and Chilton, N.B. (1994). An electrophoretic comparison of *Physaloptera* Rudolphi, 1819 (Nematoda: Physalopteridae) from two species of Australian bandicoot (Marsupialia: Peramelidae). *Systematic Parasitology* **29**, 223–228.
Okimoto, R., Chamberlin, H.M., Macfarlane, J.L. and Wolstenholme, D.R. (1991). Repeated sequence sets in mitochondrial DNA molecules of root knot nematodes (*Meloidogyne*): Nucleotide sequences, genome location and potential for host-race identification. *Nucleic Acids Research* **19**, 1619–1626.
Okimoto, R., Macfarlane, J.L., Clary, D.O. and Wolstenholme, D.R. (1992). The mitochondrial genomes of two nematodes, *Caenorhabditis elegans* and *Ascaris suum*. *Genetics* **130**, 471–498.
Okimoto, R., Macfarlane, J.L. and Wolstenholme, D.R. (1994). The mitochondrial ribosomal RNA genes of the nematodes *Caenorhabditis elegans* and *Ascaris suum*: Consensus secondary-structure models and conserved nucleotide sets for phylogenetic analysis. *Journal of Molecular Evolution* **39**, 598–613.
Orrechia, P., Paggi, L., Mattiucci, S., Smith, J.W., Nascetti, G. and Bullini, L. (1986). Electrophoretic identification of larvae and adults of *Anisakis* (Ascaridida: Anisakidae). *Journal Of Helminthology* **60**, 331–339.
Orecchia, P., Mattiucci, S., S, D.A., Paggi, L., Plotz, J., Cianchi, R., Nascetti, G., Arduino, P. and Bullini, L. (1994). Two new members in the *Contracaecum osculatum* complex (Nematoda, Ascaridoidea) from the Antarctic. *International Journal for Parasitology* **24**, 367–377.
Paggi, L., Nascetti, G., Cianchi, R., Orecchia, P., Mattiucci, S., D'Amelio, S., Berland, B., Brattey, J., Smith, J.W. and Bullini, L. (1991). Genetic evidence for three species within *Pseudoterranova decipiens* (Nematoda, Ascaridida, Ascaridoidea) in the North Atlantic and Norwegian and Barents Seas. *International Journal for Parasitology* **21**, 195–212.
Palumbi, S.R. and Baker, C.S. (1994). Contrasting population structure from nuclear intron sequences and mtDNA of humpback whales. *Molecular Biology and Evolution* **11**, 426–435.
Paran, I. and Michelmore, R.W. (1993). Development of reliable PCR-based markers linked to downy mildew resistance genes in lettuce. *Theoretical and Applied Genetics* **85**, 985–993.
Philippe, H., Chenuil, A. and Adoutte, A. (1994). Can the Cambrian explosion be inferred through molecular phylogeny? *Development* 15–25.
Poinar, G.O. (1983). *The Natural History of Nematodes*. New Jersey: Prentice-Hall.
Powers, T.O., Harris, T.S. and Hyman, B.C. (1993). Mitochondrial DNA sequence divergence among *Meloidogyne incognita*, *Romanomermis culicivorax*, *Ascaris suum* and *Caenorhabditis elegans*. *Journal of Nematology* **25**, 564–572.
Prichard, R.K. (1990). Anthelmintic resistance in nematodes: extent, recent

understanding and future directions for control and research. *International Journal for Parasitology* **20**, 515–523.

Rand, D.M. and Kann, L.M. (1996). Excess amino acid polymorphism in mitochondrial DNA: Contrasts among genes from *Drosophila*, mice, and humans. *Molecular Biology and Evolution* **13**, 737–748.

Rand, D.M., Dorfsman, M. and Kann, L.M. (1994). Neutral and non-neutral evolution of *Drosophila* mitochondrial DNA. *Genetics* **138**, 741–756.

Remme, J.H.F. (1995). The African Programme for Onchocerciasis Control: Preparing to launch. *Parasitology Today* **11**, 403–406.

Richardson, B.J., Baverstock, P.R. and Adams, M. (1986). *Allozyme Electrophoresis: A Handbook for Animal Systematics and Population Studies.* Sydney: Academic Press.

Roepstorff, A., Bjorn, H., Nansen, P., Barnes, E.H. and Christensen, C.M. (1996). Experimental *Oesophagostomum dentatum* infections in the pig: Worm populations resulting from trickle infections with three dose levels of larvae. *International Journal for Parasitology* **26**, 399–408.

Roos, M.H., Boersma, J.H., Borgsteede, F.M.H., Cornelissen, J. and Taylor, M. (1990). Molecular analysis of selection for benzimidazole resistance in the sheep parasite *Haemonchus contortus. Molecular and Biochemical Parasitology* **43**, 77–78.

Schierwater, B., Streit, B., Wagner, G.P. and DeSalle, R. (1994). *Molecular Ecology and Evolution: Approaches and Applications*, pp. 1–622: Basel, Boston & Berlin: Birkhauser Verlag.

Schinkel, A.H., Smit, J.J.M., Van Telling, O. Beijnen, J. and Wagenaar, E. (1994). Disruption of the mouse *mdr1a* P-glycoprotein gene leads to a deficiency in the blood-brain barrier and to increased sensitivity to drugs. *Cell* **77**, 491–492.

Shorrocks, B., Rosewell, J. and Edwards, K. (1990). Competition on a divided and ephemeral resource: Testing the assumptions: II. Association. *Journal of Animal Ecology* **59**, 1003–1018.

Sironi, M., Bandi, C., Sacchi, L., Di Sacco, B., Damiani, G. and Genchi, C. (1995). Molecular evidence for a close relative of the arthropod endosymbiont *Wolbachia* in a filarial worm. *Molecular and Biochemical Parasitology* **74**, 223–227.

Slade, R.W., Moritz, C. and Heideman, A. (1994). Multiple nuclear-gene phylogenies: Application to pinnipeds and comparison with a mitochondrial DNA gene phylogeny. *Molecular Biology and Evolution* **11**, 341–356.

Smith, N.H., Smith, J.M. and Spratt, B.G. (1995). Sequence Evolution of the porB Gene of *Neisseria gonorrhoeae* and *Neisseria meningitidis*: Evidence of Positive Darwinian Selection. *Molecular Biology and Evolution* **12**, 363–370.

Stevenson, L.A., Gasser, R.A. and Chilton, N.B. (1996). The ITS-2 rDNA of *Teladorsagia circumcincta*, *T. trifurcata* and *T. davtiani* (Nematoda: Trichostrongylidae) indicates that these taxa are one species. *International Journal for Parasitology* **26**, 1123–1126.

Tajima, F. (1989). Statistical method for testing the neutral mutation hypothesis by DNA polymorphism. *Genetics* **123**, 585–596.

Taylor, M.F.J., Shen, Y. and Kreitman, M.E. (1995). A population genetic test of selection at the molecular level. *Science* **270**, 1497–1499.

Thomas, W.K. and Wilson, A.C. (1991). Mode and tempo of molecular evolution in the nematode *Caenorhabditis*: Cytochrome oxidase II and calmodulin sequences. *Genetics* **128**, 269–280.

Toe, L., Merriweather, A. and Unnasch, T.R. (1994). DNA probe-based classification of *Simulium damnosum* S. L.-borne and human-derived filarial parasites in

the onchocerciasis control program area. *American Journal of Tropical Medicine and Hygiene* **51**, 676–683.
Toe, L., Tang, J., Back, C., Katholi, C.R. and Unnasch, T.R. (1997). Vector-parasite transmission complexes for onchocerciasis in West Africa. *Lancet* **349**, 163–166.
Triantaphyllou, A.C. and Esbenshade, P.R. (1990). Demonstration of multiple mating in *Heterodera glycines* with biochemical markers. *Journal of Nematology* **22**, 452–456.
Vanfleteren, J.R., van de Peer, Y., Blaxter, M.L., Tweedie, S.A.R., Trotman, C., Lu, L., van Hauwaert, M.L. and Moens, L. (1994). Molecular genealogy of some nematode taxa as based on cytochrome c and globin amino acid sequence. *Molecular Phylogenetics and Evolution* **3**, 92–101.
Viney, M.E. (1994). A genetic analysis of reproduction in *Strongyloides ratti*. *Parasitology* **109**, 511–515.
Viney, M.E., Matthews, B.E. and Walliker, D. (1992). On the biological and biochemical nature of cloned populations of *Strongyloides ratti*. *Journal of Helminthology* **66**, 45–52.
Viney, M.E., Matthews, B.E. and Walliker, D. (1993). Mating in the nematode parasite *Strongyloides ratti*: Proof of genetic exchange. *Proceedings of the Royal Society of London Series B Biological Sciences* **254**, 213–219.
Vrijenhoek, R.C. (1978). Genetic differentiation among larval nematodes infecting fishes. *Journal of Parasitology* **64**, 790–798.
Wakelin, D. (1995). Vaccines against intestinal helminths. In: *Enteric Infection 2. Intestinal Helminths* (M.J.G. Farthing, eds), pp. 287–297. Chapman and Hall.
Ward, R.D., Skibinski, D.O.F. and Woodwark, M. (1992). Protein heterozygosity, protein structure and taxonomic differentiation. In: *Evolutionary Biology, Volume 26* (M.K. Hecht, B. Wallace and R.J. MacIntyre, eds) pp. 73–159. New York: Plenum Press.
Weir, B. S. and Cockerham, C. C. (1984). Estimating F-statistics for the analysis of population structure. *Evolution* **38**, 1358–1370.
Werren, J.H. (1997a). Biology of *Wolbachia*. *Annual Review of Entomology* **42**, 587–609.
Werren, J.H. (1997b). *Wolbachia* and speciation. In: *Endless Forms: Species and Speciation* (D. Howard and S. Berlocher, eds). Oxford: Oxford University Press (in press).
Wright (1951). The genetical structure of populations. *Annals of Eugenics* **15**, 323–354.
Zarlenga, D.S., Aschenbrenner, R.A. and Lichtenfels, J.R. (1996). Variations in Mricrosatellite sequences provide evidence for population differences and multiple ribosomal gene repeats within *Trichinella pseudospiralis*. *Journal of Parasitology* **82**, 534–538.
Zimmerman, P.A., Dadzie, K.Y., De Sole, G., Remme, J., Alley, E.S. and Unnasch, T.R. (1992). *Onchocerca volvulus* DNA probe classification correlates with epidemiologic patterns of blindness. *Journal of Infectious Diseases* **165**, 964–968.
Zimmerman, P.A., Toe, L. and Unnasch, T.R. (1993). Design of *Onchocerca* DNA probes based upon analysis of a repeated sequence family. *Molecular and Biochemical Parasitology* **58**, 259–267.
Zimmerman, P.A., Katholi, C.R., Wooten, M.C., Lang Unnasch, N. and Unnasch, T.R. (1994). Recent evolutionary history of American *Onchocerca volvulus*, based on analysis of a tandemly repeated DNA sequence family. *Molecular Biology and Evolution* **11**, 384–392.
Zvelebil, M.J.J.M., Tang, L., Cookson, E., Selkirk, M.E. and Thornton, J.M. (1993). Molecular modelling and epitope prediction of gp29 from lymphatic filariae. *Molecular and Biochemical Parasitology* **58**, 145–153.

Schistosomiasis in Cattle

Jan De Bont and Jozef Vercruysse

Department of Parasitology, Faculty of Veterinary Medicine, University of Gent, Salisburylaan 133, 9820 Merelbeke, Belgium

1. Introduction 286
2. Schistosomes and Their Life Cycles 286
   2.1. Schistosome species and interactions 286
   2.2. Geographical distribution and occurrence. 290
   2.3. Life cycle of cattle schistosomes. 292
   2.4. Taxonomic features of schistosomes infecting cattle 295
3. Pathology and Pathophysiology 303
   3.1. Introduction 303
   3.2. Disease spectrum 303
   3.3. Clinical pathology and pathophysiology 305
   3.4. Pathology 310
4. Diagnosis. 315
   4.1. Diagnostic methods 315
   4.2. Measures of infection at the population level 320
5. The Epidemiology of Cattle Schistosomiasis 320
   5.1. Introduction 320
   5.2. Levels of infection in cattle populations 321
   5.3. Host–parasite relationship and immunity development 322
   5.4. Transmission 334
6. Treatment and Control. 337
   6.1. Treatment 337
   6.2. Control 339
7. Conclusion and Perspectives 345
Acknowledgements 345
References 345

1. INTRODUCTION

Schistosome infection is common in cattle in Africa and Asia. Although schistosomes may, in rare circumstances (where intensive transmission is present), act as important pathogens *per se* (examples include Kulkarni *et al.*, 1954; van Wyk *et al.*, 1974; Wang *et al.*, 1988; Chen, 1993; Markovics *et al.*, 1993), most infections in endemic areas occur at a subclinical level. However, it has been established that high rates of prevalence of subclinical infections cause significant losses due to long-term effects on animal growth and productivity and increased susceptibility to other parasitic or bacterial disease (Dargie, 1980; Pitchford and Visser, 1982; McCauley *et al.*, 1983, 1984). Despite this, schistosomes of veterinary concern have received relatively little attention compared with species affecting man, although reviews on particular species of schistosomes infecting cattle or on distinct aspects of the infection have been published (Hussein, 1973; Lawrence, 1978a; Christensen *et al.*, 1983; Kumar and de Burbure, 1986; Taylor, 1987; Agrawal and Alwar, 1992; Aradaib and Osburn, 1995; Aradaib *et al.*, 1995a; De Bont and Vercruysse, 1997). In recent years, new insights have been gained in various domains of the parasitic infections, thanks mainly to studies on the host–parasite relationships in natural infections and to investigations on potential vaccine candidates. Therefore, the aim of this chapter is to provide an updated and comprehensive review on the main features of cattle schistosomiasis.

2. SCHISTOSOMES AND THEIR LIFE CYCLES

2.1. Schistosome Species and Interactions

Schistosomes are members of the genus *Schistosoma* which belongs to the family Schistosomatidae. The adult worms are obligate parasites of the blood vascular system of vertebrates. Schistosomes are dioecious. The mature female is more slender than the male and normally carried in a ventral groove, the gynaecophoric canal, which is formed by ventrally flexed lateral outgrowths of the male body. A total of 19 different species are described worldwide (Table 1). They live in the perivesical (*S. haematobium*), nasal (*S. nasale*) or mesenteric and hepatic (remaining species) veins of the host where they feed on blood and produce nonoperculated eggs with a characteristic terminal or lateral spine. As many as 10 different species of schistosomes have been reported to naturally infect cattle: *S. mattheei*, *S. bovis* and *S. curassoni* in Africa and *S. spindale*, *S. indicum*,

S. nasale and *S. japonicum* in Asia; the other three species are primarily parasites of antelopes (*S. margrebowiei* and *S. leiperi* in Africa) or pigs (*S. incognitum* in Asia) which only occasionally infect cattle. *Schistosoma japonicum* is an important zoonosis.

There is little information on the ability of the other human pathogens *S. haematobium*, *S. mansoni* and *S. intercalatum* to infect cattle. Massoud and Nelson (1972) observed that in calves experimentally exposed to large doses of *S. haematobium* (21 000 cercariae) about 2% of the worms developed to the adult stage. However, eggs were only recovered from the tissues of one animal. They were all deformed and blackened and contained no miracidia. Saeed *et al.* (1969) successfully infected two calves with a Puerto Rican strain of *S. mansoni* (10 000 cercariae). Viable eggs were passed in the faeces and the worm recovery at autopsy was less than 1.5%. However, except for a report from Barbosa *et al.* (1962) who found adult *S. mansoni* in four of 29 cattle slaughtered in an abattoir in Brazil, there is no evidence of *S. mansoni* occurring naturally in cattle. *Schistosoma intercalatum* has never been reported from cattle in the field. A calf experimentally exposed to approximately 1750 *S. intercalatum* cercariae passed viable eggs 53 days post-infection (p.i.) but no worms were recovered during post-mortem examination 160 days p.i. (Wright *et al.*, 1972).

Some schistosome species are known to interact in areas where they coexist and instances of interspecific hybridization have been reported. For example, the cattle parasite *S. mattheei* may occasionally infect man (Pitchford, 1959; Hira, 1975; Hira and Patel, 1981; Chandiwana *et al.*, 1987a) where it has been shown to hybridize with the human species *S. haematobium* (Pitchford, 1961; Wright and Ross, 1980; Kruger *et al.*, 1986a,b). Further enzyme studies suggested that all *S. mattheei*-like human infections were the product of pairing between male *S. haematobium* and female *S. mattheei* (Kruger and Evans, 1990). However, there is little information on the ability of *S. mattheei* × *S. haematobium* hybrids to infect cattle. Pitchford (1961) isolated *S. mattheei*-like eggs from the urine of a girl and, after three generations in laboratory animals, successfully infected a calf. During a prevalence survey conducted in Zambia, 11% of 542 adult worms collected from 104 cattle showed heterozygote electrophoretic patterns, indicative of hybrids (De Bont *et al.*, 1994). Two different origins were suggested for the hybrid worms collected in Zambia. They could be F_1 *S. mattheei* × *S. haematobium* hybrids, probably originating from humans since *S. haematobium* is not known to infect cattle, or the worms could also be F_1 *S. mattheei* × *S. leiperi* hybrids. Pitchford (1974) suggested the existence of natural hybridization between *S. mattheei* and *S. leiperi* in wild ruminants, but the hybrid has never been produced experimentally. However, as both species occur in mixed natural

Table 1 *Schistosoma* species, their snail hosts, adult female length, uterine egg counts, egg size (length × width), mammalian Orders containing important definitive hosts and prepatent period (P.P.P.) (Rollinson and Southgate, 1987a; measurements based on Loker, 1983).

Species	Snail host genus	Female length (mm)	No. of uterine eggs (mean)	Egg size (μm)	Definitive hosts Orders	P.P.P. (days)
S. haematobium group						
S. mattheei (Veglia and Le Roux, 1929)	*Bulinus*	17–25	5–42 (26)	173 × 53	Artiodactyla Primates	42
S. bovis (Sonsino, 1876; Blanchard, 1895)	*Bulinus* *Planorbarius*	13–44	5–62 (29)	202 × 58	Artiodactyla	41
S. curassoni (Brumpt, 1931)	*Bulinus*	18.3–25.7	47–65 (50)	149 × 63	Artiodactyla	40
S. margrebowiei (Le Roux, 1933)	*Bulinus*	20–33.8	30–205 (130)	87 × 62	Artiodactyla	33
S. leiperi (Le Roux, 1955)	*Bulinus*	7–14.8	6–17 (12)	270 × 53	Artiodactyla	49
S. haematobium (Bilharz, 1852; Weinland, 1858)	*Bulinus*	13.5–22.5	4–56 (29)	144 × 58	Primates	56
S. intercalatum (Fisher, 1934)	*Bulinus*	13–28	12–54 (30)	175 × 62	Primates	41
S. indicum group						
S. indicum (Montgomery, 1906a)	*Indoplanorbis*	4.9–26.4	86	122 × 57	Artiodactyla	52
S. spindale (Montgomery, 1906b)	*Indoplanorbis*	7.2–16.2	4–5	382 × 70	Artiodactyla	46
S. nasale (Rao, 1933)	*Indoplanorbis*	6.9–11.7	0–2 (1)	456 × 66	Artiodactyla	77
S. incognitum (Chandler, 1926)	*Lymnaea Radix*	2.6–7.6	1	116 × 60	Rodentia Artiodactyla[a,b]	35
S. japonicum group						
S. japonicum (Katsurada, 1904)	*Oncomelania*	15–30	(161)	81 × 63	Primates Rodentia Artiodactyla[a]	34

Table 1 Continued

Species	Snail host genus	Female length (mm)	No. of uterine eggs (mean)	Egg size (μm)	Definitive hosts Orders	P.P.P. (days)
S. mekongi (Voge et al., 1978)	Neotricula	14.5–20.1	120–130	66 × 58	Primates[a]	43
S. sinensium (Pao, 1959)	Neotricula	3.3–3.8	1	105 × 45	Rodentia	—
S. malayensis (Greer et al., 1988)	Robertsiella	6.5–11.3	numerous	67 × 54	Primates	30
S. mansoni group						
S. mansoni (Sambon, 1907)	Biomphalaria	7.2–14	1	142 × 60	Primates	34
S. rodhaini (Brumpt, 1931)	Biomphalaria	3.0–10.5	1	149 × 55	Rodentia[a]	30
S. edwardiense (Thurston, 1964)	Biomphalaria	2.9–5.9	1	62 × 53	Artiodactyla	—
S. hippopotami (Thurston, 1963)	?	3.3–4.7	1	93 × 40	Artiodactyla Rodentia	—

[a] Also in Carnivora.
[b] Also in Perissodactyla.

infections in cattle in Zambia (De Bont *et al.*, 1994), clearly the opportunity for cross-mating exists. In West Africa, Southgate *et al.* (1985) and Rollinson *et al.* (1987) suggested the natural occurrence of *S. curassoni* × *S. bovis* pairings in sheep and cattle, respectively. Subsequently, viable *S. curassoni* × *S. bovis* hybrid parasites were produced in the laboratory and were maintained up until the F_4 generation. They resulted from both crossings between male *S. curassoni* and female *S. bovis*, and between female *S. curassoni* and male *S. bovis*. Comparison of these experimental hybrid lines and suspected hybrids isolated from naturally infected cattle confirmed the occurrence of natural hybridization between these two species (Rollinson *et al.*, 1990a).

Taylor (1970) produced in mice and hamsters some hybrids which have never been observed in the field. For example, the cross *S. mattheei* male × *S. bovis* female produced some eggs for three generations in mice and the offspring had morphological characteristics intermediate between those of the parental species. The cross *S. bovis* male × *S. mattheei* female was less fertile. Hybrids of *S. bovis* and *S. mattheei* were found to be particularly difficult to produce, and Taylor (1970) discussed that if hybridization occurred between the two species in areas where their distribution overlaps (e.g. in Tanzania), the offspring would be eliminated by natural selection because of their low viability and infectivity. Hybridization between *S. bovis* females and *S. haematobium* males was found to be as easy as that of *S. mattheei* females and *S. haematobium* males. It was suggested that the two species were kept isolated in nature by differences in definitive host specificity (Taylor, 1970).

2.2. Geographical Distribution and Occurrence

The geographical distribution of schistosome species infecting cattle is mainly determined by the distribution of their respective intermediate host snails. *Schistosoma bovis* is found in the Mediterranean region (Spain, Italy, Sicily, Sardinia and Corsica), the Middle East (Iran, Iraq, Israel), and is common in northern, western and eastern Africa (except Egypt), extending southwards to central Angola and southern Zaire (reviewed by Hussein, 1973; Southgate and Knowles, 1975; Pitchford, 1977). The southern range of *S. bovis* and the northern limits of *S. mattheei* are believed to overlap in Tanzania and possibly Zambia (Dinnik and Dinnik, 1965). However, the presence of *S. bovis* in Zambia could not be confirmed by De Bont *et al.* (1994).

Schistosoma mattheei is found in south-eastern Africa, from the Cape Province in South Africa northwards to Tanzania and Zambia (reviewed by Hussein, 1973; Lawrence, 1987a). Isolated reports of *S. mattheei* much further north in Nigeria (Ramsay, 1934) and Chad (Birgi and Graber, 1969) would require further investigation for confirmation. The range of *S. curassoni* overlaps that of *S. bovis*. The parasite has been found in ruminants in Senegal, The Gambia, Mauritania, Mali, Niger and Nigeria (reviewed by Rollinson *et al.*, 1990a). *Schistosoma margrebowiei* is found in northern Botswana, Namibia (eastern Caprivi Strip), Zambia, southern Zaire and Chad (reviewed by Pitchford, 1977). The distribution of *S. leiperi* includes Botswana, Namibia (eastern Caprivi), Zambia, Tanzania (Pitchford, 1976), and possibly Uganda (Malek and Ongom, 1984) and Sudan (Rollinson and Southgate, 1987).

Schistosoma spindale infections have been recorded from India, Sri Lanka, Indonesia, Malaysia, Thailand, Laos and Vietnam (reviewed by Kumar and de Burbure, 1986). The distribution of *S. indicum* is apparently confined to the Indian subcontinent (reviewed by Kumar and de Burbure, 1986). *Schistosoma nasale* is found in India, Sri Lanka, Pakistan, Bangladesh, Burma (Myanmar) (reviewed by Agrawal and Alwar, 1992), and Malaysia (Saharee *et al.*, 1984; J. Vercruysse, personal communication). *Schistosoma incognitum* has been reported from India, Thailand and Indonesia (reviewed by Agrawal and Shah, 1989). *Schistosoma japonicum* is endemic in mainland China, Taiwan, the Philippines and Central Sulawesi in Indonesia (reviewed by Chen, 1993).

In endemic areas, the occurrence of cattle schistosomes is known to be highly focal because of the underlying aggregated distribution of intermediate snail hosts and the somewhat restricted stock movement from one farm to another (Christensen *et al.*, 1983). Cattle generally move as a herd and therefore individuals from the same herd are likely to be exposed to similar ranges of cercarial challenge. This means prevalence studies at the farm level are of limited value. About 530 million head of cattle live in areas known to be endemic for cattle schistosomiasis in Africa and Asia (De Bont and Vercruysse, 1997). Only a limited number of prevalence studies have been conducted at a regional or national level, of which a few included worm detection in animals from as many different herds as possible, and were carried out in abattoirs. Reports from Zimbabwe (Condy, 1960), Zambia (De Bont *et al.*, 1994), Zaïre (Chartier *et al.*, 1990), West Africa (Diaw and Vassiliades, 1987; Rollinson *et al.*, 1990a), Sri Lanka (De Bont *et al.*, 1991a) and Bangladesh (Islam, 1975) have shown overall infection rates of between 31 and 81% in cattle populations. From this, De Bont and Vercruysse (1997) speculated that at least 165 million cattle are infected with schistosomes worldwide.

2.3. Life Cycle of Cattle Schistosomes

The paired adult worms live in the nasal (*S. nasale*) or mesenteric (other species) veins of the host where they produce eggs which contain embryonic cells (the miracidium takes about 1 week to develop after the eggs have been extruded from the female worm). The determination of daily egg production by schistosomes within a host is a very imprecise exercise (Sturrock, 1966; Basch, 1991). Estimates of daily egg production in laboratory animals were tabulated by Loker (1983) who found that the rate at which different schistosome species produce eggs correlated positively with uterine egg counts. The factors which influence the number of eggs produced by a female schistosome have been discussed by Basch (1991). In addition to intrinsic characteristics of the worms, other factors such as density dependence of fecundity and host immunity may play important roles (see Section 5.3.2.). Known estimates of daily egg productions per pair in laboratory animals are listed in Table 2. In cattle, the egg production per female per day has to our knowledge never been calculated, but estimates of numbers of eggs passed in the faeces (per female worm and per day) have been published. In calves artificially infected with *S. bovis* the number of eggs found in faeces per female worm per day declined from 106 eggs 63 days p.i. to 67 eggs 126 days p.i. (Massoud, 1973). The daily faecal egg output during the early weeks after an artificial infection with *S. mattheei* was estimated to range from 640 to 3285 eggs per female worm (Lawrence, 1973a).

Eggs laid in the blood vessels penetrate the wall of the veins and migrate to the intestinal lumen, or the nasal cavity (*S. nasale*) of the host. This passage through the vein wall and tissues of the host is aided by the release of lytic enzymes secreted by the miracidium contained in the egg, a process possibly aided by the inflammatory and immune responses of the host (Doenhoeff *et al.*, 1978; Damian, 1987). The migration takes several days to weeks, during which time nonembryonated eggs develop to maturity.

Table 2 Published estimates of daily egg production per schistosome female worm in hamsters.

Schistosoma species	Eggs per female worm per day	Reference
S. bovis	21–267	Southgate and Knowles (1975)
S. mattheei	99–268	Wright *et al.* (1972)
S. curassoni	87–340	Southgate *et al.* (1985)
S. margrebowiei	189–1450	Southgate and Knowles (1977)
S. japonicum	2980–4530	Moore and Sandground (1956)

Many eggs are carried away in the blood stream and/or trapped in the tissues, and die. Eggs which reach the lumen are either passed in the faeces (intestinal schistosomes), or disseminated while drinking or sneezing (*S. nasale*) and hatch only if they come in contact with fresh water. The passage of eggs in the urine has been observed in cattle suffering from heavy *S. mattheei* infection (Alves, 1953; McCully and Kruger, 1969; Bartsch and Van Wyk, 1977) (see Section 3.4.1).

The miracidium released from the egg represents the first free larval stage in the life cycle of schistosomes, ensuring transmission between the vertebrate and the snail. The larva is ciliated and mobile, and remains viable for up to 16–30 h (*S. bovis*; Lengy, 1962). It shows negative geotactism and is attracted by light and chemical stimuli originating from the potential snail host, which it actively penetrates. Continuation of the life-cycle depends upon the miracidium infecting an appropriate snail (Table 1). Numerous snail compatibility studies aimed at determining the natural and potential intermediate hosts of different parasite species and strains have been reported. Different parameters including infection rates, duration of infection, total cercarial production per snail, daily average cercarial production per snail and snail mortality have been used, and their validity as indicators of the host–parasite compatibility has been discussed (Frandsen, 1979; Combes, 1985; Mouahid and Théron, 1987). In an attempt to quantify the degree of compatibility, Frandsen (1979) proposed an index that involves the evaluation of total cercarial production from 100 exposed snails. The test incorporates measures of the infection rate, level and duration of cercarial production, and survival of the snails, but makes no reference to the possible effect of the miracidial dose on the parameters included in the formula. Miracidial dose has been demonstrated by several authors to affect the infection rate, snail mortality and/or cercarial production (see Lwambo *et al.*, 1987; Mouahid and Combes, 1987). Another disadvantage of the compatibility test proposed by Frandsen (1979) is the considerable amount of work it represents. Various attempts have been made to simplify the procedure without loosing too much in sensitivity (Théron, 1981; Combes, 1985). However, there still appears to be no consensus on a compatibility index which standardizes quantitative comparisons.

Current knowledge on the intramolluscan development, which includes asexual multiplication of daughter sporocysts in the digestive gland of the snail and eventually results in the emergence of numerous cercariae, will not be discussed here, as it is outside the scope of the present review (for reviews, see Jourdane and Théron, 1987; Basch, 1991; Sturrock, 1993a). However, one important aspect is the length of the prepatent period in the molluscan host, which is highly dependent upon environmental temperature, the size and age of the snail, the number of miracidia penetrating, host environmental factors, such as nutrition and crowding, and the overall compatibility between schistosome and snail. For example, in South Africa the prepatent period of *S. mattheei* in *Bulinus* snails was found to vary from a mean of 38 days in the summer (minimum 34 days) to 126 days in the winter (maximum 154 days) (Pitchford and Visser, 1969).

The free-swimming cercariae have a typical bifurcated tail and can remain viable for up to 40 h (*S. bovis*; Lengy, 1962), although their infectivity diminishes rapidly. The daily cercarial productivity, which rarely exceeds 2000 cercariae per snail (Webbe, 1965; Pitchford *et al.*, 1969; De Bont *et al.*, 1991b), is a function of the same biotic and abiotic factors as those affecting the length of prepatent period. For most species, the rhythm of cercarial emission is closely correlated with the periods of activity of the most permissive host. Schistosome parasites of man and cattle have a diurnal emission rhythm. For *S. mattheei*, nocturnal shedding also occurs during the summer (Pitchford *et al.*, 1969). Cercarial emission of *S. margrebowiei*, which shows a first peak at dawn and a second one at dusk, is perfectly adapted to the antelope which come to water pools early in the morning and late in the evening (Pitchford and Du Toit, 1976).

Infection of ruminants by cercariae is usually accomplished by penetration of the skin, although it has been shown that peroral infection while drinking may also be of importance (Fairley and Jasudasan, 1930; van Wyk *et al.*, 1974; Kassuku *et al.*, 1985). After penetrating the skin or mucosa the cercariae shed their tail and transform to schistosomula which migrate to the lungs, probably via the blood vessels. The route of migration of the schistosomula from the lungs to the hepatic portal system has been a subject of controversy for many years. From studies involving laboratory animals infected with the schistosomes of man, evidence has accumulated in favour of the intravascular route in the direction of the blood flow (reviewed by Wilson, 1987; Basch, 1991), with the schistosomula making one or several circuits of the vasculature before either being trapped in the liver or lodging in a location unfavourable for development. However, Kruger *et al.* (1969) studied the route of migration of *S. mattheei* in sheep and found no schistosomula in the left side of the heart, pulmonary vein, and aorta; these authors concluded that schistosomula moved against the bloodstream via the pulmonary artery, right ventricle and atrium, posterior vena cava, and hepatic vein into the liver. Further work is necessary to clarify the situation and determine the route of migration in cattle. The route of migration of the *S. nasale* to the veins of the nasal cavity has never been studied.

Visceral schistosomes mature in the hepatic portal veins, mate and migrate to the mesenteric veins where egg production starts. Pairing appears to be of paramount importance to the development and sexual maturation of the female worm (Popiel, 1986). Female *S. mattheei* worms from unisexual infections remain much smaller and their vitellaria are less developed than in female worms from mixed infections, but some are still capable of producing eggs (albeit infertile) (Taylor et al., 1969; Taylor, 1971). The matraclonal offspring produced after pairing S. mattheei females with S. mansoni males, two species which are not closely related, was interpreted as parthenogenetic reproduction (Taylor et al., 1969). Finally, the prepatent period in the mammalian host represents the time from infection to the first appearance of eggs in the host's excreta. Its minimum value is generally regarded as characteristic for the species (Table 1).

2.4. Taxonomic Features of Schistosomes Infecting Cattle

Schistosome species can be differentiated through such taxonomic features as morphological, life cycle or behavioural characteristics, chromosomes, host specificity or enzyme and DNA studies (Rollinson and Southgate, 1987). Some of the main features which are used for differentiating species and strains naturally infecting cattle are considered below.

2.4.1. Adult Worms

(a) *Morphological characteristics.* At autopsy, adult worm pairs are clearly visible in the mesenteric veins of infected cattle, and can easily be collected by dissection for closer examination. The size of the adult female worm varies from 2.6 (*S. incognitum*) to 34 mm (*S. margrebowiei*) (Loker, 1983) and may be a useful distinguishing character (Table 1). However, some caution should be exercised in treating the length of an adult female worm as typical because experimental studies in cattle have demonstrated that the size of adult females decreases in immune animals (Hussein et al., 1970; Preston and Webbe, 1974; Lawrence, 1977a). Lawrence (1977a) observed that, in paired worms collected soon after experimental infection with *S. mattheei*, female worms were protruding at both ends of the gynaecophoric canal of the males whereas, in later stages, most females were contained within the groove. The number and shape of eggs found *in utero* (Table 1) may also provide some indications on the identity of the female worm. For example, adult *S. margrebowiei* females normally have many more eggs (mean 130) *in utero* than adult *S. mattheei* (mean 26) or *S. leiperi* (mean 12) females (Loker, 1983).

Studies on the surface morphology using scanning electron microscopy have shown that interspecific differences occur on the dorsal and dorsolateral surfaces of the adult male worms, including the presence or absence of tubercles, and the degree of spination of the tubercles. Three general surface types are recognized: nontuberculate — *S. japonicum* (Sakamoto and Ishii, 1977) and *S. spindale* (Kruatrachue *et al.*, 1983); tuberculate with spines — *S. curassoni* (Southgate *et al.*, 1986) and *S. incognitum* (Kruatrachue *et al.*, 1982); and tuberculate without spines — *S. bovis* (Kuntz *et al.*, 1979; Southgate *et al.*, 1986), *S. mattheei* (Tulloch *et al.*, 1977) and *S. margrebowiei* (Ogbe, 1982). However, following earlier studies on *S. leiperi* in which individual male worms were found either with or without spines, Southgate *et al.* (1981) warned that intraspecific variation of surface topography may occur, and therefore argued for caution in using the structure of tubercles as a taxonomic criteria. The existence of such intraspecific variation in surface topography was later confirmed in *S. bovis* (Ngendahayo *et al.*, 1987), *S. mattheei* (Kruger *et al.*, 1986b, 1988), *S. margrebowiei* (Probert and Awad, 1987; Kruger *et al.*, 1988), and *S. nasale* (Southgate *et al.*, 1990). Probert and Awad (1987) and Southgate *et al.* (1990) observed that unpaired males lacked spined tubercles, whereas some paired males had tubercles with spines. Further studies are required to establish whether differences between isolates of a same species in the morphology of the male teguments represent intraspecific variation or are the result of examining worms at different stages of development or physiological state.

Hybridization of schistosomes has been clearly shown to induce variation in the morphology of the male teguments. For example, the tubercles of the F_1 hybrids of *S. haematobium* (heavily spined) and *S. mattheei* (spineless) show variation ranging from forms without spines to those with spines, but the majority of hybrid parasites were spined (Tchuem Tchuenté *et al.*, 1997). This supports earlier work by Kruger *et al.* (1986b) who postulated that the presence of tubercle spines on *S. mattheei* males from populations which are sympatric to *S. haematobium* in South Africa was a characteristic inherited from *S. haematobium*. Similar inheritance of spination via hybridization has been shown between *S. bovis* (spineless) and *S. curassoni* (spined) by Rollinson *et al.* (1990a). The tubercles of adult male F_1 worms possessed small numbers of stunted spines, while the examination of worms through to the F_4 generation revealed a mixed population with various degrees of spination, the majority of male worms being of intermediate form (Rollinson *et al.*, 1990a).

(b) *Enzyme studies.* Several studies have demonstrated that enzyme electrophoresis technique (starch-gel, isoelectric focusing or cellulose acetate membrane) are reliable tools for characterising schistosomes (see Rollinson and Southgate, 1985), particularly in areas where two or more species overlap in their distribution. In West Africa, detailed comparative enzyme analyses by isoelectric focusing of isolates of *S. haematobium*, *S. bovis* and *S. curassoni* have revealed different profiles for four out of seven enzymes examined: acid phosphatase (AcP), phosphoglucomutase (PGM), hexokinase (HK) and glucose phosphate isomerase (GPI) (Southgate *et al.*, 1985).

Rollinson et al. (1990a) was able to use isoelectric focusing (for AcP, PGM and glucose-6-phosphate dehydrogenase, G6PDH) to characterize worms collected from cattle in abattoirs in Senegal, The Gambia and Mali. In southern Africa, G6PDH and PGM allow the differentiation between *S. mattheei, S. bovis, S. leiperi* and *S. margrebowiei* (Ross et al., 1978; Mahon and Shiff, 1978; Southgate et al., 1981; Kruger, 1987). De Bont et al. (1994) determined the relative occurrence of different schistosome species in the Zambian cattle population by using electrophoretic techniques on worms collected from cattle in abattoirs. A number of enzymes, including G6PDH and PGM have also been employed to study intraspecific variation between isolates of a particular species, such as *S. bovis* (Southgate et al., 1980), *S. japonicum* (Fletcher et al., 1980), *S. mattheei* (Kruger, 1988, 1989) or *S. curassoni* (Rollinson et al., 1990a).

Enzyme analyses have been particularly helpful in studies on schistosome hybrids occurring in cattle. Wright and Ross (1980) used the differing isoelectric focusing patterns of G6PDH and PGM in *S. haematobium* and *S. mattheei* to study *S. mattheei* × *S. haematobium* hybrids that were laboratory-produced or isolated from human infections in South Africa. The patterns obtained from F_1 laboratory hybrids were the combination of the two distinctive sets of bands observed in the parental species. In addition, some of the worms isolated from human infections produced the same bands, providing unequivocal confirmation of the occurrence in man of natural hybridization between the two species. The same isoenzymes were later used by Kruger and Evans (1990), who suggested that all *S. mattheei* eggs passed in the urine of humans derive from *S. mattheei* females *in copula* with *S. haematobium* males. Recently, Tchuem Tchuenté et al. (1997) noted polymorphism in the phenotypes of PGM enzyme in laboratory-produced F_1 *S. mattheei* × *S. haematobium* hybrid populations, which was attributed to inter-individual variation that exists in *S. haematobium* (Wright and Ross, 1983). Rollinson et al. (1990a) produced experimental hybrids between *S. curassoni* and *S. bovis*, and confirmed their hybrid nature by enzyme analysis of AcP. Similar analyses of adult worms collected from cattle in Senegal and Mali confirmed the occurrence of natural hybridization between *S. curassoni* and *S. bovis* in West Africa (Rollinson et al., 1990a).

(c) *DNA studies.* Considerable developments have occurred in this area of biology in recent years. A comprehensive review on the progress made in understanding the relationships between species of the genus *Schistosoma* has recently been published by Rollinson et al. (1997). Particular attention was given to the detection and analysis of parasite variation, as shown by studies on ribosomal RNA genes, mitochondrial DNA and randomly amplified polymorphic DNA (RAPDs). It is not our intention here to duplicate the above review, but to present the techniques of DNA analysis which may prove useful for the rapid identification of schistosomes infecting cattle.

Using cloned rRNA probes derived from *S. mansoni* (Simpson *et al.*, 1984), Walker *et al.* (1986) produced detailed restriction site maps of the rRNA genes of six species of schistosomes with terminal spined eggs: *S. haematobium, S. curassoni, S. bovis, S. intercalatum, S. margrebowiei* and *S. mattheei*. Comparison of species showed that the transcribed spacer in *S. mattheei* and *S. margrebowiei* contained inserts (additional DNA) of 0.2 and 0.1 kb, respectively. *Schistosoma haematobium* and *S. margrebowiei* differed from other species in the nontranscribed spacer region: *S. haematobium* had a deletion of 0.5 kb which did not occur in either *S. curassoni* or *S. bovis*, whereas *S. margrebowiei* contained single or multiple inserts of 0.4 kb in the same region. No clear differences in the rRNA gene unit have emerged between *S. bovis* and *S. curassoni*. Later studies of the transcribed spacer region of various species of schistosomes, including *S. spindale* and *S. leiperi*, confirmed the potential of rRNA gene analyses for species identification (Kaukas *et al.*, 1994a; Kane and Rollinson, 1994; Kane *et al.*, 1986). Rollinson *et al.* (1990b) used probes to study the rRNA gene units of hybrids produced by experimental crosses of *S. haematobium* × *S. mattheei*, *S. mattheei* × *S. bovis* and *S. haematobium* × *S. intercalatum*. In each cross, F_1 hybrids produced a composite major banding pattern of the two parental species.

Dias Neto *et al.* (1993) described the application of RAPDs to the study of schistosomes and used the techniques to differentiate five species: *S. mansoni, S. haematobium, S. intercalatum, S. curassoni* and *S. margrebowiei*. Barral *et al.* (1993) worked on *S. mansoni, S. rodhaini, S. intercalatum, S. bovis* and *S. japonicum* and concluded that RAPD profiles are useful in distinguishing species of schistosomes. This conclusion was shared by Kaukas *et al.* (1994b) who used RAPDs to compare species of the *S. haematobium* group. Recently, Tchuem Tchuenté (1997) has observed that the RAPD profiles of laboratory-produced F_1 *S. mattheei* × *S. haematobium* hybrids were composites of the profiles of the two parental strains.

2.4.2. *Eggs*

The size, and in particular the shape, of the egg are the most commonly used criteria for differentiating between schistosome species (see Dinnik and Dinnik, 1965) — eggs can be readily obtained from nasal secretions (*S. nasale*) or faecal samples (other species). In many cases, the egg morphology, the egg-laying site, and details of the area where the animal grazes are sufficient clues to identify the parasite. However, difficulties in identification arise when species share similar egg morphologies, as in the case of *S.*

haematobium and *S. curassoni*, or when species overlap in their distribution, as in the cattle schistosomes *S. bovis* and *S. curassoni*, or *S. bovis* and *S. mattheei*.

Classical egg measurements include the total length and maximum width (Table 1) (Alves, 1949; Dinnik and Dinnik, 1965), but the breadth at a given distance from the tip of the spine and the breadth at a given distance from the blunt end of the egg have also been used (Kruger *et al.*, 1986a). Rollinson and Southgate (1987) warned that some caution should be exercised in treating a particular egg as typical and advised measuring a representative selection from a given sample. This may be difficult in cattle because of the small number of eggs which are normally found in a particular sample. Care should also be taken to measure eggs containing a fully developed miracidium. During faecal examinations in Zambia, De Bont *et al.* (1996a) observed *S. mattheei* eggs which were filled with unstructured material and vacuoles. They accounted for up to 31% of all eggs seen in faecal materials and, at measurement, were found significantly shorter and thinner than live eggs. In contrast, Touassem (1987) compared two *S. bovis* strains from Sudan and Spain maintained in mice and found no intraspecific variation in egg morphology.

Hybridization experiments have demonstrated that eggs resulting from cross-breeding are likely to be intermediate in shape (Taylor, 1970; Rollinson *et al.*, 1990a). Kruger *et al.*, (1986a) examined eggs of *S. mattheei* isolated from cattle from four different areas in South Africa and from a hybrid *S. mattheei* × *S. haematobium* from a human patient. They concluded that *S. mattheei* populations which are sympatric to *S. haematobium* possess *S. haematobium* characteristics. They suggested that these *S. mattheei* populations are infiltrated with *S. haematobium* genes via the *S. mattheei* × *S. haematobium* hybrid originating from human hosts.

2.4.3. Miracidia

Miracidia isolated from faecal materials (see Section 4.1.1.c for details on the technique) cannot be readily identified on a morphological basis, particularly in areas where mixed infections may occur. Dimensions of miracidia for some schistosome species are given by Loker (1983), they range in size from 79 × 51 µm for *S. japonicum* to 165 × 57 µm for *S. nasale*. Kruger and Hamilton-Attwell (1988) compared the morphology of the terebratoria of *S. haematobium* and *S. mattheei* populations from different locations in South Africa. It was found that the terebratorial membrane of some of the *S. haematobium* miracidia from an area with a high *S. mattheei* prevalence in humans resembled the more intricate membrane of *S. mattheei*. Kruger and Hamilton-Attwell (1988) suggested that this could be due to introgressive hybridization between *S. haematobium* and *S. haematobium* × *S. mattheei*.

2.4.4. *Cercariae*

(a) *Morphological characteristics*. Frandsen and Christensen (1984) provided general guidelines on how to harvest cercariae from infected snails, and distinguish *Schistosoma* cercariae from others emerging from African freshwater snails. Further identification of *Schistosoma* cercariae to the species level on the basis of morphological criteria is, according to these authors, not possible for the nonspecialist.

Bayssade-Dufour (1982) reported on the number and disposition of argentophilic papillae (chaetotaxy) on the cercarial surface of nine species of *Schistosoma* and produced groupings which were consistent with the *S. haematobium*, *S. mansoni* and *S. japonicum* groups. Briefly, the technique consists of the determination of three indices (AD, AL and U) after impregnation of cercariae with silver nitrate (Combes *et al*., 1976). AD and AL represent the relative distances between dorsal and lateral papillae, respectively, whereas U corresponds to the total number of papillae on the tail stem. Chaetotaxic indices have been used to distinguish the cercarie of *S. bovis*, *S. curassoni* and *S. haematobium* shed by naturally infected intermediate hosts (Ross *et al*., 1987; Bayssade-Dufour *et al*., 1989; Cabaret *et al*., 1990; Albaret *et al*., 1993).

(b) *Life cycle characteristics*. Mouchet *et al*. (1992) observed that the cercarial emergence patterns of *S. curassoni* and *S. bovis* in Niger (i.e. maximal emergence limited to the first few hours after dawn) was different from the emergence pattern of *S. haematobium* (i.e. maximal emergence in the late morning and early afternoon), and suggested using these characteristics to distinguish between bulinids (*B. umbilicatus* or *B. truncatus*) infected with either human or bovine parasites.

(c) *Enzyme studies*. Field studies involving snails may be complicated by the presence of more than one schistosome species in a same intermediate host species. Mahon and Shiff (1978) successfully used PGM in starch-gel electrophoresis to distinguish *S. mattheei* cercariae from *S. haematobium* cercariae emerging from *B. globosus* snails. Similarly, De Bont *et al*. (1991c) found differences between *S. nasale* and *S. spindale* cercariae for GPI, malate dehydrogenase (MDH), PGM and AcP. Larval schistosomes can also be detected and identified by isoelectric focusing when still associated with the digestive gland of the snail (Wright *et al*., 1979a). In Sri Lanka, cellulose acetate membrane electrophoresis was used to differentiate *S. nasale* and *S*. spindale larval stages in the digestive glands of naturally infected *Indoplanorbis exustus* snails (De Bont *et al*., 1991c). The GPI allele of *S. spindale* showed faster migration than the one of *S. nasale*, and the enzymes of both schistosomes could be distinguished from bands of activity attributable to the snail digestive gland.

(d) *DNA studies.* The techniques of analysis of rRNA gene used for adult worms can easily be applied to cercariae. As with the DNA extracted from eggs, miracidia, or sporocysts within snails, there was no difference in the patterns of hybridization obtained for cercariae and adult worms (Rollinson *et al.*, 1986, 1990b).

2.4.5. *Host Specificity*

(a) *Snail hosts.* A detailed review of the species of snails transmitting schistosomes in Africa, and of those found to have the potential to do so, has recently been published by Brown (1994). Therefore, we present here only a short summary. *Schistosoma bovis* has a very wide specificity in its snail host. In the north of its range, *S. bovis* develops in *B. truncatus* (except in Spain, where *Planorbarius metidjensis* serves as an intermediate host; Ramajo-Martin, 1978). In certain parts of East and West Africa, *S. bovis* is capable of utilizing a wide variety of *Bulinus* species as molluscan hosts, often developing naturally in snails of the *B. truncatus/tropicus, B. africanus* and *B. forskalii* groups (Southgate and Knowles, 1975). In Senegal, *B. umbilicatus* appears to be the most important natural host snail for *S. curassoni* (Southgate *et al.*, 1985; Diaw and Vassiliades, 1987), whereas *S. bovis* was observed only in *B. forskalii*. Natural infections with *S. mattheei* in southern Africa have only been recorded in members of the *B. africanus* group (*B. globosus* and *B. africanus*). *Schistosoma margrebowiei* is transmitted by *B. forskalii* and *B. scalaris* (Wright *et al.*, 1979b), and *B. tropicus* (Southgate *et al.*, 1985) in the Lochinvar Park, central Zambia. Snails naturally infected with *S. leiperi* have not yet been reported, but laboratory studies have shown that the parasite is compatible with members of the *B. africanus* group (Le Roux, 1955; Pitchford and Du Toit, 1976; Southgate *et al.*, 1981).

In the Indian subcontinent *I. exustus* appears to be the only natural intermediate host for *S. indicum, S. spindale* and *S. nasale* (Kumar and de Burbure, 1986). For *S. incognitum, Lymnaea luteola* serves as the natural intermediate host in India (Sinha and Srivastava, 1960), whereas *Radix auricularia rubiginosa* appears to be the natural host snail in Indonesia (Carney *et al.*, 1977). *Schistosoma japonicum* is transmitted by populations of *Oncomelania hupensis*, of which there are six subspecies; *O.h. nosophora* in Japan, *O.h. hupensis* in mainland China, *O.h. chiui* and *O.h. formosana* in Taiwan, *O.h. lindoensis* in Sulawesi and *O.h. quadrasi* in the Philippines (Davis, 1980; Cross *et al.*, 1984).

(b) *Definitive hosts.* Each schistosome species has its own definitive host-range. The current knowledge on which mammalian host is susceptible to which schistosome species is generally based on accidental observations of the parasites in a particular host, or of their eggs in excreta. It is of little practical interest to identify all the receptive host species for a particular schistosome. Of greater importance is the study of the relative epidemiological importance of each definitive host. The objective is then to identify the potential reservoirs of infection, which should be dealt with during the development of control strategies.

In addition to infecting cattle, *S. bovis* and *S. mattheei* occur commonly in sheep and goats, and have been recorded in horses and several species of antelope (Pitchford, 1977). Natural *S. curassoni* infections have only been observed in cattle, sheep and goats (Vercruysse *et al.*, 1984). *Schistosoma margrebowiei* and *S. leiperi* have been described from a wide range of wild herbivores and are primarily parasites of antelopes (Pitchford, 1976). They are occasionally isolated from cattle living in areas within the distribution of *Kobus* spp. (Pitchford, 1976; De Bont *et al.*, 1994). *Schistosoma indicum*, *S. spindale* and *S. nasale* are known to infect a variety of domesticated animals on the Indian subcontinent, including cattle, water buffalo, sheep, goat, and horse (Kumar and de Burbure, 1986; Agrawal and Alwar, 1992). *Schistosoma incognitum* has been recorded as a naturally acquired infection in pigs, dogs, sheep, goats and rats, and occasionally cattle (Sinha and Srivastava, 1965; Biswas, 1975). *Schistosoma japonicum* is an important zoonosis in southeast Asia. In China, mice, dogs, goats, rabbits, cattle, guinea pigs, sheep, rats, horses, and water buffaloes have all been found to harbour the infection, and more than 30 species of wild mammals have been found with natural infections (Mao and Shao, 1982). However, cattle are by far the most important reservoir of *S. japonicum* (Chen, 1993).

The occurrence in humans of schistosome species that infect cattle (other than *S. japonicum* and *S. mattheei*) is still a matter for debate. The presence of *S. bovis* eggs in human stools has been reported on several occasions (Webbe, 1982; Chunge *et al.*, 1986; Mouchet *et al.*, 1988). However, these infections appear to be slight and transient (Mouchet *et al.*, 1988). Kinoti and Mumo (1988) demonstrated that human volunteers fed infected liver passed *S. bovis* eggs in their faeces which could result in the mistaken diagnosis of a viable infection. Various reports, considered by Southgate and Knowles (1977), have suggested that *S. margrebowiei* is capable of developing in man. Again, *S. margrebowiei* eggs passed in stools could also originate from the ingestion of infected animals (Sturrock, 1993b). Grétillat (1962) reported infecting sheep with cercariae from snails that had been infected with miracidia originating from urine samples of children from Dakar, Senegal, and concluded that *S. curassoni* occurs in man. This opinion was later supported by the results of chaetotaxic studies (Albaret *et al.*, 1985). However, infection experiments involving *S. curassoni* and *S. haematobium* tend to contradict these results and suggest that it is unlikely that a zoonosis of *S. curassoni* exists in Senegal (Vercruysse *et al.*, 1984; Rollinson *et al.*, 1987).

3. PATHOLOGY AND PATHOPHYSIOLOGY

3.1. Introduction

Detailed descriptions on the pathology and pathophysiology of schistosomiasis in cattle are available mainly for *S. mattheei* (McCully and Kruger, 1969; Hussein, 1971; van Wyk *et al.*, 1974; Bartsch and van Wyk, 1977; Lawrence, 1977b,c, 1978b,c) and *S. bovis* (Hussein, 1971; Massoud, 1973; Hussein *et al.*, 1975). For *S. spindale* (Rao, 1934; Kalapesi and Puhorit, 1954; Fransen *et al.*, 1990), *S. nasale* (De Bont *et al.*, 1989a), and *S. japonicum* (Wang *et al.*, 1988) only a limited number of studies have been published. The pathology of *S. curassoni, S. leiperi* and *S. margrebowiei* has yet not been studied in cattle, however, from observations in other ruminants, it is thought to be similar to infections with *S. bovis* or *S. mattheei*.

The pathogenesis of cattle schistosomiasis (as for other animals and man) is largely caused by schistosome eggs, rather than worms. Adult worms are able to evade host immune attacks (Butterworth, 1993), although dead worms swept into the liver and lungs do give rise to focal lesions (Lawrence, 1978c). In most cases lesions caused by worms are few in number and have a limited clinical importance, compared with the lesions caused by the millions of eggs generated during the course of infection. Lesions caused by eggs are seen mainly in the intestine wall and the liver (for *S. nasale* in nasal tissues), and more rarely in other downstream parenchymal organs such as the lungs, spleen, bladder, genital tract or ectopic sites (abomasum, forestomachs). The inflammatory response around eggs is an immune-mediated event in which hypersensitivity to the egg plays a vital role (Weinstock, 1992). In early patency of primary infections tissue reactivity to eggs is low compared with later stages (Lawrence, 1978b; Saad *et al.*, 1980). It is well known that granuloma formation progresses concurrently with the development of delayed hypersensitivity (Hirata *et al.*, 1993).

3.2. Disease Spectrum

3.2.1. *Schistosome Species*

Biological features specific for each schistosome species (localization and egg productivity) and host–parasite interactions, i.e. compatibility and susceptibility, play a determining role in how the disease manifests. Visceral schistosomiasis is the most common clinico-pathological entity — except for *S. nasale*, all cattle schistosomes are located in the mesenteric veins. The ensuing pathology is intimately linked with egg production, with data on egg distribution in tissues indicating preferred sites of egg deposition. The rate at which schistosome produce eggs correlates positively with uterine egg counts (see Table 1) (Loker, 1983). Published estimates of total number of eggs produced per worm pair per day (mainly studies in hamsters) reveal a wide range of values but some clear differences exist between species (Table 2 in Section 2.3). Differences in host–parasite relationships are also known to play an important role in the pathology caused by a parasite species. For example, schistosomiasis caused by *S. mattheeei* runs an acute course in cattle, with clinical illness occurring in the early months of infection (sometimes leading to death), but in most cases followed by apparently a complete recovery. In contrast, the disease in sheep tends to run a prolonged course with very slow recovery in the animals which survive the initial acute stage. This difference explains, at least in part, the greater economic importance of *S. mattheei* in sheep as opposed to cattle in southern Africa (Lawrence, 1974).

Finally, experimental infection studies suggest that the susceptibility of cattle to primary schistosome infection differs with the parasite species to which they are exposed. Percentage worm recoveries after experimental exposure to *S. bovis* generally range between 7 and 34% (Saad *et al.*, 1980; Bushara *et al.*, 1980, 1983a,b, 1993; Aradaib *et al.*, 1995b), although Massoud (1973) reported a mean recovery rate of 62.3% in seven calves. In contrast, average worm recoveries of more than 55% have been reported after exposure to *S. mattheei* (van Wyk *et al.*, 1975, 1997; Lawrence, 1977a; De Bont *et al.*, 1997).

3.2.2. *Intensity of Infection*

It is widely accepted that symptoms, morbidity and mortality from schistosomiasis are to some extent related to the intensity of the infection, as measured by faecal egg or worm counts. However, large variations in worm burdens have been observed in cattle. Other factors such as immunity development, age, sex, genetic predisposition, nutritional effects and stresses of field conditions, including gastrointestinal nematode infection and having to graze over large tracts of land (Saad *et al.*, 1980), are certainly also playing a role in determining levels of disease.

In the great majority of cases, visceral schistosome infections in endemic areas occur in a subclinical form, characterized by a high prevalence of low to moderate worm burdens in the cattle population (Condy, 1960; Lawrence, 1978a; Vercruysse *et al.*, 1985). Recent slaughterhouse surveys conducted on *S. spindale* in Sri Lanka and *S. mattheei* in Zambia have shown that over 92% of the animals infected had less than 100 worm pairs in the mesenteric veins (De Bont *et al.*, 1991a, 1994). However, although little or no overt clinical signs may be recognized in the short term, it has been established that high prevalence rates of chronic schistosome infections do, in the long term, cause significant losses on a herd basis. These losses are caused by less easily recognizable effects on animal growth and productivity, and increased susceptibility to other parasitic and bacterial diseases (Dargie, 1980; Pitchford and Visser, 1982; McCauley *et al.*, 1984).

Occasional outbreaks of clinical intestinal schistosomiasis caused by *S. mattheei* (Strydom, 1963; Reinecke, 1970; Lawrence and Condy, 1970; Lawrence and McKenzie, 1972; van Wyk *et al.*, 1974; Lawrence, 1977b), *S. bovis* (Eisa, 1966; Hussein, 1968; Markovics *et al.*, 1993), *S. spindale* (Kalapesi and Puhorit, 1954; Kulkarni *et al.*, 1954) and *S. japonicum* (Wang *et al.*, 1988) have been reported. They are usually restricted to young livestock or adult animals undergoing relatively heavy primary infections under conditions of intensive transmission: worm burdens as high as 71 000 have been reported (van Wyk *et al.*, 1974). Recently van Wyk *et al.* (1997) conducted experiments to try to estimate the lethal cercarial dose of *S. mattheei* for unsensitized cattle. The LD_{50} could not be determined accurately, but appeared to be near 248 cercariae kg^{-1}. This conclusion differs from that of Lawrence (1977b), who estimated that the lethal level of infection was clearly greater than 341 cercaria kg^{-1}. Possible reasons for this apparent discrepancy are differences in cattle breeds and parasite strains, large variations in the development of cercariae in individual animals, and the relatively small numbers of calves used in both trials (van Wyk *et al.*, 1997). Clinical signs of schistosomiasis caused by *S. bovis* became prominent after exposure to 100 cercariae kg^{-1} (Saad *et al.*, 1980). Examples of worm counts in naturally infected animals are listed in Table 3.

Compared with visceral schistosomiasis, nasal schistosomiasis is probably a slower-developing progressive disease, leading to dyspnea due to obstruction of the nasal cavities (De Bont *et al.*, 1989a).

3.3. Clinical Pathology and Pathophysiology

The majority of clinico-pathological studies on visceral schistosomes in cattle have been made following heavy experimental infections with either *S. mattheei* or *S. bovis*, but some studies have reported during natural outbreaks. Comprehensive reviews on the subject were provided by Hussein (1973), Lawrence (1978a) and Taylor (1987). In the great majority of cases, bovine schistosomiasis is a chronic disease, characterized in the short

Table 3 Comparison of worm burdens in cattle naturally infected with *Schistosoma* spp. (1) in clinically healthy cattle from normal endemic situations and (2) in animals with acute schistosomiasis.

Schistosome species	Number of cattle perfused	Duration of exposure	Worm recovery (mean on range)	Reference
(1) Normal endemic situation				
S. mattheei	32	2–12 months	15–2651	De Bont et al. (1995b, 1997)
	3	Unknown	2660	van Wyk et al. (1974)
	2	> 7 years	3634	De Bont et al. (1995a)
S. bovis	10	10 months	835–2408	Majid et al. (1980b)
	2	>5 years	2903 & 10 053	Bushara et al. (1980)
	5	>5 years	774–6133	Bushara et al. (1983c)
(2) Acute schistosomiasis				
S. mattheei	5	Unknown	41 654–71 083	van Wyk et al. (1974)
	1	Unknown	40 679	Lawrence (1977b)
S. bovis	[a]	[a]	7000–17 000	Dargie (1980)
S. japonicum	9	55 days	480–2460	Hsü et al. (1984)
	5	2.5 months	136	Xu et al. (1993)

[a] Author did not provide details.

term by very undramatic clinical changes, but in the long term, and particularly on a herd basis, by retardation of growth. It should, however, be emphasized that in the field other parasites such as gastrointestinal nematodes, trypanosomes or liverflukes may also affect productivity, and influence the pathophysiology of schistosomiasis. In the relatively rare event of an outbreak of acute or subacute infection, overt disease is short lived. It tends only to last for about 2–3 months and, is characteristically always associated with the period following the onset of patency. During this period of overt schistosomiasis animals are anorexic, show a severe diarrhoea, and develop a marked anaemia and hypoalbuminaemia. Because of gut damage and associated protein leak, animals infected with schistosomes fall into an hypercatabolic state with respect to haemoglobin, albumin and the globulins. At the same time, they also synthesize much more of these proteins which are needed for host survival. This high turnover associated with the anorexia and reduced digestibility explains the observed substantial loss of bodyweight from which animals never really appear to recover (Dargie, 1980).

3.3.1. Schistosoma mattheei

Two different clinical syndromes have been described in *S. mattheei* infections: the 'intestinal syndrome', which occurs after heavy primary infection, and the 'chronic hepatic syndrome', which sometimes occurs after heavy reinfection. Lawrence (1976, 1977b,c,d) infected 19 calves aged 7–11 months with between 5000 and 45 000 cercariae of *S. mattheei*. Eggs were first detected in the faeces 6–7 weeks p.i., with all the calves developing the 'intestinal syndrome' (characterized by diarrhoea or dysentery, anorexia, and loss of condition) 1 week later. The severity and duration of illness was proportional to the level of infection, but recovery occurred in all animals not slaughtered for worm counts. Haemoglobin levels began to fall after 3 weeks and reached the lowest level 13 weeks p. i.. Blood packed cell volume (PCV) and red cell counts showed similar changes to those described for haemoglobin and a normocytic, normochromic anaemia developed. There was a rise in neutrophils in the period 7–10 weeks p.i., followed by a steep fall until week 15, with a gradual return to normal thereafter. Lymphocyte counts decreased sharply between weeks five and ten, and recovered slowly after 15 weeks, while albumin concentrations fell from week seven to week twenty, with a subsequent return to normal. Lawrence (1977d) observed that the most severe fall in these blood values (between 7 and 10 weeks p.i.) coincided with the early acute stage of clinical illness and the period of greatest blood loss in the faeces. The gradual improvement from 15 weeks p.i. onward coincided with the reduction in faecal egg output and the general clinical improvement of the animals. These results suggested that the major causes of anaemia, hypoalbuminaemia and lymphopaenia were blood loss and leakage of plasma from intestinal lesions caused by eggs — although anaemia began to develop during the prepatent period. Lawrence (1977d) also observed a short low peak of eosinophilia 7 weeks p.i. and a short high peak at 20–25 weeks p.i., while serum globulin levels (mainly the gamma fraction) rose sharply from the time of infection to peak 16 weeks p.i., with a subsequent fall. The elevated serum globulin levels indicate antigenic stimulation, whereas eosinophil levels are believed to reflect the degree of inflammatory response provoked by the parasite. The eosinophil counts were markedly higher in calves on a higher plane of nutrition, which was believed to reflect a more vigorous cell-mediated immune response to antigens.

The acute 'intestinal syndrome' (Lawrence, 1977b) occurred during the period of high faecal egg output in the first weeks after patency and was attributed mainly to the passage of large numbers of eggs through the intestinal wall. However, Lawrence (1977b) observed that the severity of diarrhoea began to diminish while faecal egg counts were still high, and suggested that other factors in addition to the numbers of eggs passed were responsible for the diarrhoea, namely a hyperreactive response by the host to the eggs (Lawrence, 1978b), and a shift of egg laying to the large intestine (Lawrence, 1977e). Apparently, cattle are able to overcome the effect of *S. mattheei* infection much more effectively than sheep (McCully and Kruger, 1969; Lawrence and McKenzie, 1972). Cattle which survive the early acute stage of the disease generally recover, and become partially resistant to the effects of reinfection (van Wyk and Bartsch, 1971; Lawrence, 1976). In contrast, sheep may die from schistosomiasis up to 4 years after reinfection (J.A. van Wyk, personal communication).

A completely different clinical condition was first observed by van Wyk and Bartsch (1971) after heavy experimental challenge of an ox, 18 months after primary infection. Four months after reinfection, the animal started to lose weight and showed straining, knuckling over the hind fetlocks, and signs of nervous derangement. At post-mortem examination, the authors found marked periportal hepatic fibrosis and named the condition 'hyperergic schistosomiasis'. The syndrome was recognized subsequently in a natural outbreak by van Wyk *et al.* (1974) who believed both macroscopic and microscopic hepatic lesions to be identical to the clay pipestem fibrosis of human schistosomiasis (Symmers, 1903). Later, this syndrome was reproduced experimentally in one animal and encountered in the field on two occasions by Lawrence (1977c). The experimental case has been exposed to 20 000 cercariae on four occasions at 4–6-week intervals. It went through a similar 'intestinal syndrome' to that found after a single experimental exposure, except that at 25 weeks p.i. improvements in body weight, haemoglobin concentration, and serum albumin levels were interrupted, until death. Terminal anaemia, eosinophilia and hypergammaglobulinaemia developed. Lawrence (1977c) considered the syndrome to be different from the clay pipestem fibrosis in that large numbers of eggs were not found in the portal tracts and there was no evidence of portal hypertension. The syndrome was therefore renamed the 'chronic hepatic syndrome'. Nevertheless, the progressive increase in eosinophils and serum globulins up to the time of death suggested that the condition was of immunological origin. Repeated infection could have resulted in large numbers of adult worms in the portal system, and death of these worms could have led to great antigenic stimulation. The hepatic syndrome seems to be more likely to occur if the immunity of infected cattle is broken down by challenge with large numbers of cercariae. How relevant this syndrome is in natural infections is not known.

3.3.2. Schistosoma bovis

The clinical pathology of *S. bovis* experimental infections in calves described by Saad *et al*. (1980, 1984) closely resembled the 'intestinal syndrome' observed after primary *S. mattheei* infections. Calves first became clinically ill 7 weeks after exposure to *S. bovis* cercariae and, during the following 2–3 months, showed intermittent diarrhoea, inappetence, anaemia, hypoalbuminaemia, hyperglobulinaemia and severe eosinophilia. They either lost weight or failed to maintain the rate of growth recorded in uninfected animals. These changes were broadly related to the level of infection. The development of anaemia and hypoalbuminaemia coincided closely with the start of faecal egg excretion and was maximal when faecal egg counts were highest. Saad *et al*. (1980) concluded that the high rates of red cell and albumin degradation observed in all infected animals, together with the presence of pure blood in the faeces, left little doubt that the anaemia and hypoalbuminaemia arose primarily from intestinal haemorrhage.

3.3.3. Schistosoma nasale

Preliminary accounts on the clinical pathology of *S. nasale* infections in cattle were given by Datta (1932) and Rao (1933, 1934). The condition is associated with cauliflower-like growths on the nasal mucosa causing partial obstruction of the nasal cavity and snoring sounds when breathing. Haemorrhagic and/or mucopurulent nasal discharge is a common feature. Koshy and Alwar (1974) investigated serum protein changes in naturally and experimentally infected cattle and found that infected animals were hypoalbuminaemic and hypergammaglobulinaemic, compared with uninfected controls.

3.4. Pathology

As with the information on clinical pathology, our current knowledge on the pathology of cattle schistosomiasis is, to a large extent, based on observations from animals experimentally or naturally infected with either *S. bovis* or *S. mattheei*. The pathology in experimental infections is found to be similar to the natural infections, the differences noted being mainly due to the variations in severity, and possibly duration of infection.

3.4.1. Schistosoma mattheei

The pathology of *S. mattheei* in cattle was studied by Hussein (1971) and Lawrence (1977b,c, 1978b,c). They exposed calves to between 5000 and up to 45 000 cercariae and slaughtered them serially between 7 and 107 weeks p.i. Particularly severe lesions were seen in the intestine, where there was a general appearance of hyper-reactivity and hypersensitivity, characterized by hyperplasia of the submucosal lymphoid tissue and Peyer's patches, and intense eosinophilic infiltration.

The course of infection and its related pathology followed a characteristic sequential development. In the early stage, lesions were mainly localized in the intestinal wall, as a result of massive egg-laying by the adult worms. Clinical signs in animals suffering from the acute 'intestinal syndrome' were associated mainly with the presence of haemorrhagic lesions in the mucosa of the large intestine. The lesions were undoubtedly provoked by the presence of large numbers of eggs in the mucosa, but many of the haemorrhages in the large intestine were not associated with mature eggs. It therefore seemed that some factor in addition to the presence of eggs was involved in the large intestine haemorrhages, particularly since haemorrhagic lesions did not occur to the same extent in the small intestine, in which egg counts were also high: it was suggested that the dilation and rupture of venules resulted from an early stage of immunological hyperreactivity to egg antigens.

In a later stage of the infection, the intestinal lesions resolved, following an immunological response of the host, which causes the elimination of some of the adult worms and suppression of egg-laying in those which survive. This elimination coincided with the appearance of vascular responses. They were either localized around dead worms lying in the blood vessels of the intestines, liver and lung, eventually leading to lymphoid nodule formation, or diffuse in the liver. The delay in the onset of these vascular lesions and their early appearance in heavy infections suggest that they were immunological in origin, perhaps involving the uptake of adult worm antigens by vascular endothelia. They were also observed in naturally infected animals after treatment (McCully and Kruger, 1969), which provides further support for the hypothesis that the antigen may have originated from dead and dying worms rather than from live ones.

The relative distribution of eggs within the intestinal wall showed striking changes as the infections progressed. Eggs accumulated rapidly for the first few weeks, and initially the majority were found in the mucosa, while only a small proportion was found in the submucosa. The eggs laid in the mucosa passed quickly into the lumen, while those in the submucosa persisted longer, until they were phagocytosed. As egg-laying was suppressed by the immunological response of the host, the number of eggs in the mucosa fell rapidly, while those in the submucosa remained high. Egg-laying in the muscle and adventitia commenced at a later stage and may have been caused by the adult worms being unable to penetrate to the submucosal and mucosal venules because of phlebitis in the intestinal veins. Lawrence (1978b) suggested that the relative distribution of eggs between the different layers of the intestinal wall and the presence or absence of intrahepatic vascular lesions could be of value in assessing the age of natural infections.

Post-mortem of fatal and slaughtered cases of the chronic 'hepatic syndrome' (Lawrence, 1977c) showed gross enlargement and induration of the liver. The capsule was grey, while the cut surface was yellow-brown in colour and presented a diffuse stellate pattern of grossly thickened portal tract. Histological examination revealed massive thickening of the portal veins and fibrosis. From the portal tracts fibrous tissue and small bile ducts had proliferated radially and invaded the parenchyma. There were foci of 'piecemeal necrosis' adjacent to the portal tracts.

Schistosoma mattheei, but not *S. bovis*, also causes urinary bladder lesions in cattle (Condy, 1960; McCully and Kruger, 1969; Bartsch and van Wyk, 1977; Lawrence, 1978b). Bartsch and van Wyk (1977) examined the bladders from 40 cattle heavily infected with *S. mattheei*. Thirty animals had been experimentally exposed to between 9500 and 145 000 cercariae, whereas the remaining 10 were selected from an outbreak on a farm in South Africa (details in van Wyk *et al.*, 1974). Macroscopically obvious granulomas were found in 75% of the bladders. Four types of lesions were recognized, ranging from scattered individual granulomas to widespread polypoid and granular patches and, in all four types of lesions, petechiae and ecchymoses were commonly observed. However, Bartsch and van Wyk (1977) found no correlation between the worm burden and the presence of bladder lesions in these heavy infections. Furthermore, they argued that in naturally infected animals average worm burdens are likely to be much lower than the mean threshold value (worm burden) necessary for development of these lesions.

Obwolo and Rogers (1988) observed *S. mattheei*-induced lesions in approximately 1% of the 3441 bovine uteri they examined in Zimbabwe. The nodular lesions were seen either at the mid-dorsal aspect of the uterine body or in the horns. Lesions were most severe in the myometrium and consisted of rings of multinucleated giant cells and macrophages occurring around eggs, with masses of eosinophils on the outside.

3.4.2. Schistosoma bovis

All studies clearly show that *S. bovis* is a visceral schistosome, invariably affecting the intestinal tract and liver, but sometimes involving other parts of the body (such as the lungs, pancreas, forestomachs and abomasum), particularly in heavily infected cattle (Hussein *et al.*, 1975). In these heavy infections, massive thrombosis of the portal vessels was seen, causing 'nodular sclerosis': the liver was hardened and extremely fibrosed, showing multiple, elevated, purplish or greyish nodules which consisted of greatly dilated portal veins occupied by haemorrhagic thrombi, up to 10 mm in diameter, in which masses of living and degenerated schistosomes were embedded. The pathological changes in experimental *S. bovis* infections in cattle were described in detail by Hussein (1971), who exposed calves to between 5000 and 11 000 cercariae and slaughtered them serially from 30 to 180 days p.i. The lesions caused by *S. bovis* were generally less severe than those induced by *S. mattheei* in other animals. In the calf slaughtered 30 days p.i., the liver showed intense focal accumulations of mononuclear cells and polymorphs in the parenchyma and around portal vessels. There was also severe patchy phlebitis and endophlebitis, and a little schistosomal pigment was seen in the portal tracts, but other organs were normal. Faecal egg excretion started 48 days p.i. and, in the five calves autopsied between days 60 and 180 p.i., large numbers of schistosomes were found in the portal, mesenteric and pancreatic veins, and in the intestinal submucosal and subserosal vessels. In the liver of calves autopsied either day 60 or day 71 p.i., many minute greyish granulomas were observed. Each of these consisted of one to 16 eggs surrounded by eosinophils, epithelioid cells, lymphocytes, plasma cells and fibroblasts, and giant cells. Surrounding these lesions were areas of hepatic cell degeneration, perilobular fibrosis, and marked inflammatory reactions in the portal tracts. In calves slaughtered between day 120 and day 180 p.i., some lesions were seen in the perivascular connective tissue and had started to undergo progressive fibrosis. Some showed central hyalinization and others showed a 'Hoeppli reaction', and there were numerous eggs and empty shells in small portal venules. In addition, whitish or pinkish, firm, elevated nodules 2–3 mm in diameter were seen on the surface of the liver. Microscopically, these nodules were extensive nodular and follicular lymphoid hyperplasia around degenerated adult schistosomes. The lesions were mostly intravascular.

At the earlier stages of the disease, the commonest lesions of the portal vein were partial or complete blockage of small intrahepatic veins by granulomas, and perivascular inflammatory reaction, sometimes with oedema of the perivenous tissue. The wall of some veins was slightly thickened and medial hypertrophy had commenced, whereas in other veins there was a slight intimal proliferation with hyperplasia and degeneration of the underlying media. From day 120 p.i. onwards, the veins showed irregular papilliform intimal projections, subintimal eosinophilic infiltration, and extensive medial hypertrophy and hyperplasia. These changes were seen mainly in veins of medium and larger size which also showed concentric perivascular fibrosis and 'angiomatoids'. Thrombosis of the veins was commonly present, causing severe local obstruction. Hussein (1971) thought that this medial hypertrophy was not caused by eggs, but developed secondarily to intimal lesions and obstruction of the vessels by lymphoid nodules and thrombi. In addition, the presence of degenerated parasites in the vicinity of the affected vessels suggested that the severe intimal reaction seen from the third month of infection onwards could be hypersensitive responses to foreign proteins released by the dead parasites rather than to the eggs. This was further supported by the absence of such severe reactions in small portal venules, where the eggs were likely to lodge. However, both eggs and dead schistosomes appeared to be responsible for stimulating lymphoid nodules, with dead parasites being the far most important in this reaction.

The intestinal lesions were most severe in the small intestine and could be seen macroscopically as numerous tiny granulomas in the mucosa and sometimes in the serosa, and in multiple punctate or diffuse haemorrhages. There was an acute catarrhal enteritis in the small intestine, with copious mucous exudate rich in eggs and red blood cells. Both central necrosis and 'Hoeppli reactions' were common in intestinal granulomas, and there was hypertrophy and tortuosity of the intestinal veins. Saad *et al.* (1980) later examined the relative distribution of *S. bovis* eggs both along the intestinal tract and between the different layers of the intestinal wall in experimentally infected calves and found sequential changes similar to those reported by Lawrence (1977e, 1978b) in *S. mattheei* experimental infections (see Section 3.4.1 for more details).

Hussein (1971) further described lesions in the pancreas, lungs, abomasum, fore-stomachs and lymph nodes. In the lungs, there was lymphoid hyperplasia and focal eosinophilic infiltration, and pulmonary vessels of very small diameter showed markedly thickened and hypertrophied media, with partial or complete obliteration of the lumen. These reactions were again interpreted as immunological in origin.

3.4.4. Schistosoma spindale

Rao (1934) briefly described the pathological changes observed in two calves slaughtered 86 and 118 days after exposure to 127 000 cercariae of *S. spindale*, and Kalapesi and Puhorit (1954) reported on the lesions found in the tissues of one naturally infected bull. A more detailed description of the pathology of natural infections in cattle has recently been published by Fransen *et al.* (1990). These natural infections were of much lower intensity than those obtained experimentally in studies on *S. mattheei* and *S. bovis*. Fransen *et al.* (1990) observed egg granulomas mainly in the liver and intestines, and to a much lesser extent in the lungs. Hepatic lesions were moderate, with periportal cell infiltration and periportal epithelioid cell granulomas within perilobular zones. Submucosal and mucosal granulomas accompanied by cellular changes were present in the small and large intestines. Epithelioid cell granulomas containing slender living eggs were observed in the urinary bladder of one of the 15 animals examined.

3.4.4. Schistosoma Japonicum

Schistosomiasis japonica is considered to be more severe than other schistosome infections because *S. japonicum* female worms have a much greater egg output than females of other species (see Table 2 in Section 2.3), and because eggs are deposited in large aggregates leading to intense focal intestinal reactions (Chen, 1993). Both the schistosome isolate and the host species are important determinants of the pathology induced by *S. japonicum*. However, little information is available on the pathology of natural *S. japonicum* infections in cattle. Wang *et al.* (1988) described an outbreak of bovine schistosomiasis japonica in Taiwan, in a herd of 75 beef cattle where 15 animals were found dead after showing severe diarrhoea. The most striking gross lesions were multifocal parasite egg granulomas, found mainly in the liver, lungs and intestines. Histologically, the circumoval granuloma consisted of *S. japonicum* eggs, surrounded by epithelioid cells and fibrous stroma.

3.4.5. Schistosoma Nasale

Preliminary accounts on the pathology of *S. nasale* infections in cattle were described by Datta (1932) and Rao (1933, 1934). The eggs deposited in the mucosa incite an inflammatory reaction with cellular infiltration and extensive fibrous tissue proliferation, leading to large cauliflower-like growths on the nasal mucosa. A detailed description of the pathological features in infected animals was given by De Bont *et al.* (1989a). Based on severity and localization of the macroscopic lesions, six types were recognized. Type 0: no visible lesions, but eggs present in nasal scrapings. Type 1: few noticeable pinhead-sized sessile nodular areas and minute ulcers with congested borders. Type 2: several nodular areas, 2 to 3 mm in diameter, and ulcers with congested borders. Type 3: several nodular areas, 4–5 mm in diameter, and ulcers, and general inflammation. Type 4: numerous, often confluent nodular areas and ulcers covering the entire surface and general inflammation. Type 5: numerous, often confluent nodular areas and ulcers, cauliflower-like growths and obstruction of the nasal cavity. The macroscopic lesions were generally observed in the anterior part of the nasal cavity, less than 10 cm posterior to the nasal opening; in the most severely affected animals (types 4 and 5) a few pinhead-sized nodules were also found more posteriorly. The first lesions (Type 1) appeared generally on the medial septum, on the dorsal edge of the ventral nasal concha near the cranial end of the middle meatus or, on the lateral wall of the middle meatus. Later (Types 2 to 5), they gradually spread over the whole mucosal surface of the anterior part of the cavity. The lesions tend to get more severe in older animals. Lesions Types 4 and 5 were found only in animals older than 5 years. Compared with visceral schistosomiasis, nasal schistosomiasis is probably a slower-developing progressive disease. The histopathology was similar to other schistosome infections in that granuloma formation, owing to the presence of eggs, is a common feature. The granulomas, which were often associated with 'Hoeppli reactions', began by accretion of mononuclear cells and eosinophils. The numerous, striking Hoeppli phenomena observed in cattle infected with *S. nasale* may be correlated with a strong hypersensitivity reaction of the respiratory mucosa (De Bont *et al.*, 1989a).

4. DIAGNOSIS

4.1. Diagnostic Methods

4.1.1. *Visceral Schistosomiasis*

(a) *Clinical signs.* Symptoms of haemorrhagic diarrhoea, anorexia with reduction in growth rate or loss of body weight, sometines leading to mortality (van Wyk *et al.*, 1974), are sufficiently characteristic to suggest visceral schistosomiasis in endemic areas; and may even be recognized as such by cattle owners (McCauley *et al.*, 1983). However, instances of clinical visceral schistosomiasis in endemic areas are rare. Infections are generally chronic in character, and symptoms in the majority of infected animals are insufficient to distinguish the illness from other debilitating diseases (Lawrence, 1978a; Dargie, 1980).

(b) *Faecal examination*. The diagnosis of schistosome infection is most commonly achieved through the demonstration and identification of eggs in faeces (Dinnik and Dinnik, 1965). In addition, counting these eggs is sometimes used as an indicator of the level of infection. Different faecal egg counting techniques, based on the examination of between 50 mg and 10 g of faeces, have been described and compared (Lawrence, 1970; Pitchford and Visser, 1975; Olaechea *et al.*, 1990). As discussed by Olaechea *et al.* (1990), various parameters should be considered when choosing a technique for a particular study: the sensitivity of the technique, the requirements in materials and expertise, the time consumption, and the aim of the study. Although time-consuming, faecal examination is a valuable diagnostic method for the field veterinarian dealing with groups of animals at risk, particularly because severe morbidity in cattle is generally associated with high egg excretion (see Secions 3.3.1 and 5.2). For epidemiological work, the major disadvantage of faecal egg counting methods is their decreasing sensitivity in ageing animals, which is clearly associated with the immune-related reduction of the fecundity of the female worms (see Section 5.3 for details). In Zambia, the sensitivity of the faecal egg count technique described by Lawrence (1970) declined from 73% in heifers, to 44% in adult cows (De Bont *et al.*, 1996b). As for the identification of the parasite, the morphological characteristics of the eggs may be sufficient for differentiating species in some parts of the world such as the Indian subcontinent, but may be of limited value in those places where several closely related species coexist and may interact.

(c) *Miracidial counts*. To obtain miracidia from faecal material, eggs should first be extracted by washing the faeces with a saline solution (to prevent premature hatching), either using repeated sedimentation or sieves with various pore sizes. The sediment or filtrate containing the eggs is then mixed with water to stimulate egg hatching, and the miracidia are attracted by light for collection. Miracidial hatching techniques have sometimes been preferred to faecal egg counts for the diagnosis and quantification of schistosome infections in cattle (Kruger and Heitmann, 1967; Saeed *et al.*, 1969; Kassuku *et al.*, 1986; Banerjee and Agrawal, 1989, 1990; Banerjee *et al.*, 1990; De Bont *et al.*, 1991a). Much larger faecal samples can be used, which improves the sensitivity and facilitates the detection of light or old infections. In addition, pooled samples from several animals may be examined to detect infection at the herd level. Miracidial counts may also be used in combination with faecal egg counts to measure the viability of the eggs found in the faeces (Lawrence, 1977a; De Bont et al., 1996a). However, miracidia cannot be readily identified. The existence of an immune-related decline in the viability of faecal eggs (De Bont et al., 1996a) also implies that miracidial counts may not simply be considered as a more sensitive way to measure faecal egg excretion.

(d) *Serological tests.* Many serological tests have been developed for the diagnosis of human schistosomiasis, including tests for the detection of antibodies, immune complexes and circulating schistosome antigens (reviewed by De Jonge, 1990; Feldmeier, 1993). In cattle, tests for the detection of host antibodies directed against parasite material have been developed (Du Plessis and van Wyk, 1972; Preston and Duffus, 1975; Lawrence, 1977f). The indirect fluorescent antibody (IFA) test was first employed to determine levels of serum antibodies in cattle infected with *S. mattheei* (Du Plessis and van Wyk, 1972). Smears of *S. mattheei* cercariae were used as antigen and, in positive cases, the strongest fluorescence was observed in the cuticle of the cercariae. High titres were obtained in sera from animals infected with schistosomes, whereas those from controls infected with other helminths were negative, suggesting both good sensitivity and specificity. However, no correlation was found between the IFA titres and the worm burdens. Preston and Duffus (1975) obtained similar results using an indirect haemagglutination test for the detection of *S. bovis* antibodies and concluded that the test was a more reliable screen for *S. bovis* infection than faecal examination. Lawrence (1977f) used the complement fixation (C), indirect haemagglutination (IH) and indirect immunofluorescent (IF) tests to follow for a period of up to 76 weeks, the antibody response in 30 calves experimentally infected with *S. mattheei*. Complement fixation and IF antibody titres rose to a peak at about 25 weeks p.i. and then dropped, while IH antibodies rose more slowly (during the phase of recovery) and remained high. There was a strong cross-reaction to *Fasciola gigantica* and *Paramphistomum microbothrium* in the C test whereas the IH and IF tests were found specific. Interestingly, peak IH and IF titres were proportional to the level of infection. Lawrence (1977f) used these two tests to examine a total of 52 sera samples from cattle naturally infected with *S. mattheei*. Results confirmed the experimental findings that a high IF titre was evidence of recent, heavy infection, and suggested that the IF test could be of value in the diagnosis of clinical schistosomiasis. However, the major drawback of antibody tests is that they do not distinguish between active and past infections. Therefore, they can only be used for screening recently acquired infection. In addition, except for some findings of Lawrence (1977f), there appears to be no correlation between worm burden and antibody levels.

Antigen determination in serum is now widely used for immunodiagnostic and sero-epidemiological studies of human schistosomiasis (Deelder *et* al., 1994). The circulating anodic antigen (CAA) and circulating cathodic antigen (CCA) are two antigens produced in the guts of schistosomes. They are released in the circulation of the hosts by the regular vomiting of the parasites. In laboratory animals a correlation was demonstrated between CAA and/or CCA levels in serum and worm recoveries at perfusion (Barsoum et al., 1992; Agnew et al., 1995), and it is assumed that antigen levels in serum and/or urine reflect the worm burden more accurately than egg counts (De Jonge et al., 1990). De Bont et al. (1996b) measured the CAA and CCA levels in serum samples collected, on a monthly basis over a period of 1.5 years, from a total of 44 cattle naturally exposed to *S. mattheei* on a Zambian farm; antigen determination tests, with sensitivities between 95 and 100% in heifers and adult cows, proved to be excellent methods for the diagnosis of cattle schistosomiasis. The specificity of the test was not assessed as all animals were known to be infected. A clear seasonal pattern in CAA levels was observed, with a significant increase during the second half of the dry season, when all animals are subjected to heavy physical and nutritional stress. It was concluded that, although circulating antigen determination may provide an indication of the worm burden, possible variations of the antigen clearance rate with the physiological condition of the host, e.g. nutritional stress, may complicate the interpretation of the results.

Banerjee and Agrawal (1990) tested the diagnostic values of the miracidial immobilization test (MIT), the ring precipitation test (RPT) and the cercarien Hüllen reaction (CHR) on serum samples collected at the slaughterhouse from cattle thought to be infected with *S. indicum* and/or *S. spindale* (Banerjee and Agrawal, 1990; Banerjee et al., 1990, 1991). The MIT consisted of adding five to seven miracidia of *S. incognitum* into wells containing nine different dilutions of each serum sample (ranging from 1:10 to 1:2560). The wells were examined after 20 min and the highest dilution at which immobilization of the miracidia occurred was taken as titre. If only animals positive on faecal examination were considered then a titre of 1:80 gave a 97% sensitivity and 68% specificity (Banerjee and Agrawal, 1990). The RPT consisted of adding diluted *S. spindale* antigen (prepared from worm homogenates) into wells containing different dilutions of the serum sample (ranging from 1:10 to 1:320). The wells were then examined after incubation for 24 h and the development of a clear, opaque white ring was considered as positive. Banerjee et al. (1990) found that when a titre of 1:10 was considered as positive, the RPT was 79% sensitive and 91% specific. The CHR, which occurs when living schistosome cercariae are placed in homologous antisera, involves the formation of a transparent membrane around the cercarial tegument. When tested in a similar manner as the MIT and RPT, the CHR was found to be 90% sensitive and 48% specific in cattle (Banerjee et al., 1991).

In conclusion, several serological tests have been shown to be of value in the diagnosis of cattle schistosomiasis. However, none of these tests has (to our knowledge) been used on a large scale.

(e) *Post-mortem examination*. At post-mortem, visceral schistosomiasis can easily be diagnosed by examining the mesenteric veins for the presence of paired worms. Rapid diagnosis can also be achieved by detection of schistosome eggs in scrapings of the intestinal mucosa (van Wyk, 1971) or crushed liver tissue (Banerjee and Agrawal, 1989; Banerjee *et al.*, 1990, 1991). During epidemiological studies, precise data on worm burdens (intensity of infection) or accumulation of eggs in tissues (morbidity of infection) may be required, which can only be obtained postmortem. Adult worms can be counted, either *in situ* by examining the mesenteric veins over the entire length of the intestine (De Bont *et al.*, 1991a, 1994), or after perfusion of the mesenteric veins and intrahepatic branches of the hepatic portal vein (McCully and Kruger, 1969). These techniques allow the detection of very light infections which could otherwise remain undetected with other techniques such as faecal egg or miracidial counts. In Sri Lanka for example, 21% of the animals having paired *S. spindale* worms in the mesenteric veins were negative on miracidial count and only 13% produced more than 100 miracidia, corresponding to an egg per g (e.p.g.) of more than one (De Bont *et al.*, 1991a). Finally, worms collected at slaughterhouses or after perfusion can be examined in detail. Another main advantage of including post-mortem examinations in epidemiological studies is the possibility to count eggs in tissues. Tissue egg counts are usually determined after digestion (at 37°C for 12–40 h) of known proportions of the organs in 5% KOH (Cheever, 1968; Lawrence 1977e).

4.1.2. *Nasal Schistosomiasis*

In areas where *S. nasale* is endemic, the observation of nasal granulomas, which may or not be associated with dyspnoea and snoring sounds, is highly indicative of nasal schistosomiasis. However, diagnosis should be confirmed by the presence of the typical boomerang-shaped eggs in the nasal secretions or scrapings of the nasal mucosa. In a first attempt to count eggs in nasal secretions Rao and Devi (1971) mixed 1 ml of nasal mucus with 9 ml of 10% KOH. After allowing to settle for 10–20 min at room temperature, the eggs present in the lower 1 ml were counted. Muraleedharan *et al.* (1976a) modified the technique by heating 1 g of mucus with 10% KOH and centrifuging the solution. An intradermal test for the diagnosis of subclinical nasal schistosomiasis was proposed by Jagannath *et al.* (1988). They injected 0.1 ml of a *S. japonicum* antigen to 60 infected cattle and five noninfected controls, and measured the dimensions of the wheal or erythema at the site of injection 10–20 min later. The mean dimensions of the wheal in infected animals were 1.65 cm × 1.55 cm, compared with a diameter of 3 mm in controls. The specificity of this intradermal test was not examined.

4.2. Measures of Infection at the Population Level

Measurements of schistosome infections at the population level obviously depend on the sensitivity and specificity of the diagnostic method used.

The prevalence of infection is the percentage of a cattle population which is found infected at a given time, whereas the intensity of infection is a measure of the worm burden in infected animals. In some field studies, the intensity of infection was measured by faecal egg counts (Majid et al., 1980a; Pitchford and Visser, 1982; Kassuku et al., 1986). However, it should be stressed here that even if faecal egg counts may, in some early infections, provide reasonably accurate indications of the intensity and morbidity of infection (see Section 5.3.2 for more details), they should not generally be considered as an indirect measure of the worm burden when the duration of infection is unknown (van Wyk et al., 1974; Lawrence, 1978a; Dargie, 1980; De Bont et al., 1991a, 1995a). Kassuku et al. (1986) used the egg count per g tissue in jejunal samples of 5-10 g as a measure of intensity of infection during an abattoir survey carried out in Tanzania. However, tissue egg counts (usually of the intestines and liver) are more commonly used as a measure of morbidity, the eggs of schistosomes acting as the primary agent of pathogenesis in the host.

The incidence rate is a measure of the level of transmission over a given period of time. It is the proportion of initially uninfected animals which become infected during this period of time. Levels of transmission may also be measured through examination of the snail populations at the transmission sites or by using tracer calves. These are initially uninfected animals which are mixed for a limited period of time with the herd using the transmission site of interest, and are eventually slaughtered and perfused.

5. THE EPIDEMIOLOGY OF CATTLE SCHISTOSOMIASIS

5.1. Introduction

Much of the information available on the host–parasite relationships and immunity development in cattle visceral schistosomiasis emanates from experimental infection studies using *S. mattheei* or *S. bovis*. Three main types of experimental infection studies have been carried out. Some consisted of monitoring faecal egg excretion and the eventual post-mortem examination of calves slaughtered serially after a single experimental infection (Lawrence, 1973a,b, 1977a,e, 1978a,b,c; Saad et al., 1980; Aradaib et al., 1995b). Other studies examined the response to a challenge several weeks after a single primary infection to investigate whether protective immunity could be observed after challenge (Massoud and Nelson, 1972; Lawrence 1973a, 1977a; Preston and Webbe, 1974). Finally, experimental infections have also been used to demonstrate the development of immunity in cattle exposed to long-term natural challenge (Bushara et al., 1980, 1983c).

In contrast, only a few studies have been carried out on the epidemiology of schistosomiasis in naturally infected animals. This is surprising because the host–parasite relationship in animals exposed almost daily during their entire life to moderate cercarial challenge is likely to be different from the one occurring after a single, often massive, experimental infection.

Only a few epidemiological studies have considered *S. nasale*. This is not just a reflection of the relatively lower economical importance of nasal infections but mainly because it is more difficult to study nasal infections in the definitive host. In practice, *S. nasale* worms or eggs cannot be counted in infected cattle and the only measurable parameters at the population level are the number of animals infected and the severity of macroscopic lesions in positive animals. Therefore, apart from prevalence data (Dutt and Srivastava, 1968; Achuthan and Alwar, 1973; Muraleedharan et al., 1973, 1976b; De Bont et al., 1989a), little is known about the epidemiology of nasal schistosomiasis in cattle. However, the prevalence of *S. nasale* infections and the severity of the lesions are known to increase with the age of the host/infection (Islam, 1975; De Bont et al., 1989a).

5.2. Levels of Infection in Cattle Populations

It has been postulated that at least 30% of the entire cattle population living in areas endemic for cattle schistosomiasis are infected with schistosomes, which represents at least 165 million cattle in Africa and Asia (De Bont and Vercruysse, 1997). However, little information is available on the average worm burden in cattle. Dargie (1980) estimated that the overwhelming majority of cattle infected with *S. bovis* in Sudan had less than 500 worms. Abattoir surveys have shown that over 92% of the animals infected with *S. spindale* in Sri Lanka (De Bont et al., 1991a), or *S. mattheei*, *S. leiperi* or *S. margrebowiei* in Zambia (De Bont et al., 1994), had less than 100 worm pairs in the mesenteric veins, as measured by worm counts *in situ*. No attempt was made during these abattoir surveys to count worms in those 8% of animals infected with more than 100 pairs. However, it was recognized that counts *in situ*, as opposed to those carried out after perfusion, could represent a low estimate of the intensity of infection as some worms may be lost to the hepatic portal system and liver during slaughter. Worm recoveries observed after perfusion of naturally infected but apparently healthy animals are listed in Table 3. These perfusion data confirm that the worm counts carried out *in situ* in abattoirs (De Bont *et al.*, 1991a, 1994) did represent an underestimation of the true intensity of infection, and probably the prevalence of infection in the animals examined. Much higher levels of infection have been recorded in cattle showing clinical signs of schistosomiasis (Table 3).

More data are available on the faecal egg excretion in naturally infected populations. Under normal endemic conditions, the faecal egg output rarely exceeds 50 e.p.g. (Majid *et al.*, 1980a; Pitchford and Visser, 1982; Kassuku *et al.*, 1986; De Bont *et al.*, 1995a). Pitchford *et al.* (1973) calculated the mean daily *S. mattheei* egg output from 149 cattle living in three areas near the Kruger National Park in South Africa and found it to vary between 20 800 and 80 600 eggs per animal per day. Faecal egg counts during clinical outbreaks may vary between 100 and 1000 e.p.g. (Lawrence, 1978a; Dargie, 1980; Markovics *et al.*, 1993).

5.3. Host–Parasite Relationship and Immunity Development

5.3.1. *Experimental Infection Studies*

The first detailed experimental studies were those of Lawrence (1973a,b, 1977a,e) who exposed Friesian calves, 7–11 months old, to between 5000 and 45 000 *S. mattheei* cercariae. Four of these calves were later challenged, two at 20 weeks and two at 95 weeks after the first exposure, with between 15 000 and 21 000 cercariae. Between 8 and 78 weeks p.i., 16 animals were slaughtered and schistosomes were recovered by perfusion. Ten calves were observed for a period of 75 weeks after a single infection, and the faecal egg excretion was monitored.

In the primary infections, the prepatent period was usually 6–7 weeks. Mean faecal egg counts rose to a peak of about 400 e.p.g. at 12 weeks p.i. and then declined sharply, stabilizing at low counts of about 10 e.p.g. after 40 weeks p.i. Paired adult parasites were found mainly in the small intestine and the proximal part of the large intestine. An analysis of the relative concentration of eggs in the different portions of the intestine indicated that 8–9 weeks p.i. (during the period of acute clinical illness) there was a partial shift of the parasites from the small intestine to the distal large intestine. Lawrence (1977e) postulated that schistosomes may temporarily be attracted to the distal part of the large intestine, possibly following changes in the composition of the blood in the veins draining the region. After about 18 weeks p.i. some of the parasites migrated from the intestine to the veins of the urinary bladder and stomachs. As a result, the stomachs accommodated 20% or more of the total population in heavy infections. Lawrence (1977e, 1978c) thought that this migration was probably caused by immunologically mediated inflammatory changes which made the veins of the intestine unfavourable sites for the worms. The liver harboured a large proportion of the unpaired, mainly male parasites in the early weeks after infestation. They were frequently reduced in length and were absent in calves slaughtered during the later course of infection. The percentage worm recovery from animals exposed to the primary infections was 62% at 8 weeks p.i. It then decreased exponentially, with a 50% decline by about week 35 p.i.

As the infection progressed female worms reduced in length, without any other morphological changes, while male worms remained unchanged. The number of eggs *in utero* in females recovered from the mesenteric veins ranged between 0 and 107 eggs, with a mean of 36 (Lawrence, 1977a). There was no apparent correlation between the numbers of eggs *in utero* and the level or duration of infection. Tissue egg counts in the intestine were highest just after patency but fell sharply as the infections progressed. In contrast, egg counts in the liver were relatively low and remained unchanged as the infections progressed. In early infections, eggs in the liver constituted an insignificant proportion of the total tissue egg count, but represented more than 50% of the tissue eggs in the body 80 weeks p.i. This relative increase was attributed to a larger proportion of eggs failing to lodge in the intestinal venules and being carried to the liver.

The sharp reduction in faecal egg counts between 15 and 25 weeks p.i. was mainly attributed to a reduction in the fecundity of the female worms, and only partly to the reduction in the numbers of worms. An estimation of the faecal egg output per female showed an 89% reduction 26–41 weeks p.i., compared with levels observed 15 to 18 weeks p.i. The similar declines in total tissue egg counts observed with time confirmed that the fall in faecal egg output was not caused by the retention of eggs in tissues. The viability of eggs passed in the faeces was estimated during the peak egg output by calculating the proportion of eggs which contained mature miracidia. This proportion varied between 50% and 75% and showed no significant change with time. No change was also recorded during the subsequent period of low egg output, when the viability of eggs was assessed by comparing egg counts with yields of miracidial hatching (Lawrence, 1977a).

By the time the animals were challenged, faecal egg counts from the primary infection had fallen to a low level. After challenge, three animals showed no increase in faecal egg count, while the fourth showed a significant increase for 1 week only (Lawrence, 1973b). This lack of increase in faecal egg excretion after challenge demonstrated that the reduction in fecundity of the female worms from the primary infection was not due to ageing of the worm population but was immunologically mediated. There was no reduction in establishment of parasites from the second infection, but females also had reduced length.

Different results were obtained by Preston and Webbe (1974) who also used a single primary infection with *S. mattheei* followed by a challenge. Twelve Jersey calves aged 3–4 months were divided into three groups of four animals (A, B and C). The calves of groups A and B were exposed to 500 cercariae (primary infection) and, 14 weeks later, animals of groups B and C to 5000 cercariae (challenge infection for animals of group B). All animals were slaughtered and perfused 10–11 weeks after the challenge infection. No reduction in tissue egg counts or worm recoveries were detected after challenge and the only evidence of acquired immunity was a reduction in the length of adult female worms. Preston and Webbe (1974) thought that the lack of significant protection they obtained against challenge resulted from the relatively small numbers of cercariae used for initial exposure. They also suggested that the number of cercariae used in the immunizing exposure may be more important than the number of adult worms that develop.

In the first experiment, using *S. bovis*, Massoud and Nelson (1972) exposed calves to a single primary infection of 1000 cercariae and challenged them 9 weeks later with 4000 cercariae. At autopsy, 9 weeks after challenge, there was no significant difference in the numbers of adult worms recovered in immunized animals compared with controls, but there was a significant reduction (58%) in tissue egg counts in the intestines. Later, Saad *et al.* (1980) examined the host–parasite relationships in calves slaughtered 3, 6, 9 and 12 months after a single experimental exposure to *S. bovis*, at a dose rate of either 100 or 200 cercariae kg^{-1} body weight (approximately 10 000 or 22 000 cercariae per calf). They reported results strikingly similar to those obtained by Lawrence after primary infections with *S. mattheei*, except that worm recovery rates at perfusion were much lower (10% to 14%, 3 months after infestation). Interestingly, the intensity of the *S. bovis* infections had no effect on the worm mortality rate — the worm burden decreased in a similar way by 51–60% between months three to six p.i. in animals with light and heavy infections (Saad *et al.*, 1980). Bushara *et al.* (1983b) exposed six calves aged 6–9 months (group P) to 10 000 *S. bovis* cercariae and challenged them 26 weeks p.i. with 20 000 cercariae, together with six uninfected controls. At perfusion 16 weeks after challenge, animals of group P were found to be highly resistant to the challenge. Faecal and tissue egg counts in group P were 78%–100% lower than in the controls. However, the calculated reduction in worm burden in the P group compared with the controls was only 11%. This further confirmed that the immunity stimulated in cattle by a single primary infection can induce significant reductions in the fecundity of the worms from the challenge infection, but does not protect against maturation of these worms.

It can be concluded from the above studies that experimentally infected cattle can develop an extremely effective immune response against *S. mattheei* or *S. bovis* (both in primary infections and in response to challenge infection), and that this immunity mainly acts through suppression of worm fecundity. It appears that the strength of the immunological response of the host, which results in the reduced worm fecundity, could be related to the intensity of the primary infection: low to moderate levels of infection fail to stimulate the host response. It should also be noted that little or no effect on worm burden was observed in all these experimental studies.

5.3.2. *Variation in Egg and Worm Counts in Naturally Infected Animals*

Observations on the faecal egg excretion in cattle living in areas endemic for *S. bovis* and *S. mattheei* have shown that egg counts, and prevalence rates, decrease with age of the host. In herds living in contact with cercarial-infested waters, newborn calves are likely to become infected as soon as they are allowed to join their mother for grazing. Faecal egg excretion may start as early as during the second month after birth (Majid *et al.*, 1980a), but begins more commonly between the 4 and 8 months of life (Majid *et al.*, 1980a; De Bont *et al.*, 1995a). Counts increase rapidly to reach a maximum of 70–310 e.p.g. at the age of 6–15 months, and then decrease markedly by the age of 18 months (Majid *et al.*, 1980a; Pitchford and Visser 1982; De Bont *et al.*, 1995a). In older animals, faecal egg counts remain below 20 e.p.g. and appear not to be affected by any seasonal variations in levels of transmission (Majid *et al.*, 1980a; Kassuku *et al.*, 1986; De Bont *et al.*, 1995a) or experimental challenge (Bushara *et al.*, 1980). Comparison of the numbers of eggs and miracidia g^{-1} faeces in groups of calves, heifers and adult cows on a farm infected with *S. mattheei* showed that the reduction in faecal egg excretion recorded as infection progresses was associated with a decline in the ability of the eggs to hatch (De Bont *et al.*, 1996a): 50% of the eggs from calves hatched in water, compared with only 15% of eggs from adult cows. The decline in egg viability was partly associated with morphological changes of the eggs: over twice as many smaller and vacuolated eggs were found in the faeces of heifers and adult cows (34%) compared with animals in early infection (16%).

From the very limited data available, it appears that total tissue egg counts in naturally infected animals follow, with the age of the host, a similar pattern to that of faecal egg counts. Two adult cows aged 7 years had 14 times less *S. mattheei* eggs in their tissues than two calves aged 9 months from the same farm (De Bont *et al.*, 1995a). The marked decline in tissue egg counts in older cows was accompanied by a shift of egg accumulation from the large intestine towards the liver, compared with young animals. A similar shift was reported by Bushara *et al.* (1980) after challenging adult cattle naturally infected with *S. bovis*.

In contrast to faecal egg counts, worm burdens in naturally infected animals increase with the age of the host. In Sri Lanka, an age-specific increase in prevalence and intensity of *S. spindale* infections was observed in bulls examined in an abattoir, as measured by worm counts *in situ* (De Bont *et al.*, 1991a). In a total of 174 infected animals classified in five age groups (<2, 2, 3, 4 and ≥5 years), the average worm burden increased by approximately 20% for each step up in age group. Perfusion data from young (9 months) and adult (7 years) animals from a herd in Zambia showed that adult cows had about 2.5 times more *S. mattheei* worms than calves (De Bont *et al.*, 1995a). However, worm burdens in adult animals from normal endemic areas appear to be low compared to those observed in tracer calves used for less than a year in the same areas (Majid *et al.*, 1980b; De Bont *et al.*, 1995b, 1997). In fact, evidence is accumulating to suggest that, with increasing duration of exposure to natural challenge, cattle become less susceptible to reinfection. Bushara *et al.* (1980, 1983a) exposed adult animals from endemic (putatively resistant cattle) and nonendemic (controls) areas to 70 000 *S. bovis* cercariae, and observed that mean worm counts were reduced by 30–85% in putatively resistant animals compared with controls. The development of protection against reinfection with *S. mattheei* was clearly demonstrated by the use of two groups of tracer calves on a Zambian farm (De Bont *et al.*, 1995b). A group of permanent tracers was introduced onto the farm at the beginning of the experiment. They were gradually removed and perfused after residential periods of 2, 4, 6, 8, 10 and 12 months — the aim was to examine the development of *S. mattheei* infections during the first year of natural exposure. During the same period, groups of temporary tracers were used on the farm to measure, per period of 8 weeks, the level of infection to which permanent tracers were exposed. Every 2 months, a group of temporary tracers was introduced onto the farm for a period of 2 months only. After being removed from the farm, the temporary tracers were penned on concrete flooring for 45 days to allow for maturation of recently acquired infections, and subsequently perfused. In permanent tracers there was an exponential increase in mean worm burden from 29 worms after 2 months to 974 worms after 1 year. The latter value was much lower than the cumulative number of worms picked up by temporary tracers during the same period (1799 worms), which could indicate a rapid development of resistance to reinfection in the presence of continuous challenge.

The acquired reduction in worm fecundity as infection progresses implies that a priori faecal egg counts cannot be used as an indicator of the worm and tissue egg burdens, when the duration of infection is unknown. Predictions of internal burdens made from faecal egg counts would be of great use in our understanding of schistosome infections. In endemic areas, the immune-related decline in faecal egg production is generally observed in cattle older than 18 months. Therefore, De Bont et al. (J. De Bont, D.J. Shaw and J. Vercruysse, unpublished data) examined *S. mattheei* adult female worm burdens (WP), total tissue egg counts (TEC) and faecal egg counts (e.p.g.) from 30 Friesian calves aged less than 18 months at slaughter. The calves had been naturally exposed to infection over periods of time ranging between 2 and 12 months. A significant positive relationship was found between log-transformed WP and e.p.g. Using the overall geometric mean burden of 90 WP, the overall e.p.g. to WP ratio was calculated as 0.32:1. Using the overall relationship, a faecal egg count of 10 e.p.g. during the first year of natural *S. mattheei* infection would indicate a parasite burden of 20 WP, whereas a count of 100 e.p.g. would indicate a burden of 540 WP. The slope of the overall e.p.g. = WP relationship was less than one (0.699), suggesting a decrease in egg counts as worm burdens increased. A significant positive relationship was also observed between the log-transformed TEC and e.p.g. with an overall e.p.g. to TEC ratio of 0.000063:1, calculated from the overall geometric mean of 445 000 tissue eggs. Using the overall relationship, a faecal egg count of 10 e.p.g. during the first year of natural *S. mattheei* infection would indicate a TEC of 101 000 eggs, whereas a count of 100 e.p.g. would indicate a TEC of 2.7 million eggs. The slope of the overall relationship was also less than one (0.701), suggesting a decrease in eggs observed in faeces, with increasing tissue egg counts. The overall relationship between WP and TEC was more complex, and followed a sigmoidal pattern. At low adult female worm numbers there was an exponential increase in TEC. This rate of increase began to level off around 200 WP, resulting in a plateauing of TEC at higher burdens. Therefore it appears that the decrease in e.p.g. = WP ratio as worm burdens increase is due to a larger accumulation of eggs in tissues, but there may also be a reduction in fecundity at higher worm burdens.

The results clearly show that the intensity and morbidity of *S. mattheei* infection in cattle can be predicted from faecal egg counts during the first year of moderate natural infection. However, the exact relationship between these parameters is likely to change dramatically when acquired immunity starts affecting worm fecundity. Faecal egg counts remain invariably low ($\leq$ 10 e.p.g.) in immune cattle and can no longer indicate possible variations in worm burden or tissue egg counts. Moreover, while the proportion of eggs that accumulate in the liver (not detected by faecal examination) may be negligible in very early infections (about 5% of the total count), it is known to increase significantly with time. In older infections about 40–50% of the total tissue eggs may be found in the liver (Lawrence, 1977e; Bushara et al. 1980; De Bont et al., 1995a) in which the development of considerable lesions may escape detection by faecal examination.

5.3.3. Studies on the Mechanisms of Acquired Immunity

(a) *Acquired immunity or age-dependent resistance?* The decline in faecal egg counts with age observed in herds exposed to natural challenge is not generally associated with a reduction of exposure to infection (see Section 5.4.1). Cattle come into contact with water when drinking, and all animals from the same herd will generally spend the same amount of time at the water contact point. However, cattle reach their sexual maturity around the age of 12–15 months, and it is therefore possible that the marked decline in faecal egg counts which occurs before the age of 18 months is not only related to experience of infection but also to age. Adult cattle probably differ from calves in their susceptibility to infection. The relative importance of experience and age in resistance to infection in human schistosomiasis (where resistance to reinfection also appears to exist in adults) has, in recent years, been the subject of active debate (Stelma, 1997). Gryseels (1994) argued that resistance is not solely due to immunity slowly acquired during childhood, but also to factors inherent to age itself. One possible factor is the sexual maturation of the host which could play a role by establishing or strengthening physiological barriers or shifting humoral and/or cellular immune responses (Gryseels, 1994).

To try to test whether the age-related decline in faecal egg excretion in cattle is due to age-dependent resistance and/or acquired immunity, Bushara et al. (1980) experimentally challenged (70 000 *S. bovis* cercariae) five adult animals (aged 4–10 years) from an endemic focus in Sudan (resistant cattle), together with another five cattle of similar breed and age from a nonendemic area (naive cattle). Preliminary examination and perfusion of three animals from each area had shown that the putatively resistant cattle, although infected with large numbers of adult worms, were negative on faecal examination, whereas the naive cattle were not infected. After challenge, clinical observations showed that the resistant animals were able to almost completely withstand the effects of challenge, whereas the naive cattle developed lethal infections. Prior to the death of the naive cattle, resistance was further demonstrated by differences between the two groups in terms of body weights, haematological measurements, histopathological and pathophysiological responses, and egg counts. These results clearly confirmed experimental observations (see Section 5.3.1) that cattle acquire a high degree resistance to schistosome infection.

Worm counts in the challenged resistant animals were high (about 70% of the level in controls), and Bushara et al. (1980) concluded that the main basis of naturally acquired resistance was against the fecundity of the incoming worms, rather than absolute prevention of their maturation. The development of clinical schistosomiasis in the naive animals suggests that pure age-dependent resistance to schistosome infection is unlikely in cattle. This assumption is further supported by reports of clinical cases in adults during field outbreaks in susceptible herds (Kulkarni et al., 1954; Lawrence and Condy, 1970; Lawrence 1977b; Wang et al., 1988). However, susceptibility of adults to heavy experimental (70 000 cercariae; Bushara et al., 1980) or natural (see Table 3 for data on worm counts during outbreaks) challenge does not preclude the possibility of age-dependent resistance occurring in cattle. Further research work is necessary to clarify this situation.

(b) *Development and maintenance of resistance.* In a first experiment on the mechanisms of naturally acquired resistance, Bushara et al. (1983c) surgically transplanted between 700 and 4000 adult *S. bovis* worms, from resistant cattle with very low faecal egg counts (0–8 e.p.g.) into adult naive recipients. Faecal egg excretion in recipient cattle started between 5 and 16 days after transplantation, and reached peak counts of 55–405 e.p.g. between 6 and 20 days after the operation. After the peak, counts decreased more sharply in animals with high egg counts than in those that had lower peak egg counts. All animals were perfused 46–56 days after transplantation, when between 0.1% and 78.5% of the transplanted worms were recovered. The lowest worm recovery was recorded in the animal which received the largest number of transplanted worms. The fact that the worms were able to achieve high levels of egg production when transplanted into naive recipients suggested that the earlier suppression of egg production was induced by specific immunological mechanisms in the resistant donors. Bushara et al. (1983c) also noted that the recipient animals had rapidly become immunized against the transplanted worms and that this reaction of the host appeared to be dose-dependent.

In another transfer experiment, Bushara et al. (1983b) demonstrated that some resistance could be stimulated in the absence of the migratory stages, by transplantation of adult worm pairs, and in the absence of both migratory stages and eggs, by the transplantation of adult male worms. Adult *S. bovis* worms obtained from calves perfused 14 weeks after a single experimental infection were surgically transplanted into two groups of naive recipient calves: five calves received 500 worm pairs each (WPR group), whereas another six received between 650 and 1000 male worms alone (MR group). Ten weeks after surgery, the calves were challenged with 20 000 *S. bovis* cercariae, together with six naive control calves (CC group), and all animals were perfused 16 weeks later. The WPR and MR groups had, on average, 43% and 37% fewer worms than the CC group, and the mean tissue egg counts were lower by 39–63% and 63–76%, respectively.

To test whether the adult worms surviving in naturally resistant cattle were necessary for the maintenance of resistance, Bushara et al. (1983a) cured six adult resistant cattle with a double treatment of praziquantel (group T) and challenged them, 7 weeks later, with 70 000 cercariae of S. bovis, together with six untreated groups of resistant adult cattle (group U) and six adult naive controls. All animals were perfused 16 to 18 weeks after challenge. Compared with the controls, average worm burdens in the T and U groups were reduced by 85% and 69%, and average tissue egg counts by 72–99% and 56–80%, respectively. The results confirmed that untreated, naturally infected cattle were highly resistant to experimental reinfection. Protection not only acted through a reduction of worm fecundity, but also through protection against reinfection. More importantly, the experiment showed that naturally acquired resistance was not abrogated by a treatment.

In the first experiment examining the role of serum components in immunity, Bushara et al. (1983c) injected 1000 ml of pooled sera from resistant adult cattle intraperitoneally into naive recipient calves (6–9 months old), and challenged them the next day with 7500 S. bovis cercariae. The results showed that the immune sera had a neglible effect on the numbers of worms that developed, and no significant effect on the faecal egg counts. However, from the tissue egg counts there was some evidence of a reduction in the fecundity of the worms in calves injected with immune sera. It was thought that the effect on fecundity was mediated by an immunoglobulin. Bushara et al. (1983c) hypothesized that most of the putatively suppressive immunoglobulin was catabolized by the time the worm pairs matured and that different results could have been obtained if the timing or quantity of serum had been different.

This hypothesis was later tested in two different experiments (Bushara et al., 1994). One experiment was designed to assess the effects of repeated transfers of serum on worm fecundity, by injecting immune serum during the period of worm sexual maturation, rather than before infection. The setup was similar to the one used 10 years before (Bushara et al., 1983c), except that the recipient calves received weekly injections between weeks four and 12 p.i., to a total of 4500 ml per calf. At perfusion, no significant difference in worm or in faecal or tissue egg counts was seen in the recipients. In the other experiment, the serum donors were experimentally infected calves. They were bled before infection, and at week four p.i. (corresponding to the prepatent infection), week 8 p.i. (corresponding to the increasing faecal egg excretion), and week 12 p.i. (corresponding to the beginning of decline in faecal egg counts). The four pools of sera were attributed to four different groups of recipient calves. Injections started 4 weeks after experimental infection of the recipient calves and were given weekly over a period of 7 weeks, up to a total of 2000 to 3500 ml per calf. Reductions in egg counts and worm burdens were recorded in the three groups which received serum collected after infection of the donors, compared with the control group which received serum collected before. These results provided some evidence that fecundity suppression following acquired resistance to *S. bovis* in cattle is due to at least serum-borne factors.

(c) *Concomitant immunity*. The partial protection against worm establishment observed in animals exposed to natural infection (De Bont et al., 1995a,b) could be interpreted in terms of the concomitant immunity model proposed by Smithers and Terry (1967, 1969). They observed that either a single primary exposure to normal cercariae of *S. mansoni* in rhesus monkeys, or the surgical transfer of adult *S. mansoni* worms in a naive monkey could induce good resistance against a challenge infection. In contrast, an exposure to irradiated cercariae failed to do so. They concluded that the adult worms provided the major stimulus to resistance, inducing an immunity which could readily destroy incoming schistosomula but to which the established adults were resistant. It was suggested that adult worms from the primary infection had acquired a surface coat of host molecules that allowed them to evade immune responses, while new incoming juveniles that had not yet acquired a protective coat were, at least for a short time, susceptible to the immune mechanisms generated and maintained by the adults. This implied that the concomitant immunity hypothesis regarded the schistosomula as the sole target of the protective immune response. Some support for concomitant immunity occurring in cattle is provided by the transplantation experiments of Bushara et al. (1983b). These authors found that some resistance to *S. bovis* cercarial challenge could be induced in calves surgically transplanted with either egg-producing pairs or adult male worms only.

However, there is also evidence to suggest that concomitant immunity is not the only form of acquired resistance to challenge schistosome infection. First, live vaccines with irradiated cercariae have been shown to induce a strong protective immunity in cattle (see Section 6.2.2.b. for more details). Irradiated cercariae fail to mature in the host, which implies that vaccinated animals become immune without ever being exposed to adult worms. It is also uncertain whether the continuing presence of adult worms is necessary to maintain the protection. Bushara et al. (1983a) demonstrated that naturally acquired resistance in cattle was not diminished 7 weeks after cure with praziquantel. However, further experiments are needed to determine how long after cure resistance is maintained. Second, there is now evidence to support the view that adult worms are also susceptible to immunological attack. Protective mechanisms mainly act through a reduction of the fecundity of the female worms, and possibly also through a reduction of the viability of the produced eggs (De Bont et al., 1996a). However, it is not known if the immune effector mechanisms are capable of killing adult worms. There is little information available on the longevity of adult schistosome worms in infected cattle. Lawrence (1977a) observed a rapid elimination of worms between 18 and 40 weeks after a single exposure to 20 000 cercariae of *S. mattheei*, but considered it unlikely that elimination of parasites would occur in very lightly infested cattle. In experimental *S. bovis* infections (Saad et al., 1980), worm burdens decreased by 51–60% between months three to six p.i., and a similar percentage of worms died in animals exposed to 100 or 200 cercariae kg^{-1} body weight. If the average lifespan of a worm is considered to be of 10 months, the maintenance of the relatively stable worm population observed in adult naturally infected animals reflects a dynamic equilibrium between gain and loss of parasites rather than the effect of a high degree of concomitant immunity.

(d) *Importance of the type and size of infection on immunity development.* Differences in levels of protection achieved after either experimental or natural infection suggest that the type and size of exposure to infection may play a major role on the strength of the immunological response of the host. Results from Zambia (De Bont et al., 1995b) suggest that the size of the primary challenge also affects the worm population dynamics in cattle exposed to moderate natural challenge. The level of transmission to which the herd was exposed on the farm was measured per period of 2 months by counting the number of worms picked up by temporary tracers. Two groups of permanent tracers (meant to stay on the farm for up to 1 year) were introduced onto the farm at different times. In permanent tracers introduced during a period of heavy transmission (430 worms in a temporary tracer), increased worm counts (4.6 times higher) and a greater susceptibility to reinfection were observed compared with permanent tracers introduced during a period of lighter transmission (70 worms in a temporary tracer). Therefore, worm burdens observed in adult animals may well reflect those acquired at the very early stage of infection. This stresses the importance of the level of cercarial challenge to which young calves are exposed.

5.3.4. *Studies on Heterologous Resistance*

Heterologous resistance studies were initiated by Nelson and colleagues who showed that mice and rhesus monkeys could be partially protected against a challenge with *S. mansoni* if they have been previously exposed to *S. bovis* or *S. mattheei* (Amin et al., 1968; Nelson et al., 1968; Amin and Nelson, 1969). In cattle, they tested whether calves could be protected against infection with *S. mattheei* or *S. bovis* by previously exposing them to *S. mansoni* or *S. haematobium*. In a first experiment, Hussein *et al.* (1970) exposed three calves to 10 000 *S. mansoni* cercariae and challenged them 8 weeks later with 6000 cercariae of *S. mattheei*. At perfusion, the *S. mattheei* worm loads in the three calves were reduced by 40%, 76% and 94%, respectively, compared with the mean burden in three control calves exposed to 6000 cercariae of *S. mattheei* only. Less severe pathological changes (associated to some reduction in tissue egg counts) were also observed in the calves primed with *S. mansoni*. In a later series of experiments, Massoud and Nelson (1972) used either single doses of 8000 cercariae of *Ornithobiolharzia turkestanicum* or 5000 cercariae of *S. bovis*, or three doses of 7000 cercariae of *S. haematobium* given at 4-week intervals. The following combinations were tested in groups of calves: (1) primed with *O. turkestanicum* and challenged with *S. bovis*; (2) primed with *S. bovis* and challenged with *O. turkestanicum*; (3) primed with *S. haematobium* and challenged with *S. bovis* or (4) *O. turkestanicum*. In each of the four groups reductions in worm burdens (30–42%) and tissue egg counts in the small intestine (45–83%) and large intestine (76–91%) were recorded and compared with controls.

In many areas of the world, cattle are dependent for their water on the same limited supply as the human population. Massoud and Nelson (1972) suggested that the simultaneous transmission of schistosomes of humans and cattle at the same site could be to the mutual benefit of man and livestock in reducing the effects of schistosomiasis. However, this hypothesis has never been tested in the field. These authors also suggested that *S. mansoni* or *S. haematobium* could be used as immunizing agents in cattle, particularly *S. haematobium* since it does not produce viable eggs in ruminants.

Heterologous immunity between *Schistosoma* and *Fasciola* spp. has also been experimentally demonstrated in cattle. Sirag *et al.* (1981) infected calves orally with 900 metacercariae of *F. hepatica* 10 weeks after a primary exposure to 10 000 cercariae of *S. bovis*. At necropsy 15 weeks after challenge, there was a significant reduction by 30% in the mean number of *F. hepatica* worms and less pronounced liver tissue damage in calves infected with *S. bovis*, compared with controls. Yagi *et al.* (1986) recorded a significant reduction of 94% in *S. bovis* worm establishment in calves which had been exposed 8 weeks earlier to 1000 metacercariae of *F. gigantica*. In the reverse experiment (i.e. calves first exposed to 10 000 *S. bovis* cercariae, then infected with 1000 metacercariae of *F. gigantica*), they recorded an 84% reduction in *F. gigantica* worm recoveries. Concurrent infections with schistosomes and liver flukes are very common in cattle (Chartier *et al.*, 1990). Therefore, the high levels of cross-protection obtained experimentally between the two trematode infections are likely to have important epidemiological implications, and call for further studies on the impact of heterologous resistance between schistosomes and liver flukes.

5.4. Transmission

5.4.1. *Exposure to Infection*

Knowledge of the pattern of exposure to infection is essential to an understanding of the epidemiology of cattle schistosomiasis. Large differences in the type and frequency of water contacts may exist from one place to another. Cattle are usually not free to go or graze wherever they want and their water contact activities are therefore dictated more by livestock owners and their traditions. For example, in Sri Lanka where dairy cows are usually kept indoors, it is common daily practice for farmers to wash their cattle one by one in the local pond or river (De Bont *et al.*, 1991a,c). However, in most cases, cattle will have free access to water holes while grazing, or will be led one to three times a day to an irrigation canal or river (Majid *et al.*, 1980a, Kassuku *et al.*, 1986). Van Wyk *et al.* (1974) reported that in drier parts of southern Africa, some herds may only come to water every 24–48 h. Two factors related to water contacts have a major bearing on the epidemiology of cattle schistosomiasis. The first one, is that in most cases cattle contact water only for drinking. This implies that the type of water supply used by the farmer plays an important role in the transmission of schistosomiasis. This correlation has for example been demonstrated by Pitchford *et al.* (1973) who, within a limited area of South Africa, observed an *S. mattheei* prevalence rate of 92% and a mean daily egg output of 81 000 eggs in herds obtaining water from canals and night-storage dams, compared with a prevalence of only 21% and a mean daily egg output of 21 000 eggs in herds obtaining water from troughs and with occasional access to natural streams, and no detectable infection in herds watered with piped water only. The second factor is that cattle generally move as a herd, which implies that within the same group of animals the frequency of water contacts is unlikely to change with age.

Data on the behaviour of cattle in relation to water is scarce, but it is known that they often enter and wade into water when drinking (Pitchford 1963). It is generally assumed that infection of cattle mainly occurs by cercarial penetration into the skin (Christensen *et al.*, 1983). However, outbreaks of schistosomiasis have been reported in herds which only contacted water when drinking from troughs containing large numbers of infected snails (Lawrence and Condy, 1970; van Wyk *et al.*, 1974), suggesting that infection occurred *per os*. The possible importance of the oral route of infection in ruminants has been demonstrated experimentally, either by introducing cercariae directly into the rumen or by allowing animals to drink water containing cercariae. Relatively poor results have at first been obtained when cercarial suspensions were directly deposited into the rumen with a stomach tube. Fairley and Jasudasan (1930) recovered only 290 *S. spindale* worms from a goat which had been intraruminally exposed to 98 000 cercariae 86 days previously, and Reinecke and Kruger (personal communication of 1973, cited by van Wyk *et al.*, 1974) only recovered four *S. mattheei* worms from a sheep exposed the same way to 15 000 cercariae. Van Wyk *et al.* (1974) argued that the volumes of cercarial suspensions used in these experiments were too small to prevent the possible adverse effects of the ruminal fluids on cercariae. They further stressed that ruminants, particularly those which only drink every 24–48 h may at once swallow very large volumes of water which, they argued, would dilute the ruminal content and help cercarial penetration. This hypothesis was tested in sheep and 44% *S. mattheei* cercariae developed in one sheep when directly introduced into the rumen with a large volume of water, compared with 14% in a similar infestation without dilution, and 72% in percutaneous infestation (van Wyk *et al.*, 1974). Much better recovery rates have been obtained after contact with the upper digestive tract. Fairley and Jasudasan (1930) obtained massive infections with *S. spindale* in two goats after the cercarial suspension was dropped slowly onto their tongue and buccal cavity. Kassuku *et al.* (1985) obtained relatively high *S. bovis* parasitic loads in goats allowed to drink cercarial infested water, as measured by total tissue egg counts at necropsy. The use of radio-isotopically labelled cercariae facilitated determination of penetration sites following uptake by drinking. Radioactivity could be demonstrated in the outer lips (41% of the penetrations), buccal cavity (10%), oesophagus (3%), reticulum (22%), omasum (3%) and rumen (21%).

An exception to the general rule is *S. japonicum*. Here, the intermediate snail host may spend prolonged periods of time out of water, crawling on emergent vegetation and mud surfaces. The snails have been reported to shed cercariae into drops of rain or dew, and humans and livestock have been infected by coming in contact with wet soil on river banks or wet vegetation growing near infested water bodies (Wang *et al.*, 1958; Cheng, 1971; Basch, 1986).

5.4.2. *Contamination of the habitat with cercariae*

Different factors are known to affect the degree of contamination of the habitat with cercariae. On the one hand, ecological factors affecting the biology and survival of the intermediate host population, the rate of intramolluscan development and cercarial shedding patterns are well documented (Christensen *et al.*, 1983; Jourdane and Théron, 1987; Basch, 1991). The effects of rainfall and temperature on these factors are the most consistent, resulting in clear seasonal patterns of transmission in various habitats (Malek, 1969; Pitchford *et al.*, 1974; Majid *et al.*, 1980a; Chandiwana *et al.*, 1987b; De Bont *et al.*, 1995a,b). The intramolluscan development can be interrupted at low winter temperature, and start again the following spring (Pitchford and Visser, 1965). The examination of *B. globosus* infection rates and cercarial population size, as monitored using hamster immersions showed that the transmission of *S. mattheei* in Zimbabwe was most intensive during the hot, dry season (September–November), moderate and variable during the rainy (December–February) and warm, post-rainy (March–May) seasons, and markedly reduced during the cold, dry season (June–August) (Chandiwana *et al.*, 1987b).

On the other hand, however, the capability of the definitive host to contaminate the environment with viable eggs may also affect the level of transmission. As mentioned earlier, little has been published on the behaviour of cattle at water contact sites, except that they often defaecate in water when drinking (Pitchford *et al.*, 1973). In Sri Lanka, De Bont *et al.* (1991a) used a miracidial hatching test to measure the excretion of *S. spindale* viable eggs in 85 infected cattle of different ages. The average numbers of miracidia passed per 100 g faeces decreased markedly with age from 96 miracidia per 100 g faeces in animals of 2 years and younger, to 12–16 miracidia in cattle of 3 years and older. However, these calculations are concerned with potential rather than actual contamination. Therefore, despite the age-related increase in the amount of faecal material produced (by five to ten times), as the number of eggs produced in faeces (i.e. from >100 e.p.g. to <10 e.p.g.) and their viability (i.e. from 50% to 15%) decline, the overall miracidial contamination is probably mainly from young animals in early infection.

Detection of transmission or its quantification at the water contact sites may be achieved through measurement of snail and infected snail densities or the use of sentinel rodents (Pitchford and Visser, 1962, 1965; Donnelly and Appleton, 1985; Joubert *et al.*, 1987; Chandiwana *et al.*, 1987b; De Bont *et al.*, 1995a). A better measure of the level of transmission is obtained by calculating the incidence rate of new infections, that is, the numbers of conversions (uninfected which become infected) in a particular group of cattle over a given period of time. Majid *et al.* (1980a) carried out monthly faecal examinations of a group of about 100 uninfected calves (negative on faecal egg examination) aged 1–9 months and which were allowed to move freely with the other cattle in a Sudanese village. Every month, those calves found positive were discarded and replaced by new negative calves. This process was continued over a period of 2 years and it was demonstrated that the transmission of *S. bovis* followed a clear seasonal pattern, being much higher in the hot summer months. Kassuku et al. (1986) carried out monthly faecal examinations over a period of 10 months on samples collected from a total of 29 calves entering the weaner's group on a Tanzanian farm. Miracidial counts showed that transmission of *S. bovis* on the farm occurred throughout the year, but was much higher during the drier periods before and after the rainy season. In these two studies, levels of transmission measured by faecal examination followed seasonal patterns similar to those observed during snail studies on the same farms. However, faecal examinations only provide an indirect measure of the intensity of infection. Light and single-sexed infections may escape detection, and it may therefore be necessary to use post-mortem examinations of tracer calves to obtain accurate data on transmission. In Zambia, a total of 14 tracer calves were used on a farm to measure seasonal variations in levels of *S. mattheei* transmission (De Bont et al., 1995b). They were aged 3–4 months and came from another farm where *Schistosoma* is absent. Every second month, a new group of up to three tracers was introduced and mixed with the farm herd for a period of 2 months. Schistosome infections were established in all the tracers, indicating that transmission of *S. mattheei* occurred throughout the year on the farm. Generally, worm counts after each period of 8 weeks correlated well with the respective infected snail densities recorded at the main transmission site on the farm: worm counts ranged from 71 worms during the cold dry season to 736 worms at the end of the warmer rainy season.

6. TREATMENT AND CONTROL

6.1. Treatment

Before praziquantel was launched on the market 20 years ago, numerous drugs with known schistosomicidal effect had been tested against visceral schistosome infections in cattle. Antimonials such as antimony potassium tartrate (tartar emetic), antimony pyrocatechol sodium disulphonate (stibophen or antimosan) and antimony dimercaptosuccinate (stibocaptate) were successfully used against *S. mattheei* and *S. japonicum* (McCully and Kruger, 1969; Lawrence and Schwartz, 1969; Reinecke, 1970; Shi, 1981). However, antimonials also have toxic effects, sometimes leading to the death of the animals being treated (Reinecke, 1970; Hussein, 1973; Lawrence 1978a), and are therefore not recommended for large-scale treatments. Lawrence and Schwartz (1969) found lucanthone hydrochloride (precursor of hycanthone) effective against *S. mattheei*. However, hycanthone is now suspected of being carcinogenic and is no longer used for the control of human schistomiasis. Isothiocyanates (nithiocyamin or amoscanate, nitroscanate) have been used quite widely in China for the control of *S. japonicum* in cattle (Qin *et al.*, 1982; Chen *et al.*, 1985), but again, amoscanate may cause severe side-effects in cattle (for example in Ling and Mao, 1981).

The use of trichlorphon (neguvon or metrifonate) has provided conflicting results in cattle. Administered orally at a dose rate of 50 or 75 mg kg^{-1} for four to six treatments, it was well tolerated and gave good results in five cattle infected with *S. bovis* (Dinnik, 1967), but caused either very severe side-effects or death in other animals infected with *S. mattheei* (Lawrence and Schwartz, 1969). Given by intramuscular injection (eight to 11 injections at 8–20 mg kg^{-1}, with intervals of 3–4 days), trichlorphon was shown to have some activity against *S. mattheei* infections (van Wyk *et al.*, 1974), but little or no effect against *S. bovis* (three injections at 25 mg kg^{-1}, with intervals of 3 days) (Bushara *et al.*, 1982).

Praziquantel has a high therapeutic efficacy, is easy to dose and is well-tolerated by patients. It is therefore, generally recognized as the antischistosomal drug of choice (Gönnert and Andrews, 1977; Davis, 1993). In cattle it is administered orally at a dose rate of 20–25 mg kg^{-1}, generally given twice with intervals of between 3 days and 5 weeks. Praziquantel has been used successfully against *S. bovis* (Bushara *et al.*, 1982, 1983a; Markovics *et al.*, 1985), *S. spindale* (Upatoom *et al.*, 1988) and *S. japonicum* (Hu *et al.*, 1989; Yuan, 1993).

All the antischistosomal drugs tested against visceral schistosomes in cattle have also been tested against *S. nasale*, and a detailed account of the different trials has been provided by Agrawal and Alwar (1992). They opined that antimonials were effective in treating field cases, but that side-effects and relapses were common. Treatment trials using a single oral administration of praziquantel at a dose rate of 20 mg kg^{-1} body weight gave mixed results: it was reported to be highly effective (clinical recovery and considerable reduction in the numbers of eggs in the nasal discharges) by Rahman *et al.* (1988), but unrewarding (persistence of nasal granulomas, eggs, miracidia and adult worms up to 12 weeks after treatment) by De Bont *et al.* (1989b). In the latter paper, the authors argued that the amount of granulation tissue which causes the snoring symptoms is probably so large that an idiopathic treatment appears to be incapable of bringing any symptomatic improvement.

If numerous treatment trials against schistosomes in cattle have been published, there is in contrast little or no information on how much these drugs and treatment schedules have actually been used in the field — except for China where millions of cattle have been treated within the framework of control programmes against *S. japonicum* (Taylor, 1987). In other parts of the world, the use of chemotherapy appears to be restricted to (relatively uncommon) outbreaks of clinical schistosomiasis (Lawrence, 1978a). However, Reinecke (1970) warned that indiscriminate treatment of clinically affected animals may produce more serious consequences than the disease itself. Worms paralysed or killed by treatment move towards the liver where they may cause extensive thrombosis (McCully and Kruger, 1969). Schistosomes are relatively large worms (>10 mm) and the sudden accumulation of considerable numbers of them (up to several tens of thousands) in the portal veins may cause occlusion and focal hepatic infection (McCully and Kruger, 1969). Therefore, Reinecke (unpublished report of 1964, cited by van Wyk *et al.*, 1974) recommended the use of multiple-dose treatment schedules which kill worms over an extended period (thus causing a more gradual accumulation of dead worms in the liver), compared with single-dose treatments with a highly effective drug.

6.2 Control

Obviously, the best way to prevent schistosomiasis is to keep the cattle away from potentially dangerous waters. Pitchford (1966) controlled *S. mattheei* infections in cattle on a South African farm by fencing snail-infested waters and supplying cercariae-free water in troughs.

Wherever regular contact with potentially dangerous waters cannot be avoided, various snail control measures can be applied to reduce the risk of transmission. In most habitats, snail densities and schistosome transmission are seasonal (see Section 5.4.2), and measures to control snails only need to be applied when high densities of infected snails are expected. Snail control may be achieved through environmental, chemical or biological means (reviewed by Webbe and Jordan, 1993). Environmental changes to the snail habitat, such as drainage, increased water flow or removal of water weed may be sufficient to eliminate the snails. Intermittent drying of canals, reservoirs and troughs may also be effective, provided that the snails are not capable of surviving dessication (for example, *B. umbilicatus* in West Africa can survive dry periods of 6–8 months (Diaw *et al.*, 1989)). For the control of human schistosomiasis, application of chemicals (in recent years practically only niclosamide) is the most important method used to control snails (see Mott, 1987). Niclosamide is a highly effective synthetic molluscicide and is biodegradable. However, its application requires highly skilled personnel, and whether it can be used without any adverse environmental effects is debatable (Madsen, 1990; Webbe and Jordan, 1993).

Detailed reviews of the biological methods which could be used to control snails have been provided by Jordan *et al.* (1980) and Madsen (1990). Biological methods of control include the use of other snails (e.g.

Helisoma duryi) which compete with the intermediate snail host species for the same environment, or of specific snail predators such as malacophagous fishes (De Bont and De Bont-Hers, 1952). Predators may also be used to eliminate the free larval stages of the schistosome. Muraleedharan *et al.* (1975) observed in the laboratory that the guppy fish *Poecilia reticulata* (*Lebistes reticulatus*) could eat about 90% of the *S. nasale* cercariae shed by an infected snail.

Some trematode species (such as *Echinostoma malayanum*) can be used in the control of schistosomiasis. These trematodes may not only interfere with the reproductive capacity of the intermediate snail host, they may also exert an antagonistic effect against the larval stages of the schistosomes inside the snails. Several trials to control *S. spindale* with *E. malayanum* have been carried out with success in Malaysia and Thailand (Heyneman and Umathevy, 1967; Lie *et al.*, 1970, 1971, 1974a). The failure of one trial (Lie *et al.*, 1974b) was attributed to low environmental temperatures and high turbidity of the water.

Oncomelania snails are amphibious and require both land and water control measures. However, massive efforts by the Chinese authorities achieved snail control in certain areas by labour-intensive, low-cost methods, including the individual removal and the burial of snails, and the use of molluscicides (Chi, 1975; Mao and Shao, 1982).

6.2.1. *Chemotherapy*

Chemotherapy plays a leading role in the control of human schistosomiasis (Webbe and Jordan, 1993). In cattle, it is also widely used in the Far East where *S. japonicum* is an important zoonosis. Cattle are important reservoir hosts of *S. japonicum* and it is recognized (Hsü *et al.*, 1984) that human infections could never be controlled without controlling the parasite in cattle.

However, in other parts of the world, chemotherapy is not suitable for the control of the infection in cattle (Hussein, 1980), mainly because of practical concerns (difficulties in applying treatment) and for economical reasons (to treat a 300 kg cow with praziquantel costs US$10–20, and treatments need to be repeated).

6.2.2. *Immunological Control*

(a) *Heterologous immunity.* As discussed earlier (Section 5.3.4), Massoud and Nelson (1972) suggested that immunizing cattle by exposing them to cercariae of *S. haematobium* may induce resistance to infection by other schistosomes. However, this method of control has never been applied in the field.

(b) *Irradiated vaccines.* The first attempts towards immunological control of cattle schistosomiasis focused on the use of homologous larval vaccines attenuated by irradiation. In Sudan, such a vaccine was shown to induce significant reductions in *S. bovis* infection rates, both under laboratory (Bushara *et al.*, 1978) and field conditions (Majid *et al.*, 1980b). In the laboratory, 18 calves received (in 1–3 intramuscular or subcutaneous injections) 10 000 irradiated schistsomula or cercariae (3 krad by a ^{60}Co source, at a rate of 95 rad min^{-1}) and, together with four controls, were exposed 8 weeks later to 10 000 normal *S. bovis* cercariae (Bushara *et al.*, 1978). All animals were necropsied 12–14 weeks p.i., when significant reductions of 60–80% in average faecal egg counts, total tissue egg counts and adult worm counts were recorded in immunized animals, compared with controls. No significant differences were observed in the effectiveness of the different immunization procedures. For the field trial, a total of 60 calves (half of them immunized as for the laboratory trial), were allowed to graze in an *S. bovis* endemic area for a period of 10 months (Majid *et al.*, 1980b). At perfusion, average faecal egg counts, tissue egg counts and adult worm counts in immunized animals were reduced by 82, 65 and 69%, respectively, compared with controls.

Irradiated vaccines were also shown to be effective against *S. japonicum* infection (Hsü *et al.*, 1983, 1984; Xu *et al.*, 1993). In a first trial using experimental infections, 18 calves received 1–3 doses (each dose given partly intradermally, partly intramuscularly) of about 10 000 irradiated *S. japonicum* schistosomula (24, 36 or 48 krad) and, together with five controls, were exposed 30 days after the last injections to 500 normal *S. japonicum* cercariae (Hsü *et al.*, 1983). At necropsy one month p.i., average reductions in worm counts in immunized animals varied from 55 to 87%, compared with controls. The best results were obtained with three injections of schistosomula which had been irradiated with 36 krad. Hsü and colleagues later repeated the experiment (irradiation with 38 krad, and three injections of 10 000 cercariae at 30-day intervals), this time perfusing the animals 54–57 days p.i., therefore allowing the infection to develop to the egg-producing stage (Hsü *et al.*, 1984). However, results of tissue egg counts were not presented. Worm counts in this experiment were reduced by 72% in vaccinated animals compared with controls. Xu *et al.* (1993) reported on another similar laboratory test: a single dose of cryopreserved irradiated schistosomula (either 5000 or 10 000 larvae, irradiated with 20 krad) was administered intradermally together with 1 ml of bacille Calmett-Guérin (BCG). Perfusions were carried out when all animals were found positive on a miracidial test. Worm counts in vaccinated calves were reduced by 48 to 55%, compared with controls.

To test the efficacy of the irradiated *S. japonicum* larval vaccine in the field, Hsü *et al.* (1984) transported eight vaccinated (three injections of

10 000 irradiated cercariae at 30-day intervals) and nine nonvaccinated yearling cattle into a 'heavily' endemic (infected snail density of 4–8%) area in China where they were mixed with local herds for a period of 55 days, and then perfused. Five of the nonvaccinated and three of the vaccinated animals showed symptoms of acute schistosomiasis starting on day 48. At perfusion, reductions in worm counts and in tissue egg counts (e.p.g. liver) in vaccinated animals compared with controls were 65% and 55%, respectively. Xu et al. (1993) conducted another field trial in the same Chinese province. This time, five calves immunized with a single dose of 10 000 cryopreserved-irradiated schistosomula (20 krad) and five nonvaccinated controls were exposed to natural infection for a period of 2.5 months. All animals developed severe bloody diarrhoea. Compared with controls, the average worm count in vaccinated calves was reduced only by 24%. However, tissue egg counts in the livers of vaccinated calves were reduced by 74%, compared with controls.

In conclusion, vaccines consisting of irradiated schistosomula can significantly protect cattle against schistosome infection. An economic study in Sudan indicated that the development and production of such a vaccine would yield very favourable returns from livestock production efficacy (McCauley et al., 1984). However, despite the fact that the great potentials of irradiated vaccines have been known for more than 15 years, the vaccines have never been used on a large scale. Live attenuated vaccines are difficult to produce, particularly those against *S. japonicum* because of the relatively low cercarial production by each *Oncomelania* snail. In addition, the vaccines require cryopreservation and are not easy to apply in the field (Hsü et al., 1984; Bashir et al., 1994).

(c) *Crude schistosome antigens.* Attempts to induce immunity by vaccination with *S. bovis* adult worm extracts (Aradaib et al., 1993) or whole-egg antigen (Aradaib et al., 1995c) failed to protect calves against an homologous challenge. In both trials, animals were exposed to a massive experimental infection (20 000 cercariae), and no significant difference between vaccinated and nonvaccinated calves were observed in faecal and tissue egg counts or worm recoveries.

Xu et al. (1993) demonstrated that crude freeze–thaw schistosomular antigen plus BCG is protective against *S. japonicum* in cattle. For the vaccination trial, three groups of four calves received one to three doses (30 000 nonliving schistosomula each time, at two 2-week intervals) of a freeze–thaw vaccine intradermally, together with 1 ml of BCG. All the vaccinated calves and four nonvaccinated controls, were exposed 30 days after the last injections to 500 normal *S. japonicum* cercariae. Perfusions were carried out when all animals were found positive on a miracidial test. The highest (57%) and lowest (38%) reductions in worm burden compared with controls were recorded in calves immunized with one and three doses, respectively, showing that the freeze-thaw vaccine could induce levels of protection similar to the cryopreserved-irradiated vaccine. However, Xu *et al.* (1993) observed that the freeze-thaw vaccine required fewer cercariae for preparation, was less voluminous, and would be much easier to use in the field than the cryopreserved-irradiated vaccine (for which motile organisms need to be counted before administration).

(d) *Defined protective antigens.* The current research trend for schistosomiasis control is to identify defined protective antigens that are easier to standardize and deliver than live attenuated vaccines, particularly in human infections where the use of irradiated schistosomula is not possible (see reviews by Bashir *et al.*, 1994; Capron *et al.*, 1994; Dunne *et al.*, 1995).

In cattle, the first trials using defined antigens were based on the immunization of naive calves with either the keyhole limpet haemocyanin (KLH), or the native glutathione S-transferases (GSTs) purified from *S. bovis* (Bushara *et al.*, 1993). The KLH is a commercially-available high molecular weight glycoprotein which shares a protective epitope with the protective antigen GP38 of *S. mansoni* (Grzych *et al.*, 1985, 1987). The GSTs are now leading candidates for human vaccination trials. The enzyme is present as two families of isoenzymes of molecular masses 26 and 28 kDa in all species of schistosomes examined to date. Since the first characterization of the 28GST in *S. mansoni* (Sm28GST) by Balloul *et al.* (1985, 1987a), the protein has been cloned and expressed in *Escherichia coli* and *Saccharomyces cerevisiae* (Balloul *et al.*, 1987a). The native and recombinant proteins have been shown to induce high levels of protection in various laboratory animals (Balloul *et al.*, 1987b; Boulanger *et al.*, 1991; Grezel *et al.*, 1993). The immunity induced by vaccination leads to a reduction of the worm burden and/or an impairment of the parasite fecundity (Balloul *et al.*, 1987b; Boulanger *et al.*, 1991; Xu *et al.*, 1991).

For the vaccination trial in cattle (1–2 year old), Bushara *et al.* (1993) compared the average faecal egg counts, worm burdens and tissue egg densities (e.p.g. tissue) in groups of animals which received (a) a total of 1.40 mg native *S. bovis* GST in three doses given at 2-week intervals, (b) a total of 0.48 mg native *S. bovis* GST in two doses given 15 weeks apart, and (c) a total of 2.0 mg KLH in two doses given 2 weeks apart. All animals (and a group of nonvaccinated controls) were exposed to 10 000 *S. bovis* cercariae 1 week after the last immunization, and perfused 12 weeks p.i. All three vaccination schedules induced specific antibodies. At perfusion, significant reductions in faecal egg counts (73% in groups A and B, 40% in group C) and tissue egg densities (43–86% in groups A and B, 35–72% in group C) were observed in vaccinated groups compared with controls. Interestingly, adult worm counts were not affected by vaccination. Nevertheless, the authors concluded that if similar levels of fecundity suppression could be induced in the field, considerable amelioration of the disease would be expected.

More recently, the potential of a recombinant *S. bovis*-derived GST (rSb28GST) to protect cattle against *S. mattheei* infection was tested in Zambia (De Bont *et al.*, 1997). Calves aged 4–6 months were challenged 2 weeks after the second inoculation with either 0.250 mg rSb28GST in adjuvants (vaccinated calves) or adjuvants alone (controls). In a first experiment, vaccinated and control animals were exposed to 10 000 *S. mattheei* cercariae percutaneously. All animals developed clinical schistosomiasis 7–8 weeks p.i. At perfusion 12 weeks p.i., vaccinated and control groups had averages of 887 and 541 e.p.g., 6515 and 5990 worms, and 4.2 million and 3.4 million tissue eggs, respectively. These results indicated that the immunization protocol used did not protect cattle against the massive single experimental challenge. In a second experiment, groups of vaccinated and control animals were challenged naturally over a period of 9 months on a farm where *S. mattheei* was known to be endemic. The natural infections were much lighter in intensity than were the experimental infections, as indicated by the mean faecal egg count (13 e.p.g.), worm count (139) and tissue egg count (294 000) in nonvaccinated controls. Nevertheless, in vaccinated calves, significant reductions in female worm burdens (50%), faecal egg counts (89%) and miracidial counts (93%) were recorded. Total tissue egg counts were also reduced by 42% in vaccinated animals. Serological tests demonstrated that sera from rSb28GST-immunized calves recognized the native *S. mattheei* 28GST (Smat28GST) and achieved comparable levels of inhibition of the GST activity of rSb28GST and Smat28GST, indicating the presence of cross-epitopes on these two molecules (J.M. Grzych, J. De Bont, J.L. Liu, J.L. Neyrinck, J. Fontaine, J. Vercruysse and A. Capron, unpublished data). In addition, examination of the immunological parameters showed that inhibition of the GST enzymatic activity in experimental infections was related to the presence of specific IgG antibodies, whereas in natural infections it appeared to be associated with IgA antibodies. It was concluded that the vaccination of cattle with the recombinant rSb28GST could significantly protect them against *S. mattheei* in normal endemic situations. The different results obtained after either experimental or natural exposure raise questions about the validity of using a single, relatively heavy experimental challenge to study the immunological response against schistosome infection. Future studies of transmission levels in pools visited by vaccinated cattle are desirable to test whether the spectacular reduction in the excretion of viable eggs (93%) could affect transmission.

7. CONCLUSION AND PERSPECTIVES

Although there has been little recognition of its veterinary significance, cattle schistosomiasis does cause significant losses throughout the world. Suitable drugs are not available for mass treatment in domestic stock and are unlikely to be developed in the near future. However, recent progress in identifying potentially protective parasite antigens has opened new perspectives in the control strategy against schistosomiasis. Whatever the means, it will be necessary to base the prevention and control of schistosomiasis on a profound knowledge of the epidemiology of the disease. Although there is overwhelming evidence that there is development of immunity to schistosome infection in cattle, the actual mechanism or mechanisms of protection have still to be determined. Most of the information available emanates from experimental infections where the use of large single cercarial doses may be a problem. There is a need for further studies on the dynamics of infection in naturally infected animals living under conditions of continuous moderate challenge and on the different factors which may affect the development of immunity. In addition, for human schistosomes 'natural' animal models such as *S. mattheei* and *S. bovis* in cattle can be very relevant for study of the dynamics of transmission of schistosomes, and immunity against infection in general (Hagan and Gryseels, 1994).

ACKNOWLEDGEMENTS

We wish to thank Darren Shaw for his critical reading of the manuscript.

REFERENCES

Achuthan, H.N. and Alwar, V.S. (1973). A note on the occurrence of nasal schistosomiasis in sheep and goats in Tamil Nadu (correspondence). *Indian Veterinary Journal* **50**, 1058–1059.

Agnew, A., Fulford, A.J.C., De Jonge, N., Krijger, F.W., Rodriguez-Chacon, M., Gutsmann, V. and Deelder, A.M. (1995). The relationship between worm burden and levels of a circulating antigen (CAA) of five species of *Schistosoma* in mice. *Parasitology* **111**, 67–76.

Agrawal, M.C. and Alwar, V.S. (1992). Nasal schistosomiasis: a review. *Helminthological Abstracts* **61**, 373–384.

Agrawal, M.C. and Shah, H.L. (1989). A review on *Schistosoma incognitum*, Chandler, 1926. *Helminthological Abstracts* **58**, 239–251.

Albaret, J.L., Picot, H., Diaw, O.T., Bayssade-Dufour, C., Vassiliades, G., Adamson, M., Luffau, G. and Chabaud, A.G. (1985). Enquête sur les schistosomes de l'homme et du bétail au Sénégal à l'aide des identifications spécifiques fournies par la chétotaxie des cercaires. I. Nouveaux arguments pour la validation de *S. curassoni* Brumpt, 1931, parasite de l'homme et des bovidés domestiques. *Annales de Parasitologie Humaine et Comparée* **60**, 417–434.

Albaret, J.L., Bayssade-Dufour, C. and Ngendahayo, L.D. (1993). Identification des cercaires de *Schistosoma* africains émises par *Bulinus umbilicatus*, *B. truncatus* et *B. forskalii*. *Systematic Parasitology* **26**, 209–214.

Alves, W. (1949). The eggs of *Schistosoma bovis*, *S. mattheei* and *S. haematobium*. *Journal of Helminthology* **23**, 127–134.

Alves, W. (1953). Urinary bilharziasis in an ox in Southern Rhodesia. *Transactions of the Royal Society of Tropical Medicine and Hygiene* **47**, 272.

Amin, M.B.A. and Nelson, G.S. (1969). Studies on heterologous immunity in schistosomiasis. III. Further observations on heterologous immunity in mice. *Bulletin of the World Health Organization* **41**, 225–232.

Amin, M.B.A., Nelson, G.S. and Saoud, M.F.A. (1968). Studies on heterologous immunity in schistosomiasis. II. Heterologous schistosome immunity in rhesus monkeys. *Bulletin of the World Health Organization* **38**, 19–27.

Aradaib, I.E. and Osburn, B.I. (1995). Vaccination against bovine schistosomosis: current status and future prospects: a review. *Preventive Veterinary Medicine* **22**, 285–291.

Aradaib, I.E., Abbas, B., Bushara, H.O. and Taylor, M.G. (1993). Evaluation of *Schistosoma bovis* adult worm extract for vaccination of calves. *Preventive Veterinary Medicine* **16**, 77–84.

Aradaib, I.E., Abdelmageed, E.M., Hassan, S.A. and Riemann, H.P. (1995a). A review on the diagnosis of infection in cattle of *Schistosoma bovis*: current status and future prospects. *Ciência Rural* **25**, 493–498.

Aradaib, I.E., Abbas, B., Riemann, H.P. and Osburn, B.I. (1995b). Experimental bovine schistosomiasis in zebu calves. *Ciência Rural* **25**, 99–103.

Aradaib, I.E., Omer, O.H., Abbas, B.B., Bushara, H.O., Elmalik, K.H., Saad, A.M., Osburn, B.I. and Taylor, M.G. (1995c). *Schistosoma bovis* whole egg antigen did not protect zebu calves against experimental schistosomosis. *Preventive Veterinary Medicine* **21**, 339–345.

Balloul, J.M., Pierce, R.J., Grzych, J.M. and Capron, A. (1985). *In vitro* synthesis of a 28 kilodalton antigen present on the surface of the schistosomulum of *Schistosoma mansoni*. *Molecular and Biochemical Parasitology* **17**, 105–114.

Balloul, J.M., Sondermeyer, P., Dreyer, D., Capron, M., Grzych, J.M., Pierce, R.J., Carvallo, D., Lecocq, J.P. and Capron, A. (1987a). Molecular cloning of a protective antigen of schistosomes. *Nature* **326**, 149–153.

Balloul, J.M., Grzych, J.M., Pierce, R.J. and Capron, A. (1987b). A purified 28 000 Dalton protein from *Schistosoma mansoni* adult worms protects rats and mice against experimental schistosomiasis. *Journal of Immunology* **138**, 3448–3453.

Banerjee, P.S. and Agrawal, M.C. (1989). Comparative efficacy of faecal and liver examination in determining prevalence of bovine schistosomiasis. *Journal of Veterinary Parasitology* **3**, 157–158.

Banerjee, P.S. and Agrawal, M.C. (1990). Miracidial immobilization test in bovine schistosomiasis. *Indian Journal of Animal Sciences* **60**, 628–630.

Banerjee, P.S., Agrawal, M.C. and Shah, H.L. (1990). Diagnosis of natural bovine schistosomiasis by using ring precipitation test. *Indian Journal of Parasitology* **14**, 223–226.

Banerjee, P.S., Agrawal, M.C. and Shah, H.L. (1991). Application of CHR and J-index in bovine schistosomiasis. *Indian Veterinary Journal* **68**, 1022–1026.

Barbosa, F.S., Barbosa, I. and Arruda, F. (1962). *Schistosoma mansoni*: natural infection of cattle in Brazil. *Science* **138**, 831.

Barral, V., This, P., Imbert-Establet, D., Combes, C. and Delseny, M. (1993). Genetic variability and evolution of the *Schistosoma* genome using randomly amplified polymorphic DNA markers. *Molecular and Biochemical Parasitology* **59**, 211–222.

Barsoum, I.S., Bogitsh, B.J. and Colley, D.G. (1992). Detection of *Schistosoma mansoni* circulating cathodic antigen for evaluation of resistance induced by irradiated cercariae. *Journal of Parasitology* **78**, 681–686.

Bartsch, R.C. and van Wyk, J.A. (1977). Studies on schistosomiasis. 9. Pathology of the bovine urinary tract. *Onderstepoort Journal of Veterinary Research* **44**, 73–94.

Basch, P.F. (1986). Schistosomiasis in China: an update. *American Journal of Chinese Medicine* **14**, 17–25.

Basch, P.F. (1991). *Schistosomes. Development, Reproduction, and Host Relations.* Oxford: Oxford University Press.

Bashir, M., Bickle, Q., Bushara, H.O., Cook, L., Shi Fuhui, Dian He, Huggins, M., Lin Jiaojiao, Malik, K., Moloney, A., Mukhtar, M., Yeh Ping, Xu Shoutai, Taylor, M. and Shi Yaochuan (1994). Evaluation of defined antigen vaccines against *Schistosoma bovis* and *S. japonicum* in bovines. *Tropical and Geographical Medicine* **46**, 255–258.

Bayssade-Dufour, C. (1982). Chétotaxies cercariennes comparées de dix espèces de schistosomes. *Annales de Parasitologie Humaine et Comparée* **57**, 467–485.

Bayssade-Dufour, C., Cabaret, J., Ngendahayo, L.D., Albaret, J.L., Carrat, C. and Chabaud, A.G. (1989). Identification of *Schistosoma haematobium*, *S. bovis* and *S. currassoni* by multivariate analysis of cercarial papillae indices. *International Journal for Parasitology* **19**, 839–846.

Bilharz, T.M. (1852). Fernere Beobachtungen uber das die Pfortader des Menschen bewohnende *Distomum haematobium* und sein verhaltniss zu gewissen pathologischen Bildungen aus brieflichen Mitheilungen an Professor v. Siebold vom 29. Marz 1852. *Zeitschrift für Wissenschaftliche Zoologie, Leipzig* **4**, 72–76.

Birgi, E. and Graber, M. (1969). Mollusques pulmonés d'eau douce bassomatophores vecteurs au Tchad d'affections parasitaires du bétail, leur élevage au laboratoire. *Revue d'élevage et de Médecine Vétérinaire des Pays Tropicaux* **22**, 393–408.

Biswas, G. (1975). Relative susceptibility of mammals to infection with *Schistosoma incognitum*. *Indian Journal of Animal Health*, **14**, 179–181.

Blanchard, R. (1895). Les vers du sang. In: *Les Hematozoaires de l'Homme et des Animaux* (A. Laveran, ed.), Part 2. Paris.

Boulanger, D., Reid, G.D., Sturrock, R.F., Wolowczuk, I., Balloul, J.M., Grezel, D., Pierce, R.J., Otieno, M.F., Guerret, S., Grimaud, J.A., Butterworth, A.E. and Capron, A. (1991). Immunization of mice and baboons with the recombinant Sm28GST affects both worm viability and fecundity after experimental infection with *Schistosoma mansoni*. *Parasite Immunology* **13**, 473–490.

Brown, D.S. (1994). *Freshwater Snails of Africa and their Medical Importance.* London: Taylor and Francis.

Brumpt, E. (1931). Description de deux bilharzies de mammifères africains, *Schistosoma currassoni*, sp. inquir. et *Schistosoma rodhaini* n.sp. *Annales de Parasitologie Humaine et Comparée* **9**, 325–328.

Bushara, H.O., Hussein, M.F., Saad, A.M., Taylor, M.G., Dargie, J.D., Marshall,

T.D. de C. and Nelson, G.S. (1978). Immunisation of calves against *Schistosoma bovis* using irradiated cercariae or schistosomula of *S. bovis*. *Parasitology*, **77**, 303–311.

Bushara, H.O., Majid, A.A., Saad, A.M., Hussein, M.F., Taylor, M.G., Dargie, J.D., Marshall, T.F. de C. and Nelson, G.S. (1980). Observations on cattle schistosomiasis in the Sudan, a study in comparative medicine. II. The experimental demonstration of naturally acquired resistance to *Schistosoma bovis*. *American Journal of Tropical Medicine and Hygiene* **29**, 442–451.

Bushara, H.O., Hussein, M.F., Majid, M.A. and Taylor, M.G. (1982). Effects of Praziquantel and Metrifonate on *Schistosoma bovis* infections in Sudanese cattle. *Research in Veterinary Science* **33**, 125–126.

Bushara, H.O., Majid, B.Y.A., Majid, A.A., Khitma, I., Gameel, A.A., Karib, E.A., Hussein, M.F. and Taylor, M.G. (1983a). Observations on cattle schistosomiasis in the Sudan, a study in comparative medicine. V. The effect of praziquantel therapy on naturally acquired resistance to *Schistosoma bovis*. *American Journal of Tropical Medicine and Hygiene* **32**, 1370–1374.

Bushara, H.O., Gameel, A.A., Majid, B.Y.A., Khitma, I., Haroun, E.M., Karib, E.A., Hussein, M.F. and Taylor, M.G. (1983b). Observations on cattle schistosomiasis in the Sudan, a study in comparative medicine. VI. Demonstration of resistance to *Schistosoma bovis* challenge after single exposure to normal cercariae or to transplanted adult worms. *American Journal of Tropical Medicine and Hygiene* **32**, 1375–1380.

Bushara, H.O., Hussein, M.F., Majid, M.A., Musa, B.E.H. and Taylor, M.G. (1983c). Observations on cattle schistosomiasis in the Sudan, a study in comparative medicine. IV. Preliminary observations on the mechanism of naturally acquired resistance. *American Journal of Tropical Medicine and Hygiene* **32**, 1065–1070.

Bushara, H.O., Bashir, M.E.N., Malik, K.H.E., Mukhtar, M.M., Trottein, F., Capron, A. and Taylor, M.G. (1993). Suppression of *Schistosoma bovis* egg suppression in cattle by vaccination with either glutathione S-transferase or keyhole limpet haemocyanin. *Parasite Immunology* **15**, 383–390.

Bushara, H.O., Omer, O.H., Malik, K.H.E. and Taylor, M.G. (1994). The effect of multiple transfers of immune serum on maturing *Schistosoma bovis* infections in calves. *Parasitology Research* **80**, 198–202.

Butterworth, A.E. (1993). Immunology of schistosomiasis. In: *Human Schistosomiasis* (P. Jordan, G. Webbe and R.F. Sturrock, eds), pp. 331–366. Wallingford: CAB International.

Cabaret, J., Bayssade-Dufour, C., Albaret, J.L., Ngendahayo, L.D. and Chabaud, A.G. (1990). A technique for identification of cercariae of *Schistosoma haematobium, S. currassoni, S. bovis* and *S. intercalatum*. *Annales de Parasitologie Humaine et Comparée* **65**, 61–63.

Capron, A., Riveau, G., Grzych, J-M., Boulanger, D., Capron, M. and Pierce R. (1994). Development of a vaccine strategy against human and bovine schistosomiasis. *Tropical and Geographical Medicine* **46**, 242–246.

Carney, W.P., Brown, R.J., van Peenen, P.F.D., Purnomo, I.B. and Koesharjono, C.R. (1977). *Schistosoma incognitum* from Cikurai, West Java, Indonesia. *International Journal for Parasitology* **7**, 361–366.

Chandiwana, S.K., Taylor, P. and Makura, O. (1987a). Prevalence and distribution of *Schistosoma mattheei* in Zimbabwe. *Annales de la société Belge de Médecine Tropicale* **67**, 167–172.

Chandiwana, S.K., Christensen, N.Ø. and Frandsen, F. (1987b). Seasonal patterns in the transmission of *Schistosoma haematobium, S. mattheei* and *S. mansoni* in the highveld region of Zimbabwe. *Acta Tropica* **44**, 433–444.

Chandler, A.C. (1926). A new schistosome infection in man, with notes on other human fluke infections in India. *Indian Journal of Medical Research* **14**, 179–183.

Chartier, C., Bushu, M. and Anican, U. (1990). Les dominantes du parasitisme helminthique chez les bovins en Ituri (Haut-Zaïre). II. Les associations parasitaires. *Revue d'Elevage et de Médecine Vétérinaire des Pays Tropicaux* **43**, 491–497.

Cheever, A.W. (1968). Conditions affecting the accuracy of potassium hydroxide digestion techniques for counting *Schistosoma mansoni* eggs in tissues. *Bulletin of the World Health Organization* **39**, 328–331.

Chen, D.R., Yan, J.B. and Yang, M.F. (1985). [Nitroscanate in the treatment of farm cattle with *Schistosoma japonicum*.] *Chinese Journal of Veterinary Science and Technology* **4**, 6–11.

Chen, M.G. (1993). *Schistosoma japonicum* and *S. japonicum*-like infections: epidemiology, clinical and pathological aspects. In: *Human Schistosomiasis* (P. Jordan, G. Webbe and R.F. Sturrock, eds), pp. 237–270. Wallingford: CAB International.

Cheng, T.H. (1971). Schistosomiasis in mainland China. A review of research and control programmes since 1949. *American Journal of Tropical Medicine and Hygiene* **20**, 26–53.

Chi, L.W. (1975). Mass control of *Oncomelania hupensis hupensis* snail vector for schistosomiasis in the People's Republic of China. *Veliger* **18**, 95–98.

Christensen, N.Ø., Mutani, A. and Frandsen, F. (1983). A review of the biology and transmission ecology of African bovine species of the genus *Schistosoma*. *Zeitschrift für Parasitenkunde* **69**, 551–570.

Chunge, R., Katsivo, M., Koko, P., Wamwea, M. and Kinoti, S. (1986). *Schistosoma bovis* in human stools in Kenya. *Transactions of the Royal Society of Tropical Medicine and Hygiene* **80**, 849.

Combes, C. (1985). Les transmissions vectorielles. L'analyse de la compatibilité schistosomes/mollusques vecteurs. *Bulletin de la Société de Pathologie Exotique* **78**, 742–746.

Combes, C., Bayssade-Dufour, C. and Cassone, J. (1976). Sur l'imprégnation et le montage des cercaires pour l'étude chétotaxique. *Annales de Parasitologie Humaine et Comparée* **51**, 399–400.

Condy, J.B. (1960). Bovine schistosomiasis in Southern Rhodesia. *Central African Journal of Medicine* **6**, 381–384.

Cross, J.H., Zaraspe, G., Lu, S.K., Chiu, K.M. and Hung, H.K. (1984). Susceptibility of *Oncomelania hupensis* subspecies to infection with geographic strains of *Schistosoma japonicum*. *Southeast Asian Journal of Tropical Medicine and Public Health* **15**, 155–160.

Damian, R.T. (1987). The exploitation of host immune responses by parasites. *Journal of Parasitology* **73**, 1–13.

Dargie, J.D. (1980). The pathogenesis of *Schistosoma bovis* infection in Sudanese cattle. *Transactions of the Royal Society of Tropical Medicine and Hygiene* **74**, 560–562.

Datta, S.C.A. (1932). The etiology of bovine nasal granuloma. *Indian Journal of Veterinary Science and Animal Husbandry* **2**, 131–140.

Davis, G.M. (1980). Snail hosts of Asian *Schistosoma* infecting man: evolution and coevolution. In: *The Mekong Schistosome* (J.I. Bruce, S. Sornmani, H.L. Asch and K.A. Crawford, eds). *Malacological Review*, Supplement **2**, 195–238.

Davis, A. (1993). Antischistosomal drugs and clinical practice. In: *Human Schistosomiasis* (P. Jordan, G. Webbe and R.F. Sturrock, eds), pp. 367–404. Wallingford: CAB International.

De Bont, A.F. and De Bont-Hers, M.J. (1952). Mollusc control and fish-farming in central Africa. *Nature* **170**, 323–324.

De Bont, J. and Vercruysse, J. (1997). The epidemiology and control of cattle schistosomiasis, *Parasitology Today* **13**, 255–262.

De Bont, J., van Aken, D., Vercruysse, J., Fransen, J., Southgate, V.R. and Rollinson, D. (1989a). The prevalence and pathology of *Schistosoma nasale* Rao, 1933 in cattle in Sri Lanka. *Parasitology* **98**, 197–202.

De Bont, J., van Aken, D., Vercruysse, J., Fransen, J., Southgate, V.R. and Rollinson, D. (1989b). The effect of praziquantel on *Schistosoma nasale* infections in cattle. *Journal of Veterinary Pharmacology and Therapeutics* **12**, 455–458.

De Bont, J., Vercruysse, J., van Aken, D., Southgate, V.R., Rollinson, D. and Moncrieff, C. (1991a). The epidemiology of *Schistosoma spindale* Montgomery, 1906 in cattle in Sri Lanka. *Parasitology* **102**, 237–241.

De Bont, J., Vercruysse, J., van Aken, D., Southgate, V.R. and Rollinson, D. (1991b). Studies of the relationship between *Schistosoma nasale* and *S. spindale* and their snail host *Indoplanorbis exustus*. *Journal of Helminthology* **65**, 1–7.

De Bont, J., Vercruysse, J., van Aken, D., Warlow, A., Southgate, V.R. and Rollinson, D. (1991c). Use of enzyme electrophoresis for differentiating *Schistosoma nasale* and *S. spindale* infections of *Indoplanorbis exustus* in Sri Lanka. *Systematic Parasitology* **20**, 161–164.

De Bont, J., Vercruysse, J., Southgate, V.R., Rollinson, D. and Kaukas, A. (1994). Cattle schistosomiasis in Zambia. *Journal of Helminthology* **68**, 295–299.

De Bont, J., Vercruysse, J., Sabbe, F., Southgate, V.R. and Rollinson, D. (1995a). *Schistosoma mattheei* infections in cattle: changes associated with season and age. *Veterinary Parasitology* **57**, 299–307.

De Bont, J., Vercruysse, J., Sabbe, F. and Ysebaert, M.T. (1995b). Observations on worm population dynamics in calves naturally infected with *Schistosoma mattheei*. *Parasitology* **111**, 485–491.

De Bont, J., Vercruysse, J. and Massuku, M. (1996a). Variations in *Schistosoma mattheei* egg morphology and viability according to age of infection in cattle. *Journal of Helminthology* **70**, 265–267.

De Bont, J., van Lieshout, L., Deelder, A.M., Ysebaert, M.T. and Vercruysse, J. (1996b). Circulating antigen levels in serum of cattle naturally infected with *Schistosoma mattheei*. *Parasitology* **113**, 465–471.

De Bont, J., Vercruysse, J., Grzych, J.M., Meeus, P.F.M. and Capron, A. (1997). Potential of a recombinant *Schistosoma bovis*-derived glutathione S-transferase to protect cattle against experimental and natural *S. mattheei* infection. *Parasitology* **115**, 249–255.

Deelder, A.M., Qian, Z.L., Kremsner, P.G., Acosta, L., Rabello, A.L.T., Enyong, P., Simarro, P.P., van Etten, E.C.M., Krijger, F.W., Rotmans, J.P., Fillie, Y.E., De Jonge, N., Agnew, A.M. and van Lieshout, L. (1994). Quantitative diagnosis of *Schistosoma* infections by measurements of circulating antigens in serum and urine. *Tropical and Geographical Medicine* **46**, 233–238.

De Jonge, N. (1990). *Immunodiagnosis of* Schistososoma *Infections by Detection of the Circulating Anodic Antigen*. Proefschrift Leiden. Den Haag: CIP-Gegevens Koninklijke Bibliotheek.

Dias Neto, E., Pereira de Souza, C., Rollinson, D. Katz, N., Pena, S.D.J. and Simpson, A.J.G. (1993). The random amplification of polymorphic DNA allows

the identification of strains and species of schistosome. *Molecular and Biochemical Parasitology* **57**, 83–88.
Diaw, O.T. and Vassiliades, G. (1987). Epidémiologie des schistosomes du bétail au Sénégal. *Revue d'Elevage et de Médecine Vétérinaire des Pays Tropicaux* **40**, 265–274.
Diaw, O.T., Seye, M. and Sarr, Y. (1989). Résistance à la sécheresse de mollusques du genre *Bulinus*, vecteurs de trématodoses humaines et animales au Sénégal. 2. Etude dans les conditions naturelles en zone nord-soudanienne. Ecologie et résistance de *Bulinus umbilicatus* et *B. senegalensis. Revue d'Elevage et de Médecine Vétérinaire des Pays Tropicaux* **42**, 177–187.
Dinnik, N.N. (1967). The effect of neguvon on *Schistosoma bovis* in naturally infected cattle. *Veterinary Medical Review, Leverkussen* **1**, 76–78.
Dinnik, J.A. and Dinnik, N.N. (1965). The Schistosomes of domestic ruminants in Eastern Africa. *Bulletin of Epizootic Diseases of Africa* **13**, 341–359.
Doenhoff, M.J., Mussallam, R., Bain, J. and McGregor, A. (1978). Studies on the host-parasite relationship in *Schistosoma mansoni* infected mice: the immunological dependence of parasite egg excretion. *Immunology* **35**, 771–778.
Donnelly, F.A. and Appleton, C.C. (1985). Observations on the field transmission dynamics of *Schistosoma mansoni* and *S. mattheei* in southern Natal, South Africa. *Parasitology* **91**, 281–290.
Dunne, D.W., Hagan, P. and Abath, F.G.C. (1995). Prospects for immunological control of schistosomiasis. *The Lancet* **345**, 1488–1492.
Du Plessis, J.L. and van Wyk, J.A. (1972). Studies on schistosomiasis. 3. Detection of antibodies against *Schistosoma mattheei* by the indirect immuno-fluorescent method. *Onderstepoort Journal of Veterinary Research* **39**, 179–180.
Dutt, S.C. and Srivastava, H.D. (1968). Studies on *Schistosoma nasale* Rao, 1933. II. Molluscan and mammalian hosts of the blood-fluke. *Indian Journal of Veterinary Science* **38**, 210–216.
Eisa, A.M. (1966). Parasitism — a challenge to animal health in the Sudan. *Sudan Journal of Veterinary Science and Animal Husbandry* **7**, 85–94.
Fairley, N.H. and Jasudasan, F. (1930). Studies in *Schistosoma spindale*. Part II. The definitive hosts of *S. spindale* with special reference to alimentary infections in ruminants. *Indian Medical Research Memoirs* **17**, 11–15.
Feldmeier, H. (1993). Diagnosis. In: *Human Schistosomiasis* (P. Jordan, G. Webbe and R.F. Sturrock, eds), pp. 271–303. Wallingford: CAB International.
Fisher, A.C. (1934). A study of the schistosomiasis of the Stanleyville District of the Belgian Congo. *Transactions of the Royal Society of Tropical Medicine and Hygiene* **28**, 277–306.
Fletcher, M., Woodruff, D.S., LoVerde, P.T. and Asch, H.L. (1980). Genetic differentiation between *Schistosoma mekongi* and *S. japonicum*: an electrophoretic study. In: *The Mekong Schistosome* (J.I. Bruce, S. Sornmani, H.L. Asch and K.A. Crawford, eds). *Malacological Review*, Supplement **2**, 113–122.
Frandsen, F. (1979). Studies on the relationship between *Schistosoma* and their intermediate hosts. I. The genus *Bulinus* and *Schistosoma haematobium* from Egypt. *Journal of Helminthology* **53**, 15–29.
Frandsen, F. and Christensen, N.Ø. (1984). An introductory guide to the identification of cercariae from African freshwater snails with special reference to cercariae of trematode species of medical and veterinary importance. *Acta Tropica* **41**, 181–202.
Fransen, J., De Bont, J., Vercruysse, J., Van Aken, D., Southgate, V.R. and

Rollinson, D. (1990). Pathology of natural infections of *Schistosoma spindale* Montgomery, 1906, in cattle. *Journal of Comparative Pathology* **103**, 447–455.

Gönnert, R. and Andrews, P. (1977). Praziquantel, a new broad spectrum antischistosomal agent. *Zeitschrift für Parasitenkunde* **52**, 129–150.

Greer, G.J., Ow-Yang, C.K. and Hol-Sen Young (1988). *Schistosoma malayensis* n.sp.: a *Schistosoma japonicum*–complex schistosome from Peninsular Malaysia. *Journal of Parasitology* **74**, 471–480.

Grezel, D., Capron, M., Grzych, J.M., Fontaine, J., Lecocq, J.P. and Capron, A. (1993). Protective immunity induced in rat schistosomiasis by a single dose of the Sm28GST recombinant antigen: effector mechanisms involving IgE and IgA antibodies. *European Journal of Immunology* **23**, 454–460.

Grétillat, S. (1962). Une nouvelle zoonose, la 'Bilharziose Ouest Africaine' à *Schistosoma curassoni* Brumpt, 1931 commune à l'homme et aux ruminants domestiques. *Comptes Rendus Hebdomadaires des Séances de l'Académie des Sciences de Paris* **255**, 1805–1807.

Gryseels, B. (1994). Human resistance to *Schistosoma* infections: age or experience? *Parasitology Today* **10**, 380–384.

Grzych, J.M., Capron, M., Lambert, P.H., Dissous, C., Torres, S. and Capron, A. (1985). An anti-idiotype vaccine against experimental schistosomiasis. *Nature* **316**, 74–76.

Grzych, J.M., Dissous, C., Capron, M., Torres, S., Lambert, P.H. and Capron, A. (1987). *Schistosoma mansoni* shares a protective carbohydrate epitope with keyhole limpet haemocyanin. *Journal of Experimental Medicine* **165**, 865–878.

Hagan, P. and Gryseels, B. (1994). Schistosomiasis research and the European Community. *Tropical and Geographical Medicine* **46**, 259–268.

Heyneman, D. and Umathevy, T. (1967). A field experiment to test the possibility of using double infection of host snails as a possible biological control of schistosomiasis. *Medical Journal of Malaya* **21**, 273.

Hira, P.R. (1975). Observations on *Schistosoma mattheei* Veglia and Le Roux 1929 infections in man in Zambia. *Annales de la Société Belge de Medecine Tropicale* **55**, 633–642.

Hira, P.R. and Patel, B.G. (1981). Transmission of schistosomiasis in a rural area in Zambia. *Central African Journal of Medicine* **27**, 244–249.

Hirata, M., Takushima, M., Kage, M. and Fukuma, T. (1993). Comparative analysis of hepatic, pulmonary and intestinal granuloma formation around freshly laid *Schistosoma japonicum* eggs in mice. *Parasitology Research* **79**, 316–321.

Hsü, S.Y.L., Hsü, H.F., Xu, S.T., Shi, F.H., He, Y.X., Clarke, W.R. and Johnson, S.C. (1983). Vaccination against bovine schistosomiasis japonica with highly X-irradiated schistosomula. *American Journal of Tropical Medicine and Hygiene* **32**, 367–370.

Hsü, S.Y.L., Xu, S.T., He, Y.X., Shi, F.H., Shen, W., Hsü, H.F., Osborne, J.W. and Clarke, W.R. (1984). Vaccination of bovines against schistosomiasis japonica with highly irradiated schistosomula in China. *American Journal of Tropical Medicine and Hygiene* **33**, 891–898.

Hu, S.G., Li, J.S., Yang, X.J. and Wang, J.X. (1989). [Treatment of infection caused by *Schistosoma japonicum* with praziquantel injections. II. Treatment of experimentally infected yellow cattle.] *Chinese Journal of Veterinary Medicine* **15**, 17–18.

Hussein, M.F. (1968). Observations on the pathology of natural and experimental bovine schistosomiasis. *Transactions of the Royal Society of Tropical Medicine and Hygiene* **62**, 9.

Hussein, M.F. (1971). The pathology of experimental schistosomiasis in calves. *Research in Veterinary Science* **12**, 246–252.
Hussein, M.F. (1973). Animal schistosomiasis in Africa: a review of *Schistosoma bovis* and *Schistosoma mattheei*. *The Veterinary Bulletin* **43**, 341–347.
Hussein, M.F. (1980). Prospects for the control of *Schistosoma bovis* infection in Sudanese cattle. *Transactions of the Royal Society of Tropical Medicine and Hygiene* **74**, 559–560.
Hussein, M.F., Saeed, A.A. and Nelson, G.S. (1970). Studies on heterologous immunity in schistosomiasis. 4. Heterologous immunity in cattle. *Bulletin of the World Health Organization* **42**, 745–749.
Hussein, M.F., Tartour, G., Imbabi, S.E. and Ali, K.E. (1975). The pathology of naturally-occurring bovine schistosomiasis in the Sudan. *Annals of Tropical Medicine and Parasitology* **69**, 217–225.
Islam, K.S. (1975). Schistosomiasis in domestic ruminants in Bangladesh. *Tropical Animal Health and Production* **7**, 244.
Jagannath, M.S., Sano, M., Rahman, S.A., Prabhakar, K.S., D'Souza, P.E. and Prem, G. (1988). An intradermal test for the diagnosis of *Schistosoma nasale* Rao, 1933 infection in cattle. *Indian Veterinary Journal* **65**, 273.
Jordan, P., Christie, J.D. and Unrau, G.O. (1980). Schistosomiasis transmission with particular reference to possible ecological and biological methods of control. *Acta Tropica* **37**, 95–135.
Joubert, P.H., Hamilton-Attwell, V.L. and Kruger, F.J. (1987). The occurrence of *Schistosoma mattheei* in the south-western Transvaal. *Onderstepoort Journal of Veterinary Research* **54**, 603–605.
Jourdane, J. and Théron, A. (1987). Larval development: eggs to cercariae. In: *The Biology of Schistosomes: from Genes to Latrines* (D. Rollinson and A.J.G. Simpson, eds). pp. 83–113. London: Academic Press.
Kalapesi, R.M. and Puhorit, B.L. (1954). Observations on histopathology of morbid tissues from a case of natural infection with *Schistosoma spindalis* in a bovine. *Indian Veterinary Journal* **30**, 336–340.
Kane, R.A. and Rollinson, D. (1994). Repetitive sequences in the ribosomal DNA internal transcribed spacer of *Schistosoma haematobium*, *Schistosoma intercalatum* and *Schistosoma mattheei*. *Molecular and Biochemical Parasitology* **63**, 153–156.
Kane, R.A., Ridgers, I.L., Johnston, D.A. and Rollinson, D. (1996). Repetitive sequences within the first internal transcribed spacer of ribosomal DNA in schistosomes contain a Chi-like site. *Molecular and Biochemical Parasitology* **75**, 265–269.
Kassuku, A.A., Nansen, P. and Christensen, N.Ø., (1985). A comparison of the efficiency of the percutaneous and per-oral routes of infection in caprine *Schistosoma bovis* infections. *Journal of Helminthology* **59**, 23–28.
Kassuku, A.A., Christensen, N.Ø., Monrad, J., Nansen, P. and Knudsen, J. (1986). Epidemiological studies on *Schistosoma bovis* in Iringa Region, Tanzania. *Acta Tropica* **43**, 153–163.
Katsurada, F. (1904). *Schistosoma japonicum*, a new parasite of man, by which an endemic disease in various areas of Japan is caused. *Annotationes Zooloogicae Japanenses* **5**, 146–160.
Kaukas, A., Johnston, D.A., Kane, R.A. and Rollinson, D. (1994a). Restriction enzyme mapping of ribosomal DNA of *Schistosoma spindale* and *S. leiperi* (Digenea) and its application to interspecific differentiation. *Systematic Parasitology* **27**, 13–17.

Kaukas, A., Dias Neto, E., Simpson, A.J.G., Southgate, V.R. and Rollinson, D. (1994b). A phylogenetic analysis of *Schistosoma haematobium* group species based on randomly amplified polymorphic DNA. *International Journal for Parasitology* **24**, 285–290.

Kinoti, G.K. and Mumo, J.M. (1988). Spurious human infection with *Schistosoma bovis*. *Transactions of the Royal Society of Tropical Medicine and Hygiene* **82**, 589–590.

Koshy, T.J. and Alwar, V.S. (1974). Electrophoretic studies in nasal schistosomiasis in bovines. *Cheiron* **3**, 114.

Kruatruchue, M., Riengrojpitak, S., Sahaphong, S. and Upatham, E.S. (1982). Scanning electron microscopy of adult *Schistosoma incognitum*. *South East Asian Journal of Tropical Medicine and Public Health* **13**, 163–173.

Kruatruchue, M., Riengrojpitak, S., Upatham, E.S. and Sahaphong, S. (1983). Scanning electron microscopy of the tegumental surface of adult *Schistosoma spindale*. *South East Asian Journal of Tropical Medicine and Public Health* **14**, 281–288.

Kruger, F.J. (1987). Enzyme electrophoresis of South African *Schistosoma mattheei* and *S. haematobium*. *Onderstepoort Journal of Veterinary Research* **54**, 93–96.

Kruger, F.J. (1988). Further observations on the electrophoretic characterization of South African *Schistosoma mattheei* and *S. haematobium*. *Onderstepoort Journal of Veterinary Research* **55**, 67–68.

Kruger, F.J. (1989). Enzyme polymorphism in *Schistosoma mattheei* from cattle in the Eastern Transvaal Lowveld. *Journal of Helminthology* **63**, 191–196.

Kruger, F.J. (1990). Frequency and possible consequences of hybridization between *Schistosoma haematobium* and *S. mattheei* in the Eastern Transvaal Lowveld. *Journal of Helminthology* **64**, 333–336.

Kruger, F.J. and Evans, A.C. (1990). Do all human urinary infections with *Schistosoma mattheei* represent hybridization between *S. haematobium* and *S. mattheei*? *Journal of Helminthology* **64**, 330–332.

Kruger, F.J. and Hamilton-Attwell, V.L. (1988). Scanning electron microscope studies of miracidia suggest introgressive hybridization between *Schistosoma haematobium* and *S. haematobium* × *S. mattheei* in the Eastern Transvaal. *Journal of Helminthology* **62**, 141–147.

Kruger, S.P. and Heitman, L.P. (1967). Studies on bilharzia I. The development of an apparatus to hatch miracidia. *Journal of the South African Veterinary Medical Association* **38**, 191–196.

Kruger, S.P., Heitman, L.P., van Wyk, J.A. and McCully, R.M. (1969). The route of migration of *Schistosoma mattheei* from the lungs to the liver in sheep. *Journal of the South African Veterinary Medical Association* **40**, 39–43.

Kruger, F.J., Schutte, C.H.J., Visser, P.S. and Evans, A.C. (1986a). Phenotypic differences in *Schistosoma mattheei* ova from populations sympatric and allopatric to *S. haematobium*. *Onderstepoort Journal of Veterinary Research* **53**, 103–107.

Kruger, F.J., Hamilton-Attwell, V.L. and Schutte, C.H.J. (1986b). Scanning electron microscopy of the teguments of males from five populations of *Schistosoma mattheei*. *Onderstepoort Journal of Veterinary Research* **53**, 109–110.

Kruger, F.J., Hamilton-Attwell, V.L., Tiedt, L., Visser, P.S. and Joubert, P.H. (1988). Notes on the occurrence of tubercular spines in *Schistosoma margrebowiei* and *Schistosoma mattheei*. *Onderstepoort Journal of Veterinary Research* **55**, 187–189.

Kulkarni, H., Rao, S.R. and Chaudhari, P.G. (1954). Unusual outbreaks of schistosomiasis in bovines due to *Schistosoma spindalis* associated with heavy mortality in Bombay State. *Bombay Veterinary College Magazine* **4**, 3–15.

Kumar, V. and de Burbure, G. (1986). Schistosomes of animals and man in Asia. *Helminthological Abstracts* (Series A) **55**, 469–480.

Kuntz, R.E., Davidson, D.L., Huang, T.C. and Tulloch, T.S. (1979). Scanning electron microscopy of the integumental surfaces of *Schistosoma bovis*. *Journal of Helminthology* **53**, 131–132.

Lawrence, J.A. (1970). Examination of ruminant faeces for schistosome eggs. *Rhodesian Veterinary Journal* **1**, 49–52.

Lawrence, J.A. (1973a). *Schistosoma mattheei* in cattle: variations in parasite egg production. *Research in Veterinary Science* **14**, 402–404.

Lawrence, J.A. (1973b). *Schistosoma mattheei* in cattle: the host–parasite relationship. *Research in Veterinary Science* **14**, 400–402.

Lawrence, J.A. (1974). *Schistosoma mattheei* in sheep: the host–parasite relationship. *Research in Veterinary Science* **17**, 263–264.

Lawrence, J.A. (1976). *Schistosoma mattheei* in the ox: clinical aspects. *Rhodesian Veterinary Journal* **7**, 48–51.

Lawrence, J.A. (1977a). *Schistosoma mattheei* in the ox: observations on the parasite. *Veterinary Parasitology* **3**, 291–303.

Lawrence, J.A. (1977b). *Schistosoma mattheei* infestation in the ox: the intestinal syndrome. *Journal of the South African Veterinary Association* **48**, 55–58.

Lawrence, J.A. (1977c) *Schistosoma mattheei* in the ox: the chronic hepatic syndrome. *Journal of the South African Veterinary Association* **48**, 77–83.

Lawrence, J.A. (1977d) *Schistosma mattheei* in the ox: clinical pathological observations. *Research in Veterinary Science* **23**, 280–287.

Lawrence, J.A. (1977e). *Schistosoma mattheei* in the ox: distribution of the parasite in the host. *Veterinary Parasitology* **3**, 305–315.

Lawrence, J.A. (1977f). *Schistosoma mattheei* in the ox: the serological response. *Research in Veterinary Science* **23**, 288–292.

Lawrence, J.A. (1978a). Bovine schistosomiasis in Southern Africa. *Helminthological Abstracts* **47**, 261–270.

Lawrence, J.A. (1978b). The pathology of *Schistosoma mattheei* infection in the ox. 1. Lesions attributable to the eggs. *Journal of Comparative Pathology* **88**, 1–14.

Lawrence, J.A. (1978c). The pathology of *Schistosoma mattheei* infection in the ox. 2. Lesions attributable to the adult parasites. *Journal of Comparative Pathology* **88**, 15–29.

Lawrence, J.A. and Condy, J.B (1970). The developing problem of schistosomiasis in domestic stock in Rhodesia. *Central African Journal of Medicine* **16** (Supplement to No. 7), 19–22.

Lawrence, J.A. and McKenzie, R.L. (1972). Schistosomiasis in farm livestock. *Rhodesia Agricultural Journal* **69**, 79–83.

Lawrence, J.A. and Schwartz, W.O.H. (1969). Treatment of *Schistosoma mattheei* infection in cattle. *Journal of the South African Veterinary Medical Association* **40**, 129–136.

Lengy, J. (1962). Studies on *Schistosoma bovis* (Sonsino, 1876) in Israel. I. Larval stages from egg to cercaria. *Bulletin of the Research Council, Israel* **10E**, 1–36.

Le Roux, P.L. (1933). A preliminary note on *Bilharzia margrebowiei*, a new parasite of ruminants and possibly man in Northern Rhodesia. *Journal of Helminthology* **11**, 53–62.

Le Roux, P.L. (1955). A new mammalian schistosome (*Schistosoma leiperi* sp. nov.), from herbivora in Northern Rhodesia. *Transactions of the Royal Society of Tropical Medicine and Hygiene* **49**, 293–294.

Lie, K.J., Kwo, E.H. and Owyang, C.K. (1970). A field trial to test the possible control of *Schistosoma spindale* by means of interspecific trematode antagonism. *Southeast Asian Journal of Tropical Medicine and Public Health* **1**, 19–28.

Lie, K.J., Kwo, E.H. and Owyang, C.K. (1971). Further trial to control *Schistosoma spindale* by trematode antagonism. *Southeast Asian Journal of Tropical Medicine and Public Health* **2**, 237–243.

Lie, K.J., Schneider, C.R., Sornmani, S., Lanza, G.R. and Impand, P. (1974a). Biological control by trematode antagonism. I. A successful field trial to control *Schistosoma spindale* in Northeast Thailand. *Southeast Asian Journal of Tropical Medicine and Public Health* **5**, 46–59.

Lie, K.J., Schneider, C.R., Sornmani, S., Lanza, G.R. and Impand, P. (1974b). Biological control by trematode antagonism. II. Failure to control *Schistosoma spindale* in a field trial in Northeast Thailand. *Southeast Asian Journal of Tropical Medicine and Public Health* **5**, 60–64.

Ling, X.F. and Mao, G.Y. (1981). [Reaction of farm cattle to amoscanate treatment for control of schistosomiasis japonica.] *Chinese Journal of Veterinary Medicine* **7**, 20–21.

Loker, E.S. (1983). A comparative study of the life-histories of mammalian schistosomes. *Parasitology* **87**, 343–369.

Lwambo, N.J.S., Upatham, E.S., Kruatrachue, M. and Viyanant, V. (1987). The host-parasite relationship between the Saudi Arabian *Schistosoma mansoni* and its intermediate and definitive hosts. 1. *S. mansoni* and its local snail host *Biomphalaira arabica*. *Southeast Asian Journal of Tropical Medicine and Public Health* **18**, 156–165.

Madsen, H. (1990). Biological methods for the control of freshwater snails. *Parasitology Today* **6**, 237–241.

Mahon, R.J. and Shiff, C.J. (1978). Electrophoresis to distinguish *Schistosoma haematobium* and *S. mattheei* cercariae emerging from *Bulinus* snails. *Journal of Parasitology* **64**, 372–373.

Majid, A.A., Marshall, T.F. de C., Hussein, M.F., Bushara, H.O., Taylor, M.G., Nelson, G.S. and Dargie, J.D. (1980a). Observations on cattle schistosomiasis in the Sudan, a study in comparative medicine. I. Epizootiological observations on *Schistosoma bovis* in the White Nile province. *American Journal of Tropical Medicine and Hygiene* **29**, 435–441.

Majid, A.A., Bushara, H.O., Saad, A.M., Hussein, M.F., Taylor, M.G., Dargie, J.D., Marshall, T.F. de C. and Nelson, G.S. (1980b). Observations on cattle schistosomiasis in the Sudan, a study in comparative medicine. III. Field testing of an irradiated *Schistosoma bovis* vaccine. *American Journal of Tropical Medicine and Hygiene* **29**,452–455.

Malek, E.A. (1969). Studies on bovine schistosomiasis in the Sudan. *Annals of Tropical Medicine and Parasitology* **63**, 501–513.

Malek, E.A. and Ongom, V.L. (1984). *Schistosoma leiperi* Le Roux, 1955 from a bushbuck in Uganda. *Journal of Parasitology* **70**, 821–822.

Mao, S.P. and Shao, B.R. (1982). Schistosomiasis control in the People's Republic of China. *American Journal of Tropical Medicine and Hygiene* **31**, 92–99.

Markovics, A., Perl, S., Chaimovitz, M., Klopfer, U. and Pipano, E. (1985). Efficacy of Praziquantel in *Schistosoma bovis* infections. *Israel Journal of Medical Science* **21**, 712.

Markovics, A., Perl, S., Orgad, U. and Pipano, E. (1993). Outbreaks of schistosomiasis (*Schistosoma bovis*) in cattle and sheep. *Israel Journal of Veterinary Medicine* **48**, 123–125.

Massoud, J. (1973). Parasitological and pathological observations on *Schistosoma bovis* Sonsino, 1876, in calves, sheep and goats in Iran. *Journal of Helminthology* **47**, 155–164.

Massoud, J. and Nelson, G.S. (1972). Studies on heterologous immunity in schistosomiasis. 6. Observations on cross-immunity to *Ornithobilharzia turkestanicum, Schistosoma bovis, S. mansoni,* and *S. haematobium* in mice, sheep, and cattle in Iran. *Bulletin of the World Health Organisation* **47**, 591–600.

McCauley, E.H., Tayeb, A. and Majid, A.A. (1983). Owner survey of schistosomiasis mortality in Sudanese cattle. *Tropical Animal Health and Production* **15**, 227–233.

McCauley, E.H., Majid, A.A. and Tayeb, A. (1984). Economic evaluation of the production impact of bovine schistosomiasis and vaccination in the Sudan. *Preventive Veterinary Medicine* **2**, 735–754.

McCully, R.M. and Kruger, S.P. (1969). Observations on bilharziasis of domestic ruminants in South Africa. *Onderstepoort Journal of Veterinary Research* **36**, 129–162.

Montgomery, R.E. (1906a). Observations on bilharziosis among animals in India. I. *Journal of Tropical Veterinary Science* **1**, 14–46.

Montgomery, R.E. (1906b). Observations on bilharziosis among animals in India. II. *Journal of Tropical Veterinary Science* **1**, 138–174.

Moore, D.V. and Sandground, J.H. (1956). The relative egg producing capacity of *Schistosoma mansoni* and *Schistosoma japonicum*. *American journal of Tropical Medicine and Hygiene* **5**, 831–840.

Mott, K.E. (1987). Schistosomiasis control. In: *The Biology of Schistosomes: from Genes to Latrines* (D. Rollinson and A.J.G. Simpson, eds). pp. 431–450. London: Academic Press.

Mouahid, A. and Combes, C. (1987). Genetic variability of *Schistosoma bovis* cercarial production according to miracidial dose. *Journal of Helminthology* **61**, 89–94.

Mouahid, A. and Théron, A. (1987). *Schistosoma bovis*: Variability of cercarial production as related to the snail hosts: *Bulinus truncatus, B. wrighti* and *Planorbarius metidjensis*. *International Journal for Parasitology* **17**, 1431–1434.

Mouchet, F., Develoux, M. and Magasa, M.B. (1988). *Schistosoma bovis* in human stools in Republic of Niger. *Transactions of the Royal Society of Tropical Medicine and Hygiene* **82**, 257.

Mouchet, F., Théron, A., Brémond, P., Sellin, E. and Sellin, B. (1992). Pattern of cercarial emergence of *Schistosoma curassoni* from Niger and comparison with three sympatric species of schistosomes. *Journal of Parasitology* **78**, 61–63.

Muraleedharan, K., Kumar, S.P., Hedge, K.S. and Alwar, V.S. (1973). Incidence of *Schistosoma nasale* Rao, 1933 infection in sheep. *Indian Veterinary Journal* **50**, 1056–1057.

Muraleedharan, K., Kumar, S.P. and Hedge, K.S. (1975). Predatory activity of the guppy *Lebistes reticulatus* (Peters, 1859) on cercariae and miracidia of *Schistosoma nasale* Rao, 1933. *Indian Veterinary Journal* **52**, 763–768.

Muraleedharan, K., Kumar, S.P. and Hedge, K.S. (1976a). An efficient egg counting technique for nasal schistosomiasis. *Indian Veterinary Journal* **53**, 143–146.

Muraleedharan, K., Kumar, S.P., Hedge, K.S. and Alwar, V.S. (1976b). Studies on the epizootiology of nasal schistosomiasis of bovines. 1. Prevalence and incidence of infection. *Mysore Journal of Agriculture Sciences* **10**, 105–117.

Nelson, G.S., Amin, M.A., Saoud, M.F.A. and Teesdale, C. (1968). Studies on heterologous immunity in schistosomiasis. 1. Heterologous schistosome immunity in mice. *Bulletin of the World Health Organization* **38**, 9–17.

Ngendahayo, L.D., Bayssade-Dufour, C., Albaret, J.L., Diaw, O.T., Deiana, S., Southgate, V.R., Ross, G.C., Luffau, G. and Chabaud, A.G. (1987). Morphologie des téguments de *Schistosoma bovis*; variations selon l'hôte vertébré; comparaison avec *S. curassoni*. *Annales de Parasitologie Humaine et Comparée* **62**, 530–541.

Obwolo, M.J. and Rogers, S.E. (1988). Schistosomal lesions in the bovine uterus. *Journal of Comparative Pathology* **98**, 501–505.

Ogbe, M.G. (1982). Scanning electron microscopy of tegumental surfaces of adult and developing *Schistosoma margrebowiei* Le Roux, 1933. *International Journal for Parasitology* **12**, 191–198.

Olaechea, F.V., Christensen, N.Ø., and Henriksen, S.A. (1990). A comparison of the filtration, concentration, and thick smear techniques in the diagnosis of *Schistosoma bovis* infection in cattle and goats. *Acta Tropica* **47**, 217–221.

Pao, T.C. (1959). Description of a new schistosome *Schistosoma sinensium* sp. nov. (Trematoda: Schistosomatidae) from Szechuan province. *Chinese Medical Journal, Peking* **78**, 278.

Pitchford, R.J. (1959). Cattle schistosomiasis in man in the eastern Transvaal. *Transactions of the Royal Society of Tropical Medicine and Hygiene* **53**, 285–290.

Pitchford, R.J. (1961). Observations on a possible hybrid between the two schistosomes *S. haematobium* and *S. mattheei*. *Transactions of the Royal Society of Tropical Medicine and Hygiene* **55**, 44–51.

Pitchford, R.J. (1963). Some brief notes on schistosomes occurring in animals. *Journal of the South African Veterinary Association* **34**, 613–618.

Pitchford, R.J. (1966). Findings in relation to schistosome transmission in the field following the introduction of various control measures. *South African Medical Journal* **40** (Supplement), 1–16.

Pitchford, R.J. (1974). Some preliminary observations on schistosomes occurring in antelope in central southern Africa. *Rhodesian Veterinary Journal* **4**, 57–61.

Pitchford, R.J. (1976). Preliminary observations on the distribution, definitive hosts and possible relation with other schistosomes, of *Schistosoma margrebowiei*, Le Roux, 1933 and *Schistosoma leiperi*, Le Roux, 1955. *Journal of Helminthology* **50**, 111–123.

Pitchford, R.J. (1977). A check list of definitive hosts exhibiting evidence of the genus *Schistosoma* Weinland, 1858, acquired naturally in Africa and the Middle East. *Journal of Helminthology* **51**, 229–252.

Pitchford, R.J. and Du Toit, J.F. (1976). The shedding pattern of three little-known African schistosomes under outdoor conditions. *Annals of Tropical Medicine and Parasitology* **70**, 181–187.

Pitchford, R.J. and Visser, P.S. (1962). Results of exposing mice to schistosomiasis by immersion in natural water. *Transactions of the Royal Society of Tropical Medicine and Hygiene* **56**, 294–301.

Pitchford, R.J. and Visser, P.S. (1965). Some further observations on schistosome transmission in the eastern Transvaal. *Bulletin of the World Health Organization* **32**, 83–104.

Pitchford, R.J. and Visser, P.S. (1969). The use of behaviour patterns of larval schistosomes in assessing the bilharzia potential of non-endemic areas. *South African Medical Journal* **43**, 983–995.

Pitchford, R.J. and Visser, P.S. (1975). A simple technique for quantitative estimation of helminth eggs in human and animal excreta with special reference to *Schistosoma* sp. *Transactions of the Royal Society of Tropical Medicine and Hygiene* **69**, 318–322.

Pitchford, R.J. and Visser, P.S. (1982). *Schistosoma mattheei* Veglia & Le Roux, 1929, egg output from cattle in a highly endemic area in the Eastern Transvaal. *Onderstepoort Journal of Veterinary Research* **49**, 233–235.

Pitchford, R.J., Meyling, A.H., Meyling, J. and Du Toit, J.F. (1969). Cercarial shedding patterns of various schistosome species under outdoor conditions in the Transvaal. *Annals of Tropical Medicine and Parasitology* **63**, 359–371.

Pitchford, R.J., Visser, P.S., Du Toit, J.F., Pienaar, U. de V. and Young, E. (1973). Observations on the ecology of *Schistosoma mattheei*, Veglia & Le Roux, 1929, in portion of the Kruger National Park and surrounding area using a new quantitative technique for egg output. *Journal of the South African Veterinary Association* **44**, 405–420.

Pitchford, R.J., Visser, P.S., Pienaar, U. de V. and Young, E. (1974). Further observations on *Schistosoma mattheei*, Veglia & Le Roux, 1929, in the Kruger National Park. *Journal of the South African Veterinary Association* **45**, 211–218.

Popiel, I. (1986). The reproductive biology of schistosomes. *Parasitology Today* **2**, 10–19.

Preston, J.M. and Duffus, W.P.H. (1975). Diagnosis of *Schistosoma bovis* infection in cattle by an indirect haemagglutination test. *Journal of Helminthology* **49**, 9–17.

Preston, J.M. and Webbe, G. (1974). Studies on immunity to reinfection with *Schistosoma mattheei* in sheep and cattle. *Bulletin of the World Health Organization* **50**, 566–568.

Probert, A.J. and Awad, A.H.H. (1987). Scanning electron microscopy of the tegument of adult *S. margrebowiei* Le Roux, 1933 with particular reference to the structure of the tubercles. *Parasitology* **95**, 491–498.

Qin, L.R., Liu, Z.L., Chen, Y.Z., Ma, L.H., Zeng, X.G. and Xu, M.G. (1982). [Treatment of *Schistosoma japonicum* infection in farm cattle with nithiocyanamin.] *Chinese Journal of Veterinary Medicine* **8**, 2–6.

Rahman, S.A., Sano, M., Jagannath, M.S., Prabhakar, K.S., D'Souza, P.E. and Prem, G. (1988). Efficacy of praziquantel against *Schistosoma nasale* infection in cattle. *Tropical Animal Health and Production* **20**, 19–22.

Ramajo-Martin, V. (1978). Observaciones acerca de la receptividad de diversas poblaciones de *Planorbarius metidjensis, Bulinus* (B.) *truncatus* y *Biomphalaria glabrata* a *Schistosoma bovis* de Espana. *Revista Ibérica de Parasitologia* **38**, 537–549.

Ramsay, G.W. ST L. (1934). A study on schistosomiasis and certain other helminthic infections in Northern Nigeria. *West African Medical Journal* **8**, 2–10.

Rao, M.A.N. (1933). Bovine nasal schistosomiasis in the Madras Presidency with a description of the parasite. *Indian Journal of Veterinary Science and Animal Husbandry* **3**, 29–38.

Rao, M.A.N. (1934). A comparative study of *Schistosoma spindalis*, Montgomery, 1906, and *Schistosoma nasalis*, n. sp. *Indian Journal of Veterinary Science and Animal Husbandry* **4**, 1–28.

Rao, P.V.R. and Devi, T.I. (1971). Nasal schistosomiasis in buffaloes. *Indian Journal of Animal Health* **10**, 185–188.

Reinecke, R.K. (1970). The epizootiology of an outbreak of bilharziasis in Zululand. *Central African Journal of Medicine* **16**, 10–12.

Rollinson, D. and Southgate, V.R. (1985). Schistosomes and snail populations: genetic variability and parasite transmission. In: *Ecology and Genetics of*

Host–Parasite Interactions (D. Rollinson and R.M. Anderson, eds). Linnean Society Symposium Series Vol. 11, pp. 91–109. London: Academic Press.

Rollinson, D. and Southgate, V.R. (1987). The genus *Schistosoma*: A taxonomic appraisal. In: *The Biology of Schistosomes: from Genes to Latrines* (D. Rollinson and A.J.G. Simpson, eds). pp. 1–49. London: Academic Press.

Rollinson, D., Walker, T.K. and Simpson, A.J.G. (1986). The application of recombinant DNA technology to problems of helminth identification. *Parasitology* **91**, S53–S71.

Rollinson, D., Vercruysse, J., Southgate, V.R., Moore, P.J., Ross, G.C., Walker, T.K. and Knowles, R.J. (1987). Observations on human and animal schistosomiasis in Senegal. In: *Helminth Zoonoses* (S. Geerts, V. Kumar and J. Brandt, eds). pp. 119–131. The Hague: Martinus Nijhoff.

Rollinson, D., Southgate, V.R., Vercruysse, J. and Moore, P.J. (1990a). Observations on natural and experimental interactions between *Schistosoma bovis* and *S. curassoni* from West Africa. *Acta Tropica* **47**, 101–114.

Rollinson, D., Walker, T.K., Knowles, R.J. and Simpson, A.J.G. (1990b). Identification of schistosome hybrids and larval parasites using rRNA probes. *Systematic Parasitology* **15**, 65–73.

Rollinson, D., Kaukas, A., Johnston, D.A., Simpson, A.J.G. and Tanaka, M. (1997). Some molecular insights into schistosome evolution. *International Journal for Parasitology* **27**, 11–28.

Ross, G.C., Southgate, V.R. and Knowles, R.J. (1978). Observations on some isoenzymes of strains of *Schistosoma bovis, S. mattheei, S. margrebowiei*, and *S. leiperi*. *Zeitschrift für Parasitenkunde* **57**, 49–56.

Ross, G.C., Bayssade-Dufour, C., Southgate, V.R., Albaret, J.L., Ngendahayo, L.D. and Chabaud, A.G. (1987). Relationships between cercarial indices of *Schistosoma haematobium, S. bovis* and *S. curassoni* from Senegal and the isoenzyme genotypes of the adult worms. *Annales de Parasitologie Humaine et Comparée* **62**, 507–515.

Saad, A.M., Hussein, M.F., Dargie, J.D., Taylor, M.G. and Nelson, G.S. (1980). *Schistosoma bovis* in calves: the development and clinical pathology of primary infections. *Research in Veterinary Science* **28**, 105–111.

Saad, A.M., Hussein, M.F., Bushara, H.O., Dargie, J.D. and Taylor, M.G. (1984). Erythrokinetics and albumin metabolism in primary experimental *Schistosoma bovis* infections in zebu calves. *Journal of Comparative Pathology* **94**, 249–262.

Saeed, A.A., Nelson, G.S. and Hussein, M.F. (1969). Experimental infection of calves with *Schistosoma mansoni*. *Transactions of the Royal Society of Tropical Medicine and Hygiene* **63**, 456–458.

Saharee, A.A., Sani, R.A., Sheikh-omar, A.R. and Greer, G.J. (1984). A case of bovine nasal schistosomiasis. *Kajian Veterinar* **16**, 33–36.

Sakamoto, K. and Ishii, Y. (1977). Scanning electron microscope observations on adult *Schistosoma japonicum*. *Journal of Parasitology* **63**, 407–412.

Sambon, L.W. (1907). Remarks on *Schistosoma mansoni*. *Journal of Tropical Medicine and Hygiene* **10**, 303–304.

Shi, S.K. (1981). [Clinical observation on the treatment of 100 cases of bovine schistosomiasis with Sb-58.] *Chinese Journal of Veterinary Medicine* **7**, 16–18.

Sirag, S.B., Christensen, N.Ø., Nansen, P., Monrad, J. and Frandsen, F. (1981). Resistance to *Fasciola hepatica* in calves harbouring primary patent *Schistosoma bovis* infections. *Journal of Helminthology* **55**, 63–70.

Simpson, A.J.G., Dama, J.B., Lewis, F.A. and McCutchan, T.F. (1984). The arrangements of ribosomal RNA genes in *Schistosoma mansoni*. Identification of polymorphic structural variants. *European Journal of Biochemistry* **139**, 41–45.

Sinha, P.K. and Srivastava, H.D. (1960). Studies on *Schistosoma incognitum*, Chandler, 1926. II. On the life-cycle of the blood fluke. *Journal of Parasitology* **46**, 629–641.

Sinha, P.K. and Srivastava, H.D. (1965). Studies on *Schistosoma incognitum*, Chandler, 1926. III. On the host specificity of the blood fluke. *Indian Veterinary Journal* **42**, 335–341.

Smithers, S.R. and Terry, R.J. (1967). Resistance to experimental infection with *Schistosoma mansoni* in rhesus monkeys induced by the transfer of adult worms. *Transactions of the Royal Society of Tropical Medicine and Hygiene* **61**, 517–533.

Smithers, S.R. and Terry, R.J. (1969). Immunity in schistosomiasis. *Annals of the New York Academy of Sciences* **160**, 826–840.

Sonsino, P. (1876). On a new parasite of cows, *Bilharzia bovis*. *Rendiconti dell'Accademia della Scienze Fisiche e Matematiche, Napoli* **15**, 84–87.

Southgate, V.R. and Knowles, R.J. (1975). Observations on *Schistosoma bovis* Sonsino, 1876. *Journal of Natural History* **9**, 273–314.

Southgate, V.R. and Knowles, R.J. (1977). On *Schistosoma margrebowiei* Le Roux, 1933: The morphology of the egg, miracidium and cercaria, the compatibility with species of Bulinus and development in *Mesocricetus auratus*. *Zeitschrift für Parasitenkunde* **54**, 233–250.

Southgate, V.R., Rollinson, D., Ross, G.C. and Knowles, R.J. (1980). Observations on an isolate of *Schistosoma bovis* from Tanzania. *Zeitschrift für Parasitenkunde* **63**, 241–249.

Southgate, V.R., Ross, G.C. and Knowles, R.J. (1981). On *Schistosoma leiperi* Le Roux, 1955: scanning electron microscopy of adult worms, compatibility with species of *Bulinus*, development in *Mesocricetus auratus* and isoenzymes. *Zeitschrift für Parasitenkunde* **66**, 63–81.

Southgate, V.R., Rollinson, D., Ross, G.C., Knowles, R.J. and Vercruysse, J. (1985). On *Schistosoma curassoni, S. haematobium* and *S. bovis* from Senegal: developments in *Mesocricetus auratus*, compatibility with species of *Bulinus* and their enzymes. *Journal of Natural History* **19**, 1249–1267.

Southgate, V.R., Rollinson, D. and Vercruysse, J. (1986). Scanning electron microscopy of the tegument of adult *Schistosoma curassoni*, and comparison with male *S. bovis* and *S. haematobium* from Senegal. *Parasitology* **93**, 433–442.

Southgate, V.R., Rollinson, D., De Bont, J., Vercruysse, J., van Aken, D. and Spratt, J. (1990). Surface topography of the tegument of adult *Schistosoma nasale* Rao, 1933 from Sri Lanka. *Systematic Parasitology* **16**, 139–147.

Stelma, F.F. (1997). *Immuno-epidemiology, Morbidity and Chemotherapy in a Community Recently Exposed to* Schistososoma mansoni *Infection*. PhD thesis, Department of Parasitology, University of Leiden.

Strydom, H.F. (1963). Bilharziasis in sheep and cattle in the Piet Retief District. *Journal of the South African Medical Association* **34**, 69–72.

Sturrock, R.F. (1966). Daily egg output of schistosomes. *Transactions of the Royal Society of Tropical Medicine and Hygiene* **60**, 139–140.

Sturrock, R.F. (1993a). The intermediate hosts and host-parasite relationships. In: *Human Schistosomiasis* (P. Jordan, G. Webbe and R.F. Sturrock, eds). pp. 33–85. Wallingford: CAB International.

Sturrock, R.F. (1993b). The parasites and their life cycles. In: *Human Schistosomiasis* (P. Jordan, G. Webbe and R.F. Sturrock, eds). pp. 1–32. Wallingford: CAB International.

Symmers, Wm. St.C (1903). Note on a new form of liver cirrhosis due to the presence of the ova of bilharzia haematobia. *Journal of Pathology and Bacteriology* **9**, 237–239.

Taylor, M.G. (1970). Hybridisation experiments on five species of African schistosomes. *Journal of Helminthology* **44**, 253–314.

Taylor, M.G. (1971). Further observations on the sexual maturation of female schistosomes in single-sex infections. *Journal of Helminthology* **45**, 89–92.

Taylor, M.G. (1987). Schistosomes of domestic animals: *Schistosoma bovis* and other animal forms. In: *Immunology, Immunoprophylaxis and Immunotherapy of Parasitic Infections* (E.J.L. Soulsby, ed), vol 2, pp. 49–90. Boca Raton: CRC Press.

Taylor, M.G., Amin, M.B.A. and Nelson, G.S. (1969). 'Parthenogenesis' in *Schistosoma mattheei*. *Journal of Helminthology* **43**, 197–206.

Tchuem Tchuenté, L.A., Southgate, V.R., Jourdane, J., Kaukas, A. and Vercruysse, J. (1997). Hybridization between *Schistosoma haematobium* and *S. mattheei*: viability of hybrids and their development in sheep. *Systematic Parasitology* **36**, 123–131.

Théron, A. (1981). Dynamics of larval populations of *Schistosoma mansoni* in *Biomphalaria glabrata*. I. Rhythmic production of cercariae in monomiracidial infections. *Annals of Tropical Medicine and Parasitology* **75**, 71–77.

Thurston, J.P. (1963). Schistosomes from *Hippopotamus amphibius*. L.I. The morphology of *Schistosoma hippopotami* sp. nov. *Parasitology* **53**, 49–54.

Thurston, J.P. (1964). Schistosomes from *Hippopotamus amphibius* L.II. The morphology of *Schistosoma edwardiense* sp. nov. *Parasitology* **54**, 67–72.

Touassem, R. (1987). Egg polymorphism of *Schistosoma bovis*. *Veterinary Parasitology*, **23**, 185–191.

Tulloch, G.S., Kuntz, R.E., Davidson, D.L. and Huang, T.C. (1977). Scanning electron microscopy of the integument of *Schistosoma mattheei* Veglia and Le Roux, 1929. *Transactions of the American Microscopical Society* **96**, 41–47.

Upatoom, N., Horchner, F. and Leidl, K. (1988). Therapy against *Schistosoma spindale* infestation in cattle and buffalo. *Veterinary Medical Review* **59**, 171–174.

van Wyk, J.A. (1971). Rapid diagnosis of bilharzia. *Journal of the South African Veterinary Medical Association* **42**, 378.

van Wyk, J.A., (1983). The importance of animals in human schistosomiasis in South Africa. *South African Medical Journal* **63**, 201–204.

van Wyk, J.A. and Bartsch, R.C. (1971). Staggers syndrome in experimental hyperergic (?) schistosomiasis. *Journal of the South African Veterinary Medical Association* **42**, 337.

van Wyk, J.A., Bartsch, R.C., van Rensburg, L.J., Heitmann, L.P. and Goosen, P.J. (1974). Studies on schistosomiasis. VI. A field outbreak of bilharzia in cattle. *Onderstepoort Journal of Veterinary Research* **41**, 39–49.

van Wyk, J.A., Heitmann, L.P. and van Rensburg, L.J. (1975). Studies on schistosomiasis. VII. A comparison of various methods for the infestation of sheep with *Schistosoma mattheei*. *Onderstepoort Journal of Veterinary Research* **42**, 71–74.

van Wyk, J.A., van Rensburg, L.J. and Heitmann, L.P. (1997). *Schistosoma mattheei* infection in cattle: the course of the intestinal syndrome, and an estimate of the lethal dose of cercariae. *Onderstepoort Journal of Veterinary Research* **64**, 65–75.

Veglia, F. and Le Roux, P.L. (1929). On the morphology of a schistosome (*Schistosoma mattheei* sp. nov.) from the sheep in the Cape Province. *15th Annual Report, Director of Veterinary Services, Union of South Africa 1929*, pp. 335–346.

Vercruysse, J., Southgate, V.R. and Rollinson, D. (1984). *Schistosoma curassoni*

Brumpt, 1931, in sheep and goats in Senegal. *Journal of Natural History* **18**, 969–976.
Vercruysse, J., Southgate, V.R. and Rollinson, D. (1985). The epidemiology of human and animal schistosomiasis in the Senegal River Basin. *Acta Tropica* **42**, 249–259.
Voge, M., Bruckner, D. and Bruce, J.I. (1978). *Schistosoma mekongi* sp. n. from man and animals compared with four geographic strains of *Schistosoma japonicum*. *Journal of Parasitology* **64**, 577–584.
Walker, T.K., Rollinson, D. and Simpson, A.J.G. (1986). Differentiation of *Schistosoma haematobium* from related species using cloned ribosomal RNA gene probes. *Molecular and Biochemical Parasitology* **20**, 123–131.
Wang, P.H., Liu, S.H., Hua, T.S. and Chang, S. (1958). A newly discovered mode of infection of schistosomiasis japonica. *Chinese Medical Journal* **77**, 569.
Wang, J.S., Wu, F.M., Tung, K.C. and Chan, J.P. (1988). An outbreak and epidemiology of bovine schistosomiasis japonica in Changhua County of Taiwan. *Journal of the Chinese Society of Veterinary Science* **14**, 289–296.
Webbe, G. (1965). Transmission of bilharziasis. 2. Production of cercariae. *Bulletin of the World Health Organization* **33**, 155–244.
Webbe, G. (1982). The parasites. In: *Schistosomiasis. Epidemiology, Treatment and Control* (P. Jordan and G. Webbe, eds), pp. 1–15. London: William Heinemann Medical Books.
Webbe, G. and Jordan, P. (1993). Control. In: *Human Schistosomiasis* (P. Jordan, G. Webbe and R.F. Sturrock, eds), pp. 405–451. Wallingford: CAB International.
Weinland, D.F. (1858). *Human Cestoides*. Cambridge: Metcalf and Co.
Weinstock, J.V. (1992). The pathogenesis of granulomatous inflammation and organ injury in schistosomiasis: interactions between the schistosome ova and the host. *Immunological investigations* **21**, 455–475.
Wilson, R.A. (1987). Cercariae to liver worms: development and migration in the mammalian host. In: *The Biology of Schistosomes: from Genes to Latrines* (D. Rollinson and A.J.G. Simpson, eds), pp. 115–146. London: Academic Press.
Wright, C.A. and Ross, G.C. (1980). Hybrids between *Schistosoma haematobium* and *S. mattheei* and their identification by isoelectric focusing of enzymes. *Transactions of the Royal Society of Tropical Medicine and Hygiene* **74**, 326–332.
Wright, C.A. and Ross, G.C. (1983). Enzyme analysis of *Schistosoma haematobium*. *Bulletin of the World Health Organization* **61**, 307–316.
Wright, C.A., Southgate, V.R. and Knowles, R.J. (1972). What is *Schistosoma intercalatum* Fisher, 1934? *Transactions of the Royal Society of Tropical Medicine and Hygiene* **66**, 28–64.
Wright, C.A., Rollinson, D. and Goll, P.H. (1979a). Parasites in *Bulinus senegalensis* (Mollusca: Planorbidae) and their detection. *Parastology* **79**, 95–105.
Wright, C.A., Southgate, V.R. and Howard, G.W. (1979b). Observations on the life cycle of *Schistosoma margrebowiei* in Zambia. *Journal of Natural History* **13**, 499–506.
Xu, C.B., Verwaerde, C., Grzych, J.M., Fontaine, J. and Capron, A. (1991). A monoclonal antibody blocking the *Schistosoma mansoni* 28-kDa glutathione S-transferase activity reduces female worm fecundity and egg viability. *European Journal of Immunology* **21**, 1801–1807.
Xu, S., Shi, F., Shen, W., Lin, J., Wang, Y., Lin, B., Qian, C., Ye, P., Fu, L., Shi, Y., Wu, W., Zhang, Z., Zhu, H. and Guo, W. (1993). Vaccination of bovines against Schistosomiasis japonica with cryopreserved, irradiated and freeze-thaw schistosomula. *Veterinary Parasitology* **47**, 37–50.

Yagi, A.I., Younis, S.A., Haroun, E.M., Gameel, A.A., Bushara, H.O. and Taylor, M.G. (1986). Studies on heterologous resistance between *Schistosoma bovis* and *Fasciola gigantica* in Sudanese cattle. *Journal of Helminthology* **60**, 55–59.

Yuan, H.C. (1993). Epidemiological features and control strategies of schistosomiasis japonica in China. *Chinese Medical Journal* **106**, 563–568.

Index

Note: page numbers in *italics* refer to figures and tables

acid phosphatase, schistosome 296, 297
acquired immune deficiency syndrome *see* AIDS
active cutaneous anaphylaxis (ACA) reaction 156
adequate clinical response (ACR) 5
adrenocorticotrophic hormone (ACTH) 153
African Programme for Onchocerciasis Control 267
agriculture 137
AIDS 64
   Pneumocystis carinii 93
albumin 307, 309
alleles per locus (A) 229, *230–1*
allopatry 253
alveolar epithelial cells
   P. carinii attachment 73–4
   type II 74
alveolar macrophages in PCP 74
aminoalcohols 26–7
4-aminoquinolines 4–6, 17
   novel 31–2
   resistance 21–4
amodiaquine 5
   enhanced efflux from host–parasite complex 23
Ancylostoma caninum 153
Anisakis 252
Anisakis simplex 262
anthelmintics 138, 262–3
   metabolic pathways 265
   resistance 262–5, *266*, 267–9
      allele frequency 265, *266*, 267
      alleles 269
      dominance 268
      genetics 265, *266*, 267–8
      immigration 269
      membrane receptor proteins 265
      migration patterns 269
      p-glycoprotein 264–5
      population biology 268–9
      ß-tubulin 264
anthroponosis 137
anti-MSG monoclonal antibodies 70–1, 74
antibiotics 34–5
antibody
   antigen binding 161
   T cells in worm expulsion 159–60
antibody response
   secretory 159
   Trichinella spiralis 155–61
   trichinosis 156–9, 162
antifols 32–4
   resistance 14
      biochemical/genetic basis 25–6
      models 16
      rodent models 19
   see also dihydrofolate reductase inhibitors
antigens
   circulating anodic/cathodic 318
   cross-reacting 192
   crude schistosome 342–3
   defined protective 343
   DvA-1 polyprotein 272
   evolution 272
   surface 272
antihistamines 187–8

antimalarial drugs
  cross-resistance 7
  development of new 28–37
  parasite response testing 2
  phytochemicals 31
  resistance 2–10, *11*, 12–28
  rodent studies 17
antimony compounds 337
antischistosomal drugs 337–9
antiserotonins 187–8
antithymocyte serum 162
Artabotrys uncinatus 30
Arteflene 30
artemether 8, 9, 12, 29
  combinations 39
  rodent models 19
Artemisia annua 2, 12, 28
artemisinin 2, *11*, 12, 28
  combinations 38–9
  gametocytocidal action 13
  recrudescence rate 30, 38
  rodent models 18–19
  synergism 38–9
artesunate 12, 29
  combinations 39
  rodent models 19
ascariasis in China 110–11, *112–13*, 114
  control 140
  cross-infection 132
  distribution 114, *115–17*, 118–19, 120, *121*
    agricultural methods 128, 130
    climate 126
    demographic factors 120, 122–3, *124*, 125
    environmental factors 125–6
    factors influencing 120, 122–3, *124*, 125–8, *129*, 130–3
    geographic factors 125–6
    income levels 127–8
    occupation 128, *129*
    pig host 131–3
    rural 125
    socio-economic factors 127–30
    topographic factors 125–6
    urban 125
    water supply 128
  drug resistance 138
  infestation rate 114, *115–17*
  intensity 133, *134*, 135, *136*, 140
  origin of human association 136–8
  pigs 131–3, 136, 140
  polyparasitism 130, *131*

  prevalence 110, 114, *115–17*, 118–20
    age 120–2
    gender differences 122–3
    humans 132–3
    nationality 123, *124*, 125
    pigs 132–3
  public health problem 110, 140
  sources of information 114
  traditional medicine 138–9
  worm burden 135
  see also *Ascaris lumbricoides*
ascarids
  allopatry 253
  allozyme surveys 250, 252
  F_1 hybrids 252
  hybrid sterility 252
  sympatry 247, 250, 252–3
Ascaris 221, 223
  allele frequency differences 247
  cross-infection 132, 248
  eggs 133
  genomic DNA 132
  host specificity 248, *250*
  ITS spacers 247, 248
  mating barriers 248
  microspatial genetic variation 242–3
  mtDNA
    diversity 224
    sequences 247, *249*
  nucleotide diversity 234
  rDNA polymorphism 241
  repetitive DNA 236–7
  sibling species divergence 247–8
  spatial clustering of infective eggs 243
  spatial patterns of transmission 242–3
  transmission cycles 132
  worm burden 243
Ascaris lumbricoides 110
  human contact 137–8
  origins of human infection 136
  polyparasitism in China 130
  population genetics 242–3
  speciation 138
  sympatry 247
  see also ascariasis in China
Ascaris suum 110
  mtDNA genes 227, *228*, 229
  population genetics 242–3
  speciation 138
  sympatry 247
  trickle infections 270

aspartic haemoglobinase 36
atovaquone 10, *11*
  combinations 10, *11*, 12, 40
  proguanil combination 10, *11*, 12
  resistance 28
  rodent models 19
  synergism 60
azithromycin 34–5

B cells
  anti-adult *Trichinella spiralis* effect 161
  growth and CD4$^+$, OX22$^-$ *198*
  worm expulsion 170
B immunoblasts 164
benflumetol 39
benzimidazole resistance 263, 264
  allele 267, 268, 269
bisbenzylisoquinoline alkaloids 31
bisquinolines
  novel 31–2
  resistance studies 17, *18*
bone marrow cells
  population for *T. spiralis* rejection 191
  role in trichinosis 193–4
Brugia 272

c-kit 184–6
Caenorhabditis elegans
  COI gene 227, 229
  ivermectin resistance 263–4
  mtDNA genes 227, *228*, 229
cancer chemotherapy 64
CD4 93
CD4$^+$
  adult *T. spiralis* rejection 197
  effector cell dissemination 166
  trichinosis infection rejection 165
CD4$^+$,OX22$^+$ cells 186
CD4$^+$,OX22$^-$ cells 160, 163, 164, 186
  B cell growth *198*
  cell population 169
  cytokine secretion 168
  dissemination sequence 165
  enterocyte transport of Ig *196*
  IL-4 production 166, 168
cell surface antigen switch mechanism 67
cercariae
  lethal dose 305
  schistosomiasis in cattle 293, 294, 334–7
cercarian Hüllen reaction test 318
*cg*2 gene 58

CG2 protein 59
chaetotaxy 300
chelating agents 35–6
  parasite-selective 29
chemistry, combinatorial 29
chemotherapy
  ascariasis 138–9
  schistosomiasis 338–9, 340
China 110–11, *112–13*
  regions *118*
chloramphenicol 35
chloroquine
  cross-resistance with primaquine 13
  enhanced efflux from host–parasite complex 22–3
  pharmacological properties 6
  resistance 3
    biochemical/genetic basis 21–4
    genetic basis 58–9
    models 16
    P. falciparum 4–6
    P. vivax 13–14
    point mutations 24
    reversal 9–10
    reversal by verapamil 37
    rodent models 17, 19
    reversing agent combination 6
    synergism 39
  therapeutic use 6, 13–14
  transmission in *P. falciparum* resistance areas 58
  WHO clinical response assay 4–5
  WHO *in vitro* procedure 5
chloroquine–haemin complex 36
chlorpheniramine 38
chlorproguanil 33, 34
chlorpromazine 38
cholera toxin 185
choline transport 36
chronotherapy 22
ciprofloxacin 34
circulating anodic/cathodic antigens 318
clindamycin 35
complement fixation test 317
conserved recombination junction element (CRJE) *79*
Contracaecum 252
cross-resistance, antimalarial drugs 7
cycloguanil
  resistance 26
  rodent models 20

cyproheptadine 38
cysteine protease 36
cytochrome oxidase II (COII) gene 223-4
cytokines 6
  adult *T. spiralis* rejection 199
  malaria 28
  mast cell effects 184
  see also interferon-γ; interleukins

dapsone 7
  chlorproguanil combination 34
  proguanil combination 33-4
delayed hypersensitivity reaction
  schistosomiasis in cattle 303
  T. spiralis 190
  trichinosis 154, 162-3
desferrioxamine *20*, 21, 35
desipramine 38
DHFR gene
  Fansidar resistance 59-60
  point mutations 59
  WR 99210 action 60
DHFR-TS enzyme complex 15, 16, 25
DHFR-TS gene 7, 37
DHPS gene 59-60
Dictyocaulus viviparus 272
dihydroartemisinin 12
dihydrofolate reductase
  Pneumocystis carinii 68
  see also DHFR gene
dihydrofolate reductase inhibitors 6-7
  resistance 25-6
  see also antifols
dihydroorotate 28
dihydropteroate synthase 25
dispersal, gender biased 241
diversity estimates, nematode parasites 257
DNA probes, drug response 5
doxycycline 35, 39
drug combinations 37-40
drug resistance 2
  allele frequency 265, *266*, 267
  biochemical basis 21-8
  dominance 268
  Fansidar 59-60
  genes 263, 264-5
  genetic basis 21-8
  modulators 37-8
  nematodes 150, 220, 262-5, *266*, 267-9
  Plasmodium falciparum 4-10, *11*, 12-13, 26, 30

Plasmodium vivax 3, 13-14
  pyrimethamine 14, 15, 25, 26
dual antibody hypothesis 151-2
DvA-1 polyprotein antigen 272

E64 36
early treatment failure (ETF) 5
Echinocephalus overstreeti 262
Echinostoma malayanum 340
endoperoxides, synthetic 30-1
Entamoeba histolytica 130
Enterobius vermicularis 130
enterocytes 170
  IgE transport 197, *198*
  T. spiralis invasion 161
eosinophilia 168, 169
eosinophils and *T. spiralis*
  IL-5 in growth and maturation 182
  new-born larvae killing 182-3
  protection 181-3
    stage-specificity 182
  response 180-3
etaquine see WR 238, 605

Fansidar 6-7, 14, 34
  resistance 59-60
Fasciola heterologous immunity with
  Schistosoma 333-4
Fcγ receptor 160
fecundity, *T. spiralis* effect 200
Fenozan B07 *11*, 30, 31
  synergism 39
fertilizers, ascariasis in China 128, 130
fibronectin 73, 96
filariasis in China 111
5-fluoroorotate 12, 15, 33
  combinations 40
fluoxetine 38
folate metabolism, compounds acting on 40
FV-1 gene 179

G25 36
galactose 71
genetic drift 246
genetic variation, microspatial 242-3, 244
Giardia lamblia, polyparasitism in China 130
Globodera 240
globule leukocyte response 162
globulins, schistosomiasis in cattle 307, 309
glucose 71
glucose phosphate isomerase 296

glucose-6-phosphate dehydrogenase 297
glutathione peroxidase (gp29) sequences 272
glutathione peroxidases 182
glutathione S-transferase (GST) 343–4
    recombinant 344
glycoprotein A antigen 75, *77*
glycosyl phosphatidylinositol anchorage 95
granulomas
    formation in schistosomiasis 303, 314, 315
    nasal 319
growth, stunted 110, 194, 200
guanosine triphosphate (GTP) 25

haem polymerase 23
haemoglobin, schistosomiasis in cattle 307, 309
Haemonchus contortus 234
    benzimidazole resistance 264
    ivermectin resistance 264
    levamisole resistance 263, 267, 268
    ß-tubulin introns 234
Haemonchus placei 241
haemozoin 23
halofantrine 7, 9–10
    combinations 39
    metabolite 9
    resistance
        biochemical/genetic basis 26–7
        Pfmdr1 amplification 24
        synergism 39
        therapy 9
Hardy–Weinberg equilibrium 244, 252
Heligmosomoides polygyrus 150
helminth infections
    allergic response 154
    host response elements 154
    primates 137
hepatic syndrome, chronic 309, 311
Heterodera, isozyme markers 272
Heterodera avenae 244
Heterodera schachtii 240
heterozygosity per locus (H_e) 229, *230–1*, *232*
hexokinase, schistosome 296
histamine, mast cell function 186–8
Hoeppli reactions 315
host response, *Trichinella spiralis* 157, 173–9
Howardula 223
    mtDNA diversity 224
    sibling species 257
Howardula aoronymphium 221

humans
    infection sharing with domesticated animals 137
    schistosomiasis 287, 302–3
hycanthone 337–8
hypergammaglobulinaemia, selective 156
hypersensitivity reactions, trichinosis 154
Hypodontus macropi 254

I-A gene 176, 177
I-E gene 176, 177
IgA 172
    trichinosis 159
IgE 169, 172
    adult *T. spiralis* rejection 161, 198–9
    enterocytes in transport 197, *198*
    protection by worm rapid expulsion 160
    trichinosis 158, 159
IgG1 response 169
IgG 172
    adult *T. spiralis* rejection 199
    protection by worm rapid expulsion 160
    response in trichinosis 156–7, 158
IgM 172
    response in trichinosis 156
Ihe-1 gene 179
immediate hypersensitivity, *T. spiralis* 154, 155
immune serum transfer, trichinosis 150
immunoglobulins
    Trichinella spiralis 172
    trichinosis 155–6, 158–9
immunosuppression, *T. spiralis* 170–3
immunosuppressive therapy 64
indirect fluorescent antibody (IFA) test 317
indirect haemagglutination test 317
indirect immunofluorescence test 317
infection, saturated 133
inflammation
    allergic 195
    non-specific in trichinosis 190–5
inflammatory cells, *T. spiralis* rejection 153
interferon-γ 166, 167, 168
interleukin-1 (IL-1) 167
interleukin-2 (IL-2) 166–7
interleukin-3 (IL-3) 166, 167
    mast cell effects 184, 186
    T. spiralis 184, 186
interleukin-4 (IL-4) 160, 161, 166, 167, 168
    adult *Trichinella* rejection 167, 170, 197, 198–9

interleukin-5 (IL-5) 166, 167
eosinophil growth and maturation 182
internal transcribed spacers (ITS) 235, 241
   Ascaris 247, 248
   cryptic species 253–4
intestinal epithelial cells 196–7
intestinal mast cell proteases (IMCP) 184
intron diversity estimates 235
iron chelators 35–6
isothiocyanates 338
isozyme markers 272
ivermectin resistance 263–4, 265

Japanese B encephalitis virus 171

kala-azar 111
Kampuchean gem miners 2–3, 8
keyhole limpet haemocyanin (KLH) 343

lactose intolerance 110
late treatment failure (LTF) 5
leak-lesion hypothesis 189, 195, 198, 199
levamisole resistance 263, 267, 268
lipid metabolism, compounds influencing 36
lucanthone hydrochloride 337–8
lymphocytes, intraepithelial 194
lysophospholipase 36

major basic protein 181
major histocompatibility antigen (MHC) 173–4
   class II molecules 176–7
   T. spiralis
      rejection 175
      variation 174–8
major surface glycoproteins (MSG)
   biochemistry 72–3
   C-terminal region 78
   cDNA 76, 78
      cloning 75, 76
      probes 80
   composition 71–2
   conserved cysteines 94–5
   conserved recombination junction element (CRJE) *79*
   fibronectin binding site 73
   genes 75–6, *77*, 78, *79*, 80–9, *90*, 91–7
      chromosomal organization 85–9
      complexity 93
      karyotype hybridization analysis *81*
      multiplicity 80–1

   Southern blot analysis *81*, 82
   splicing 84–5
   subtelomeric region 86
   tandem 84
   tandem clusters *82*
   telomeric clones 86–8
   telomeric DNA *trans*-action 88–9
   telomeric repertoire 86–8
   telomeric UCS 85–6
   transcript expression 84–5
   transcription start sites 84, *85*
   upstream conserved sequence 78, *79*, 80, 82–4, 85–6, 88
   upstream conserved sequence attachment 83, 84
   upstream conserved sequence copy number 84
genomic fragments 80
glycoprotein A antigen 75, 77
glycosyl phosphatidylinositol anchorage 95
isoelectric variation 72–3
isoform expression 80
mRNA 83–4, 91
N-linked carbohydrates 73
open reading frames (ORFs) 80, *82*
P. carinii 70–5
polypeptide switching 89
protein features 93–6
pulmonary surfactant binding 74
purification 73
RACE clones 78, 80, 82
recombination 89, 91
repetitive sequence elements 75–6
silent genomic repertoire 80–1, *82*
subtilisin-like protease 96–7
subtypes 76, *77*
switching 96, 97
threonine residues 94
UCS 89, 91
upstream conserved sequence 78, *79*, 80, 82–4
variable regions 76, 78
malaria
   2% relapse technique 17, 19
   cerebral 6, 29, 35
   China 111
   control 2–3
   drug resistant 2–3
   endemic areas 2–3
   in vitro resistance models 15–16

malaria (*Continued*)
  intraerythrocytic schizogony 22
  merozoite latency 22
  new drug development 28–37
  nonimmune travellers 5
  pigment 23
  plant remedies 31
  prevalence *3*
  rodent models 16–21
  TNF 28
  see also Plasmodium spp.
Malarone 10, *11*, 12
Maloprim 7
Mannich bases 31–2
mannose 71
mast cells
  adult *T. spiralis* rejection 198
  allergic inflammation 195
  cytokine effects 184
  histamine 186–8
  IL-3 effects 184, 186
  intestinal 162
  nematode infection protection 183–9
  prostaglandins 186–7
  serotonin 186–9
  T. spiralis infection 183–9
mastocytosis 168, 169
  T. spiralis infection 183
mating barriers
  Ascaris 248
  cryptic species 257
mating patterns, nematode parasites 244–5, 270–2
Mazamastrongylus odocoili 238, 240–1
medicine, traditional 139
mefloquine 2, 7–10
  combinations 39
  quinine cross-resistance 8–9
  resistance 8, 18
    biochemical/genetic basis 26–7
    Pfmdr1 amplification 24
  sulfadoxine and pyrimethamine (MSP) combination 8
Meloidogyne 221
  mtDNA diversity 224
  mtDNA sequence variation 240
  population history 258–9, *260*
  sibling species 258–9, *260*
Meloidogyne javanica 227, *228*, 229
membrane receptor proteins 265
menoctone 10

merozoite latency 22
methylene blue 31
microsatellites 236–7
  markers 272
miracidia, schistosomiasis in cattle 293, 316–17, 337
miracidial immobilization test 318
monophyly, reciprocal 246
mtDNA clock 223
mtDNA genes, nematode parasites *226*, 227, *228*, 229
multilocus enzyme electrophoresis (MLEE) 229
myeloperoxidase 181
myosins 96

N-acetyl-D-glucosamine 71
Nacobbus aberrans 254
naphthoquinones 10, 12
  resistance 28
  rodent models 19
natural selection, population genetic structure (PGS) 242
ND4 gene 223, 224, 227, 229
nematode parasites
  alleles per locus (A) 229, *230–1*
  allozyme variation 250, 252
  anthelmintic control 262–3
  antigen evolution 272
  benzimidazole resistance 263, 264
  chemotherapy programmes 267
  COII gene 223–4
  cryptic species 246
    coexisting 255
    distribution 257
    genetic variation 257
    mating barriers 257
  direct life cycles 232–3
  diversity estimates 257
  domestic livestock 240–1
  drug resistance 138, 150, 220, 262–5, *266*, 267–9
    genes 263, 264–5
  dual antibody hypothesis 151–2
  effective population size (N_e) 233
  endotherms 224
  exotherms 224
  gender biased dispersal 241
  gene flow 237–8, *239*, 240
    barriers 247
  genetic drift 246

nematode parasites (*Continued*)
    genetic markers 220
        drug resistance 262–5, *266*, 267–9
        intrahost dynamics 279
        larval stage identification 262
        parasite transmission 243–4
    geographical differentiation 237–8, *239*, 240
    geographical population structure *239*
    heterozygosity 220
        per locus (H_e) 229, *230–1*, *232*
    heterozygote excess 244–5
    host co-evolution 272
    immunity in rodents 151
    immunology 272
    indirect life cycles 232–3
    intron diversity estimates 235
    ITS 241
    ivermectin resistance 263–4
    larval stage identification 262
    levamisole resistance 263
    life history 260, *261*, 262
        traits 240–1
    loci differences 241–2
    maternal markers 270
    mating patterns 244–5, 270–2
    microsatellite markers 272
    microspatial genetic variation 242–3, 244
    migration patterns 269
    mitochondrial DNA 221, *222*, 223–4, *225–6*, 227, *228*, 229
        A+T bias 224, 227
        evolution rate 223–4
        genes *226*, 227, *228*, 229
        mutation rate 235
        recombination 229
        structural evolution 227, *228*, 229
        variation 241
    ND4 gene 223, 224, 227
    nuclear genome 229, *230–1*, 232–7
        allozyme variation 229, *230–1*, 232–3
        microsatellites 236–7
        nuclear sequence diversity 233–5
        RAPD 236–7
        rDNA 234
        repetitive DNA 236–7
        ribosomal DNA spacers 235–6
    phylogenetic trees 223
    plant parasitic 238, 240
        microspatial genetic variation 244
    population
        admixture 221, 223
        biology 268–9
        genetic structure (PGS) 237–8, *239*, 240–1, 240–5
        proportion of polymorphic loci (P) 229, *230–1*
        RAPD 238, 240
        rDNA genes 241
        resistance alleles 269
        scnDNA 241–2
        selection hypothesis 232
        sequence
            diversity 221, *222*, 223–4, *225–6*
            types 220
            variation 246
        sibling species 245–8, *249*, 250, *251*, 252–5, *256*, 257–8
            early stage of divergence 247–8, *249*, 250
            genetic divergence 253–5
            late stage of divergence 250, 252–3
            morphological divergence 253–5
            population history 258–9, *260*
            ubiquity 255, *256*, 257–8
        sociobiology 270–2
        species splitting 246
        substitution bias 224, 227
        surface antigens 272
        transmission patterns 242–4
        trickle infections 270, *271*
        ß-tubulin isotypes 267
        variation 221–37
        wild animals 240–1
neutrophils
    T. spiralis 189–90
    TNF-α 190
niclosamide 339
night soil 128, 130
Nippostrongylus brasiliensis 150
nitric oxide 6
norfloxacin 34
nuclear genes, single copy 237
nucleotide diversity 234

Onchocerca volvulus 238
    hybridization patterns 248
    molecular markers of pathogenesis 248, 250
    ocular pathogenesis 248
    population history 258, *259*
    sympatry 250

onchocerciasis, forest/savannah forms 248, 250
open reading frames (ORFs), MSG 80, *82*
Ostertagia, polymorphisms 254
Ostertagia ostertagi 241, 242
oxidant compounds 30–1

p-glycoprotein 264–5
Pailin (Kampuchea) 3
pamaquine 31
paraphyly 246
Parascaris 252
parasitaemia, falciparum 7
Parasitodiplogaster 224
parthenogenesis 260
passive cutaneous anaphylaxis (PCA) reaction 156
Pcmdr1 24
penfluridol 38
pepstatin A 36
peroxidases 181
Peyer's patch cells 163
Pfmdr1 gene 15, 16, 37–8
  amplification 24, 27
Pfmdr2 gene 37–8
phosphatidylcholine 36
phosphoglucomutase, schistosome 296, 297
phospholipid synthesis inhibition 29
phosphorylcholine 192
phytochemicals 31
pigs
  ascariasis in China 131–3, 136, 140
  domestication 137
  infection sharing with humans 137
piperaquine 9
plasmalepsins 60–1
Plasmodium berghei 10, 12, 16, 17, 18
  ANKA 25
  multi-drug resistant 30
  N (A) strain *20*, 21
  1,2,4-trioxane treatment 30
Plasmodium chabaudi 17, *20*, 21
  Pcmdr1 24
  subcurative treatment 58
Plasmodium cynomolgi 17
Plasmodium falciparum 2
  antibiotics 34–5
  artemisinin resistance 12–13
  chloroquine resistance 3, 4–6, 9–10
    genetic basis 58–9

chloroquine sensitivity with mefloquine/quinine resistance 10
cross-resistance 7, 8–9
  proguanil/pyrimethamine 25
cycloguanil resistance 26
dihydrofolate reductase inhibitor resistance 6–7
drug resistance 4–10, *11*, 12–13, 30
etaquine 61
folate-resistance model 58
genome 40
halofantrine resistance 7, 9–10
HD3 line 58, 59
in vitro resistance models 15–16
iron requirements 35
K1 line 15
mefloquine resistance 7–10
multi-drug resistant 30
naphthoquinone resistance 10, 12
point mutation 26
pyrimethamine resistance 4, 26
quinine resistance 7–10
rodent models 16
sesquiterpene lactone resistance 12–13
sulfadoxine resistance 26
1,2,4-trioxane treatment 30
Plasmodium malariae 15
Plasmodium ovale 15
Plasmodium vinckei petteri 17
Plasmodium vinckei vinckei 17
Plasmodium vivax 61
  antifol resistance 14
  chloroquine resistance 3, 13–14
  cross-resistance of chloroquine/primaquine 13
  drug resistance 13–14
  pyrimethamine resistance 14
Plasmodium yoelii 12, 16
Plasmodium yoelii ssp. NS 16, 17, 18, 19
Pneumocystis carinii 64
  45–55 kDa antigen complex 73
  AIDS 93
  antigenic variation 93
  attachment 73–4
  cell surface antigen switch mechanism 67
  cyst cell wall 71
  DHFR gene 68
  genes 68
    detection 69
  genetic control of antigenic variation 89, *90*, 91–3

Pneumocystis carinii (*Continued*)
  genetic heterogeneity with UCS locus 92
  genome 68
  host response 74–5
  immune response 74
  intracystic bodies 65
  karyotypes 68
  life cycle 64–6
  macrophage binding 75
  major surface glycoproteins (MSGs) 70–5
    antigenic diversity 92–3
    conserved cysteines 94–5
    genes 75–6, 77, 78, 79, 80–9, 90, 91–7
    genetic control of antigenic variation 89, 90, 91–3
    glycosyl phosphatidylinositol anchorage 95
    mRNA 91
    ORF-3 96–7
    polypeptide switching 89
    protein features 93–6
    recombination 89, 91
    sequence comparison of genes 91
    subtilisin-like protease 96–7
    switching mechanism 96, 97
    UCS 89, 91
  molecular biology 67–8
  molecular taxonomy 67
  pathobiology 73–5
  pulmonary surfactant interaction 74
  resting form 65
  RNA analysis 67
  SDS-PAGE 71, 72
  switch model 89, 90
  telomere-associated recombination model 89
  trophozoites 73
  TS gene 68
  ultrastructure 66
  vegetative form 65
Pneumocystis carinii pneumonia (PCP) 64, 66
  alveolar macrophages 74
  clinical course 69
  diagnosis 68–9
  pathogenesis 73
  reinfection 66
  relapse 66
point mutations 24, 26
polymerase chain reaction, PCP diagnosis 68–9
polyparasitism, ascariasis in China 130, 131
polyphyly 246
population genetic structure (PGS)
  allele frequency 237
  life history traits 240–1
  loci differences 241–2
  microspatial 242–4
  natural selection 242
  nematode parasites 237–8, 239, 240–5
  RAPD markers 237
  single copy nuclear genes 237
  transmission patterns 242–4
praziquantel 337, 338
primaquine 14, 31
  cross-resistance with chloroquine 13
  synergism 39
prochlorperazine 38
proguanil 6, 7, 33
  atovaquone combination 10, 11, 12
  combinations 40
  dapsone combination 33–4
  pyrimethamine cross-resistance 25
  synergism 33, 60
proportion of polymorphic loci (P) 229, 230–1
prostaglandins, mast cell function 186–7
protease inhibitors 36–7
proteinases, parasite 36–7
PS-15 11, 33
Pseudoterranova, allozyme variation 252
Pseudoterranova decipiens 238
pulmonary surfactant 74
pyonaridine 11
  resistance 18
pyrimethamine 6
  drug combinations 6–7
  proguanil cross-resistance 25
  resistance 4
    P. falciparum point mutation 26
    P. malariae 15
    P. vivax 14
pyrimidine metabolism 32–4
pyronaridine 17–18, 32
pyroninophilic cells 162

qing hao 2
quinine 7–10
  resistance 26–7
    Pfmdr1 amplification 24

R4 237
Radophilus 240

random amplified polymorphic DNA
    (RAPD) 236-7
  plant parasitic nematodes 238, 240
  schistosome 297-8
rDNA spacers 235-6
refugia 269
relapse technique, 217, 19
repetitive DNA 236-7
resistance-reversing agents 37-8
restriction fragment length polymorphism
    (RFLP) 235
reversing agent, chloroquine combination 6
ribose 71
rifampicin 34
ring precipitation test 318
Ro 42-1611 30
rodent models of malaria 16-21
Romanomermis culicivorax 227, *228*, 229
Rugopharynx 254

Saccharomyces cerevisiae 15-16
Schistosoma bovis 286, *288*, 290
  cercariae 294
  concomitant immunity 331
  defined protective antigens 343-4
  eggs 312
    production 292
  experimental infection 324-5
  immunity 325
  immunoglobulin suppression 330-1
  intestinal lesions 313
  irradiated vaccines 341
  liver damage 312-13
  miracidium 293
  pathology 309, 312-14
  portal vessel lesions 312, 313
  resistance 329-30
  worm burden 324
Schistosoma curassoni 286, *288*, 290
Schistosoma haematobium 286, 287, *288*
  heterologous immunity 340
Schistosoma indicum 287, 287, 291
Schistosoma intercalatum 287, *288*
Schistosoma japonicum 111, 287, *288-9*, 291
  irradiated vaccines 341-2
  pathology 314
Schistosoma leiperi 287, *288*, 291
Schistosoma mansoni 287, *289*
  concomitant immunity 331
  matraclonal offspring 295
Schistosoma margrebowiei 288, 291

  cercariae 294
Schistosoma mattheei 286, 287, *288*, 290-1
  cercariae 294
  chronic hepatic syndrome 309, 311
  clinical syndromes 307-9
  eggs 292, 293, 308, 310, 311
    faecal counts 323, 327
    tissue counts 323, 325-6, 327
  GST antigens 344
  hyperergic schistosomiasis 308-9
  immune response 325
  infection
    experimental 322-4
    intensity 327-8
  intestinal lesions 310-11
  intestinal syndrome 307-8
  intramolluscan development 294
  migration route in sheep 294
  morbidity 327-8
  pathology 310-12
  protection against reinfection 326
  reproduction 294-5
  urinary bladder lesions 311-12
  uterine lesions 312
  vascular lesions 310-11
  worm burden 312, 327
Schistosoma nasale 286, 287, *288*, 291,
    309-10
  egg production 292
  migration route 294
  pathology 314-15
Schistosoma spindale 286, *288*, 291
  pathology 314
schistosomes
  definitive host 301-3
  interspecific hybridization 289-90
  species 286, 287, *288-9*, 290
  visceral 294-5
schistosomiasis
  chemotherapy 338-9, 340
  drinking water 333, 339
  human infection 287, 302-3
schistosomiasis in cattle 286
  acquired immunity 328-32
  adult worms 292, 295-8
    DNA studies 297-8, 301
    enzyme studies 296-7, 300-1
    host immune attack 303
    host specificity 301-3
    immunological attack 331-2
    morphology 295-6

schistosomiasis in cattle (*Continued*)
   size 295, 323
  antigen determination 317–18
  antischistosomal drugs 337–9
  bodyweight loss 307
  cercariae 293, 294, 300–1, 334–7
   contamination of habitat 335–7
  cercarian Hüllen reaction test 318
  chaetotaxy 300
  chemotherapy 338–9, 340
  chronic hepatic syndrome 309, 311
  circulating anodic/cathodic antigens 318
  clinical signs 304, 305
  complement fixation test 317
  control 339–44
   immunological 340–4
   programmes 338–9
   trematodes 340
  crude schistosome antigens 342–3
  defined protective antigens 343
  delayed hypersensitivity 303
  diagnosis 315–20
  disease spectrum 303–5
  drinking water 333, 334, 339
  eggs 298–9
   age-related decline 328
   counts in naturally infected animals 325–8
   faecal counts 316, 320, 323
   faecal excretion 322
   inflammatory immune response 303
   production 292
   productivity 303
   tissue counts 319, 320, 323
   transport 293
   viability 325
  endemic 291, 321
  epidemiology 320–37
  faecal egg excretion 322
  geographical distribution 290–1
  glutathione S-transferase 343–4
  granulomas
   formation 303, 314, 315
   nasal 319
  gut damage 307
  Hoeppli reactions 315
  host
   definitive 301–3, 336
   sexual maturity 328
  host–parasite relations 303–4, 320, 322–34
  human schistosomes 287
  hybridization 287, 290, 296, 297, 299
  hyperergic 308–9
  immune response 325
  immunity
   concomitant 331–2
   development 322–34, 332
   heterologous 333, 340
   serum components 330
  immunoglobulin suppression 330–1
  incidence rate 320
  indirect fluorescent antibody (IFA) test 317
  indirect haemagglutination test 317
  indirect immunofluorescence test 317
  infection
   duration 327
   experimental 322–5
   exposure 328, 334–5
   incidence rate 336–7
   intensity 304–5
   levels 321–2
   measures at population level 320
   natural 325–8
   resistance 328–32
   route 334–5
   size 332
   type 332
  influence of other disease 307
  intestinal 305
  intestinal haemorrhage 309
  intestinal syndrome 307–8
  intramolluscan development 293–4
  irradiated vaccines 341–2
  lethal cercarial dose 305
  life cycle 292–5
  localization 303
  miracidia 293, 299–300
   counts 316–17, 337
  miracidial immobilization test 318
  nasal 319–20
  occurrence 290–1
  pairing 294–5
  pathogenesis 303
  pathology 303–5, *306*, 307–15
   clinical 305, 307–10
  pathophysiology 305, 307–10
  perfusion studies 322
  prevalence 291
  protection against reinfection 326
  RAPD 298
  resistance 328–32

schistosomiasis in cattle (*Continued*)
　development 329–31
　heterologous 332–4
　maintenance 329–31
　ring precipitation test 318
　rRNA genes 298
　ruminant infection 294
　schistosome species 303–4
　schistosomular antigen 342–3
　snails 301–3
　　density 337
　　habitat 339
　　vector 293
　susceptibility 304
　sympatry 299
　taxonomy 295–303
　tissue egg counts 319, 320
　transmission 334–7
　treatment 337–9
　tubercles 295–6
　visceral 304, 312
　　clinical signs 315–16
　　diagnosis 315–19
　　faecal examination 316
　　miracidial counts 316–17
　　post-mortem examination 319
　　serological tests 317–19
　water supply 333, 339
　worm burden *306*, 321, 324, 332
　　naturally infected animals 326
　worm counts 329
　　snail density 337
　worm fecundity 327
SDS–PAGE 71, 72
selection hypothesis, nematode parasites 232
serotonin
　antagonists 188
　mast cell function 186–9
sesquiterpene lactones 12–13, 19
　development 29–30
sialic acid 71
sibling species
　allopatric host-associated 253
　allozyme divergence, late stage 250, 252–3
　early stage of divergence 247–8, *249*, 250
　genetic divergence 253–5
　genetic markers 245–8, *249*, 250, *251*,
　　252–5, *256*, 257–8
　morphological divergence 253–5
　population history 258–9, *260*
　ubiquity 255, *256*, 257–8

snails
　schistosomiasis in cattle 293, 301–3, 337
　　control 339–40
sociobiology, nematode parasites 270–2
sodium dodecyl sulphate-polyacrylamide gel
　　see SDS–PAGE
sphingomyelin synthase 36
stem cell factor (SCF) receptor 183–4, 185–6
strongyles 138
Strongyloides 223
　heterogonic life cycle 244–5
　homogonic life cycle 244
　parthenogenetic female 260
　sexual exchange 260, *261*
Strongyloides ratti 150, 221, 236
　actin locus 244, 245
　allozyme variation 238
　heterozygote excess 244
　life cycle *261*
stunting
　ascariasis 110
　T. spiralis effect 194, 200
subtilisin-like protease 96–7
sulfadoxine 6, 59
　resistance 26
sulfalene 6
sulphonamides, rodent models 19
superoxide dismutase 182
surface antigens, nematode parasites 272
switch model, *Pneumocystis carinii* 89, *90*
sympatry
　ascarids 247, 250, 252–3
　schistosome 299

T cells
　activated 163
　adult *T. spiralis* rejection 200
　antibody in worm expulsion 159–60
　B cell anti-adult *T. spiralis* effect 160–1
　homing 164
　intraintestinal migration pathways 164
　phenotypic polarization 166
　protection against *T. spiralis* 162–70
　trichinosis role 194
Tajima's D statistic 223
TAS (transposable element of Ascaris) 237
Teladorsagia circumcincta 225–6, 238
Telodorsagia, polymorphisms 254
terebratorial membrane, schistosome
　　miracidial 299–300
tetracycline 35

tetrandine 31
T$_h$2 cytokines 166, 167, 169
thiostrepton 34
thymidylate synthase (TS) 68
  blocking 33
thymus, *T. spiralis* immunity 162
transferrin 35
trematodes, schistosomiasis control 340
Trichinella, rDNA 236
Trichinella spiralis 150
  adult rejection 197–9
  allergic inflammation 195
  anti-newborn larvae immunity 200
  antibody
    in protection 159–61
    responses 155–61
    role 200–1
  B cells 161
  bone marrow cell
    population for rejection 191
    role 193–4
  c-kit 184–6
  CD4$^+$ 176
  concurrent infections 192–3
  cross-immunity 191
  cross-reacting antigens 192
  cross-reactivity 192
  delayed hypersensitivity response 154, 190
  enterocyte transport of IgE 197
  eosinophil response 180–3
  eosinophil-mediated protection 181–3
  expulsion 159, 160
    adult 160–1
    damage absence 197
    intestinal mucus 195–6
    lack of damage 192, 194–5
  FV-1 gene 179
  granulocytic cell population response 180–90
  H-2 region 176, 177
  histamine 187–8
  host mediators 173
  host response 157
    strength 199–200, 201
    variation 173–9
  I-A gene 176, 177
  I-E gene 176, 177
  Ia antigen recognition 176
  Ihe-1 gene 179
  IL-3 184, 186
  IL-4 in rejection 167, 170
  immediate hypersensitivity 154, 155
  immune response enhancement by concurrent infection 193
  immunity 200–1
    local gut 152–3
    in rodents 150–1
    transfer 154
  immunoglobulins 172
  immunosuppression 170–3
  intestinal epithelial cells 196–7
  intestinal immunity 153–4
  intestinal mast cell proteases 184
  Japanese B encephalitis virus 171
  larval antigen 172
  larval immunization 151, 152
  LD$_{50}$ values 151
  mast cells in infection 183–9
  mastocytosis 183
  MHC 173–4
    genes 175, 176
    in rejection 175
  MHC-linked variation 174–8
  mucus trapping 196
  muscle larvae 175
    antigen response 177
  neutrophils 189–90
  newborn larvae killing 181, 182–3
  non-MHC genes 175
  non-MHC-linked variation 178–9
  non-specific inflammation 190–5
  parasite-derived immunosuppressive substances 171–2
  prostaglandins 187–9
  protection mechanisms 190–7
  re-establishment in intestine 199–200
  reinfection 152
  rejection 152, 153
    delay 173
    terminal effector 195
  resistance 175
  serotonin 187–8
  serum immunoglobulin levels 155–6
  splenocytes 172
  stem cell factor (SCF) receptor 183–4, 185–6
  T cell role in protection 162–70
  T$_h$2-type response 173
  thymus in immunity 162
  Ts-1 gene 177–8, 179
  Ts-2 gene 177–8, 179

Trichinella spiralis (*Continued*)
   worm dose/worm expulsion relationship 172–3
   worm expulsion 152
trichinosis 150–5
   ACTH treatment 153
   cytokine production 166–9
   delayed hypersensitivity reaction 162–3
   host fecundity 194, 200
   hypersensitivity reactions 154
   immune serum transfer 150
   immunoglobulins 155–6, 158–9
   infection rejection 165
   intestinal mucus 195–6
   secretory antibody response 159
   specific antibody response 156–9
   stunting 194, 200
trichlorphon 338
trichostrongylids
   heterozygote deficits 255
   life history traits 240
   morphs 254
   PGS 240
   phenograms 223
   polymorphisms 254
   reproductively isolated host 254–5
   sR allele 255
Trichostrongylus colubriformis 255
   levamisole resistance 263, 268
Trichuris muris 150
Trichuris trichiura, polyparasitism in China 130
trickle infection regimes 270, *271*
1,2,4-trioxanes 30
tropomyosins 96
Trypanosoma brucei, variant surface glycoprotein (VSG)
   gene rearrangement 88–9, 91
   glycosyl phosphatidylinositol anchorage 95
Ts-1 gene 177–8, 179
Ts-2 gene 177–8, 179
TTAGGG sequence 86

tubercles, schistosome 295–6
β-tubulin 264
   introns 234
tumour necrosis factor (TNF) 6
   malaria 28
tumour necrosis factor-α (TNF-α) 190

UCS-MSG gene 91
upstream conserved sequence (UCS) 78, *79*, 80, 82–4, 85–6
   expression site 88
   promoter activity 85
   telomeric 85–6
   TTAGGG sequence 86

vaccines, irradiated 341–2
variant surface glycoprotein (VSG) 88–9, 91, 95
verapamil
   chloroquine resistance reversal 37, 38
   resistance reversal 27
vitamin A, impaired absorption 110
vitronectin 73
VLA-4 integrin 164

water supply
   ascariasis in China 128
   schistosomiasis in cattle 333, 334, 339
Wolbachia bacteria 257
World Health Organization (WHO)
   chloroquine clinical response assay 4–5
   chloroquine *in vitro* procedure 5
   malaria control 2
worm expulsion
   adult 160–1
   rapid 160
WR 99,210 *11*, 33
   parasite DHFR 60
WR 238,605 *11*, 14, 61

yingzhaosu 28
   development of derivatives 29–30